NEUROEPIDEMIOLOGY

HANDBOOK OF CLINICAL NEUROLOGY

Series Editors

MICHAEL J. AMINOFF, FRANÇOIS BOLLER, AND DICK F. SWAAB

VOLUME 138

ELSEVIER

AMSTERDAM BOSTON HEIDELBERG LONDON NEW YORK OXFORD
PARIS SAN DIEGO SAN FRANCISCO SINGAPORE SYDNEY TOKYO

NEUROEPIDEMIOLOGY

Series Editors

MICHAEL J. AMINOFF, FRANÇOIS BOLLER, AND DICK F. SWAAB

Volume Editors

CATERINA ROSANO, M. ARFAN IKRAM, AND MARY GANGULI

VOLUME 138

3rd Series

ELSEVIER

AMSTERDAM BOSTON HEIDELBERG LONDON NEW YORK OXFORD
PARIS SAN DIEGO SAN FRANCISCO SINGAPORE SYDNEY TOKYO

ELSEVIER
Radarweg 29, PO Box 211, 1000 AE Amsterdam, Netherlands
The Boulevard, Langford Lane, Kidlington, Oxford OX5 1GB, United Kingdom
50 Hampshire Street, 5th Floor, Cambridge, MA 02139, United States

British Library Cataloguing-in-Publication Data
A catalogue record for this book is available from the British Library

Library of Congress Cataloging-in-Publication Data
A catalog record for this book is available from the Library of Congress

ISBN: 978-0-12-802973-2

For information on all Elsevier publications
visit our website at https://www.elsevier.com/

Publisher: Shirley Decker-lucke
Acquisition Editor: Mara Conner
Editorial Project Manager: Kristi Anderson
Production Project Manager: Sujatha Thirugnana Sambandam
Cover Designer: Alan Studholme

Typeset by SPi Global, India

Handbook of Clinical Neurology 3rd Series

Available titles

Vol. 79, The human hypothalamus: basic and clinical aspects, Part I, D.F. Swaab, ed. ISBN 9780444513571
Vol. 80, The human hypothalamus: basic and clinical aspects, Part II, D.F. Swaab, ed. ISBN 9780444514905
Vol. 81, Pain, F. Cervero and T.S. Jensen, eds. ISBN 9780444519016
Vol. 82, Motor neurone disorders and related diseases, A.A. Eisen and P.J. Shaw, eds. ISBN 9780444518941
Vol. 83, Parkinson's disease and related disorders, Part I, W.C. Koller and E. Melamed, eds. ISBN 9780444519009
Vol. 84, Parkinson's disease and related disorders, Part II, W.C. Koller and E. Melamed, eds. ISBN 9780444528933
Vol. 85, HIV/AIDS and the nervous system, P. Portegies and J. Berger, eds. ISBN 9780444520104
Vol. 86, Myopathies, F.L. Mastaglia and D. Hilton Jones, eds. ISBN 9780444518996
Vol. 87, Malformations of the nervous system, H.B. Sarnat and P. Curatolo, eds. ISBN 9780444518965
Vol. 88, Neuropsychology and behavioural neurology, G. Goldenberg and B.C. Miller, eds. ISBN 9780444518972
Vol. 89, Dementias, C. Duyckaerts and I. Litvan, eds. ISBN 9780444518989
Vol. 90, Disorders of consciousness, G.B. Young and E.F.M. Wijdicks, eds. ISBN 9780444518958
Vol. 91, Neuromuscular junction disorders, A.G. Engel, ed. ISBN 9780444520081
Vol. 92, Stroke – Part I: Basic and epidemiological aspects, M. Fisher, ed. ISBN 9780444520036
Vol. 93, Stroke – Part II: Clinical manifestations and pathogenesis, M. Fisher, ed. ISBN 9780444520043
Vol. 94, Stroke – Part III: Investigations and management, M. Fisher, ed. ISBN 9780444520050
Vol. 95, History of neurology, S. Finger, F. Boller and K.L. Tyler, eds. ISBN 9780444520081
Vol. 96, Bacterial infections of the central nervous system, K.L. Roos and A.R. Tunkel, eds. ISBN 9780444520159
Vol. 97, Headache, G. Nappi and M.A. Moskowitz, eds. ISBN 9780444521392
Vol. 98, Sleep disorders Part I, P. Montagna and S. Chokroverty, eds. ISBN 9780444520067
Vol. 99, Sleep disorders Part II, P. Montagna and S. Chokroverty, eds. ISBN 9780444520074
Vol. 100, Hyperkinetic movement disorders, W.J. Weiner and E. Tolosa, eds. ISBN 9780444520142
Vol. 101, Muscular dystrophies, A. Amato and R.C. Griggs, eds. ISBN 9780080450315
Vol. 102, Neuro-ophthalmology, C. Kennard and R.J. Leigh, eds. ISBN 9780444529039
Vol. 103, Ataxic disorders, S.H. Subramony and A. Durr, eds. ISBN 9780444518927
Vol. 104, Neuro-oncology Part I, W. Grisold and R. Sofietti, eds. ISBN 9780444521385
Vol. 105, Neuro-oncology Part II, W. Grisold and R. Sofietti, eds. ISBN 9780444535023
Vol. 106, Neurobiology of psychiatric disorders, T. Schlaepfer and C.B. Nemeroff, eds. ISBN 9780444520029
Vol. 107, Epilepsy Part I, H. Stefan and W.H. Theodore, eds. ISBN 9780444528988
Vol. 108, Epilepsy Part II, H. Stefan and W.H. Theodore, eds. ISBN 9780444528995
Vol. 109, Spinal cord injury, J. Verhaagen and J.W. McDonald III, eds. ISBN 9780444521378
Vol. 110, Neurological rehabilitation, M. Barnes and D.C. Good, eds. ISBN 9780444529015
Vol. 111, Pediatric neurology Part I, O. Dulac, M. Lassonde and H.B. Sarnat, eds. ISBN 9780444528919
Vol. 112, Pediatric neurology Part II, O. Dulac, M. Lassonde and H.B. Sarnat, eds. ISBN 9780444529107
Vol. 113, Pediatric neurology Part III, O. Dulac, M. Lassonde and H.B. Sarnat, eds. ISBN 9780444595652
Vol. 114, Neuroparasitology and tropical neurology, H.H. Garcia, H.B. Tanowitz and O.H. Del Brutto, eds. ISBN 9780444534903
Vol. 115, Peripheral nerve disorders, G. Said and C. Krarup, eds. ISBN 9780444529022
Vol. 116, Brain stimulation, A.M. Lozano and M. Hallett, eds. ISBN 9780444534972
Vol. 117, Autonomic nervous system, R.M. Buijs and D.F. Swaab, eds. ISBN 9780444534910
Vol. 118, Ethical and legal issues in neurology, J.L. Bernat and H.R. Beresford, eds. ISBN 9780444535016
Vol. 119, Neurologic aspects of systemic disease Part I, J. Biller and J.M. Ferro, eds. ISBN 9780702040863
Vol. 120, Neurologic aspects of systemic disease Part II, J. Biller and J.M. Ferro, eds. ISBN 9780702040870
Vol. 121, Neurologic aspects of systemic disease Part III, J. Biller and J.M. Ferro, eds. ISBN 9780702040887
Vol. 122, Multiple sclerosis and related disorders, D.S. Goodin, ed. ISBN 9780444520012
Vol. 123, Neurovirology, A.C. Tselis and J. Booss, eds. ISBN 9780444534880

Foreword

Epidemiology is an ancient discipline. It can be argued that Hippocrates (460–370 BC) was the first epidemiologist, not just because of the appellation of some of his works (*Epidemic*), but above all because he was one of the first to consider that diseases are not God-sent supernatural phenomena, but that they have natural, identifiable causes, establishing, for instance, that malaria occurs more frequently in swampy areas. Epidemiology was at first mainly focused on communicable diseases, but later became an essential tool in our understanding of all diseases, particularly those affecting the nervous system, and it has always been a part of their description. For instance, the magisterial "Huntington's chorea" chapter written close to 50 years ago (in 1968) by George Bruyn for volume 6 of the *Handbook of Clinical Neurology* (HCN) included a section on epidemiology, even though this term does not appear in the index of the volume. For several reasons, including the introduction of large-population studies, many of which are longitudinal and still ongoing, and the entry of epidemiology into fields such as genetics and molecular epidemiology, neuroepidemiology has become a discipline of its own. We are therefore proud to present the first HCN volume entirely dedicated to neuroepidemiology.

The volume deals with a variety of neuroepidemiology-related topics organized into three sections. The first deals with the principles and foundation of epidemiology and neurology research. It includes a description of the methods involved and their applications. Emphasis is given to population neuroscience, including a thorough discussion of genetics. The section also deals with the application of epidemiology to imaging, neuropsychology, and -omes technologies. The second section discusses the epidemiology of diseases primarily affecting the nervous system, such as neurodegenerative diseases including dementia, stroke, epilepsy, trauma, amyotrophic lateral sclerosis, multiple sclerosis, migraine, and peripheral neuropathies. The final section deals with the epidemiology of conditions originating outside the nervous system. This could have been endless. Two emblematic examples are included: cancer and the side-effects of its treatment, as well as sickle-cell anemia, the most common serious genetic disease in childhood.

We have been fortunate to have as volume editors three distinguished scholars, Caterina Rosano of the Graduate School of Public Health, University of Pittsburgh, M. Arfan Ikram of the Department of Epidemiology, Erasmus Medical Center, Rotterdam, and Mary Ganguli of the Department of Psychiatry, Epidemiology and Neurology, University of Pittsburgh. All three have been at the forefront of neuroepidemiology research for many years. They have assembled a truly international group of authors with acknowledged expertise and together they have produced this authoritative, comprehensive, and up-to-date volume. Its availability electronically on Elsevier's Science Direct site as well as in print format should ensure its ready accessibility and facilitate searches for specific information.

We are grateful to the volume editors and to all the contributors for their efforts in creating such an invaluable resource. As series editors we read and commented on each of the chapters with great interest. We are therefore confident that both clinicians and researchers in many different medical disciplines will find much in this volume to appeal to them.

And last but not least, it is always a pleasure to acknowledge and thank Elsevier, our publisher – and, in particular, Michael Parkinson in Scotland, and Mara Conner and Kristi Anderson in San Diego – for their unfailing and expert assistance in the development and production of this volume.

Michael J. Aminoff
François Boller
Dick F. Swaab

Preface

This book is aimed at clinical neurologists seeking to expand their knowledge of clinical epidemiology and to further their understanding of quantitative research approaches to clinical issues. One of our goals is to help prepare investigators from a variety of disciplines to conduct independent epidemiological studies. Such studies will discern the etiologic factors, underlying mechanisms, and prognostic markers that influence the onset, progression, and recurrence of disorders and diseases affecting the nervous system. Both clinicians new to research and those already engaged in research will find useful information within these chapters.

The etiology of many important disorders of the central nervous system remains elusive. For some diseases, like stroke, the most important determinants have been identified, yet better prevention and therapeutic approaches are needed to reduce the continued high incidence and mortality. This is of especially high relevance in some racial, ethnic and socioeconomic groups. Emerging technologies in recent years have provided exciting new opportunities to define these conditions, understand their etiology, and respond with effective prevention measures and novel treatments. Specifically, the application of such technologies can facilitate the characterization of the phenotype earlier in the course of the disease, improve causal inference in observational epidemiological studies, and enable accurate monitoring of response to therapy.

Yet, evolving technologies alone are not sufficient to advance our understanding of the etiology and pathogenesis of neurological disorders. It is essential for contemporary clinicians and researchers to understand how studies should be designed, with carefully selected, sized, and characterized samples, in which the technology can be applied appropriately. Once the data are collected, it is also important for the neurologist to understand in principle how they should be analyzed and interpreted.

Our new volume on neuroepidemiology addresses these challenges by paying special attention to methodological approaches. The book describes the most recent methodologies to define and quantify the burden of CNS disease, and that help to understand the underlying mechanisms of disease. Neuroimaging and molecular methods receive particular emphasis due to their expanding use in neuroepidemiology both for clinical and basic research. The book also includes descriptions of those neurological conditions that are secondary to other diseases, and whose incidence is increasing because of longer survival rates.

We have organized the book into three parts. In Part 1, the focus is on methodologies, and specifically on the multimodal assessment of neurological conditions, epidemiological methods, and advanced statistical analyses. Knowledge of methods to investigate the nervous system is key to defining the clinical manifestations of neurological diseases and to identify related risk factors and biomarkers. In addition, knowledge of study design and statistical methods is necessary to quantify disease frequency through careful ascertainment of cases within the population at risk, and to provide morbidity and mortality rates. Moreover, the rapid advance of technology continues to yield even greater amounts of data, which require careful data mining approaches. A section is dedicated to a description of advanced analytical methods and their application in clinical studies. Such methods can be applied to predict risk of developing neurological diseases over the life course, as well as variability of clinical manifestations, survival, and response to treatment.

In Parts 2 and 3, we describe the epidemiology of neurological disorders. Each chapter provides extensive and up-to-date information on observational studies and clinical trials, including study design considerations for therapeutic studies of neuroepidemiology. Part 2 focuses on the epidemiology of disorders primarily affecting the nervous system, whereas Part 3 examines the neurocognitive consequences of some conditions originating outside the nervous system.

We are delighted that the volume is part of the prestigious *Handbook of Clinical Neurology* series, where it will serve as a reference book, but we hope as well that this volume will be useful as a general text to help the individual reader gain up-to-date knowledge of the epidemiology of neurological disease and a thorough appreciation of the methodological skills required to conduct research in this exciting field.

Caterina Rosano
M. Arfan Ikram
Mary Ganguli

Contributors

S. Abramovici
University of Pittsburgh Comprehensive Epilepsy Center (UPCEC), University of Pittsburgh Medical School, Pittsburgh, PA, USA

E. Albanese
Department of Psychiatry, University of Geneva, Geneva, Switzerland

A. Bagić
University of Pittsburgh Comprehensive Epilepsy Center (UPCEC), University of Pittsburgh Medical School, Pittsburgh, PA, USA

C.K. Barha
Aging, Mobility, and Cognitive Neuroscience Laboratory, Department of Physical Therapy, Faculty of Medicine and Djavad Mowafaghian Centre for Brain Health, University of British Columbia Vancouver, BC, Canada

B.F. Boeve
Department of Neurology, Mayo Clinic, Rochester, MN, USA

D. Bos
Departments of Radiology and Epidemiology, Erasmus MC, University Medical Center Rotterdam, Rotterdam, The Netherlands

M.C. Carlson
Department of Mental Health, Johns Hopkins Bloomberg School of Public Health, Baltimore, MD, USA

M. de Groot
Departments of Radiology and Medical Informatics, Erasmus MC, University Medical Center Rotterdam, Rotterdam, The Netherlands

M. Ganguli
Departments of Psychiatry, Neurology, and Epidemiology, University of Pittsburgh, Pittsburgh, PA, USA

A.J. Gardner
Hunter New England Local Health District Sports Concussion Program; Priority Research Centre for Stroke and Brain Injury, School of Medicine and Public Health, University of Newcastle, Callaghan, NSW, Australia

D.S. Goodin
Multiple Sclerosis Center at the University of California, San Francisco, San Francisco, CA, USA

R. Hanewinckel
Department of Epidemiology, Erasmus MC University Medical Center, Rotterdam, The Netherlands

M.A. Ikram
Department of Epidemiology, Erasmus MC University Medical Center, Rotterdam, The Netherlands

M.E. Jacob
Department of Epidemiology, University of Pittsburgh, Pittsburgh, PA, USA

P. Jean-Pierre
Cancer Neurocognitive Translational Research Laboratory and Department of Psychology, University of Notre Dame, Notre Dame, and Indiana University Melvin and Bren Simon Cancer Center, Indianapolis, IN, USA

Y.L. Johnson
Department of Psychology, University of Mississippi, Oxford, MS, USA

P.J. Koudstaal
Department of Neurology, Erasmus MC University Medical Center, Rotterdam, The Netherlands

D. Lacomis
Departments of Neurology and Pathology, University of Pittsburgh Medical Center, Pittsburgh, PA, USA

T. Liu-Ambrose
Aging, Mobility, and Cognitive Neuroscience Laboratory, Department of Physical Therapy, Faculty of

Medicine and Djavad Mowafaghian Centre for Brain Health and Brain Research Centre, University of British Columbia Vancouver, BC, Canada

G. Logroscino
Department of Neurology, Mayo Clinic, Rochester, MN, USA and Neurodegenerative Diseases Unit, Department of Basica Medicine, Neuroscience and Sense Organs, University of Bari, Italy

A.M. Malek
Department of Public Health Sciences, Medical University of South Carolina, Charleston, SC, USA

B.C. McDonald
Indiana University Melvin and Bren Simon Cancer Center and Center for Neuroimaging, Department of Radiology and Imaging Sciences, Indiana University School of Medicine, Indianapolis, IN, USA

L.S. Nagamatsu
Exercise, Mobility and Brain Health Laboratory, School of Kinesiology, Faculty of Health Sciences, Western University, London, Ontario, Canada

P. Palta
Department of Epidemiology, University of North Carolina at Chapel Hill, Chapel Hill, NC, USA

T. Paus
Rotman Research Institute and Departments of Psychology and Psychiatry, University of Toronto, Toronto, Canada and Child Mind Institute, New York, NY, USA

K.R. Peck
Department of Psychology, University of Mississippi, Oxford, MS, USA

M.L.P. Portegies
Department of Epidemiology, Erasmus MC University Medical Center, Rotterdam, The Netherlands

C. Reitz
Taub Institute for Research on Alzheimer's Disease and the Aging Brain, Gertrude H. Sergievsky Center, Department of Neurology and Department of Epidemiology, Columbia University, New York, NY, USA

C. Rosano
Graduate School of Public Health, University of Pittsburgh, Pittsburgh, PA, USA

S.F. Sacuiu
Department of Neuropsychiatry, Sahlgrenska University Hospital and Department of Psychiatry and Neurochemistry, University of Gothenburg Institute of Neuroscience and Physiology, Gothenburg, Sweden

R. Savica
Department of Neurology, Mayo Clinic, Rochester, MN, USA

T.A. Smitherman
Department of Psychology, University of Mississippi, Oxford, MS, USA

B. Snitz
Department of Neurology, University of Pittsburgh, Pittsburgh, PA, USA

J. Strouse
Division of Hematology, Department of Medicine and Division of Pediatric Hematology/Oncology, Department of Pediatrics, Duke University, Durham, NC, USA

E.O. Talbott
Department of Epidemiology, University of Pittsburgh Graduate School of Public Health, Pittsburgh, PA, USA

G. Tosto
Taub Institute for Research on Alzheimer's Disease and the Aging Brain and Gertrude H. Sergievsky Center, Columbia University, New York, NY, USA

P.A. van Doorn
Department of Neurology, Erasmus University Medical Centre, Rotterdam, The Netherlands

M.W. Vernooij
Departments of Radiology and Epidemiology, Erasmus MC, University Medical Center Rotterdam, Rotterdam, The Netherlands

R. Zafonte
Department of Physical Medicine and Rehabilitation, Harvard Medical School, Spaulding Rehabilitation Hospital; MassGeneral Hospital for Children Sport Concussion Program and Red Sox Foundation and Massachusetts General Hospital Home Base Program, Brigham and Women's Hospital, Boston, MA, USA

Contents

Section 1

Principles and foundations of epidemiology and neurology research

Handbook of Clinical Neurology, Vol. 138 (3rd series)
Neuroepidemiology
C. Rosano, M.A. Ikram, and M. Ganguli, Editors
http://dx.doi.org/10.1016/B978-0-12-802973-2.00001-X

Chapter 1

Epidemiology for the clinical neurologist

M.E. JACOB[1] AND M. GANGULI[2]*

[1]*Department of Epidemiology, University of Pittsburgh, Pittsburgh, PA, USA*

[2]*Departments of Psychiatry, Neurology, and Epidemiology, University of Pittsburgh, Pittsburgh, PA, USA*

Abstract

Epidemiology is a foundation of all clinical and public health research and practice. Epidemiology serves seven important uses for the advancement of medicine and public health. It enables community diagnosis by quantifying risk factors and diseases in the community; completes the clinical picture of disease by revealing the entire distribution of disease and presenting meaningful population averages from representative samples; identifies risk factors for disease by detecting and quantifying associations between exposures and disease and evaluating causal hypotheses; computes individual risk to identify high-risk groups to whom preventive interventions can be targeted; evaluates historic trends that monitor disease over time and provide clues to etiology; delineates new syndromes and disease subtypes not previously apparent in clinical settings, helping to streamline effective disease management; and investigates the effects of health services on population health to identify effective public health interventions. The clinician with a grasp of epidemiologic principles is in a position to critically evaluate the research literature, to apply it to clinical practice, and to undertake valid clinical epidemiology research with patients in clinical settings.

The neurologist is called to the emergency department to see a patient with acute-onset, left-sided weakness. While conducting the neurologic examination, she obtains the patient's history and learns that he has a long history of hypertension and also of heavy smoking and alcohol intake. Soon, she is initiating treatment, ordering the appropriate investigations, and explaining to the patient and family the diagnosis of a stroke, the likely causes, and the prognosis. This standard clinical practice is not based solely on her anecdotal experience with her own previous patients or even her knowledge of the clinical features of stroke. Also entering into her clinical assessment, and her conversation with her patient and family, is her awareness of the "distribution and determinants" of stroke in large numbers of people, i.e., the epidemiology of stroke (Davis et al., 1987; Shinton and Beevers, 1989; Sacco et al., 1999).

As in the above example, and as we will show throughout this chapter, epidemiology is a foundation of clinical and public health practice. The thoughtful clinician with a grasp of epidemiologic principles is able to critically evaluate research that is being disseminated and also to incorporate findings into clinical practice. The clinical epidemiologist, who undertakes epidemiologic research among patients in the clinical setting, is in a unique position to advance knowledge through scientific research on vital clinical questions.

WHAT IS EPIDEMIOLOGY?

The word epidemiology is derived from the Greek terms *epi* meaning "on" or "upon," *demos* meaning "people," and *logos,* meaning "study." Epidemiology is classically defined as: (1) the study of the distribution and determinants of health-related states and events in populations; and (2) the application of this study to the prevention and control of health problems. The field of epidemiology, as we know it, originated in a landmark study of a cholera epidemic by John Snow in England, in the 1850s. By establishing that the disease was being spread

*Correspondence to: Mary Ganguli, MD, MPH, WPIC, 3811 O'Hara Street, Pittsburgh PA 15213, USA. Tel: +1-412-647-6516, E-mail: GanguliM@upmc.edu

through London via ingestion of contaminated water from the now infamous Broad Street pump, Snow revolutionized the understanding of the pathogenesis of cholera. Since then, epidemiologic research has led to game-changing public health measures to improve the health of populations, ranging from smallpox vaccination that eradicated the disease, to multidrug antiretroviral treatment which has transformed HIV-AIDS from a fatal disease to a manageable chronic disease.

Beyond investigating communicable diseases, epidemiology has also moved on to understanding chronic disease at the population level, applying the same principles. Technologic advancement has led to the emergence of subspecialty fields like molecular epidemiology and geospatial epidemiology. However, translating lab discoveries into measurable health benefits in the human population is not straightforward. While basic and clinical neurobiologic research has resulted in immense progress in our understanding of the human brain, it has become evident that we need to look beyond small convenience samples of patients to larger representative population samples to investigate the pathogenesis of diseases and conditions affecting the central nervous system. This motivation underlies the emerging field of "population neuroscience," which aims to marry the knowledge base and skill sets of neuroscientists with those of population scientists (Paus, 2010; Falk et al., 2013).

USES OF EPIDEMIOLOGY

In 1955, Jerry Morris published an article entitled "Uses of epidemiology," which he later expanded into a textbook on epidemiology. The seven "uses" Morris listed (Table 1.1) remain remarkably relevant today, and we will use them as an outline to review the basic principles of epidemiology.

Although a narrow focus on epidemiologic methods, particularly statistical methods, can potentially take away from the broader purposes of epidemiology, we will briefly outline the key methods corresponding to each "use" of epidemiology.

Table 1.1

The J.N. Morris "seven uses of epidemiology"

1. Make a community diagnosis
2. Supplement the clinical picture
3. Identify causal/risk factors
4. Compute individual risk
5. Chart historic trends
6. Delineate new syndromes
7. Evaluate health services

Making a community diagnosis

Epidemiology investigates health at the population level. A community or population diagnosis identifies the magnitude and distribution of diseases present in the community, i.e., the public health burden of disease. By helping to prioritize problems and identify at-risk subpopulations, community diagnosis is a prerequisite for formulating health policy and planning public health programs. A complete picture of the health of the community can be obtained only by collecting information on a comprehensive list of variables, including sociodemographic, health, and environmental factors related to the presence of disease. However, to the clinical epidemiologist, typically focused on a single disease, the pertinence of the "community diagnosis" is in understanding the burden of that disease in the population. Required reporting or voluntary registries for a disease can make community diagnosis easier, but since most neurologic disorders are not reportable, only population-based studies can provide estimates of disease burden.

Here we will focus on the epidemiologic measures most commonly used to quantify the health of populations and allow for comparisons among them.

MEASUREMENT OF DISEASE IN POPULATIONS: PREVALENCE VS. INCIDENCE

We quantify the magnitude of a disease in the population is in two primary ways – prevalence and incidence.

PREVALENCE

Prevalence is the proportion of diseased persons in a defined population at a given point in time, or during a short, fixed period of time, often expressed as a percentage or per 1000. It represents the public health burden of the disease at that time point or during that period. Prevalence is measured by means of a one-time cross-sectional study or survey.

$$\text{Prevalence} = \frac{\text{Number of existing cases of a given disease in a defined population}}{\text{Total number of persons in that population}}$$

INCIDENCE

Incidence is the rate at which new cases of a given disease develop in a population over a specified time period, among people who at the beginning of that period were free of that disease (i.e., at potential risk of developing that disease in the future). Incidence represents the risk of developing the disease for persons in that population. It stands to reason that incidence can only be measured in a prospective (cohort) study which assesses study participants repeatedly over time.

$$\text{Incidence rate} = \frac{\text{Number of new cases in a given population over a specified period of time}}{\text{Number of persons who were disease-free at the beginning of that period}}$$

Incidence could be understood and expressed very simply as, e.g., an annual rate. However, it is usually calculated in terms of "person-years of follow-up," which takes into account the different lengths of time that different individuals in the cohort were followed, by simply summing up the disease-free periods for which each individual was observed.

Whether calculating prevalence or incidence, it is critical to understand that everyone in the numerator must also be in the denominator, and everyone in the denominator must have a chance of being in the numerator.

Prevalence and incidence are most meaningfully reported not for overall populations but for specific demographic groups, most commonly as age-specific estimates. When the prevalence of dementia is shown for age groups 65–74, 75–84, and 85+ years, it becomes immediately obvious that prevalence of dementia increases with age. In interpreting this as an age effect, another possibility to be considered is that of cohort effects, described later.

What is the relationship between prevalence and incidence?

Incidence

The rate at which new disease develops in a given population depends on many factors, including that population's age, sex, racial/ethnic characteristics, life expectancy, stability, and distribution of risk factors. The estimation of incidence further depends on the accuracy of screening and diagnostic methods and the earliest stage of the disease that these methods can detect.

Prevalence

The proportion of diseased people in the population at a given time clearly depends on the rate at which new disease develops in that population, but incidence is only one of two key components of prevalence. The missing link is duration of disease, i.e., how long people live with that disease in that population, until either they recover from the disease or die with the disease.

Thus,

$$\text{Prevalence} = \text{Incidence} \times \text{duration}$$

The duration of disease depends on factors which remove the diseased person from the population – the recovery rate from the disease (as in curable and self-limiting diseases) and the death rate with, or due to, the disease (as in most chronic diseases).

Thus, with no change in incidence, prevalence could be reduced by either a high case fatality rate or high recovery rate; conversely, prevalence would be increased by any factors that prolong life with the disease. Factors associated with prevalent disease may simply reflect their association with duration, rather than with incidence. This is one reason why we cannot prove causality from associations between disease and various potential explanatory factors ("exposures") in cross-sectional studies (see more on this topic under Observational studies, later in this chapter). Earlier detection of disease can appear to lengthen survival and increase prevalence estimates even if mortality and recovery rates remain unchanged.

Mortality

Mortality rate due to a given cause, such as stroke, head trauma, or Alzheimer's disease, is a common and dramatic way of describing the impact of a disease in the population. In cohort studies where participants are followed for considerable lengths of time, disease-specific death rates could be calculated. Clinical trials may use survival/mortality as their primary outcome and compare mortality rates between the intervention and control populations.

$$\text{Cause-specific death rate} = \frac{\text{Number of deaths occurring from that cause during a specified period}}{\text{Total population during that period}}$$

The above death rate is not particularly informative, as it does not take into account the age or gender composition of the population and thus cannot be used to compare the health status of different populations with different demographic characteristics.

Age-specific and sex-specific death rates for specific causes are more commonly reported and calculated as follows:

$$\text{Age-sex-specific death rate} = \frac{\text{Number of deaths occurring from that cause in a specified age/sex group in a specified period}}{\text{Total population in that age-sex group during that period}}$$

Completing the clinical picture

Diseases are typically first described in specialized clinical settings. However, because of the various selection factors that govern access to these settings, the patients seen there are not typical of people with the given disease in the population at large. Alois Alzheimer first described, in the early 1900s, the pathology and clinical symptoms of a single case of presenile dementia (later called Alzheimer's disease) in a 51-year-old woman admitted to a psychiatric hospital in Germany. From then on, the condition was assumed to be an unusual disease of the middle-aged (Cipriani et al., 2011). In the 1960s, however, Martin Roth and colleagues in England conducted a population-based neuropathologic study and convincingly established that Alzheimer's disease was in fact also a relatively common disease among the elderly (Roth et al., 1966). In another example, Leo Kanner described 11 children with autism in his 1943 paper entitled "Autistic disturbances of affective contact," and perceived an association between higher parental educational attainment and the occurrence of autism (Kanner, 1968). Kanner's specialized case load of patients had generated a selection bias and the apparent association was proved wrong in subsequent population-based epidemiologic studies (Kogan et al., 2009).

The importance of sampling

While recognizing that anecdotal evidence is woefully inadequate to inform clinical practice, we also realize that research on entire populations is not feasible. Scientific principles underpinning clinical care are derived from research on adequately sized samples that are representative of the population which is to receive the care. This target population could be defined in geographic terms (e.g., a region in Europe) or as a group with a certain disease (e.g., epilepsy) or exposure (e.g., factory workers exposed to asbestos).

While few readers of this chapter may do their own sampling, it is important for all readers to gain a basic grasp not just of how sampling is performed, but also how the sampling approach can influence the results of a study and the inferences that can be drawn from it.

Random sampling

The purpose of drawing a random sample from the target population is to ensure that each person in the population has an equal chance of being selected. It is perfectly legitimate to study the prevalence of epilepsy in a neurology clinic sample, or of stroke in a nursing-home sample, recognizing that these prevalence estimates will be considerably higher than the prevalence estimates from a general practice clinic, or the community at large. Further, since clinic patients with epilepsy are not randomly drawn from all people with epilepsy in the community, the inherent selection bias in the clinic sample will limit the extent to which inferences about characteristics of clinic patients can be generalized to all individuals with epilepsy in the community.

Unbiased estimates of population averages can be obtained only by applying random sampling techniques to the reference population. This method requires having a comprehensive list (such as a census) from which the sample is drawn using a table of random numbers or random-number-generating software. In simple random sampling, participants for the study are drawn in such a way that, every time a person is selected, every other person in the population has the same probability of being selected; chance alone determines the probability of any individual being selected. Other variations exist, such as systematic random sampling (selecting every *n*th person on a list) and stratified random sampling (selecting randomly within strata defined by age group or other characteristic). Random sampling techniques are consistently employed in survey analysis.

Many excellent observational studies and clinical trials are undertaken in convenience samples of patients or volunteers who both fulfill strict eligibility criteria for that study and also are willing to undergo all study procedures. While these selection factors are usually essential to testing that study's hypotheses, it must also be remembered that they also limit our ability to generalize the study's inferences to the individuals with the same disease in the larger population.

The word "generalizability" is interpreted differently in different circles and is sometimes used as a default criticism of any study, since no study is fully representative of, or generalizable to, all populations. If a study's selection factors themselves are associated with the variables being investigated, then they will influence the results and undermine internal validity. Such factors may be inadvertent or unavoidable, such as when older adults with an age-associated disease are less likely to volunteer for research than middle-aged adults with the same disease. Alternatively, they may be under the control of the investigator, e.g., if individuals with stroke are explicitly excluded from participation in a study of Alzheimer's disease, making it impossible for that study to later examine the relationship between stroke and Alzheimer's disease. For external validity to be present, the sample should be representative of the population from which it purports to be drawn, i.e., the selection process should be as unbiased as possible. While the investigator does want the study results to be replicated in other samples, we do not expect every study to have the same result. For example, we do not expect that results from a study of head trauma outcomes in female nuns would fully generalize to a study of male military personnel, given that nuns and soldiers have different exposures and head trauma rates. However, we would hope that the results from the sample of nuns would be true of the larger population of nuns from which the sample was drawn, and perhaps generalize to other samples of nuns. Further, if an association between a given exposure and a given disease were to be found both in the nuns and also in the soldiers, that finding would have greater likelihood of reflecting a true phenomenon (Kukull and Ganguli, 2012).

Sample size

Whether reviewing an article about a study or designing a new study, the reader should understand how the results can be influenced by the number of participants in the study. The larger the sample that can be drawn, the more likely it is to be representative of the population from which it is drawn. Smaller samples are more likely to be influenced by variability in the sampling process, because, unlike a batch of laboratory rats, human populations are rarely homogeneous. Since it is difficult and expensive to recruit a sample that is both large and random, a key factor in study design is the sample size calculation. Here, the principle is to calculate the minimum number of participants required to estimate prevalence or incidence, or more often, to test a hypothesis such as a significant difference in some characteristic or exposure between groups within the sample.

When calculating sample size, the following values are considered.

p-value

The *p*-value is the probability that the observed association is due to chance. Even with random sampling, there remains a possibility of random error; the sample can demonstrate differences or associations that are not true of the parent population. We conventionally restrict the probability of this error to no more than alpha $=0.05$ (i.e., if the study was repeated multiple times, the same result would be obtained 95% of the time). This $p=0.05$ threshold minimizes the chance of incorrectly interpreting a chance finding as genuine (alpha error or type 1 error).

Power

The power of the study is the probability of correctly detecting a difference between two groups in the sample, if a difference truly exists in the population from which the sample is drawn. Setting power at the conventional 80–95% range minimizes the possibility of beta error or type 2 error, the probability of missing a true difference.

Effect size

The magnitude of the expected difference between two groups in a study is stated in terms of an estimate of the effect size (along with the *p*-value and confidence interval). The effect size that we expect to see in our study is factored into the calculation of the sample size; the smaller the effect to be detected, the larger the sample (greater the power) that is required to detect it. The expected effect size may be chosen based on previous studies or on the investigator's clinical judgment; it should be large enough to be clinically relevant but not so large as to be implausible.

Making inferences from the sample to the population

We re-emphasize here that the purpose of drawing a representative sample from a population is to be able to infer that the results from the sample are true of the population from which it was drawn. Specifically, we want the mean (average) value of some attribute (e.g., height) measured in the sample to also be the mean of that attribute in the larger population. Similarly, we want the proportion of the sample that has a certain characteristic (e.g., college education) to be similar to the proportion with that characteristic in the population. In turn, the effect size that we determine for an exposure factor, for example, a new drug, should represent the true effect in the population. That is, we want the sample value to be a good estimate of the population value, within some confidence limits.

Normal distribution and confidence limits

We will assume that readers are familiar with the concepts of normal distribution (bell-shaped curve, Gaussian curve) and of central tendency (e.g., the mean), and dispersion (e.g., the standard deviation, or SD, around the mean). Many of the statistical tests used to examine differences between groups assume that the variables of interest are normally distributed in the population. While this is often not the case, most tests will be sufficiently robust if the sample is large enough. The rule of thumb here is a sample size of at least 30, with no fewer than 5 individuals in each group or subgroup being compared.

The standard error (SE) is the standard deviation of the sample distribution. In a normal distribution, 95% of sample observations fall within 1.96 SE of the population mean, which is why 95% confidence intervals are provided along with prevalence, incidence, and effect size estimates. The 95% confidence limits indicate that there is 95% probability that the actual population value is within those limits obtained from the sample, i.e., the range of values for the true effect in the population. It is the corollary of the *p*-value cutoff and alpha error level of 0.05 that we described earlier. The narrower the confidence interval around an estimate, the more precise that estimate will be; larger samples generally provide narrower confidence intervals. Thus, in a study with appropriate sample size and random sampling procedures, the mean values, effect sizes, and confidence limits provide a reasonably accurate picture of the reference population.

Identifying causal/risk factors

Epidemiologists use a range of study designs to evaluate associations among social, environmental, or biologic variables. Broadly, these study designs can be classified as observational or experimental (Table 1.2). In observational studies, the researcher only observes participants, collecting data on the variables required to test the hypothesis in question. In experimental studies, the researcher intervenes in some way to try and influence an outcome.

Observational studies

Descriptive studies

A descriptive study describes a disease in a population in terms of its magnitude and distribution but does not test any associations. For example, the tabulation of the mortality rate due to stroke in a country over several decades is used to describe the mortality trend over time. These data are usually available from centers for health statistics in different countries. It is always possible to speculate about the causes of a trend, but analytic studies are required to evaluate the factors associated with the trend and potentially establish its causes. Purely descriptive studies are rare; a descriptive analysis is often the preliminary stage of an epidemiologic study that eventually tests associations. Case studies and case series (such as the example of Alzheimer's and Kanner's original patients) provide important but limited descriptive data regarding a disease; the distribution of the disease in the population is not evident. These clinical case descriptions often lead to larger observational studies of the phenomena.

Ecologic studies

Ecological studies are those in which the units of observation are not individuals but rather populations of a region. Associations are tested between summary measures of populations, often collected for other purposes. For example, an ecologic study suggested the apparent cardioprotective effect of wine consumption by correlating per capita consumption of alcohol with coronary heart disease mortality rates in different countries (Criqui and Ringel, 1994). While intriguing and potentially useful for hypothesis generation, such studies can result in erroneous conclusions because of the phenomenon of ecologic fallacy. This is a bias that occurs when the associations observed at the population level do not represent the association at the individual level. For example, Emile Durkheim's 1897 treatise on suicide demonstrated that suicide rates were higher in Protestant countries than in Catholic nations; he inferred that greater social control among Catholics was responsible for their lower suicide rates (Selkin, 1983). However, the Protestant countries were different from the Catholic countries in many ways besides religion (confounding), which could not be adjusted for in that study. Also, the predictor and the outcome were measured for countries and not for individuals (the suicides were not linked to an individual's faith but attributed to the predominant faith in that country) and this gives rise to an aggregate bias. Thus confounding and aggregate bias contribute to ecologic fallacies and incorrect inferences (Freedman, 2001).

Table 1.2

Classification of epidemiologic studies

Observational
Descriptive
Analytic
Ecologic
Cross-sectional
Case-control
Cohort
Experimental
Randomized controlled trials
Community trials
Field trials

Cross-sectional studies

Cross-sectional studies or surveys measure both the exposure and outcome in a sample of the population at a point in time. Ideally, the sample should be randomly selected from the population. Here, a matter of concern is the proportion of selected individuals who refuse to participate, since they are almost certainly dissimilar in some way from those who consent. The larger the refusal rate, the greater the likelihood of response bias within the sample. Analysis of such studies should always report the number of eligible individuals who were initially selected and approached and what proportion of them enrolled in the study.

Surveys of representative samples capture the prevalence of disease in the population being studied. It is also possible to test for associations of prevalent disease with potential risk factors, but not possible to know whether the exposure preceded the effect. Since temporality of association is a strong criterion for causality, cross-sectional studies cannot prove causality but help to generate causal hypotheses. Cross-sectional surveys of representative samples are useful in the assessment of healthcare needs of the population and are often used by countries and regions for this purpose. Repeated surveys can provide important information regarding health trends.

Case-control studies

Case-control studies usually collect data at a single point in time, as in a cross-sectional study, but are conceptually longitudinal in that they collect exposure data from the past. They are ideally used to investigate rare diseases with prevalence too low to be cost-effectively detected by random sampling from a population. Typically, they compare a group of patients with identified disease (cases) to a group without the disease (controls). Cases are usually recruited from clinical settings or by advertisement, from a broadly defined population (e.g., residents of a given city). Controls are selected from the same broadly defined population so as to be similar to the cases except for not having disease. Exposure data are obtained from both cases and controls, often by direct questioning, examination, or laboratory tests. Association between exposure and outcome in case-control studies is quantified using the odds ratio, which for rare diseases is a reasonable estimate of the relative risk.

Selection of suitable controls is an important aspect of case-control studies because if there is a systematic difference between cases and controls in any aspect other than the disease itself (selection bias), a true association between the exposure and disease may be missed or a spurious association observed. Random selection of non-diseased controls from the reference population would be ideal, but is often not feasible. Common control groups that are used include friends, relatives, or neighbors of patients, patients from the same hospital with other diseases, all of which can potentially introduce different types of biases. It is also essential to collect data in the same way from cases and controls, rather than, e.g., obtaining the exposure data directly from the controls but from the family members of the cases. In case-control studies an additional potential problem is that an individual classified as a disease-free control today may develop the disease at a later time in the future; if the exposure under study is a gene, the study will suffer from misclassification error.

Cohort studies

Cohort studies are longitudinal studies where a representative group of people in the community are followed up prospectively. A good example is a disease-free group that is followed to identify incident cases of disease as they arise. At the beginning of the study (and during repeated follow-up assessments), exposure status (often to multiple exposures) is measured and the cohort is closely examined for the development of disease. As new cases of disease are identified in this study, calculation of incidence rates is possible. Associations between initial exposure and subsequent disease are measured using relative risk. Besides following disease-free cohorts for incidence of disease, cohorts of participants with a given disease, e.g., stroke or Parkinson's, could be followed up for investigating outcomes such as rate of progression, development of complications, or mortality.

While cohort studies are ideal for studying causal factors of diseases and disease outcomes, they are major undertakings involving much expenditure and follow-up. Recruitment of randomly sampled participants, and subsequent retention of the cohort over many years, is labor-intensive and challenging. The internal validity of a cohort study depends partly on retention of the cohort, as attrition introduces new biases.

Many cohort studies like the Baltimore Longitudinal Study of Aging are composed of volunteer participants and have provided valuable information on associations between exposures and disease (Shock et al., 1984); however, they cannot provide information on incidence rates in the population at large.

Nested case-control studies

Nested case-control studies are case-control studies embedded within cohort studies. Cases and controls are chosen from the same cohort; cases are the participants who were initially disease-free but developed the disease, i.e., became incident cases, during follow-up. Nested case-control studies improve accuracy of information of exposure as data were systematically collected prior to the occurrence of disease, and not subject to length bias (prevalence bias) which we will describe later.

Experimental studies

Experimental epidemiology involves intervention in a group of people – it might be the addition or the removal of a factor, for example, addition of a dietary supplement or a weight loss program with the objective of reducing body mass. The effects of the intervention are then assessed by comparing outcomes between a group that received the intervention and a group that did not. Ethical considerations play an enhanced role in the design and implementation of these studies. Randomized controlled trials, field trials, and community trials are experimental studies.

Randomized controlled trials

Randomized controlled trials are experimental studies where participants are randomly allocated to receive or not receive the intervention. Randomization ensures that at the beginning of the study the intervention and control group are comparable and that the selection to receive treatment is not biased. Clinical trials are randomized controlled trials of treatment options, e.g., a specific drug for patients with a specific disease.

Field trials

Field trials are experimental studies conducted on healthy people in the field, i.e., the community-living population. Field trials are commonly conducted for testing the effect of vaccines.

Community trials

Community trials provide the intervention to communities rather than to individuals. This is particularly useful when the intervention focuses on changes in group behavior. An example is the Stanford five-city project which provided prevention measures to reduce cardiovascular disease risk (Farquhar et al., 1985).

Measures of association

When potential causal/risk factors are identified in epidemiologic studies, their associations with disease are typically estimated using relative risk and odds ratio. The key to interpreting these estimates is that a ratio of 1.0 means that the risk or odds is identical between the two groups being compared (e.g., cases and controls). For a given exposure, a ratio greater than 1.0 (e.g., 1.5) means that the risk of disease is higher (in this example, by 50%) in the exposed group compared to the unexposed group. Conversely, a ratio less than 1.0 (e.g., 0.5) means that the risk in the exposed group is lower than in the unexposed group. In addition to the ratio itself, the reader should pay close attention to the 95% confidence interval around the ratio; if the ratio is higher or lower than 1.0, but the confidence interval includes 1.0, then the observed increase or decrease in risk is not significantly different between the groups being compared.

In the examples below we demonstrate the calculation of the ratio, but not of the 95% confidence intervals, which are excessively complex for our illustration purposes.

Odds ratio

Odds ratio is the measure of association between exposure and outcome in case-control studies and cross-sectional studies. It is the ratio of the odds of exposure among cases to the odds of exposure among controls. The odds ratio is a good approximation of the risk ratio when the disease is rare and when the controls are representative of the general population in terms of the exposure.

Here is a hypothetic example of how the odds ratio is calculated in a case-control study. A case-control study examined 100 cases of multiple sclerosis (MS) and 200 controls and tested for the presence of Th1 cytokine in cerebrospinal fluid (CSF) samples in all cases and controls. The results are given in Table 1.3.

Table 1.3

Results from a hypothetic case-control study examining the association between multiple sclerosis and Th1 cytokine in cerebrospinal fluid

Test for Th1	Cases	Controls	Total
Positive	62 (*a*)	24 (*b*)	86
Negative	38 (*c*)	176 (*d*)	214
Total	100	200	300

$$\text{Odds ratio for exposure among cases versus controls} \\ = a^{*}d/b^{*}c = (62^{*}176)/(24^{*}38) = 12$$

This indicates that the MS cases were about 12 times more likely to be positive for the marker than the controls. As the exposure odds and disease odds are equivalent, this odds ratio can also be interpreted as follows – individuals with a positive CSF test for Th1 cytokine have a 12 times higher odds for developing MS than individuals whose CSF is negative for the marker.

Relative risk or risk ratio

Relative risk or risk ratio is the ratio of incidence among the exposed to the incidence among the unexposed. As this involves the use of incidence, relative risks can be calculated only in cohort studies and clinical trials.

Here is a hypothetic example of how relative risk is calculated in a cohort study. A cohort study followed 5000 older adults for 10 years to study the association between lifestyle habits and the development of cardiovascular disease. The findings for smoking and stroke are given in Table 1.4.

$$\text{Relative risk} = \frac{\text{Incidence among the exposed}}{\text{Incidence among the unexposed}} \\ = (100/850)/(150/4150) = 0.12/0.04 = 3$$

This relative risk can be interpreted as follows: smokers have a three times higher risk of developing stroke when compared to nonsmokers.

Table 1.4

Results from a hypothetic cohort study examining the association between smoking and stroke

Smoking	Stroke	No stroke	Total
Yes	100 (*a*)	750 (*b*)	850
No	150 (*c*)	4000 (*d*)	4150
Total	250	4750	5000

In observational studies with multiple potential confounding factors, regression models are employed which adjust for these factors (i.e., including these factors as covariates in the model, and, where appropriate, also including interaction terms between covariates). Results are presented as adjusted risk ratios and odds ratios along with 95% confidence intervals and *p*-values to quantify associations.

Timing and duration of exposure

This is a critical and often overlooked aspect of risk factor studies. An observational study may show that individuals with a certain lifetime exposure have a lower risk (i.e., incidence) of developing a certain disease. This result should not be interpreted as necessarily meaning that short-term administration of that exposure will prevent that disease, or even treat the disease once it has already manifested. The Cache County Study in Utah showed that women who had taken estrogen supplements to treat symptoms of menopause subsequently had a lower incidence of Alzheimer's disease than women who had not taken estrogen (Zandi et al., 2002). This discovery led to a randomized clinical trial of opposed (combined with progestin) and unopposed estrogen in older women; in the trial, women taking opposed estrogen showed an elevated risk of dementia within the Women's Health Initiative Study (Shumaker et al., 2003). The trial was seen as having discredited the observational study, but in fact the two studies had tested different hypotheses. The observational study had shown a protective effect among women taking estrogens for at least 10 years, and having done so at least 10 years before the onset of dementia. The trial had been conducted in women who were older and perhaps already in the preclinical stages of Alzheimer's disease. The Cache County study later demonstrated that observational findings in older women replicated those in the trial (Shao et al., 2012).

Sources of error in epidemiologic studies

Bias

Bias is the result of systematic error in the design and conduct of the study, such that the observed results in the sample will be different from the true results. Bias occurs due to flaws in the method of selection of study participants or in the process of gathering information regarding exposure and disease. This systematic error is different from random error due to sampling variability, which results from the use of a sample to estimate parameters for the reference population. We will discuss two broad categories of bias: selection bias and information bias.

Selection bias

Selection bias occurs when there are systematic differences between members of the population selected for the study and those who are not. For example, cases of Alzheimer's disease recruited from a research clinic were found to be more likely to carry the *APOE*4* genotype than cases captured by population surveillance within the same area (Tsuang et al., 1996). It was subsequently determined that participants at the clinic registry were younger with earlier disease onset and more advanced Alzheimer's, all characteristics associated with carrying the *APOE*4* allele. There was thus an inadvertent selection bias in the clinic sample which led to a biased overestimate of the relative risk.

Prevalence bias (length bias) is a kind of selection bias that occurs because, at any given point in time, the prevalent cases are those who have survived the longest. Prevalence bias can distort associations between risk factors and diseases. In the 1990s, several case-control studies demonstrated a protective association between smoking and Alzheimer's disease (Kukull, 2001). It was later understood that smokers who developed Alzheimer's were dying earlier than nonsmokers with Alzheimer's because of other diseases associated with smoking. This resulted in inflated numbers of smokers among controls and a reduced number of smokers among those with Alzheimer's, leading to the apparent protective association. This is the phenomenon of competing risks, in which the exposure factor (smoking) is associated with more than one event (death, Alzheimer's) and the occurrence of one event (death) will prevent the other (Alzheimer's) from being observed in the study.

Attrition bias is an important bias in longitudinal studies, since individuals lost to follow-up over the course of the study are likely to be different from those who remain under observation until the outcome or the end of the study. For example, those who drop out of a study because they die are probably those who were more severely ill than those who survived and remained in the study; those who drop out of a weight loss study are very often those for whom the intervention is not effective. Treating these dropouts as random would bias the results. Since some degree of attrition is inevitable, statistical methods are available to evaluate and address attrition bias to varying extents.

Information/measurement bias

Information bias occurs when the measurement and classification of the exposure and outcome are inaccurate. Recall bias is a type of information bias common in case-control studies where the cases (or their families) are more likely to recall a prior exposure than the controls. Many previous case-control studies showed an association between Alzheimer's disease and head trauma (Mortimer et al., 1991) but could not be replicated in a prospective study, where exposure was determined before the onset of dementia (Chandra et al., 1989). If the investigator who is measuring the outcome is aware of the exposure status, this can influence the measurement; the resulting inaccuracy is termed observer bias. To avoid this bias, measurements are performed in a blinded fashion.

Confounding

Confounding occurs when a certain exposure *A* (the confounder) is associated with both the exposure/risk factor being studied *B* (the exposure) and with the disease *C* (outcome), and its effect has not been separated out. The researcher erroneously concludes that the exposure *A* is associated with the disease *C*, whereas in fact the association is spurious. An apparent protective effect of antioxidant supplement consumption against cognitive impairment might be confounded by education, if more highly educated individuals perform better on a cognitive test and are also more likely to buy nutritional supplements (Mendelsohn et al., 1998), as demonstrated in Figure 1.1.

Confounding can be controlled at the design stage using randomization, matching, and restriction. Randomization, the random allocation of participants to intervention and control groups, is standard in experimental studies and ensures that the confounder variables are equally distributed among the intervention and control groups. Matching for confounder variables between cases and controls is employed in case-control studies. For example, if age is a confounder, cases and controls are selected to be matched on age. Restriction is the method of excluding participants using exclusion criteria such that certain confounders are eliminated. At the analysis stage, confounding can be controlled by stratification (separate analysis for participants with and without the confounding exposure) or multivariate modeling, where confounders are adjusted for (i.e., included as covariates) in the statistical models.

Even after using multiple methods to adjust for confounding, some amount of confounding is often persistent because certain confounders are not known or have simply not been measured – this is termed unmeasured or residual confounding. If there are multiple confounders which are known but not measured, it might be best not to attempt the analysis at all, as confounding can gravely distort associations.

Establishing causality in epidemiologic studies

Causal inference is the term used for the process of determining whether an observed association truly reflects a cause-and-effect relationship. Establishing causation is complicated; in theory, we can only establish causality if we examine the same group of individuals with and without the exposure simultaneously (the counterfactual framework) and examine for the onset of disease. This is impossible in the real world; the randomized controlled trial comes closest to achieving this kind of a scenario. However, not all exposures can be randomized, e.g., we cannot randomize individuals to smoke/not smoke, or to experience or not experience head trauma. Causality is most often established by triangulation of evidence from multiple animal and human studies.

Computing individual risk

The risk to an individual can be computed only by first studying the experience of populations and computing population averages. The most defensible estimates of human health risks due to exposures are from epidemiologic studies rather than laboratory experiments.

The measures of association described earlier quantify the increase or decrease in the probability for disease in exposed individuals compared to the unexposed, but it is important to understand that they do not quantify the probability itself. After computing measures of association between exposure and outcome, adjusted for confounding factors, most often by regression analysis, it

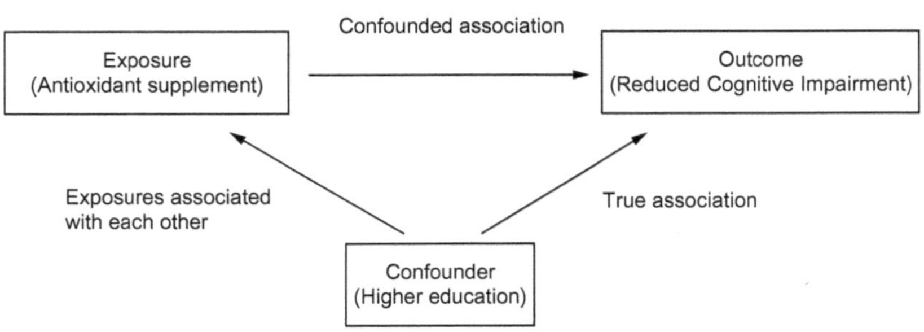

Fig. 1.1. Higher education as a confounder in the association between antioxidant supplement and reduced cognitive impairment.

Table 1.5

Schematic representation of results from a screening test

	Screen positive	Screen negative	Total
Disease	True positive (TP)	False negative (FN)	All diseased (TP+FN)
No disease	False positive (FP)	True negative (TN)	All nondiseased (FP+TN)
Total	All screen positive TP+FP	All screen negative FN+TN	

is possible to compute the individual probability of the outcome for each participant in the study based on exposure factors. In this era of personalized medicine and mobile phone apps, there is great potential for these predictive probabilities to be used on a larger scale for determination of individual risks for the purpose of clinical interventions as well as health education. For example, data from the Framingham Heart Study have been used to develop a risk assessment tool to calculate a person's 10-year risk of having a heart attack (D'Agostino et al., 2008). Other multivariate risk scores have been developed to predict the risk of developing diabetes and stroke. Risk assessment, however, is fraught with uncertainty due to limitations in the available exposure data and the limitations of statistical modeling. Much care needs to be taken to develop accurate models.

In the above paragraph, we considered the calculation of risk for disease in an individual who is free of the disease. This is important for taking preventive measures. For a clinician, a more common scenario is the assessment and diagnosis of disease in a symptomatic patient. Epidemiologic methods come in handy in determining whether a test can detect disease accurately. The concept of screening tests and their capacity to diagnose disease is important for understanding accuracy of tests.

Screening tests

Strictly speaking, from a public health perspective, a screening test is one that is applied to all persons in the population (irrespective of symptoms) to identify those who are likely to have the disease. When a screening test detects that a person has a strong probability of a given disease, a detailed assessment using a "gold-standard" test is necessary to diagnose the disease. A Pap smear screens for early cervical cancer among women, but a positive Pap smear does not confirm the diagnosis; it merely identifies the women who require a cervical biopsy and histopathologic examination of cervical tissue for the gold-standard diagnosis of cervical cancer.

When the Geriatric Depression Scale is used to screen a population for depression symptoms, some individuals may screen positive (i.e., obtain a high score) without having a major depressive disorder (as per the gold-standard diagnosis). Conversely, some individuals with major depression may score low on the same test. The validity of the screening tests is described in terms of specificity, sensitivity, positive predictive value, and negative predictive value.

In Table 1.5,

Sensitivity of a screening test
= true positives/all diseased; how likely a positive test is to capture all individuals with disease

Specificity of a screening test
= true negative/all nondiseased; how likely a negative test is to eliminate all individuals without disease

Positive predictive value
= true positive/all screen positive; how likely the patient with the positive test is to have the disease

Negative predictive value
= true negative/all screen negatives; how likely the patient with the negative test is to not have the disease

Sensitivity and specificity are fixed properties of the test. Positive predictive value and negative predictive value vary according to the prevalence of the disease in the population. In a low-prevalence population, where there are very few true positives compared to false positives, positive predictive value decreases and negative predictive value will increase. Thus, the same screening test may perform differently in different populations.

Charting historic trends

Epidemiologic analyses of data make it possible to examine trends in the health of populations over time. The epidemiologic transition, i.e., the change in morbidity and mortality patterns from predominantly infectious causes to predominantly chronic disease-related causes, was identified by evaluation of secular trends. An analysis of stroke incidence over time in the Atherosclerosis Risk in Communities study revealed that stroke incidence had decreased from 1987 to 2011 (Koton et al., 2014). Charting historic trends may provide clues to the etiology of disease. A recent decrease in the incidence of Alzheimer's disease has been attributed to the better control of vascular risk factors and vascular diseases like hypertension (Schrijvers et al., 2012).

Cohort effects

Cohort effects are variations over time, in one or more characteristics, among groups of individuals defined by some shared experience such as year or decade of birth, or years of a specific exposure. Any given population comprises multiple subcohorts with different rates of exposures and outcomes. This makes the overall population heterogeneous and can mask or distort effects which might be present in smaller, more homogeneous, constituent subcohorts. For example, an apparent relationship between aging and cognitive impairment within an American population as a whole may in fact reflect not an age effect but a cohort effect. The earlier-born cohort (now aged 85+ years) grew up during the Depression Era and many boys dropped out of school at age 12 to work in the coal mines. Their poor cognitive impairment in their 80s might be the result of their early adverse educational or environmental exposures compared to their children's generation (now aged 65–74 years), and not merely a function of "age." Within each birth cohort, there may be no age effect. In addition to age and cohort effects, there can be period effects due to events or developments at a specific time, e.g., a nuclear radiation exposure, or introduction of a new therapeutic class of drugs.

Additional factors to be kept in mind when assessing trends over time include changes in the age composition of the population, and changes in screening and diagnostic criteria. If any of these factors are in play, they can produce changes in incidence and prevalence which do not in fact indicate a true trend due to, e.g., improved control of the disease.

Delineating new syndromes

Clinical syndromes like parkinsonism and Guillain–Barré syndrome were first identified in clinical settings and appeared in the scientific literature initially as case studies or case series. Subsequently, epidemiologic studies were used to delineate different subtypes and develop case definitions. For example, acute motor axonal neuropathy was recognized as an important subtype of the Guillain–Barré syndrome as a result of a study on a large sample of patients in China (McKhann et al., 1993). Psychologic and behavioral disturbances, largely disregarded in clinical dementia research, were recognized as common features of dementia in a population-based study of dementia (Lyketsos et al., 2000). Each subtype of a disease may have distinct pathogenesis, clinical features, and response to therapy, which may not be recognized unless a clinical epidemiology approach is taken. Epidemiology has also aided in the recognition and prioritization of important phenomena which may not present directly to the clinician, as in subclinical cardiovascular disease (Chaves et al., 2004). Epidemiologic studies have also helped in identifying that apparently disparate phenomena are linked, as in the case of metabolic syndrome (Reaven, 1997), and the frailty syndrome (Fried et al., 2001).

Evaluating health services

Epidemiology evaluates the impact of healthcare on population health. Concrete evidence for changes in disease burden in the population mostly comes from review of epidemiologic research. A review of research findings on stroke mortality over the past decades in the USA concluded that the decline in stroke mortality could be attributed to a lower incidence rate of stroke as well as lower case fatality of stroke (Lackland et al., 2014). The declining incidence was mostly attributed to better control of hypertension. Such an evaluation of medical care is possible only with data from longitudinal population-based studies that estimate incidence rates and mortality rates. Epidemiologic surveillance of lab samples was used to detect a decline in incidence of Japanese encephalitis following a vaccination campaign in a district in India (Ranjan et al., 2014). Epidemiology is critical for estimating the actual community-level impact of health services and programs.

MISUSES OF EPIDEMIOLOGY

Thus far, we have discussed at some length the "uses" of epidemiology. In closing we would like to briefly propose a few "misuses" of epidemiology as well, indicating common practices which reflect a lack of understanding of the epidemiologic principles we have described:

1. Using the term "epidemiologic" to mean a specific study design rather than a perspective on understanding and examining disease in populations
2. Making directional inferences from cross-sectional data
3. Generalizing results from biased nonrepresentative samples, or failing to examine information on the study population to allow for contextual inferences
4. Assuming observed risk factor associations are causal; i.e., failing to recognize that an observed association is merely a signal, and does not in itself explain the underlying mechanism

5. Basing recommendations and interventions (or intervention trials) on observational data without first understanding the required timing and duration of the exposure
6. Reporting and publicizing obscure associations without considering biologic plausibility or underlying mechanisms
7. Defining exposures and outcomes interchangeably and too broadly to be useful.

SUMMARY

Using the classic "seven uses of epidemiology" as a framework, we have provided a brief overview of epidemiologic principles and methods as relevant to clinicians and clinical researchers. These include community diagnosis (prevalence, incidence, mortality, cohort effects), completing or supplementing the clinical picture, establishing risk relationships, the proper use of different study designs and methods to address different research questions, delineating new syndromes, establishing trends over time, and evaluating health services. Armed with a grasp of these principles, the clinical neurologist will be in a better position to critically read the scientific literature, to apply epidemiologic knowledge to clinical and public health practice, to avoid common pitfalls in the interpretation of epidemiologic data, and, if desired, to initiate clinical epidemiology studies.

References

Chandra V, Kokmen E, Schoenberg BS et al. (1989). Head trauma with loss of consciousness as a risk factor for Alzheimer's disease. Neurology 39: 1576–1578.

Chaves PH, Kuller LH, O'Leary DH et al. (2004). Subclinical cardiovascular disease in older adults: insights from the Cardiovascular Health Study. Am J Geriatr Cardiol 13: 137–151.

Cipriani G, Dolciotti C, Picchi L et al. (2011). Alzheimer and his disease: a brief history. Neurol Sci 32: 275–279.

Criqui MH, Ringel BL (1994). Does diet or alcohol explain the French paradox? Lancet 344: 1719–1723.

D'Agostino Sr RB, Vasan RS, Pencina MJ et al. (2008). General cardiovascular risk profile for use in primary care: the Framingham Heart Study. Circulation 117: 743–753.

Davis PH, Dambrosia JM, Schoenberg BS et al. (1987). Risk factors for ischemic stroke: a prospective study in Rochester, Minnesota. Ann Neurol 22: 319–327.

Falk EB, Hyde LW, Mitchell C et al. (2013). What is a representative brain? Neuroscience meets population science. Proc Natl Acad Sci U S A 110: 17615–17622.

Farquhar JW, Fortmann SP, Maccoby N et al. (1985). The Stanford Five-City Project: design and methods. Am J Epidemiol 122: 323–334.

Freedman DA (2001). Ecological Inference and Ecological Fallacy. In: NJ Smelser, PB Baltes (Eds.), International Encyclopedia of Social and Behavioral Sciences, Elsevier, Amsterdam, New York.

Fried LP, Tangen CM, Walston J et al. (2001). Frailty in older adults: evidence for a phenotype. J Gerontol A Biol Sci Med Sci 56: M146–M156.

Kanner L (1968). Autistic disturbances of affective contact. Acta Paedopsychiatr 35: 100–136.

Kogan MD, Blumberg SJ, Schieve LA et al. (2009). Prevalence of parent-reported diagnosis of autism spectrum disorder among children in the US, 2007. Pediatrics 124: 1395–1403.

Koton S, Schneider AC, Rosamond WD et al. (2014). Stroke incidence and mortality trends in US communities, 1987 to 2011. JAMA 312: 259–268.

Kukull WA (2001). The association between smoking and Alzheimer's disease: effects of study design and bias. Biol Psychiatry 49: 194–199.

Kukull WA, Ganguli M (2012). Generalizability: the trees, the forest, and the low-hanging fruit. Neurology 78: 1886–1891.

Lackland DT, Roccella EJ, Deutsch AF et al. (2014). Factors influencing the decline in stroke mortality: a statement from the American Heart Association/American Stroke Association. Stroke 45: 315–353.

Lyketsos CG, Steinberg M, Tschanz JT et al. (2000). Mental and behavioral disturbances in dementia: findings from the Cache County Study on Memory in Aging. Am J Psychiatry 157: 708–714.

McKhann GM, Cornblath DR, Griffin JW et al. (1993). Acute motor axonal neuropathy: a frequent cause of acute flaccid paralysis in China. Ann Neurol 33: 333–342.

Mendelsohn AB, Belle SH, Stoehr GP et al. (1998). Use of antioxidant supplements and its association with cognitive function in a rural elderly cohort: the MoVIES Project. Monongahela Valley Independent Elders Survey. Am J Epidemiol 148: 38–44.

Morris JN (1955). Uses of epidemiology. Br Med J 2: 395–401.

Mortimer JA, van Duijn CM, Chandra V et al. (1991). Head trauma as a risk factor for Alzheimer's disease: a collaborative re-analysis of case-control studies. EURODEM Risk Factors Research Group. Int J Epidemiol 20 (Suppl 2): S28–S35.

Paus T (2010). Population neuroscience: why and how. Hum Brain Mapp 31: 891–903.

Ranjan P, Gore M, Selvaraju S et al. (2014). Decline in Japanese encephalitis, Kushinagar District, Uttar Pradesh, India. Emerg Infect Dis 20: 1406–1407.

Reaven GM (1997). Banting Lecture 1988. Role of insulin resistance in human disease. 1988 Nutrition 13. 65; discussion 64, 66.

Roth M, Tomlinson BE, Blessed G (1966). Correlation between scores for dementia and counts of 'senile plaques' in cerebral grey matter of elderly subjects. Nature 209: 109–110.

Sacco RL, Elkind M, Boden-Albala B et al. (1999). The protective effect of moderate alcohol consumption on ischemic stroke. JAMA 281: 53–60.

Schrijvers EM, Verhaaren BF, Koudstaal PJ et al. (2012). Is dementia incidence declining? Trends in dementia incidence since 1990 in the Rotterdam Study. Neurology 78: 1456–1463.

Selkin J (1983). The legacy of Emile Durkheim. Suicide Life Threat Behav 13: 3–14.

Shao H, Breitner JC, Whitmer RA et al. (2012). Hormone therapy and Alzheimer disease dementia: new findings from the Cache County Study. Neurology 79: 1846–1852.

Shinton R, Beevers G (1989). Meta-analysis of relation between cigarette smoking and stroke. BMJ 298: 789–794.

Shock NW, Greulich RC, Andres R et al. (1984). Normal Human Aging: The Baltimore Longitudinal Study of Aging, US Government Printing Office, Washington, DC.

Shumaker SA, Legault C, Rapp SR et al. (2003). Estrogen plus progestin and the incidence of dementia and mild cognitive impairment in postmenopausal women: the Women's Health Initiative Memory Study: a randomized controlled trial. JAMA 289: 2651–2662.

Tsuang D, Kukull W, Sheppard L et al. (1996). Impact of sample selection on APOE epsilon 4 allele frequency: a comparison of two Alzheimer's disease samples. J Am Geriatr Soc 44: 704–707.

Zandi PP, Carlson MC, Plassman BL et al. (2002). Hormone replacement therapy and incidence of Alzheimer disease in older women: the Cache County Study. JAMA 288: 2123–2129.

Handbook of Clinical Neurology, Vol. 138 (3rd series)
Neuroepidemiology
C. Rosano, M.A. Ikram, and M. Ganguli, Editors
http://dx.doi.org/10.1016/B978-0-12-802973-2.00002-1

Chapter 2

Population neuroscience

T. PAUS*

Rotman Research Institute and Departments of Psychology and Psychiatry, University of Toronto, Toronto, Canada and Child Mind Institute, New York, NY, USA

Abstract

Population neuroscience endeavors to identify influences shaping the human brain from conception onwards, thus generating knowledge relevant for building and maintaining brain health throughout the life span. This can be achieved by studying large samples of participants drawn from the general population and evaluated with state-of-the-art tools for assessing (a) genes and their regulation; (b) external and internal environments; and (c) brain properties. This chapter reviews the three elements of population neuroscience (principles, tools, innovations, limitations), and discusses future directions in this field.

Population neuroscience endeavors to identify influences shaping the human brain from conception onwards, thus generating knowledge relevant for building and maintaining brain health throughout the lifespan. Practitioners of population neuroscience are cognizant of three key challenges inherent in their pursuits:

1. An infinite combination of factors influencing the brain from within (genes and their regulation) and the outside (social and physical environment);
2. The presence of developmental cascades that carry such influences from one time point to the next (e.g., prenatal to postnatal), from one organ to another (e.g., cardiometabolic to brain), and from one level of organization to a different one (e.g., behavior to gene regulation, and *vice versa*);
3. Structural and functional complexity of the human brain.

The three challenges can be met by studying large samples of participants drawn from the general population and evaluated with state-of-the-art tools for assessing (1) genes and their regulation; (2) external and internal environments; and (3) brain properties – all done in an integrative fashion and across the lifespan (Fig. 2.1).

Unlike clinical (case-control) studies, population neuroscience does not focus on patients. An ideal (i.e., representative) sample includes a mix of healthy individuals, individuals in preclinical stages of a disease, and those with a fully blown disorder, with numbers corresponding to the population prevalence of different conditions (and their antecedents) at a given age. Broad sampling of environments and genomes is essential to identify the key influences shaping brain capacity (and health) under different circumstances.

In this chapter, I will describe briefly basic concepts pertinent for the assessment of: (1) genes and their regulation; (2) physical and social environment; and (3) brain structure and function. For a detailed treatment of these topics, see Paus (2013). I will conclude by describing some of the current population neuroscience studies and their possible extensions in the future.

GENES AND GENE REGULATION

We differ from each other by thousands to millions of variants in the DNA sequence (Manolio et al., 2009). These variants include single-nucleotide polymorphisms (SNPs), copy number variants, as well as copy number neutral inversions and translocations. As shown in Figure 2.2, there is a negative relationship between allele frequency and effect size: typically, common (minor allele frequency, MAF $> 5\%$) variants are associated with small effect sizes (polygenic complex traits) while

*Correspondence to: Tomáš Paus, M.D., Ph.D., Rotman Research Institute, 3560 Bathurst Street, Toronto, Ontario, M6A 2E1, Canada. Tel: +1-416-785-2500, Fax: +1-416-785-4230, E-mail: tpaus@research.baycrest.org

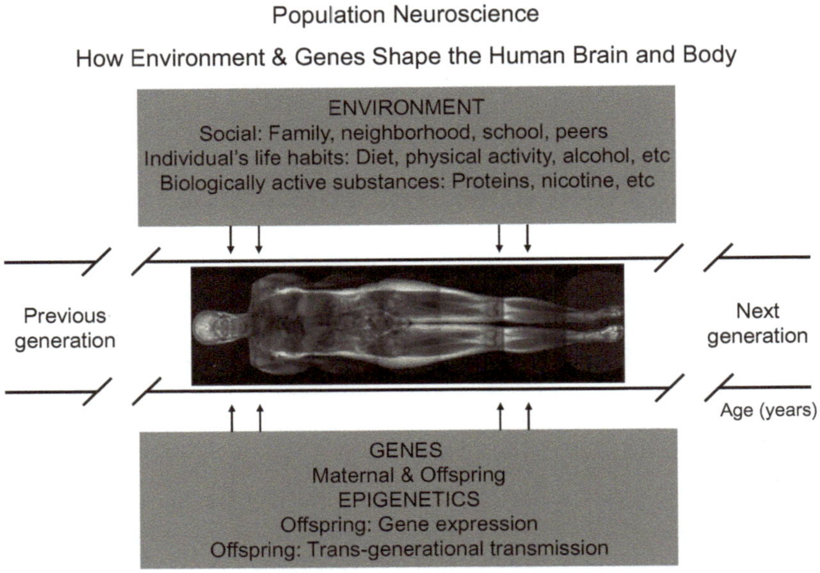

Fig. 2.1. Population neuroscience: assessing (1) genes and their regulation; (2) external and internal environments; and (3) brain properties across the lifespan and generations.

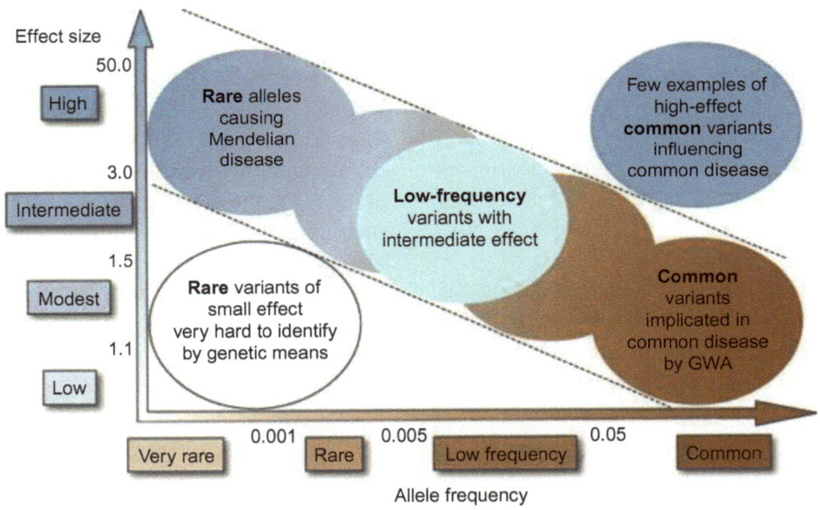

Fig. 2.2. Feasibility of identifying genetic variants by risk – allele frequency and strength of effect size (odds ratio) of genetic effects on a trait. GWA, genomewide association. (Reproduced from Manolio et al., 2009, with permission from Nature Publishing Group.)

the known rare variants (MAF < 0.5%) have large effect sizes (monogenic Mendelian disorders). Therefore, it is not surprising that discovering new associations between common SNPs and complex traits requires large samples. For example, the Enhancing NeuroImaging Genetics through Meta-Analysis (ENIGMA) consortium identified SNPs associated with subcortical brain volumes, such as those of the hippocampus and putamen, using a meta-genomewide association study) (GWAS) involving about 30 000 individuals (Stein et al., 2012a; Hibar et al., 2015). Similarly, a GWAS carried out by the Psychiatric Genomics Consortium identified 128 SNPs (108 independent loci) associated with schizophrenia by comparing 36 989 cases with 113 075 controls (Schizophrenia Working Group of the Psychiatric Genomics Consortium, 2014). As expected, however, the amount of phenotypic variance explained by these SNPs is small. A top SNP identified in the ENIGMA-based meta-GWAS explained between 0.2% and 0.5% of variance in a subcortical brain volume (Hibar et al., 2015). Polygenic scores calculated from the 108 genome-wide significant loci explained 3.4% variation on the liability scale to schizophrenia (Schizophrenia Working Group of the Psychiatric Genomics Consortium, 2014). These numbers are consistent with the highly polygenic nature of complex traits.

Would considering all common SNPs explain more variance in a given phenotype? This is indeed the case, as demonstrated for 49 different traits using the Genome-wide Complex Trait Analysis (Yang et al., 2011, 2013). We have shown that an overall pattern of genotypic variations across ~500 000 SNPs explains up to 50% of phenotypic variations in brain structure (Toro et al., 2015) and function (Dickie et al., 2014) (Fig. 2.3). It is unlikely, however, that each of the 500 000 SNPs contributes equally, each adding 0.0001% to the total of 50% of variance explained; not all genomic locations are created equal. This brings us to genome biology and the knowledge gained from projects such as the Encyclopedia of DNA Elements, ENCODE (Kellis et al., 2014).

Genomes can be partitioned into a number of functional domains. To start with, all SNPs included in the 1000 Genomes dataset can be annotated (classified) as belonging to one of the following six partitions: coding region (0.9% of all SNPs), untranslated region (UTR; 0.9%), promoter (2.6%), DNase I hypersensitivity site (DHS; 16.4%), intron (28.6%) and intergenic region (50.5%) (see Table S4 in Gusev et al., 2014). Genetic variants located in the different partitions may be over- or underrepresented among SNPs associated with a given trait. For example, across 11 common diseases, the top hits are highly enriched in the coding regions (13.8-fold enrichment), and less so in the DHS regions (5.1 × enrichment). At the same time, SNP-based heritability estimates are the highest for the DHS partition (79% SNP heritability), as compared with the coding regions (8% SNP heritability) (Gusev et al., 2014). As shown in Figure 2.4, the high contribution of SNPs located in the DHS partition cannot be attributed solely

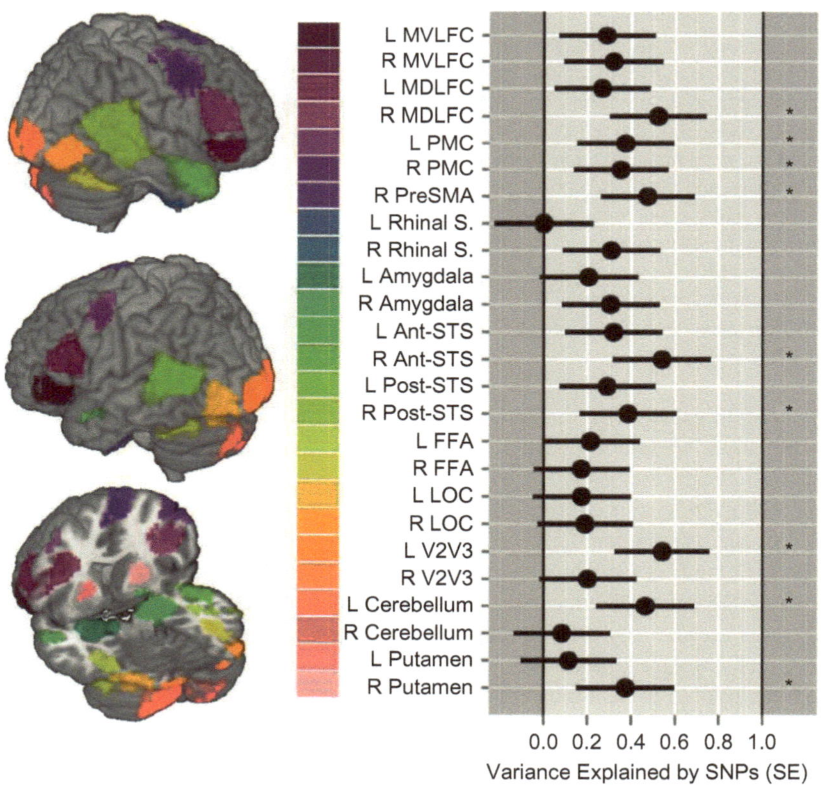

Fig. 2.3. Single-nucleotide polymorphism (SNP)-based estimate of heritability of the brain response to ambiguous faces. Left: Locations of the 25 functional regions of interest (ROI) defined for the dynamic ambiguous face vs. control contrast. Right: Proportion of variance in mean percent blood oxygen level-dependent signal change (%BCS) explained by common genetic variance for each ROI when viewing facial expressions in 1620 unrelated adolescents. Error bars indicate the standard error (SE) of the estimate. Asterisks indicate those estimates significant at an alpha 0.05 (uncorrected). Vertical gridlines are at intervals of 0.2. MVLFC, mid-ventrolateral frontal cortex; MDLFC, mid-dorsolateral frontal cortex; PMC, premotor cortex; PreSMA, pre supplementary motor area; STS, superior temporal sulcus; FFA, fusiform face area; LOC, lateral occipital cortex; L, left; R, right. (Reproduced from Dickie et al., 2014.)

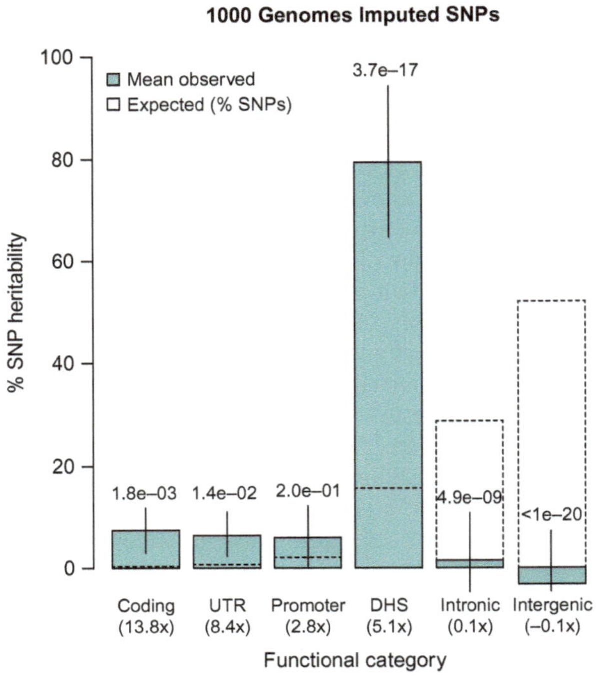

Fig. 2.4. Functional partitioning of single-nucleotide polymorphism (SNP)-based heritability across 11 traits. Joint estimates of the percentage of h2g from six functional components are shown in filled bars (meta-analyzed over 11 traits). The null expectation, equal to the percentage in each category, is shown by dashed, unfilled bars, and *p*-values report the difference from this expectation. Fold enrichment relative to the null expectation is shown in parentheses below each category. Error bars show 1.96 × standard error (SE) after adjustments for shared controls. UTR, untranslated region; DHS, DNase I hypersensitivity site. (Reproduced from Gusev et al., 2014.)

to its size. It is likely that SNPs located in this transcriptionally active partition mark regulatory variants that influence gene expression. There are many other functional annotations that allow one to examine possible mechanisms underlying genotype–phenotype associations discovered by GWAS, such as sites rich in histone modifiers (Zhang et al., 2015) or superenhancers (Pott and Lieb, 2015); histones are the key components of chromatin, regulators of its structure and, in turn, accessibility of DNA for transcription. For example, Finucane and colleagues (2015) combined four histone marks (H3K4me1, H3K4me3, H3K9ac, H3K27ac) that were assayed across many different cell types (e.g., brain, liver, pancreas, muscle) to identify genomic locations (SNPs) associated with cell-specific modifiers of histone H3. After grouping these marks into 10 cell-type groups (e.g., central nervous system (CNS) and immune tissues), they observed specific patterns of enrichment for SNPs associated with 15 out of the 17 complex traits and diseases (Fig. 2.5). Some of the enrichment patterns were consistent with the tissues known to be involved in a given disorder (e.g., CNS for schizophrenia or immune tissues for rheumatoid arthritis), while other patterns revealed additional involvements of tissues not considered to be the primary ones (e.g., CNS for body mass index and for age at menarche). Thus, knowledge of genome biology is beginning to guide the interpretation of GWAS-based findings. The key term here is regulation of gene expression, either globally or in a tissue-specific manner.

In addition to *in vitro* assays of gene regulation, one can also use the classic GWAS approach to search for genetic variants associated with interindividual variations in gene expression: expression quantitative trait loci (eQTL) (Albert and Kruglyak, 2015). Here, the level of gene expression represents a trait, with expression levels of 20 000+ genes measured in each sample either through expression microarrays or RNA sequencing. A GWAS approach is then used to identify associations between SNPs and expression levels: eQTLs. Given the current sample size limitations, most of the discovered eQTLs are local and act in *cis*: a genetic variant is near the regulated gene (<2 Mbp) and affects its expression in an allele-specific manner (i.e., acts on the gene located on the same physical chromosome, not on its copy on the homologous chromosome). Such GWAS-based analyses of gene expression have been carried out in a number of different tissues, including the human brain (Ramasamy et al., 2014). In most cases, levels of gene expression (in blood or other tissues) have been measured at "baseline" and compared across various conditions/diseases.

One interesting exception is a recent study that identified genetic variants associated with changes in gene expression (in blood cells) in response to dexamethasone, a glucocorticoid receptor agonist (Arloth et al., 2015). The authors used these eQTLs to calculate a polygenic score indexing the strength of an individual's response to a hypothetic stressor: the higher the score, the stronger the stress response (Arloth et al., 2015). We have tested this approach by calculating a polygenic "stress" score in a sample of adolescents and demonstrated its (positive) association with the cortisol response to math challenge (Wong and Paus, unpublished observation). This example illustrates yet another utility of identifying genetic variants associated with gene regulation: construction of a gene panel indexing genetic variations relevant for a family of specific biologic processes and, subsequently, using such a panel to test the importance of these processes for a given system level trait. Furthermore, we have been developing a similar approach using methylation levels at CpG sites in the gene body, taking advantage of the widespread availability of data on DNA methylation generated by

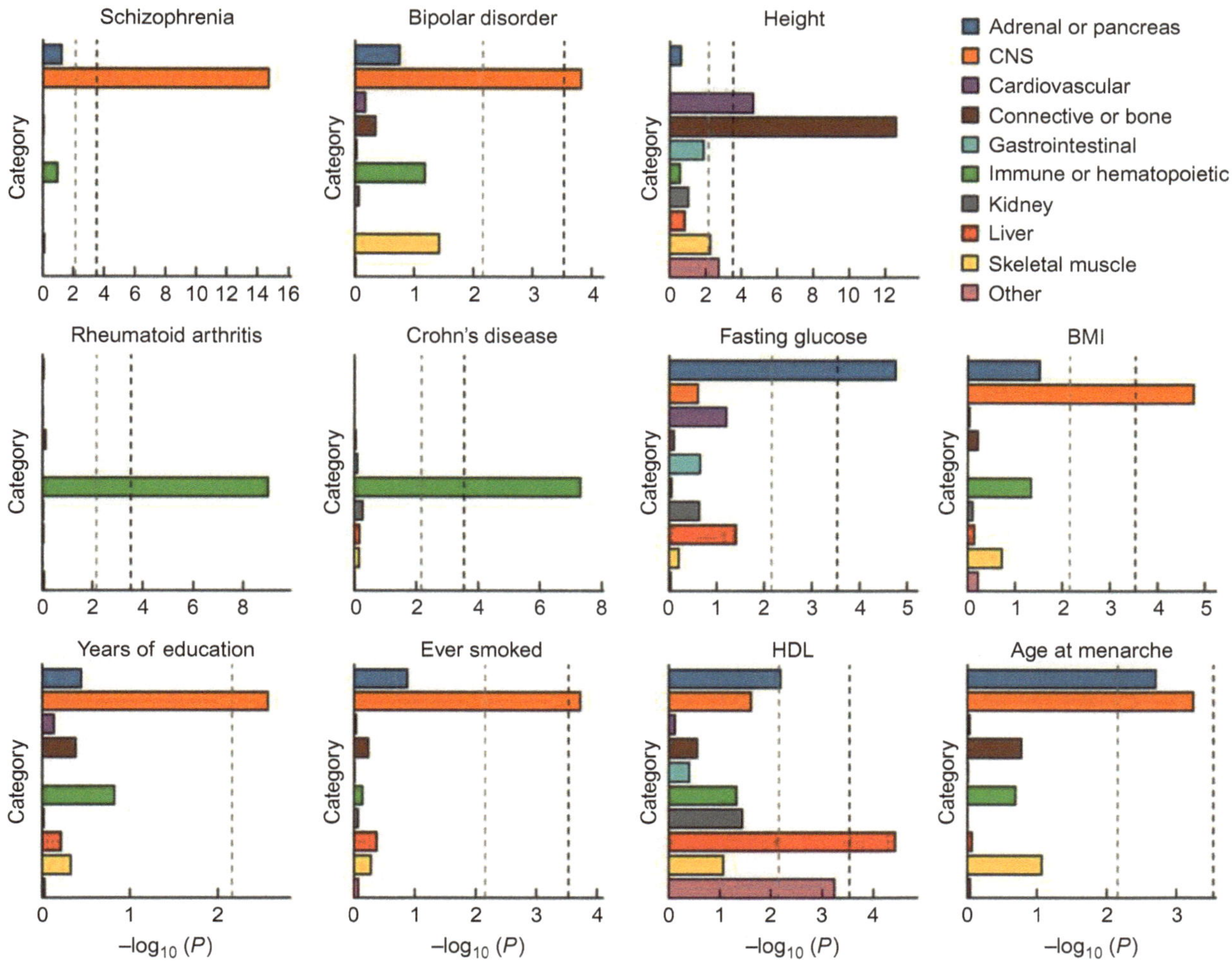

Fig. 2.5. Single-nucleotide polymorphisms associated with 11 traits (indicated above each plot) show enrichment for genomic locations associated with cell-specific modifiers of histone H3. Cell-type groups are color-coded (see legend in the top right corner). The black dashed lines at $-\log 10\ (p) = 3.5$ is the cutoff for Bonferroni significance. The gray dashed lines at $-\log 10\ (p) = 2.1$ is the cutoff for False Discovery Rate < 0.05. CNS, central nervous system; BMI, body mass index; HDL, high-density lipoprotein. (Reproduced from Finucane et al., 2015, with permission from Nature Publishing Group.)

microarrays covering about 450 000 such sites across the genome (Bibikova et al., 2011). It appears that polyepigenic scores calculated for a panel of genes may provide a useful index of gene transcription effectiveness in a given individual (French and Paus, unpublished observation). As shown in Figure 2.6, polyepigenic score based on the (normalized) methylation levels of gene bodies across ~150 growth factor genes predicts the median expression of this group of genes in the human brain. This is consistent with the known positive relationship between gene body methylation and gene expression (Kulis et al., 2013).

In summary, current advances in high-throughput genomics, transcriptomics, and epigenomics, as well as related progress in genome biology and bioinformatics, make it possible to gain detailed insights into factors (and mechanism) shaping the human brain from within an individual. Let us now turn our attention to the individual's environment.

PHYSICAL AND SOCIAL ENVIRONMENT

We are both "recipients" and "creators" of our environment (Kendler et al., 2003). There are countless variations and permutations of physical and social environments that surround us in space and time. These "external" environments contribute – together with genes – to many "internal" environments that exist in our body: on the body surfaces (e.g., microbes on our skin and in the gut) and in the circulating blood (e.g., micronutrients, toxins, inflammatory molecules).

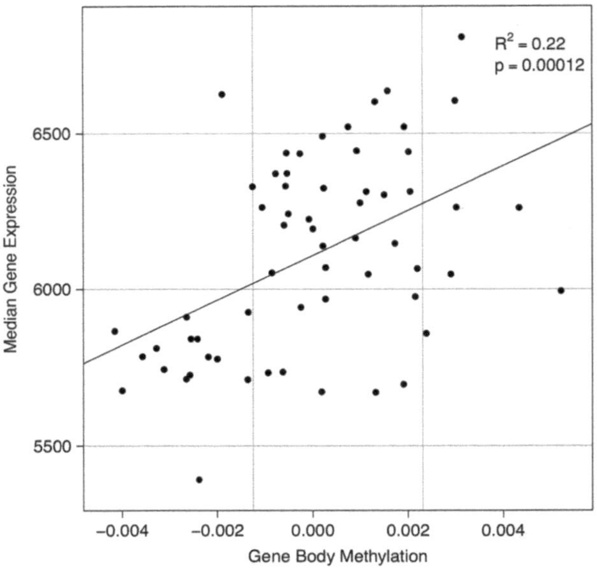

Fig. 2.6. Correlation between methylation levels in gene bodies of the growth-factor genes (adjusted for age, neuron fraction, and genomewide methylation) and their expression in postmortem male brain tissue from the BrainSpan atlas (samples = 61, donors = 7 males, 59 degrees of freedom, $t = 4.12$). Gene expression is ranked on an increasing scale so that high expression corresponds to high rank. (French and Paus, unpublished observation.)

Measuring various "internal" environments is straightforward: one needs a biologic sample (stool, urine, blood) and an appropriate high-throughput assay for assessing a particular part of the metabolome, a catalog containing at least ~8500 endogenous and ~40 000 exogenous compounds (Wishart, 2011). Efforts are on the way to use – in a consistent and harmonized manner – biologic samples for monitoring exposures of individuals to various chemicals, be it through the air, water, food, or various consumer products (Collins et al., 2014; Exley et al., 2015). In 2015, the US National Institute of Environmental Health Sciences founded the Children's Health Exposure Analysis Resource that will provide access to a state-of-the-art infrastructure for the analysis of biologic samples collected in health studies of children (biomonitoring) and suitable for creating an overall "exposome" based on a catalogue of various environmental exposures acting in a developmental context.

Assessing "external" environments – rather than its biologic markers – is challenging. Asking a series of questions using a standard survey is the most common way to collect information about the individual's physical and social environment. For example, the PhenX Toolkit contains a number of questions about different social environments (e.g., family, neighborhood, workplace, school, residence), as developed by the Social Environments Working Group (https:/www.phenxtoolkit.org/index.php). Although valuable, there are two main disadvantages of the survey-based approach: (1) participant's time (many hours required to cover multiple domains); and (2) self-reported nature of the collected information and, therefore, possible rating biases and errors. Next, we review an alternative approach that uses aggregate data to characterize the physical and social environments in which an individual spends most of his/her time (e.g., home, workplace).

Geographic Information Systems

Geographic information systems (GIS) allow data to be mapped in space and time, providing multiple layers of information about the individual's physical and social environment at a given time point. Tools such as ArcView by the Environmental Systems Research Institute (http:/www.esri.com/) enable investigators to organize, analyze, and visualize aggregate data relative to the geospatial coordinates (longitude and latitude) of a cohort member.

Aggregate data

Aggregate data come from a variety of sources (Table 2.1), and are organized geographically at the level of administrative units such as census tracts; in metropolitan areas, census tracts typically contain between 2000 and 8000 individuals. Census tracts can be subdivided into smaller geospatial units; in Canada, for example, dissemination areas are the smallest standard units (~600 individuals) used to report data collected during a census or by local government.

Cohort member

Location of a cohort member may also vary in its granularity; depending on the study and measures taken to protect confidentiality of study participants at a given level of data access, this can be as precise as the exact street address or a postal code, or as coarse as a city district, census tract, county, state or country. Of course, the appropriate level of granularity for linking the individual and aggregate data depends on the type of a cohort (e.g., same city vs. many countries) and questions being asked. For example, we have investigated relationships between green space and subjective well-being (and self-reported health) in a sample of 31 000 participants from the Ontario Health Study living in Toronto, Ontario, Canada (Kardan et al., 2015). The postal codes of their homes were mapped into ~3200 dissemination areas (Fig. 2.7A); information about tree canopy in each of these dissemination areas came from reports generated by the city of Toronto

Table 2.1

Aggregate-level descriptors of social and physical environments

Variable	Data source	World Bank
Demographic		
Population (total and densities)	Census	
Proportion by age group	Census	x
Proportion by sex	Census	x
Proportion by ethnicity	Census	x
Proportion by race	Census	
Per capita marital status	Census	x
Recent mobility/migration status	Census	x
Proportion by religion	Census	
Proportion by mother tongue	Census	x
Households		
Average household size	Census	x
Total housing units	Census	x
Proportion of homes rented	Census	x
Type of dwelling	Census	x
Housing construction material	Not available	x
Age of dwelling	Not available	x
Availability of utilities (by type)	Not available	x
Size of dwelling	Not available	x
Mortgage status and repayment amount	Not available	x
Up-to-date bill payment status	Not available	x
Possession of durable goods (e.g., TV, cell phone)	Not available	X
Household energy use	National survey*	x
Water consumption and conservation	National survey*	x
Water quality concerns	National survey*	
Composting and recycling behavior	National survey*	
Types of owned recreational vehicles (e.g., motor boats)	National survey*	
Involvement in outdoor activities	National survey*	
Household daily expenses (total amount)	Not available	x
Household food expenditures (total amount and quantity by food type)	Not available	x
Household nonfood expenditures (e.g., clothing, household products, entertainment)	Not available	x
Socioeconomic		
Proportion by education level	Census	x
Proportion by major field of study		x
Household cost of education (e.g., tuition, transportation, textbooks, accommodation)	Not available	x
Per capita income by age/sex	Census	x
Household income	Census	x
Proportion of income below poverty level	Census	x
Unemployment rate	Census	x
Proportion by age/sex in labor force	Census	x
Proportion by occupation (national classification)	Census	x
Proportion of social housing	Municipal	
House prices	Municipal	x
Per capita benefits/social security received	Not available	x
Per capita employer method of payment	Not available	x
Per capita loans, credits, savings, taxes, and transfers	Not available	x

Continued

Table 2.1

Continued

Variable	Data source	World Bank
Health and wellbeing		
Rate of chronic conditions	National survey†	x
Mental health status	National survey†	
Mental health service use and experience	National survey†	
Smoking behavior	National survey†	x
Work stress	National survey†	
Amount of physical activity	National survey†	
Cancer screening rates	National survey† and municipal	
Fertility rates	National survey† and municipal	
Proportion of teenage births	National survey‡ and municipal	
Proportion of births by maternal education level	National survey‡ and municipal	
Proportion of stillbirths	National survey‡ and municipal	
Death rates	National survey‡ and municipal	
Disability rates	National survey§	x
Location of ambulance stations	Municipal	
Substance use		
Alcohol outlet density	County	
Tobacco outlet density	County	
Alcohol sale rules (wet/moist/dry)	County	
Proportion of drinkers and binge drinkers	County	x
Proportion of smokers	National survey†	x
Alcohol use and dependency	National survey†	
Substance use and dependency	National survey†	
Crime		
Number of traffic tickets	Municipal	
Proportion of crimes by neighborhood	Municipal	
Built environment and schools		
Number of active and cleared building permits	Municipal	
Number of development projects	Municipal	
Location and boundaries of heritage districts	Municipal	
School boundaries/location	Municipal	
School population	Municipal	
Transportation		
Proportion using private transport	Census	
Proportion using public transport	Census	
Intersection pedestrian and traffic volume by time of day	Municipal	
Road volume	Municipal	
Subway stop density	Municipal	
Traffic light and camera locations	Municipal	
Trails, bike lanes, and road length and location	Municipal	
Household/per capita transportation expenses	Not available	x
Physical environment and land use		
Forest and land cover	Municipal	
Street tree data	Municipal	
Days of extreme heat alerts	Municipal	
Area of green space by neighborhood	Municipal	
Area of tree cover by neighborhood	Municipal	
Area of water by neighborhood	Municipal	
Area of agriculture by neighborhood	Municipal	
Area of roads by neighborhood	Municipal	

Table 2.1

Continued

Variable	Data source	World Bank
Area of paved surfaces by neighborhood	Municipal	
Type size and location of trees	Municipal	
Zoning bylaws by neighborhood	Municipal	
Location of parks and recreation facilities	Municipal	
Agriculture		
Total number of farms	Census	x
Farm type (major crop/livestock production)	Census	x
Proportion of farm injuries	Census	
Size of farm land area	Census	x
Total income generated from farm activity	Census	x
Farm activity-related expenses	Census	x
Farm ownership status (owned, leased, rented)	Census	x
Pollution sources		
Pollutant and carcinogenic scores by neighborhood	Municipal	
Household pesticide and fertilizer use	National survey*	x
Household disposal of hazardous and electronic waste	National survey*	
Local beach water quality	Municipal	
Pollutant and carcinogen scores by neighborhood	Municipal	
Number of polluting facilities	Municipal	

First column contains variables grouped into various domains, second column indicates potential sources of data for a given variable, and third column indicates whether a given variable is included in household surveys used by the World Bank. Census, based on census administered by the government of Canada; national survey, based on *Household and the Environment Survey (Canada), †Canadian Community Health Survey, ‡national databases on birth, stillbirth, and death (Canada) and §Canadian Survey on Disability; municipal, based on the City of Toronto database (http:/goo.gl/AB0DCu); household survey, based on a survey administered by the World Bank in Bulgaria (2003).

(Fig. 2.7B) and by QuickBird satellite imagery (Fig. 2.7C). Thus, this cross-sectional study linked measures of "exposures" (green space) generated by others at an aggregate level with "outcomes" (wellbeing) collected at an individual level in a particular cohort (within a large metropolitan area).

Moving beyond a single time point, one can also examine relationships between environment and health over time. For example, in the Multi-Ethnic Study of Atherosclerosis, investigators assessed physical and social neighborhood (mostly availability of healthy food and resources for physical activity) and incidence of type 2 diabetes mellitus over a 5-year period in ~5000 participants living in six US cities (Christine et al., 2015). For the "exposures," they used two sources of information: (1) locations of food stores likely to use healthy food and those of commercial recreational establishments; and (2) survey-based rating of the participants' neighborhood with regard to the availability of healthy foods and walking environment. The two types of information were linked to the location of participants' homes using a 1-mile buffer around each home. The design of this study shows another interesting feature of GIS-based assessments of physical and social environment, namely the possibility of using historic (time-varying) records about a given environment (e.g., "walkability" of the built environment) and an emergence of health problems over the study period.

Existing records gathered and administered by local and state governments represent a rich source of information about physical and social environments of individuals participating in many large-scale studies of brain health. Furthermore, the emergence of tools such as Google Street View and social media (Twitter, Facebook) provides new opportunities for characterizing the physical and social environment at an aggregate level. Google Street View allows investigators to assess different features of the built environment using panoramic street-level images of local neighborhoods taken mostly by camera-equipped cars. These geo-coded images can be rated for various features, such as signs of physical disorder (e.g., litter, graffiti), physical decay (e.g., poor conditions of sidewalks), type of stores, or street walkability (Odgers et al., 2012; Less et al., 2015); there are some limitations of this approach, however (Curtis et al., 2013).

Twitter provides a rich source of information about the interests of its users; a short message (a tweet, up

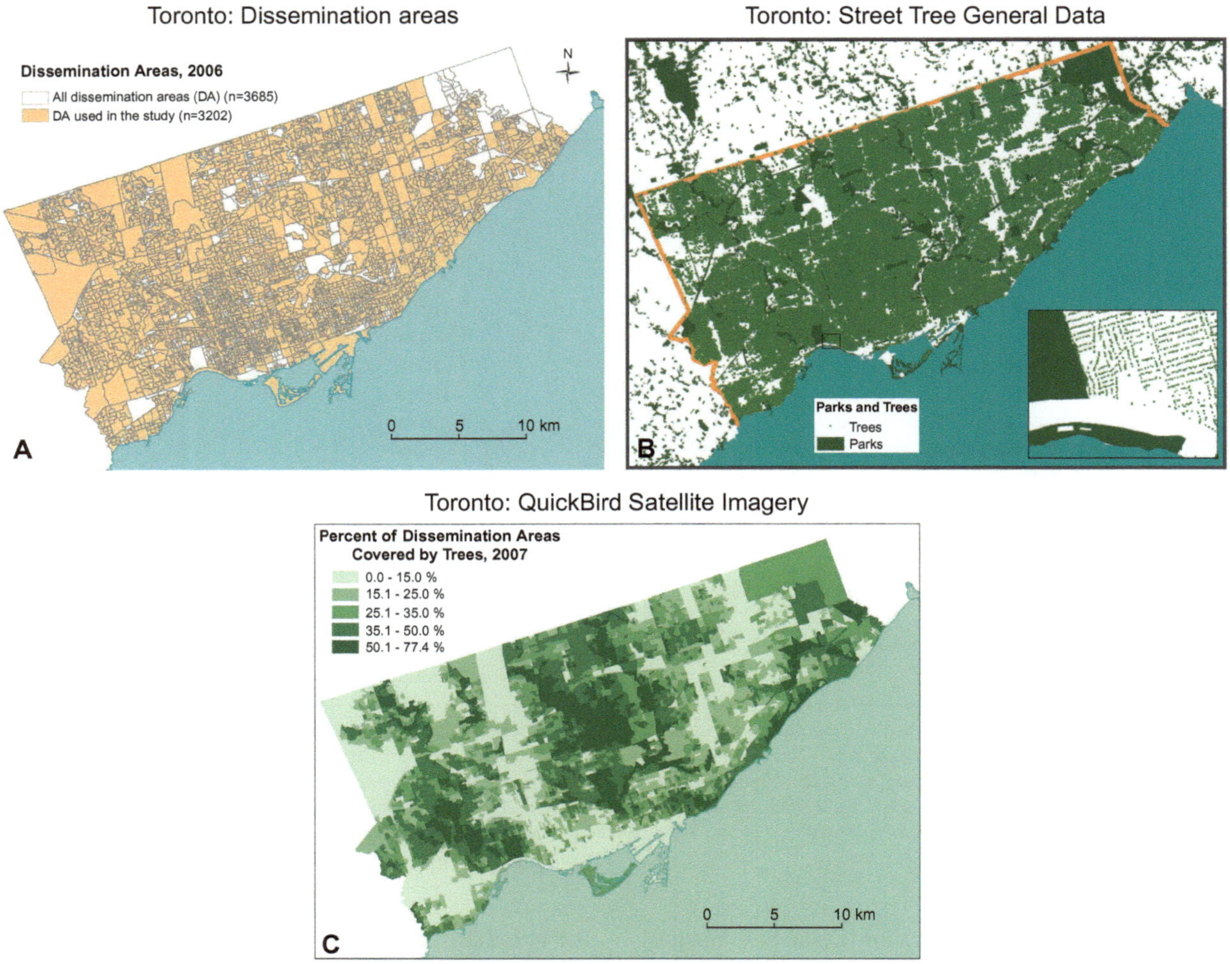

Fig. 2.7. (**A**) Map of dissemination areas in the city of Toronto (2006). The colored regions show the dissemination areas included in the original study. (**B**) The greenspace map of the city of Toronto constructed from the individual tree information contained in the Street Tree General Data dataset. This image is shown at much lower resolution compared with the real image; the dissociation between individual trees and other areas is clearly perceivable for the zoomed-in area. Parks are shown in dark green. (**C**) The greenspace map of the city of Toronto constructed from the geographical information system (GIS) polygon dataset called Forest and Land Cover. The levels are shown in units of 10–15% for display purposes. These figures were created using Environmental Systems Research Institute's (ESRI) ArcGIS software v. 10.2. (Reproduced from Kardan et al., 2015.)

to 140 characters in length) sent by a user can be tagged by a hashtag identifying explicit topics of the message. Although users do not follow any rules when generating hashtags, one can group these into topic clusters (Table 2.2; Weng and Menczer, 2015). In addition to hashtag-based analyses (or in the absence of hashtags), a plain-text analysis can be used to identify (and classify) words and phrases constituting tweeted messages. Importantly, users are able to attach geo-coordinates or place names while tweeting; in one estimate, 91% of all geo-tagged tweets contained coordinates derived from the global positioning system (GPS, built in in smartphones) and 9% had place names (Blanford et al., 2015). Therefore, topics of the tweets (e.g., interest in conservative politics or food) or the interrelationships between topics (topic networks) or between users (social networks) can be mapped in space and time, thus creating aggregate data characterizing social structure of the different administrative units, such as dissemination areas described above. It is of note that Twitter provides a variety of tools for data mining through its subsidiary Gnip, Twitter's enterprise API (application programming interface) platform (https://gnip.com/sources/twitter/).

Facebook – an online social network – provides another window into the social activities of its users. Various Facebook activities, such as wall postings, likes, or status messages, can be used to classify users into different categories with regard to their interests, such

Table 2.2

Examples of topic clusters in the hashtag co-occurrence network

Cluster	Example #hashtags
Technology	google, microsoft, supercomputers, ibm, wikipedia, startuptip, topworkplace
Politics	tcot, p2, top, usgovernment, dems, owe, politics, teaparty
Lifestyle	pizza, pepsi, health, vacation, caribbean, ford, honda, volkswagen, hm, timberland
Twitter-specific	followme, followfriday, friday, justretweet, instanfollower, rt
Mobile devices	apple, galaxy3, note2, iosapp, mp3player, releases

Reproduced from Weng and Menczer (2015).

as watching TV (Chunara et al., 2013), or their individual characteristics, such as the Big Five personality traits (Park et al., 2015). These can be, in turn, related to the geographic locations of Facebook users (e.g., postal code), thus creating additional layers of information about social features of a given administrative unit.

The above examples illustrate the power of digital ecology, an objective way of assessing social environment by measuring behavior through the individual's use of social media and mobile technology.

Finally, smartphones and other mobile technology equipped with a GPS device, such as accelerometers, heart rate monitors or even sweat-based monitors of metabolites and electrolytes (Gao et al., 2016), provide new opportunities for sampling individuals' behavior and physiology, and evaluating this individual-based metric against physical and social context using aggregate data. For example, Hurvitz and colleagues (2014) used mobile technology (accelerometers, GPS devices, and/or a multimodal sensor) to create "LifeLogs" containing data derived from these devices that are, in turn, visualized in geo-space with "LifeLog Views" (Fig. 2.8).

In summary, technologic advances are expanding our ability to assess the individual's internal and external environments with an increasing breadth and precision. Together with aggregate data mapped in well-defined geospatial units with high granularity, these tools allow us to relate the individual's envirome to his/her exposome and phenome.

BRAIN STRUCTURE AND FUNCTION

As stated in the opening paragraph, population neuroscience is concerned with factors shaping the human brain and – by extension – cognition, behavior, and mental health. Magnetic resonance imaging (MRI) is the key phenotyping tool allowing interindividual variations to be assessed in a large number of structural and functional properties of the human brain throughout the lifespan. In this section, I will review the most common MRI-based metrics and highlight some of the key challenges associated with the interpretation of MRI findings. Tools for assessing cognition, behavior, and mental health are described elsewhere (Paus, 2013).

The noninvasive nature of MRI, its versatility *vis-à-vis* imaged features of brain structure and function, and the widespread availability of MR scanners make it an ideal tool for mapping the human brain throughout the lifespan. MRI is a remarkably flexible tool for measuring a wide array of functional and structural properties of the human brain and other organs (Fig. 2.9). The same scanner can be used for acquiring different types of images by changing the exact parameters of an MR "scan": the temporal sequence of radiofrequency pulses, gradients, and readouts. Thus, in one MRI session, we can run several scans to assess a number of structural and functional properties of the human brain (Table 2.3).

Brain structure

Regarding brain structure, multimodal MRI allows derivation of a variety of measures using automated, image analysis algorithms (or "pipelines").

T1-WEIGHTED IMAGES

T1-weighted images have been the richest and most flexible source of data for morphometric analyses. Over the past 20 years, a number of computational pipelines have been developed by the neuroimaging community, such as FMRIB Software Library (FSL) from the FMRIB laboratory at the University of Oxford (http:/www.fmrib.ox.ac.uk/fsl) and FreeSurfer from the Martinos Center for Biomedical Imaging (http:/surfer.nmr.mgh.harvard.edu/). These pipelines enable us to extract a large number of brain features, or phenotypes, in a fully automatic fashion. There are two general classes of these features: (1) voxel- or vertex-wise features derived for each *x, y,* and *z* location (e.g., gray- and white-matter "density" maps, cortical thickness, cortical folding); and (2) volumetric

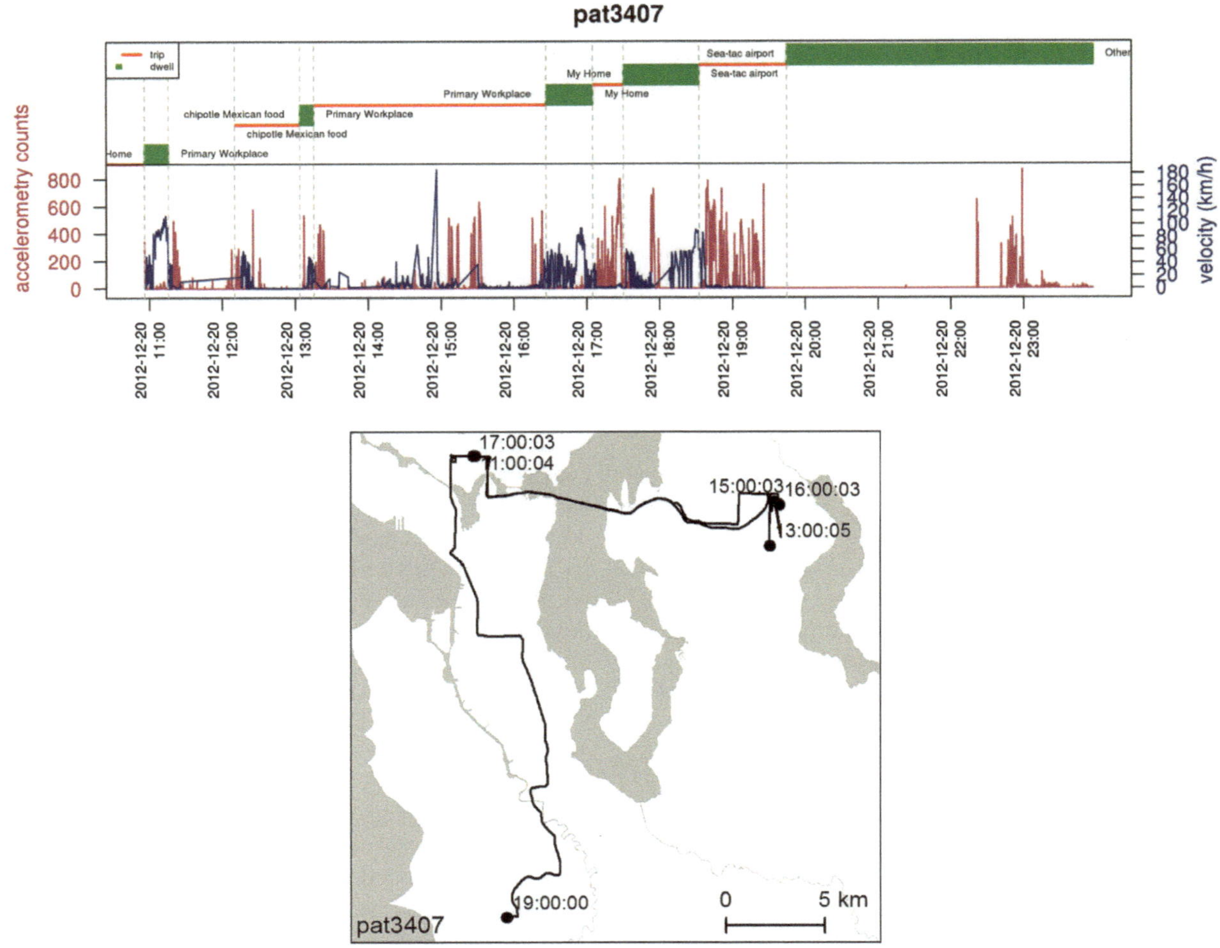

Fig. 2.8. One-day LifeLogs collected from a participant. The upper panel shows self-reported data on place (red) and trip (green) from the travel diary along with objective accelerometry counts (magenta) and global positioning system velocity (blue). The lower panel maps travel patterns for the same individual on the same day, with markers shown at hourly intervals. (Reproduced from Hurvitz et al., 2014.)

measures (e.g., volumes of gray matter in the frontal lobes, volume of the hippocampus, area of the corpus callosum).

Diffusion tensor imaging

Using diffusion tensor imaging (DTI), one can estimate several parameters of water diffusion, such as mean diffusivity and fractional anisotropy (FA). FA reflects the degree of water diffusion directionality; voxels containing water that moves predominantly along a single direction have higher FA. In white matter, FA is believed to depend on the microstructural features of fiber tracts, including the relative alignment of individual axons, how tightly they are packed (which affects the amount of interstitial water; see above), myelin content, and axon diameter. Values of FA can be calculated using various tools, such as FDT (http:/www.fmrib.ox.ac.uk/fsl) or ExploreDTI (http:/www.exploredti.com).

Magnetic transfer imaging

Magnetic transfer (MT) imaging is another MR technique employed in studies of the structural properties of white matter. Contrast in MT images reflects the interaction between free-water protons and protons bound to macromolecules (McGowan, 1999). The macromolecules of myelin are the dominant source of the MT signal in white matter (Kucharczyk et al., 1994). Postmortem studies revealed a significant positive correlation between myelin content and magnetization transfer ratio (MTR) (Schmierer et al., 2004, 2008). The fastest way to acquire magnetization transfer data is to use a dual acquisition, with and without an MT saturation pulse; MTR is calculated as the percent signal change between the two acquisitions (Pike, 1996).

Other MR acquisitions suitable for the characterization of structural properties in general, and those of white

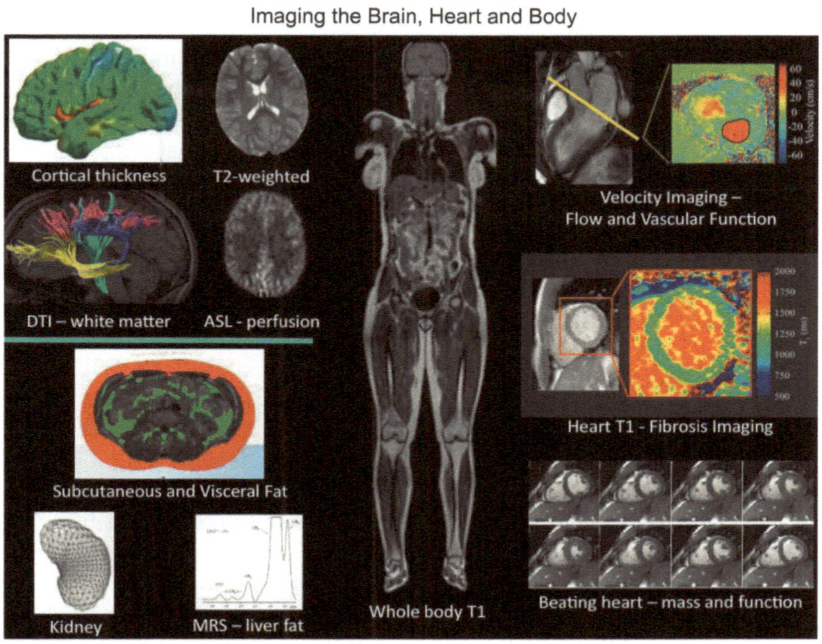

Fig. 2.9. Versatility of magnetic resonance imaging for assessing structural and functional properties of the human brain, heart, and body. ASL, arterial spin labeling; DTI, diffusion tensor imaging; MRS, magnetic resonance spectroscopy.

Table 2.3

An example of a 60-minute magnetic resonance imaging (MRI) protocol enabling one to characterize a number of structural and functional properties of the human brain

MRI sequence	Time (minutes)	Structure and physiology
T1-weighted	10	Volumes, thickness, folding, shape, tissue density
T2-weighted	4	White-matter hyperintensities (number, volume, location)
Diffusion tensor imaging	12	Fractional anisotropy, mean diffusivity, track delineation
Magnetization transfer	8	Myelination index
Arterial spin labeling	5	Perfusion
Resting-state functional	8	Spontaneous cerebral networks; functional connectivity; physiologic fluctuations
Paradigm-based functional	6–10	Brain response associated with specific stimuli/tasks; functional connectivity

Reproduced from Paus (2013), with permission from Springer Science and Business Media.

matter in particular, include quantitative T_1, T_2, and myelin water fraction imaging using, for example, the DESPOT methods (Deoni et al., 2005, 2008; Deoni, 2011), and the NODDI (neurite orientation distribution and density imaging) technique, the latter being suitable for mapping axonal properties in the human brain (Zhang et al., 2012). Conceptually, structural properties of white matter are important for the conduction of action potentials and for axonal transport (Paus et al., 2014).

Brain function

When mapping brain function, the most common MR signal used as an indirect index of brain "activity" is the blood oxygenation-level dependent (BOLD) contrast, arising from the disproportionate increase of blood flow into the region engaged functionally at a given moment. In most functional MRI (fMRI) studies, the degree of engagement of a particular brain region is measured simply by tracking the BOLD signal over time and calculating the difference between time segments that differ from each other with regard to the presence or absence of a particular stimulus or, in more general terms, a difference in a particular well-defined behavioral state. This calculation is carried out again voxel by voxel, typically after a number of preprocessing steps, including corrections for head motion, alignment of (functional) BOLD images with (structural) T1-weighted images acquired in the same session, and the convolution of the timeline of relevant events (e.g., presence of a stimulus) during the scan with a hemodynamic response function, which models the temporal delay (~6 seconds) between the neural activation and the BOLD response. There are a number of software packages suitable for analyzing fMRI time series (e.g., SPM, www.fil.ion.ucl.ac.uk/spm/software/spm8/ and FSL, fsl.fmrib.ox.ac.uk/fsl/fslwiki/). The main outputs of such analyses are statistical maps that indicate in which regions (voxels) the BOLD signal measured during one state differs from that in another state.

Although voxel-wise analyses of fMRI data allow for an unbiased search for condition- or group-related differences throughout the brain, restricting the "search space" to a set of specific brain regions can be useful in many instances. In particular, such a region-of-interest (ROI) approach is suitable for testing specific hypothesis in large population-based studies in which one models the brain response in the context of a variety of genetic and environmental influences (Fig. 2.3). Owing to the use of the same standardized stereotaxic space in the majority of imaging studies, there are various ways in which such ROIs can be defined *a priori*, both functionally and anatomically. The former can be achieved through meta-analyses of published reports (Eickhoff et al., 2009) or by constructing probabilistic maps using large datasets (Tahmasebi et al., 2012), while the latter can take advantage of various anatomic parcellations of the human brain (e.g., the Automatic Anatomical Labelling atlas (Tzourio-Mazoyer et al., 2002), probabilistic maps of cytoarchitecture (Eickhoff et al., 2007), or meta-analyses of structural imaging studies (Cheung et al., 2010)).

The network structure of our brains implies constant functional interactions across various spatially segregated modules. Functional imaging studies provide a unique opportunity to quantify such interregional interactions by virtue of measuring BOLD signals across the entire brain simultaneously. The primary tool for exploring so-called functional connectivity has been a statistical analysis of concurrent fluctuations in the BOLD signal over time, with or without the engagement of a participant in a particular paradigm. Various methods have been introduced to measure functional and effective connectivity, from simple interregional correlations to multivariate techniques (reviewed in Friston, 2011).

Over the years, MRI – together with the other brain-mapping techniques – has provided a wealth of data about the structural and functional organization of the human brain, in health and disease. In my view, there are two key challenges facing both the producers and consumers of this ever-growing literature: (1) reproducibility and (2) interpretation.

Reproducibility

Reproducibility of imaging-derived measures is paramount, especially when trying to identify factors shaping the human brain. Various metrics derived from (multimodal) structural images show high test–retest reliability. For example, Wonderlick and colleagues (2009) have evaluated test–retest reliability (two sessions, 2 weeks apart), and the influence of several acquisition parameters (same 3 T scanner) for a number of morphometric measures derived from T1-weighted images by FreeSurfer. They found that the reliability – estimated with intraclass correlation coefficients (ICCs) – was "excellent" (ICCs > 0.75) for most measures; with the exception of globus pallidus, all ICC values were above 0.95. The test–retest reliability of DTI-based measures appears to vary across the measures and fiber tracts. Wang et al. (2012) found excellent reliability for the mean length of the corpus callosum and the uncinate fasciculus, and fair (ICC 0.4–0.75) reliabilities for fractional anisotropy in most fiber tracts. Test–retest reliability of task-based fMRI data has been characterized as "fair" in adults and adolescents, and "poor" (ICC < 0.4) in children; it was lower in regions with weak "activation," as revealed by group t-maps (Caceres et al., 2009; Koolschijn et al., 2011; Plichta et al., 2012). Test–retest reliability of the fMRI response varies across regions engaged by a given task (Caceres et al., 2009). The relatively low within-subject test–retest reliability of functional data is likely related to a number of factors, including the fact that the BOLD signal is an indirect measure of brain activity, its measurement is affected by a number of noise-generating factors (e.g., head motion, physiologic "noise" related to respiration and cardiac cycle) and by the state of the participant during scanning. The latter factors, such as interindividual and session-by-session (longitudinal) variations in task-related behavior (performance, attention) and general state of arousal (anxiety, sleepiness), are difficult to control, thereby adding error to these measurements.

Interpretation

Provided we are able to obtain reliable measurements, the next challenge is that of interpretation. Let us illustrate the challenge of interpreting imaging-based findings using two examples: (1) directionality of structure–function relationships; and (2) the ambiguity in our use of the term "connectivity."

Directionality of structure–function relationships

When considering directionality of structure–function relationships, it is often assumed to be unidirectional: from structure to function. This is not always the case, however.

An ever-growing literature on experience-induced (or experience-associated) changes in brain morphology, such as those in gray-matter density, cortical thickness, or fractional anisotropy, attests to the possibility of the opposite directionality: from function (experience) to structure. These experience-related structural changes have been observed in adult brains in multiple situations, including musical training (Gaser and Schlaug, 2003), profession (taxi drivers; Maguire et al., 2000), acquisition of a second language (Mechelli et al., 2004; Stein et al., 2012b), studying for medical exams (Draganski et al., 2006), or learning a new visuomotor skill (Draganski et al., 2004; Boyke et al., 2008; Driemeyer et al., 2008). The last example (learning how to juggle three balls in the air) is particularly informative given its longitudinal (pre- and posttraining) design. Thus, it is clear that cumulative functional engagement of a particular neural network shapes its structure.

In some cases, however, structure does indeed influence function. A rather unexpected example of such a structure-to-function pathway is the relationship between posttraumatic stress disorder (PTSD) and hippocampal volume, with the smaller hippocampus being often attributed to the known neurotoxic effects of trauma-induced stress. But, in fact, smaller hippocampus volumes may be a trait indicating an individual's predisposition for developing PTSD after being exposed to trauma. This interpretation was put forward following an observation that PTSD is more likely to develop after an exposure to combat trauma in soldiers whose monozygotic co-twins – not exposed to combat – have a smaller hippocampus (Gilbertson et al., 2002).

Functional and structural connectivity

The second example concerns functional and structural connectivity. Over the past 30+ years, we have seen the term "connectivity" used in reference to multiple features of the functional and structural organization of the human brain. A quick PubMed search – using the terms "connectivity" and "brain," and limited to "humans" as the species – yielded 10 853 citations (as of January 25, 2016). In the neuroimaging literature of the past 20 years, studies of "functional connectivity" have come to dominate the field: 5945 citations represent a subset of the 10 853 citations, now restricted by adding the terms "functional" and "imaging." In most of these studies, the authors used various statistical techniques to identify and describe coordinated patterns of variation in brain activity measured with electroencephalography (magnetoencephalography) or indexed indirectly by regional cerebral blood flow or the BOLD signal. But such correlation-based connectivity is not the same as the true neural connectivity between two spatially discrete populations of neurons, as revealed by classic tract-tracing studies (Bohland et al., 2009) and used to construct wiring diagrams of the monkey brain (e.g., CoCoMac: Stephan et al., 2001). For obvious reasons, the same approach cannot be used to map the neural connectivity of the human brain. Thus, it is important to keep in mind that, in most cases, the term "functional connectivity" is used only as a metaphor; its meaning is determined by the analytic tool used to derive the measure associated with the term in a given study.

If we go back to our PubMed search and replace the term "functional" by "structural," a total of 1492 citations now emerge. Most of these studies use the term "structural connectivity" in the context of the macrostructural or microstructural properties of white matter derived (mostly) from T1-weighted and diffusion tensor images, respectively. DTI-based tractography has been embraced as a way to overcome the absence of knowledge about neural connectivity in the human brain; the hope is that it can approximate the gold standard – anatomic connectivity revealed by neural tracers in the monkey brain (see above). This is unlikely, however. DTI-based tractography provides a representation of the main fiber tracts, which may change depending on the algorithm used to derive it (Chung et al., 2011); it is not equivalent to the neural connectivity obtained with tracers transported from cell to cell by axonal transport.

In summary, MRI represents a powerful technology for mapping brain structure and function across the lifespan. Combining multimodal approaches with automatic image analysis pipelines enables investigators to derive multiple structural and functional characteristics, or quantitative traits. But caution should be exercised both in terms of the test–retest reliability of these measures and their neurobiologic interpretations.

FUTURE DIRECTIONS

To date, most of the population neuroscience work has been done in high-income countries and carried out in the context of observational studies. This is the case for the great majority of sites participating in the ENIGMA consortium (Thompson et al., 2014) and the Cohorts for Heart and Aging Research in Genomic Epidemiology (CHARGE) consortium (Psaty et al., 2009). These two consortia brought together various case-control and cohort studies employing brain imaging and genetics in disease-oriented research carried out mostly in North America, Europe, Australia, and Japan. There are many reasons, however, for extending this work in two directions: (1) middle-income countries; and (2) interventions.

The World Bank classifies countries based on gross national income (GNI) per capita.

> *As of 1 July 2015, low-income economies are defined as those with a GNI per capita, calculated using the World Bank Atlas method, of $1,045 or less in 2014; middle-income economies are those with a GNI per capita of more than $1,045 but less than $12,736; high-income economies are those with a GNI per capita of $12,736 or more. Lower-middle-income and upper-middle-income economies are separated at a GNI per capita of $4,125 (Anonymous, 2016).*

Here are a few examples of countries in each of the four categories: (1) low income (<$1045): Afghanistan, Malawi; (2) lower-middle income ($1046–4125): Bangladesh, India, Indonesia, Nicaragua, Nigeria, Pakistan, Zambia; (3) upper-middle income ($4126–12 735): Brazil, China, Malaysia, Mexico, Paraguay, South Africa, Turkey; and (4) high income (>$12 736): Argentina, Canada, Chile, Japan, Russian Federation, Uruguay, USA, Venezuela, all Western and Central European countries. There are many reasons to facilitate population neuroscience studies in middle-income countries. First, 7/10 most populous countries have middle-income economies (China, India, Indonesia, Brazil, Pakistan, Bangladesh, Nigeria: a total population of ~3.3 billion in 2008). Second, the contribution of noncommunicable diseases, including psychiatric and cardiometabolic disorders, to the global disease burden shows an increasing trend in middle-income countries (Murray et al., 2012; Global Burden of Disease Study 2013 Collaborators, 2015). This is likely due to an increasing lifespan but also to lifestyle changes associated, for example, with obesity (Poskitt, 2014). Third, a broad range of physical and social environments provides opportunities for disentangling complex relationships between exposures and outcomes with regard to the potential causes and confounders. For example, finding a similar relationship between breastfeeding and the child's general intelligence in middle- and high-income countries strengthens its interpretation as being causal given the different economic and/or cultural reasons for breastfeeding (Brion et al., 2011). These are but a few reasons for expanding population neuroscience studies in middle-income countries. In Brazil, for example, two such studies are already under way: one in the metropolitan areas of Porto Alegre and São Paulo (Salum et al., 2015) and another in a rural community of Baependi (Taporoski et al., 2015).

Overall, our understanding of forces shaping the human brain would be greatly enhanced by taking a global approach that embraces diversity of physical and social environments as a driver of knowledge. Diverse environments are likely to generate diverse pathways ensuring healthy brain development. When discovered and understood, such "alternative solutions" brought about by particular circumstances can be, in turn, harnessed to advance child development in other contexts. This brings us to the second direction in which to develop further population neuroscience: intervention studies.

Many noncommunicable diseases are related to our lifestyle. A point in case: Alzheimer's disease (AD). Based on a systematic review of epidemiologic evidence, Barnes and Yaffe (2011) estimated that seven modifiable risk factors: midlife obesity, diabetes, midlife hypertension, physical inactivity, smoking, depression, and low education – account for up to 51% of the population-attributable risk of AD in the world (Table 2.4). Poor health behavior (e.g., physical inactivity, poor diet) can be viewed as distal risk factors of AD, acting in part

Table 2.4

Alzheimer's disease cases attributable to potentially modifiable risk factors worldwide

Risk factor	Population prevalence	Relative risk (95% CI)	PAR (confidence range)	Number of cases attributable (thousands; confidence range)
Diabetes mellitus	6.40%	1.39 (1.17–1.66)	2.4% (1.1–4.1)	826 (365–1374)
Midlife hypertension	8.90%	1.61 (1.16–2.24)	5.1% (1.4–9.9)	1746 (476–3369)
Midlife obesity	3.40%	1.60 (1.34–1.92)	2.0% (1.1–3.0)	678 (387–1028)
Depression	13.20%	1.90 (1.55–2.33)	10.6% (6.8–14.9)	3600 (2295–5063)
Physical inactivity	17.70%	1.82 (1.19–2.78)	12.7% (3.3–24 0)	4297 (1103–8122)
Smoking	27.40%	1.59 (1.15–2.20)	13.9% (3.9–24 7)	4718 (1338–8388)
Low education	40.00%	1.59 (1.35–1.86)	19.1% (12.3–25.6)	6473 (4163–8677)
Combined (maximum)			50.70%	17 187 028*

Reproduced from Barnes and Yaffe (2011).

*Absolute number.

PAR, population-attributable risk; CI, confidence intervals.

via proximal cardiometabolic risk factors, such as obesity, type 2 diabetes mellitus, and hypertension. Over time, these proximal and distal risk factors affect adversely "brain maintenance," the preserved chemistry, structure, and function in healthy aging brains (Nyberg et al., 2012), and "brain reserve" (Stern, 2012). Not surprisingly, health trajectories culminating in AD begin decades before the full-blown disease emerges. In fact, developmental cascades that increase the probability of AD begin at conception. For example, we have shown that maternal smoking during pregnancy is associated with smaller cerebral cortex (Paus et al., 2012) and adolescent obesity (Syme et al., 2010). Adolescent obesity is, in turn, associated with metabolic syndrome (Syme et al., 2008), lower executive functions (Schwartz et al., 2013), and variations in structural brain properties (Schwartz et al., 2014), all during adolescence. Clearly, the best way to prevent AD is to prevent these and other risk factors from emerging, thus building and maintaining brain reserve.

Low-cost interventions aimed at modifying health behaviors can be readily embedded in cohort studies. There are several advantages of doing so. First, the same mobile technology described above in the context of monitoring individuals' behaviour can be used to facilitate behavior change. For example, a recent Tobacco, Exercise and Diet Messages (TEXT ME) trial showed the effectiveness of this mode of support (four text messages per week for 6 months) in adults with coronary heart disease; the intervention group showed reduction of smoking, increase in physical activity, lowering of body mass index, systolic blood pressure, and low-density lipoprotein cholesterol (Chow et al., 2015). When possible, such automatic texting can be enhanced by online interactions with health coaches or through moderated peer interactions via social media. Interactive computer-based interventions (online chats, bidirectional e-mails) have been pursued mostly in the context of weight loss; a systematic review of such (14 studies, 2537 participants) concluded that they are effective (Wieland et al., 2012). Second, baseline assessments can be used to personalize subsequent interventions (e.g., focus on diet) and use the initial results to motivate changes in health behaviors. For example, providing information about a specific risk gene for AD (*APOE*) to individuals with a family history of AD appears to be 2.7 times more effective in changing their health behaviors (1 year later) than providing risk estimates based on family history alone (Chao et al., 2008). Similarly, providing genotype-based personalized (vs. general) dietary advice was judged more useful by the participants when considering their diet (Nielsen and El-Sohemy, 2012). Such motivational enhancements can, of course, extend beyond genotypes and include information about a number of state-of-the-art health indicators derived in the course of a cohort study (Table 2.5).

Third, a state-of-the-art phenotyping infrastructure of a cohort study would not only serve as an ideal platform for evaluating primary and secondary outcomes of an intervention but would also provide possible insights into processes mediating its effectiveness. Furthermore, the broad nature of population-based cohorts with regard to other assessments would allow one to identify characteristics differentiating responders from nonresponders. This knowledge would be helpful for modifying the tested intervention and designing new intervention strategies.

In summary, supporting the growth of population neuroscience in middle-income countries and embedding interventions in large-scale studies would: (1) enhance our understanding of the basic processes underlying the influence of physical and social environments (and

Table 2.5

Enhanced health information that can be provided to participants in a population neuroscience study

Domain	Information
Brain	Hippocampal volume
Cognition	Executive score, memory score
Body composition	BMI, fat/lean mass, visceral fat
Biochemistry	Glucose, hemoglobin A1c, insulin, cholesterol, LDL-chol, HDL-chol, TG
Cardiovascular	BP, BP variability, BP response to mental stress
Diet	Macro- and micronutrients; glycemic index; fatty acid index
Physical activity	Activity index (metabolic equivalent of task)
Mood	Depression and anxiety

BMI, body mass index; LDL-chol, low-density lipoprotein cholesterol; HDL-chol, high-density lipoprotein cholesterol; BP, blood pressure; TG, triglycerides.

genes) on the human brain; and (2) provide a useful test bed for investigating strategies for preventing noncommunicable diseases in the general population.

CONCLUSIONS

Technology drives progress in science. In the context of population neuroscience, this is most apparent in the use of MRI for multimodal *in vivo* assessments of brain structure and function, and in the availability of high-throughput assays for genotyping (including whole-genome sequencing), epigenotyping, and other omics-based analyses of biologicals samples. Information and communication technology is entering this field both as a mobile tool for assessing social environment and monitoring individuals' behaviour in everyday life, as well as a tool for supporting behavior change. The widespread presence of mobile technology in today's society makes it well suited for preventing common diseases and increasing population health. The highly integrative approach embraced by population neuroscience provides an opportunity for tackling the multiplicity of influences shaping the human brain in an extremely interconnected fashion throughout the life course. As noted by Walter Fontana, a senior author of a recent paper on factors influencing the life span of *Caenorhabditis elegans*: "there is profound interdependence in the physiology of an organism, and changes in one physiological aspect affect all others to determine life span" (Harvard Medical School, 2016).

ACKNOWLEDGMENTS

The work described in this chapter was made possible by our funders, including the Canadian Institutes of Health Research and the National Institutes of Health (USA). I serve as the Ann and Max Tanenbaum Chair in Population Neuroscience at the University of Toronto and the Dr. John and Consuela Phelan Scholar at the Child Mind Institute in New York. I am grateful to my students, fellows, and colleagues for their contributions made in the course of our studies of the adolescent brain. I very much appreciate the collaborative spirit of my academic colleagues associated with a number of cohorts, including the IMAGEN Study, ALSPAC, Northern Finland Birth Cohort, and the Porto Alegre-São Paulo High Risk Cohort Study for the Development of Childhood Psychiatric Disorders. My work on population neuroscience would not be possible without Dr. Zdenka Pausova. Over more than 20 years, Zdenka has provided me the inspiration and knowledge necessary for embarking on studies in genetics and epigenetics. Together, we built the Saguenay Youth Study, which provides the template for most of the ideas and concepts described here. Finally, I thank my graduate student Nadine Parker for creating Table 2.1.

References

Albert FW, Kruglyak L (2015). The role of regulatory variation in complex traits and disease. Nat Rev Genet 16: 197–212.

Anonymous (2016). The World Bank: New Country Classifications [Online]. The World Bank. Available: http://data.worldbank.org/news/new-country-classifications-2015.

Arloth J, Bogdan R, Weber P et al. (2015). Genetic differences in the immediate transcriptome response to stress predict risk-related brain function and psychiatric disorders. Neuron 86: 1189–1202.

Barnes DE, Yaffe K (2011). The projected effect of risk factor reduction on Alzheimer's disease prevalence. Lancet Neurol 10: 819–828.

Bibikova M, Barnes B, Tsan C et al. (2011). High density DNA methylation array with single CpG site resolution. Genomics 98: 288–295.

Blanford JI, Huang Z, Savelyev A et al. (2015). Geo-located tweets. Enhancing mobility maps and capturing cross-border movement. PLoS One 10: e0129202.

Bohland JW, Wu C, Barbas H et al. (2009). A proposal for a coordinated effort for the determination of brainwide neuroanatomical connectivity in model organisms at a mesoscopic scale. PLoS Comput Biol 5: e1000334.

Boyke J, Driemeyer J, Gaser C et al. (2008). Training-induced brain structure changes in the elderly. J Neurosci 28: 7031–7035.

Brion MJ, Lawlor DA, Matijasevich A et al. (2011). What are the causal effects of breastfeeding on IQ, obesity and blood pressure? Evidence from comparing high-income with middle-income cohorts. Int J Epidemiol 40: 670–680.

Caceres A, Hall DL, Zelaya FO et al. (2009). Measuring fMRI reliability with the intra-class correlation coefficient. Neuroimage 45: 758–768.

Chao S, Roberts JS, Marteau TM et al. (2008). Health behavior changes after genetic risk assessment for Alzheimer disease: the REVEAL Study. Alzheimer Dis Assoc Disord 22: 94–97.

Cheung C, Yu K, Fung G et al. (2010). Autistic disorders and schizophrenia: related or remote? An anatomical likelihood estimation. PLoS One 5: e12233.

Chow CK, Redfern J, Hillis GS et al. (2015). Effect of lifestyle-focused text messaging on risk factor modification in patients with coronary heart disease: a randomized clinical trial. JAMA 314: 1255–1263.

Christine PJ, Auchincloss AH, Bertoni AG et al. (2015). Longitudinal associations between neighborhood physical and social environments and incident type 2 diabetes mellitus: the Multi-Ethnic Study of Atherosclerosis (MESA). JAMA Intern Med 175: 1311–1320.

Chunara R, Bouton L, Ayers JW et al. (2013). Assessing the online social environment for surveillance of obesity prevalence. PLoS One 8: e61373.

Chung HW, Chou MC, Chen CY (2011). Principles and limitations of computational algorithms in clinical diffusion tensor MR tractography. AJNR Am J Neuroradiol 32: 3–13.

Collins A, Koppen G, Valdiglesias V et al. (2014). The comet assay as a tool for human biomonitoring studies: the ComNet project. Mutat Res Rev Mutat Res 759: 27–39.

Curtis JW, Curtis A, Mapes J et al. (2013). Using Google Street View for systematic observation of the built environment: analysis of spatio-temporal instability of imagery dates. Int J Health Geogr 12: 53.

Deoni SC (2011). Correction of main and transmit magnetic field (B0 and B1) inhomogeneity effects in multicomponent-driven equilibrium single-pulse observation of T1 and T2. Magn Reson Med 65: 1021–1035.

Deoni SC, Peters TM, Rutt BK (2005). High-resolution T1 and T2 mapping of the brain in a clinically acceptable time with DESPOT1 and DESPOT2. Magn Reson Med 53: 237–241.

Deoni SC, Rutt BK, Arun T et al. (2008). Gleaning multicomponent T1 and T2 information from steady-state imaging data. Magn Reson Med 60: 1372–1387.

Dickie EW, Tahmasebi A, French L et al. (2014). Global genetic variations predict brain response to faces. PLoS Genet 10: e1004523.

Draganski B, Gaser C, Busch V et al. (2004). Neuroplasticity: changes in grey matter induced by training. Nature 427: 311–312.

Draganski B, Gaser C, Kempermann G et al. (2006). Temporal and spatial dynamics of brain structure changes during extensive learning. J Neurosci 26: 6314–6317.

Driemeyer J, Boyke J, Gaser C et al. (2008). Changes in gray matter induced by learning–revisited. PLoS One 3: e2669.

Eickhoff SB, Paus T, Caspers S et al. (2007). Assignment of functional activations to probabilistic cytoarchitectonic areas revisited. Neuroimage 36: 511–521.

Eickhoff SB, Laird AR, Grefkes C et al. (2009). Coordinate-based activation likelihood estimation meta-analysis of neuroimaging data: a random-effects approach based on empirical estimates of spatial uncertainty. Hum Brain Mapp 30: 2907–2926.

Exley K, Cano N, Aerts D et al. (2015). Communication in a human biomonitoring study: focus group work, public engagement and lessons learnt in 17 European countries. Environ Res 141: 31–41.

Finucane HK, Bulik-sullivan B, Gusev A et al. (2015). Partitioning heritability by functional annotation using genome-wide association summary statistics. Nat Genet 47: 1228–1235.

Friston KJ (2011). Functional and effective connectivity: a review. Brain Connect 1: 13–36.

Gao W, Emaminejad S, Nyein HY et al. (2016). Fully integrated wearable sensor arrays for multiplexed in situ perspiration analysis. Nature 529: 509–514.

Gaser C, Schlaug G (2003). Brain structures differ between musicians and non-musicians. J Neurosci 23: 9240–9245.

Gilbertson MW, Shenton ME, Ciszewski A et al. (2002). Smaller hippocampal volume predicts pathologic vulnerability to psychological trauma. Nat Neurosci 5: 1242–1247.

Global Burden of Disease Study 2013 Collaborators (2015). Global, regional, and national comparative risk assessment of 79 behavioural, environmental and occupational, and metabolic risks or clusters of risks in 188 countries, 1990–2013: a systematic analysis for the Global Burden of Disease Study 2013. Lancet 386: 2287–2323.

Gusev A, Lee SH, Trynka G et al. (2014). Partitioning heritability of regulatory and cell-type-specific variants across 11 common diseases. Am J Hum Genet 95: 535–552.

Harvard Medical School (2016). 'Lifespan machine' probes cause of aging: Findings suggest that aging has no single mechanism. Science Daily. Available online at www.sciencedaily.com/releases/2016/01/160127141412.htm.

Hibar DP, Stein JL, Renteria ME et al. (2015). Common genetic variants influence human subcortical brain structures. Nature 520: 224–229.

Hurvitz PM, Moudon AV, Kang B et al. (2014). Emerging technologies for assessing physical activity behaviors in space and time. Front Public Health 2: 2.

Kardan O, Gozdyra P, Misic B et al. (2015). Neighborhood greenspace and health in a large urban center. Sci Rep 5: 11610.

Kellis M, Wold B, Snyder MP et al. (2014). Defining functional DNA elements in the human genome. Proc Natl Acad Sci U S A 111: 6131–6138.

Kendler KS, Gardner CO, Prescott CA (2003). Personality and the experience of environmental adversity. Psychol Med 33: 1193–1202.

Koolschijn PC, Schel MA, De rooij M et al. (2011). A three-year longitudinal functional magnetic resonance imaging study of performance monitoring and test-retest reliability from childhood to early adulthood. J Neurosci 31: 4204–4212.

Kucharczyk W, Macdonald PM, Stanisz GJ et al. (1994). Relaxivity and magnetization transfer of white matter lipids at MR imaging: importance of cerebrosides and pH. Radiology 192: 521–529.

Kulis M, Queiros AC, Beekman R et al. (2013). Intragenic DNA methylation in transcriptional regulation, normal differentiation and cancer. Biochim Biophys Acta 1829: 1161–1174.

Less EL, Mckee P, Toomey T et al. (2015). Matching study areas using Google Street View: a new application for an emerging technology. Eval Program Plann 53: 72–79.

Maguire EA, Gadian DG, Johnsrude IS et al. (2000). Navigation-related structural change in the hippocampi of taxi drivers. Proc Natl Acad Sci U S A 97: 4398–4403.

Manolio TA, Collins FS, Cox NJ et al. (2009). Finding the missing heritability of complex diseases. Nature 461: 747–753.

McGowan JC (1999). The physical basis of magnetization transfer imaging. Neurology 53: S3–S7.

Mechelli A, Crinion JT, Noppeney U et al. (2004). Neurolinguistics: structural plasticity in the bilingual brain. Nature 431: 757.

Murray CJ, Vos T, Lozano R et al. (2012). Disability-adjusted life years (DALYs) for 291 diseases and injuries in 21 regions, 1990–2010: a systematic analysis for the Global Burden of Disease Study 2010. Lancet 380: 2197–2223.

Nielsen DE, El-sohemy A (2012). A randomized trial of genetic information for personalized nutrition. Genes Nutr 7: 559–566.

Nyberg L, Lovden M, Riklund K et al. (2012). Memory aging and brain maintenance. Trends Cogn Sci 16: 292–305.

Odgers CL, Caspi A, Bates CJ et al. (2012). Systematic social observation of children's neighborhoods using Google Street View: a reliable and cost-effective method. J Child Psychol Psychiatry 53: 1009–1017.

Park G, Schwartz HA, Eichstaedt JC et al. (2015). Automatic personality assessment through social media language. J Pers Soc Psychol 108: 934–952.

Paus T (2013). Population Neuroscience. Springer-Verlag, Berlin.

Paus T, Bernard M, Chakravarty MM et al. (2012). KCTD8 gene and brain growth in adverse intrauterine environment: a genome-wide association study. Cereb Cortex 22: 2634–2642.

Paus T, Pesaresi M, French L (2014). White matter as a transport system. Neuroscience 276: 117–125.

Pike GB (1996). Pulsed magnetization transfer contrast in gradient echo imaging: a two-pool analytic description of signal response. Magn Reson Med 36: 95–103.

Plichta MM, Schwarz AJ, Grimm O et al. (2012). Test-retest reliability of evoked BOLD signals from a cognitive-emotive fMRI test battery. Neuroimage 60: 1746–1758.

Poskitt EM (2014). Childhood obesity in low- and middle-income countries. Paediatr Int Child Health 34: 239–249.

Pott S, Lieb JD (2015). What are super-enhancers? Nat Genet 47: 8–12.

Psaty BM, O'donnell CJ, Gudnason V et al. (2009). Cohorts for Heart and Aging Research in Genomic Epidemiology (CHARGE) consortium: design of prospective meta-analyses of genome-wide association studies from 5 cohorts. Circ Cardiovasc Genet 2: 73–80.

Ramasamy A, Trabzuni D, Guelfi S et al. (2014). Genetic variability in the regulation of gene expression in ten regions of the human brain. Nat Neurosci 17: 1418–1428.

Salum GA, Gadelha A, Pan PM et al. (2015). High risk cohort study for psychiatric disorders in childhood: rationale, design, methods and preliminary results. Int J Methods Psychiatr Res 24: 58–73.

Schizophrenia working group of the psychiatric genomics Consortium (2014). Biological insights from 108 schizophrenia-associated genetic loci. Nature 511: 421–427.

Schmierer K, Scaravilli F, Altmann DR et al. (2004). Magnetization transfer ratio and myelin in postmortem multiple sclerosis brain. Ann Neurol 56: 407–415.

Schmierer K, Wheeler-kingshott CA, Tozer DJ et al. (2008). Quantitative magnetic resonance of postmortem multiple sclerosis brain before and after fixation. Magn Reson Med 59: 268–277.

Schwartz DH, Leonard G, Perron M et al. (2013). Visceral fat is associated with lower executive functioning in adolescents. Int J Obes (Lond) 37: 1336–1343.

Schwartz DH, Dickie E, Pangelinan MM et al. (2014). Adiposity is associated with structural properties of the adolescent brain. Neuroimage 103C: 192–201.

Stein JL, Medland SE, Vasquez AA et al. (2012a). Identification of common variants associated with human hippocampal and intracranial volumes. Nat Genet 44: 552–561.

Stein M, Federspiel A, Koenig T et al. (2012b). Structural plasticity in the language system related to increased second language proficiency. Cortex 48: 458–465.

Stephan KE, Kamper L, Bozkurt A et al. (2001). Advanced database methodology for the collation of connectivity data on the macaque brain (CoCoMac). Philos Trans R Soc Lond B Biol Sci 356: 1159–1186.

Stern Y (2012). Cognitive reserve in ageing and Alzheimer's disease. Lancet Neurol 11: 1006–1012.

Syme C, Abrahamowicz M, Leonard GT et al. (2008). Intra-abdominal adiposity and individual components of the metabolic syndrome in adolescence: sex differences and underlying mechanisms. Arch Pediatr Adolesc Med 162: 453–461.

Syme C, Abrahamowicz M, Mahboubi A et al. (2010). Prenatal exposure to maternal cigarette smoking and accumulation of intra-abdominal fat during adolescence. Obesity (Silver Spring) 18: 1021–1025.

Tahmasebi AM, Artiges E, Banaschewski T et al. (2012). Creating probabilistic maps of the face network in the adolescent brain: a multicentre functional MRI study. Hum Brain Mapp 33: 938–957.

Taporoski TP, Negrao AB, Horimoto AR et al. (2015). Shared genetic factors of anxiety and depression symptoms in a Brazilian family-based cohort, the Baependi heart study. PLoS One 10e0144255.

Thompson PM, Stein JL, Medland SE et al. (2014). The ENIGMA consortium: large-scale collaborative analyses of neuroimaging and genetic data. Brain Imaging Behav 8: 153–182.

Toro R, Poline JB, Huguet G et al. (2015). Genomic architecture of human neuroanatomical diversity. Mol Psychiatry 20: 1011–1016.

Tzourio-mazoyer N, Landeau B, Papathanassiou D et al. (2002). Automated anatomical labeling of activations in SPM using a macroscopic anatomical parcellation of the MNI MRI single-subject brain. Neuroimage 15: 273–289.

Wang JY, Abdi H, Bakhadirov K et al. (2012). A comprehensive reliability assessment of quantitative diffusion tensor tractography. Neuroimage 60: 1127–1138.

Weng L, Menczer F (2015). Topicality and impact in social media: diverse messages, focused messengers. PLoS One 10e0118410.

Wieland LS, Falzon L, Sciamanna CN et al. (2012). Interactive computer-based interventions for weight loss or weight maintenance in overweight or obese people. Cochrane Database Syst Rev 8CD007675.

Wishart DS (2011). Advances in metabolite identification. Bioanalysis 3: 1769–1782.

Wonderlick JS, Ziegler DA, Hosseini-varnamkhasti P et al. (2009). Reliability of MRI-derived cortical and subcortical morphometric measures: effects of pulse sequence, voxel geometry, and parallel imaging. Neuroimage 44: 1324–1333.

Yang J, Lee SH, Goddard ME et al. (2011). GCTA: a tool for genome-wide complex trait analysis. Am J Hum Genet 88: 76–82.

Yang J, Lee T, Kim J et al. (2013). Ubiquitous polygenicity of human complex traits: genome-wide analysis of 49 traits in Koreans. PLoS Genet 9e1003355.

Zhang H, Schneider T, Wheeler-kingshott CA et al. (2012). NODDI: practical in vivo neurite orientation dispersion and density imaging of the human brain. Neuroimage 61: 1000–1016.

Zhang T, Cooper S, Brockdorff N (2015). The interplay of histone modifications – writers that read. EMBO Rep 16: 1467–1481.

Handbook of Clinical Neurology, Vol. 138 (3rd series)
Neuroepidemiology
C. Rosano, M.A. Ikram, and M. Ganguli, Editors
http://dx.doi.org/10.1016/B978-0-12-802973-2.00003-3

Chapter 3

Advanced epidemiologic and analytical methods

E. ALBANESE*
Department of Psychiatry, University of Geneva, Geneva, Switzerland

Abstract

Observational studies are indispensable for etiologic research, and are key to test life-course hypotheses and improve our understanding of neurologic diseases that have long induction and latency periods. In recent years a plethora of advanced design and analytic techniques have been developed to strengthen the robustness and ultimately the validity of the results of observational studies, and to address their inherent proneness to bias. It is the responsibility of clinicians and researchers to critically appraise and appropriately contextualize the findings of the exponentially expanding scientific literature. This critical appraisal should be rooted in a thorough understanding of advanced epidemiologic methods and techniques commonly used to formulate and test relevant hypotheses and to keep bias at bay.

INTRODUCTION

Our understanding of the causes and physiopathology of several disorders and diseases of the nervous system is still limited. Epidemiologic evidence contributes to improve our understanding of how modifiable risk/protective factors may relate to the clinical expression of neurologic syndromes, the underlying neuropathology, neuroinflammation (Heneka et al., 2015), and neurovascular damage (Zlokovic, 2005; Jellinger, 2008). This provides important insights into the manifestation of disease in community-dwelling individuals. Further, this evidence is of fundamental importance to inform the definition of theoretic frameworks and the formulation of clinical and research hypotheses. These hypotheses can then be tested and verified in experimental studies to prove causality, advance knowledge, and promote translation from basic science to clinical practice and public health interventions (Launer, 2007). Therefore, the role played by bias, confounding, and chance in epidemiologic studies should be carefully considered at the outset. The interpretation of the results should be informed by a critical appraisal of the methodologic characteristics and quality of the study. Critical appraisal is an analytic and structured evaluation used to establish the level of confidence in the results (internal validity), and to determine the applicability of study findings beyond the specific research context (external validity). With critical appraisal we consider potential sources of errors, and we determine whether the information is reliable and useful for us. Critical appraisal is relevant in both observational and experimental studies, and is at the core of causal inference.

Relevance of critical appraisal in observational epidemiology

Descriptive epidemiology is used to estimate the number of people affected by a given disease, or with relevant health characteristics, including symptoms and signs, at a population level. Prevalence and incidence are the two key measures used in descriptive epidemiology, and constitute fundamental information to inform public health and policy decisions. Even though descriptive studies may appear to have a simple design, some methodologic aspects need to be considered. A number of neurologic diseases are degenerative in nature, chronic, or may not be treated at present. Therefore, their prevalence, which is the product of incidence and duration, is influenced by the number of new cases in the population

*Correspondence to: Prof. Emiliano Albanese, MD PhD, Department of Psychiatry (WHO CC for Mental Health Research and Training), University of Geneva and Swiss School of Public Health (SSPH+), 70c rue du Grand-Pré, Geneva, Switzerland (CH) 1202. Tel: +41-22-372-5554, E-mail: emiliano.albanese@unige.ch

at risk (i.e., incidence) and by survival rather than recovery. However, although length of survival after diagnosis may increase if case fatality rate is reduced, it is also influenced by earlier detection of the disease (typically of milder cases), which in turn can inflate prevalence even when mortality rates remain unchanged. On the one hand longer survival and changes in diagnostic methods may increase or inflate prevalence. On the other hand they may lead to a greater heterogeneity amongst the enumerated cases. This heterogeneity can have relevant implications for internal and external validity and in both descriptive and analytic studies.

Relevance of critical appraisal in analytic epidemiology

Another key use of epidemiology is the identification of risk and protective factors to inform prevention, and to provide clues to etiology and disease mechanisms and contribute to better delineate new syndromes. In etiologic research an important distinction is made between observational and experimental studies. In observational studies, which include case-control and cohort designs, amongst others, disease occurrence is investigated with respect to exposure to putative causes. These typically include measurable characteristics that capture the existence and/or the extent of exposure to intrinsic (including genetics and biologic features) and extrinsic factors (including social, environmental, and lifestyles), which may in turn be associated to the presence or absence of a disease and its clinical features (symptoms and signs), and/or to the underlying neuropathology. Observational studies are often used in etiologic research. The results may inform clinical and public health practice, particularly prevention, and prompt new hypotheses about the etiopathologic mechanisms of diseases, but cannot prove causality. Conversely, causal inference can be established in experimental studies to a markedly higher degree of confidence. This is because the associations, or risk estimates, evidenced in experimental studies are less prone to confounding, bias, and chance (see Chapter 1). Confounding is minimized in randomized controlled trials (RCTs), because randomization matches groups by all known, and unknown, characteristics except the intervention. Bias mainly depends on how RCTs are conducted and on their methodologic quality. Key design features include the appropriateness of randomization, blinding of participants, investigators and data analysts, and concealment of allocation to the intervention. Formal quality assessment of RCTs is used to explore the internal validity of the results of primary studies and to investigate and interpret the heterogeneity across study results in meta-analyses and systematic reviews. Yet, RCTs are often conducted in controlled experimental conditions that limit external validity. In addition, not infrequently, the manipulation of potential causes or protective factors to test differential effects on health outcomes can be impractical or unethical.

Observational studies remain an indispensable instrument for etiologic research. Their inherent proneness to bias and confounding cannot be eliminated, but it can be substantially reduced and controlled for during conduction of the study and statistical analysis. In recent years a plethora of advanced design and analytic techniques have been developed to strengthen the robustness and ultimately the validity of the results of observational studies.

THE SCOPES OF NEUROEPIDEMIOLOGY

Neuroepidemiology investigates risk and protective factors that may exert their effect directly on the underlying neuropathologies of diseases or that may influence the clinical expression of signs and symptoms in the presence of damage. In the former case, modulation of risk exposure is ultimately concerned with avoidance of disease, and can be conceived as primary prevention. In the latter case the modification of the risk profile may instead delay (or advance) the actual clinical onset of the disease, which would correspond to secondary prevention. There is very likely a complex interplay between the expression of clinical symptoms and the underlying neuropathology of disease, and empiric evidence may not be easy to reconcile. Here we illustrate the case of dementia, and the transformative relevance of endorsing a life-course approach to improve its study.

The relationship between the manifest clinical symptoms of dementia during life and the underlying pathologic hallmarks of Alzheimer's disease (AD) measured postmortem varied markedly in a large representative sample of community-dwelling elders, particularly at older ages (Savva et al., 2009). More recent studies have used functional and structural brain imaging techniques to quantify beta-amyloidosis and neurodegeneration in individuals without cognitive impairment. Both features of brain damage were common and their presence increased with age (Jack et al., 2010). This confirms that dementia is not a normal part of aging, not only because the neuropathology is not inexorable, but also because in its presence the clinical expression of the disease may greatly vary between individuals or be absent through late life and until death. A large body of evidence needs to be considered to decipher this variability.

It has been suggested that the modulation and variability of the clinical expression of symptoms in dementia may be explained by brain (structural) and cognitive (functional) reserve, which is influenced by a number of factors (Stern, 2012). These factors may operate from *in utero* (i.e., intrauterine environment, placental function, and maternal nutrition), through early life (i.e., socioeconomic circumstances and nutrition), adolescence, and young adulthood (i.e., educational and occupational achievements; social life, and lifestyle), to midlife (i.e., cardiovascular risk profile) and late life (i.e., preservation of healthy lifestyle, of cognitive stimulating activities, and social interactions). Psychosocial factors would ward off the development and clinical manifestations of the disease, and vascular risk factors (i.e., high blood pressure, cholesterol, diabetes, and obesity) have consistently been shown to increase the risk of late-onset dementia (Prince et al., 2014). However, further evidence suggests that the effects of psychosocial factors on cognitive function markedly vary with age, and declining values of vascular risk factors after midlife are often associated with dementia, possibly in relation to the underlying disease processes. This corpus of evidence suggested that dementia may have a lifelong trajectory in a manner comparable to several other chronic diseases of late life, with risks clustering around specific developmental epochs and often accumulating in their effects, sometimes with relatively long latent periods before onset of clinical syndromes. A life-course approach to late-onset dementia has been called for in order to reconcile this complex corpus of evidence (Launer, 2005).

In the next sections, we will illustrate the main concepts and methods of life-course epidemiology and its relevance in the context of neurologic diseases and disorders, highlighting the existing methodologic challenges. Then we will describe some sources of errors and bias that may reduce the internal and external validity of epidemiologic studies, and illustrate some of the analytic and methodologic approaches and techniques used to prevent these errors, reduce their effect, and improve the robustness of findings.

LIFE-COURSE EPIDEMIOLOGY

Epidemiologic evidence plays a critical role to inform the formulation of complex theoretic models and relevant hypotheses. Observational epidemiology prompts the formulation of specific hypotheses that may be tested in experimental studies, and further refined after testing

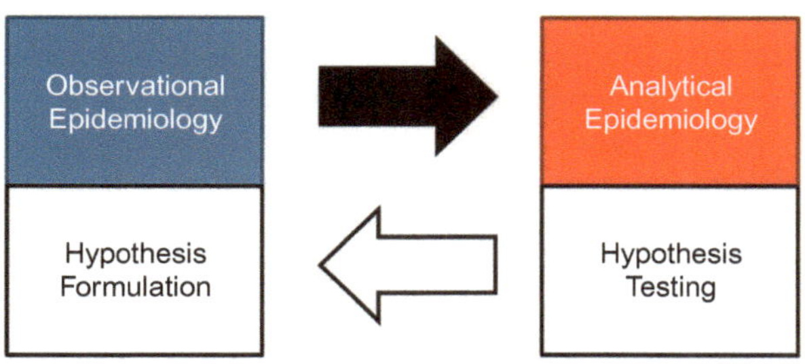

Fig. 3.1. From hypothesis formulation to testing and back.

(Fig. 3.1). A conceptual framework is needed to maximize this circular process. In a framework mutated from oncology that distinguishes factors into those that initiate, promote, proliferate, and advance the disease, better model specifications are needed to address the implications of a life-course approach to neurologic diseases.

Definitions and key concepts

The study of how risk and protective factors dynamically interact throughout life to influence later disease outcomes is referred to as life-course epidemiology (Kuh et al., 2003). Life-course epidemiology is a theoretic model and its endorsement to the study of neurologic diseases is grounded on the biologic plausibility of a large body of epidemiologic evidence. This evidence supports the notion that certain risk and protective factors may exert their effects with variable degree of strength at different time periods and with potential additive effects through cumulative exposure during the life course. Also relevant to life-course methods is the appreciation that most exposures and potential causes of disease are unlikely independent, and that exposure status may be influenced by the disease, after, but also before, its overt clinical onset. The latter is often referred to as "reverse causality," because the direction of the association under study "reverts" with respect to the one originally hypothesized. In other words, reverse causality refers to the circumstance where a putative exposure or cause is in fact an early marker of the disease itself.

In several diseases of the nervous system, for instance tumors, cardiovascular disease, and neurodegenerative diseases like AD, Parkinson's, amyotrophic lateral sclerosis, the neuropathology likely accumulates gradually and may precede by up to several decades the onset of the clinical symptoms. This underlying pathology and the preclinical, very mild symptoms of the disease may have direct and indirect effects on putative risk and protective factors. Progressive brain damage may influence blood pressure, body weight, and social activities and lifestyle, including diet and exercise, up to decades before the overt symptomatology of the ensuing disease.

In prospective observational studies the absence of clinical symptoms at the time of exposure assessment and allowing for long follow-ups before the outcome is ascertained may only reduce the likelihood that the observed associations may be disease-confounded, because one cannot be entirely sure of having measured the exposure status in disease-free individuals. Issues related to reverse causality are of critical importance in observational studies and should be carefully considered in light of prespecified, hypothesis-driven explanatory mechanisms of the expected and observed associations between presumed causes and disease. Reverse causality should also be resolved before the conduction of experimental studies to avoid the intervention under study modulating a marker (or consequence) of the disease instead of a risk factor or a cause.

Main purpose(s)

The relevance of timing and accumulation of exposure to potential causes throughout the life course and reverse causality are the main contentions that several neurologic diseases are best studied using a life-course approach. The ultimate goal of a life-course approach is to advance etiologic research. This is achieved testing, separately and in conjunction, hypotheses of different pathways between lifelong exposure to putative risk and protective factors and relevant health outcomes, including disease biomarkers, diagnosis, and clinical signs and symptoms measured prospectively. The development and specification of life-course theoretic models, the formulation of relevant hypotheses, and the choice of the appropriate analytic strategies and techniques are the key components of life-course epidemiology. These components broadly depend on the concepts of: causal pathways, timing of exposure, and the possible explanatory mechanisms.

Causal pathways

One of the central concepts in life-course epidemiology is accumulation of risk. This is closely related to the notion of dose–response relationship that may occur when changes (increments) in the level of exposure to a putative cause are associated with an increasing probability of observing the effect under study (prevalence/incidence of the disease). The number of insults, their duration, and severity may all contribute to augment the "dose" of exposure. Accumulation of risk occurs over time and through the possible coexistence of various and different types of exposures.

With the obvious exception of diseases with an autosomal-dominant genetic transmission with complete penetrance (like Huntington disease), for most neurologic diseases causation is multifactorial. Most causes are necessary rather than sufficient, and may cause damage to biologic systems and organs either independently or clustering around one or more predisposing (e.g., age, sex, and genetic traits) or enabling factors (e.g., low-socioeconomic status) (Figs 3.2 and 3.3). For instance, a life-course approach to the study of the complex relationship between adiposity and dementia has been called for with a focus on accumulation of risk, and trajectories of body weight throughout life (see below) (Gustafson, 2008). Obesity is strongly influenced by sociodemographic and genetic characteristics, and excess body weight may have detrimental effects on the structure

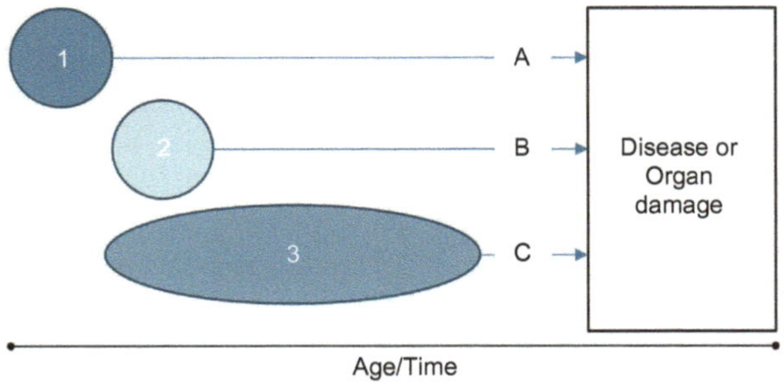

Fig. 3.2. Three independent risk factors are related to a later outcome (i.e., disease or organ damage) through three independent pathways (A, B, and C). The longer duration of exposure (circle 3) and the dose of exposure (more intense color, 1) are also relevant.

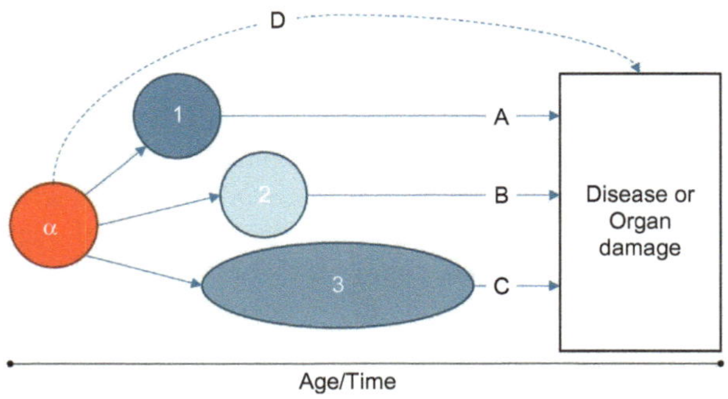

Fig. 3.3. The three risk factors (1, 2, and 3) may be clustered by a common factor (α) which may or not be related to the later outcome (dotted path, D).

and function of the brain through vascular (i.e., ischemic and toxic events, thickening of the carotid artery, endothelial dysfunction, peripheral resistance and arterial stiffness, ventricular hypertrophy), endocrine (i.e., excess secretion of sex hormones, adipokines, and cytokines, including interleukins, neurotrophins, and growth factors) and dysmetabolic mechanisms (i.e., hyperinsulinemia, advanced glycosylation end products) (Gustafson, 2006). Further, obesity in childhood tends to track into adult life, and the duration and severity of obesity may increase brain damage in a dose–response manner, and similarly increase the risk of diabetes and high blood pressure, which in turn contribute to dementia and cardiovascular disease risks too (Prince et al., 2014). The former can be seen as an example of independent cumulative life-course model (Fig. 3.2), the latter of a clustered cumulative model (Fig. 3.3).

Closely linked to risk accumulation is the concept of lifelong trajectory of an individual's dimension like a social, physiologic, or psychologic feature. In this context, body weight measured over the life course becomes of interest, because it would capture an individual's trajectory, and whether and to what extent this has deviated from a normative trajectory of body weight that is assumed to be

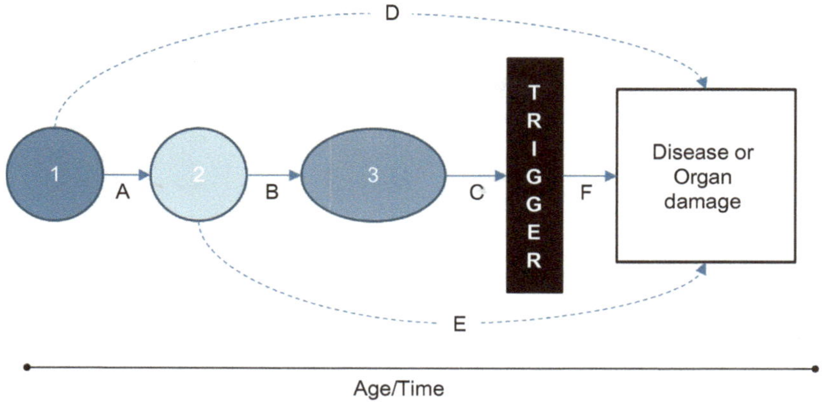

Fig. 3.4. Chain of probabilistic events (A, B, and C) through risk factors (1, 2, and 3) with a necessary final (or trigger) factor for the final link with the outcome (F). Independent effects of the factors in the chain may or not exist (dotted paths D and E).

optimal for health. In addition to the cumulative dimension, the study of trajectories allows to explore whether an individual's relative (transition) or radical (turning point) change in trajectory with respect to the norm is related to a change in disease risk. The effects of weight gain and particularly of weight loss in those who are obese are typical examples of transitions and turning points in individuals' body weight trajectories that receive great attention in research, and which have important implications for prevention. For instance, weight gain at various ages in adult life was not associated with midlife cognitive decline and cognitive reserve in the 1946 British Birth Cohort study (Albanese et al., 2012). However, the limited evidence from RCTs suggests that weight loss in overweight and obese people can be beneficial for cognitive function (Siervo et al., 2011).

Finally, a chain-of-risk model is a particular multifactorial causal pathway in which the accumulation of risk occurs through a sequence of probabilistic, temporally ordered events in which a risk factor leads to a subsequent risk factor in the chain. Each factor in the chain may or not have an independent and direct effect on disease risk, and the subsequent factors in the chain may either mediate or modify the indirect effect in an additive or multiplicative manner. Moreover, it can also be hypothesized that a chain of risk may culminate in a final and necessary cause without which disease cannot occur because only this final link in the chain can trigger the onset of the disease (Fig. 3.4). An example of this is encephalitis due to herpes simplex virus type 1 (HSV-1), in which the trigger effect that precipitates the disease is the centripetal travel of HSV-1 from the ganglion of the trigeminal nerve into the central nervous system (CNS). More than 50% of adults are infected by HSV-1 in the USA, but the proportion of those who will develop encephalitis is extremely low. Although this is not fully understood, there is a chain of risks that starts with predisposing conditions to infection, followed by the initial infection, the later reactivation of the dormant neurotropic virus to its pathway into the CNS.

Timing

The chronologic age of an individual and the year and location of birth are the two central notions around which revolve most of the concepts and measures of life-course epidemiology that capture timing and duration of exposure. Timing and duration can be measured in various ways. These measures may be used to test the aforementioned accumulation hypothesis and/or the existence of periods in life when effects may be stronger (sensitive periods) or strongest (critical period) compared to other periods in the life course.

Birth cohort effects and context are defined by the year and location of birth and are used to delineate broad characteristics of large groups of individuals who have collectively shared the exposure to the same environmental circumstances. Geographic or year-of-birth variations in adult disease occurrence can be explained by differential exposure during different epochs in life, since as early as conception. For instance, the fetal origin of adult disease (FOAD) hypothesis stemmed from the observation of significant geographic variations in mortality due to cardiovascular disease across England and Wales that could not be explained by genetic inheritance or risky lifestyle in adult life (Barker, 2007). Namely, the FOAD hypothesis posits that undernutrition *in utero* leads to permanent changes in the structure and function of organs and body systems and their metabolism. These adaptive changes make the best of the present conditions and constitute an anticipatory strategy to also best adapt to the "expected" environment at birth, of which the mother's hormones and nutrients that cross the placenta provide cues. When a substantial mismatch exists

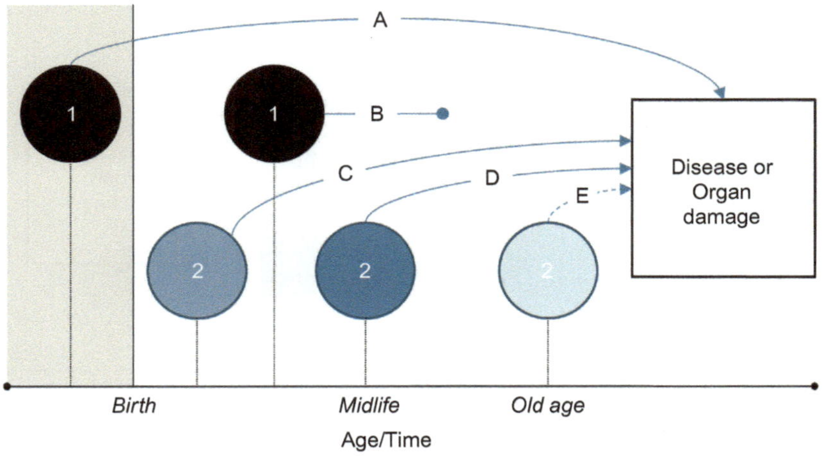

Fig. 3.5. The critical-period model is illustrated in black: exposure to risk factor 1 before birth increases risk of disease (path A); exposure after birth does not (interrupted path B). The sensitive period model is in blue: the effect is strongest for exposure to risk factor 2 in midlife (path D), but exposure in early life is associated with later disease too (path C); exposure in old age has the weakest association (path E).

between the predicted environment and the actual environmental conditions during early childhood, the permanent changes *in utero* may become maladaptive. Moreover, the ability to adapt to the environment (developmental plasticity; see below) may also be reduced and the subsequent risk of disease is further increased through a greater sensitivity to exposure throughout the life course, which is exacerbated by the "mismatched" environment. Low birthweight has been used as a proxy of this unfavorable intrauterine environment, and it has been shown to be associated to greater risk of coronary heart disease in late life in prospective population-based studies (Barker et al., 1989). In addition, growth during early childhood has been shown to increase risk of later coronary events independently of weight at birth, supporting the FOAD hypothesis that exposure during the early years of development after birth may also play a role in modulating risk of cardiovascular disease in adulthood (Barker et al., 2005).

In first instances of exploration of the FOAD hypothesis with respect to brain disease, smaller birth size was prospectively associated with cognitive impairment and brain damage in late life in a prospective population-based study (Muller et al., 2014b). A further additive effect on brain atrophy was also found in the same cohort for the subsequent exposure to vascular risk factors in midlife (Muller et al., 2016).

The FOAD is also an example of the so-called critical period (or latency) model of disease. A critical period is usually a short period of time in life, particularly during rapid development, when exposure may cause an adverse health outcome up to several decades later. Outside this time window, the structure, function, and metabolism of body systems cannot be modified, thus there cannot be any effect of this exposure on disease risk. When the disease mechanisms are well known, the critical-period model may even fully explain disease occurrence. Different from critical periods, during sensitive periods an exposure has a stronger effect on subsequent disease risk compared to other periods in the life course when excess risk is weaker but may still exist (Fig. 3.5).

For instance, folate deficiency during pregnancy and the association between excess body weight and hypertension are good examples of the critical and sensitive period model, respectively.

The first trimester, and particularly the third week of pregnancy, can be considered a critical period during which severe folate deficiency in pregnant women causes neuronal tube defects. As pregnancy advances, low levels of folic acid may increase the risk of giving birth to low-birthweight and premature infants, but can no longer cause neuronal tube defects.

In the British 1946 Birth Cohort, those who became overweight or obese earlier in life had on average a higher blood pressure in midlife compared to those who became obese later (Wills et al., 2010). This may suggest that adolescence and early adulthood may be sensitive periods during which the effect of excess body weight on midlife hypertension is stronger compared to later exposure. However, because too few participants who became overweight or obese lost any weight by midlife in this cohort, it was not possible to disentangle whether cumulative exposure to overweight, the sensitive periods earlier in life, or both best explained the observed associations (Wills et al., 2010). Further, as highlighted above, reverse causality may render the appreciation of sensitive periods particularly challenging. For instance, because a decline in blood pressure may begin up to two decades before the clinical onset of dementia (Stewart et al., 2009), the stronger effect on dementia risk of exposure to hypertension in mid compared to late life may be in part disease-confounded rather than due to sensitive periods of hypertension exposure in midlife. Epidemiologic evidence suggests that higher systolic and lower diastolic blood pressure in late life are both independently associated with focal and diffuse vascular brain damage, but only in those with a history of hypertension in midlife (Muller et al., 2014a).

It is an essential requisite of a putative cause to precede the effect (i.e., the disease). The interval between the exposure and the beginning of the disease is termed the induction period, which can be equal to, but is more likely shorter than, the latency period, which is demarcated by when the disease is detected or diagnosed. The causes of disease initiation are usually extremely difficult to investigate in humans. In fact, because the variation in time between the induction and latency periods is often unknown, the factors that modulate the clinical expression of the disease may not be distinguishable from the causes of the disease itself. In addition, both induction and latency periods are likely very long in chronic diseases, which seemingly contribute to weaken the observable associations. For instance, the neuropathology underlying AD and other neurologic diseases has an insidious onset and likely accumulates gradually over several years (induction period) (Jack et al., 2010). At present biomarkers may be used to detect and characterize the presence of brain damage but only when symptoms and signs become apparent, which may be up to several decades after the disease initiation, and too late to revert pathologic cascades.

Induction and latency periods tend to be little investigated, particularly in world regions that undergo rapid epidemiologic (Yusuf et al., 2001) and nutritional transition (Popkin, 2001), and where population aging is particularly pronounced. This is problematic because the rapidly increasing proportion of older adults in these settings will lead to an epidemic rise of age-related chronic diseases in the next decades, which may be detrimentally influenced by unfavorable conditions in early life, for which interventions may be needed quickly. Indeed, despite the fact that undernutrition in childhood coupled with increasing exposure to mismatched environmental conditions (i.e., Westernized, calorie-dense diets and lifestyles) have known long-term effects on adult health, greater attention is still devoted to the short-term effects on survival and poor health.

Explanatory mechanisms

A life-course approach that considers factors that may act throughout life to influence distal disease initiation and onset implies the formulation of specific research hypotheses. The biologic plausibility of these research questions and hypotheses should be framed within the boundaries of precise explanatory mechanisms. These mechanisms should be consistent with a causal role of extrinsic, modifiable factors, which would modulate disease risk through complex interactions with intrinsic, inherited factors.

Plasticity refers to the ability to adapt to changing environmental conditions through the modification of the structure and function of organs and biologic systems. While rapid, reversible, functional adaptation is used to maintain homeostasis, the changes of plasticity are usually irreversible. As discussed above in the context of the FOAD, the more specific term developmental plasticity is used for those irreversible changes in metabolic systems and organs' structure and function that are induced by suboptimal conditions (mainly nutritional and dysmetabolic) experienced during fetal life and the first years of childhood development. Epigenetic processes are likely implicated in developmental plasticity, and the changes might even be transmitted across generations (Gluckman et al., 2009). However, permanent changes to body functions and structures may also occur outside development, typically in response to damage and as a consequence of repair (without *restitutio ad integrum*), and with age. Life-course epidemiology seeks to explore the implications of the outcomes of plasticity at different ages, in the short, mid, and long term, and with respect to a varying environment, to ultimately explain both beneficial and detrimental effects on health outcomes.

Plasticity and developmental plasticity are relevant mechanisms in neuroepidemiology. For instance, optimal nutrition in the last weeks of pregnancy is key for the maturation and development of the central and peripheral nervous system, and 95% of brain growth is completed by 5 years of age. This explains why several neurologic disorders in adult life are strongly influenced by a suboptimal or abnormal development of the CNS during pregnancy, particularly in the third trimester (Rees and Inder, 2005). These early changes likely interact to cause disease (either additively or multiplicatively) with structural and functional damage that accumulates in the brain during the life course. The effects of early changes in brain structure and function can be further modulated by cognitive reserve, which may compensate for the effects of suboptimal brain development on later cognitive outcomes (Muller et al., 2014b). Similarly, optimal brain development may contribute to brain (or structure) reserve and ward off the effects of brain damage accumulated during the life course above and beyond cognitive (functional) reserve.

The cognitive and brain reserve hypothesis (described above) is an example of developmental plasticity that may act through resilience – namely, resilience to adverse events or circumstances through positive adaptive mechanisms facilitated by the greater number of neurons and synapses formed in early childhood (brain reserve) and/or the greater complexity and efficiency of neuronal networks that may be procured by education, occupation, and social and intellectually stimulating activities during the life course (cognitive reserve) (Stern, 2012). Interestingly, although resilience is a positive adaptive mechanism, an excess of it may lead to allostatic load, which consists in the persistent physiologic adaptation to physical and psychologic stressors (McEwen, 1998). Allostatic load causes sustained stress response and the solicitation of the hypothalamic–pituitary–adrenal axis, which stimulates the secretion of glucocorticoids (mainly cortisol). Cortisol has a number of effects on organs and metabolisms and may cause direct damage to the brain, particularly to regions rich in cortisol receptors that are also implicated in its feedback regulation, like the hippocampus (Lupien et al., 1998). The allostatic load mechanism may explain some recent findings from a large prospective cohort study that showed a significant, independent, 25-year-long association between the high propensity to cope actively in response to life challenges and difficulties, which may increase sustained stress, with worse cognitive function in midlife (Albanese et al., 2016).

Susceptibility (or vulnerability) is an explanatory mechanism closely related to the chain-of-risks model of causation which consists in the harmful adaptation to adverse events because of predisposing (intrinsic) factors or previous exposure. Susceptibility mounts progressively over time and assumes the existence of multiple causes that add up to gradually lead to a higher proneness to disease. Not all factors that contribute to susceptibility are causes. Some of these are mediators that are associated with the risk factor of interest and the disease, but, different from confounders, they also lie on the causal pathway. Thus, mediators do not provide alternative explanations of the association between an exposure and disease, even though they are associated with both. This is easily understood when considering that a mediator must temporally follow a putative cause, while a confounder may pre-exist.

Methods in life-course epidemiology

Study designs and type of data needed

The first requisite to conduct a life-course study is the availability of the appropriate data. Longitudinal data of a population-based, representative cohort followed prospectively are indispensable. Birth cohort studies, in which individuals are sampled at birth, are rare but are the ideal design, because a large number of assessments and measures are taken at fixed intervals in childhood, adolescence, early, mid- and late adulthood, and through old age, that is, across the entire life course. Birth cohort studies are extremely expensive, and are very difficult to conduct and maintain, not least to limit attrition of participants who in essence accept to live their lives as "guinea pigs." Birth cohort studies have several advantages though, including the availability of genetic data, and of repeated measures of a large number and variety of anthropometric, psychologic, sociodemographic, and lifestyle characteristics, and health measures and diagnoses that are collected using standardized procedures. Few studies of this kind are available in the world, and only one or two have lasted long enough to follow up participants into old age (Kuh et al., 2011).

An alternative to birth cohort studies to obtain data across adult life and into old age has been the systematic leveraging of existing cohort studies set up decades ago, which have been revitalized by hypothesis-driven new waves of data collection conducted tracking the same participants who were part of the original cohort. The already available and newly collected data are combined to create a continuum across adult life and into old age (Harris et al., 2007).

Record linkage methods have also been extensively used to acquire information and data routinely collected for clinical, medical, or other reasons in populations or large groups of people, for instance, members or clients of health insurances or health integrated delivery systems and organizations (Somers and Fund, 1971). In all these cases, data on mortality from national or regional registries may be linked and cause of death reported on death certificates may be used as proxies of disease occurrence at the time of death.

Formulating hypothesis

In biomedical research statistical analysis of data should be dictated by a solid, *a priori* hypothesis. This is particularly important in epidemiology. Cohort studies consist of the collection of a wealth of information on large and representative samples of individuals followed up prospectively over time. Secondary analysis of the data of existing cohorts is extremely valuable, not least to maximize the important financial and resource investments made to conduct these types of studies. However, data mining and other exploratory techniques are highly contraindicated and should be discouraged. Publication bias is inevitable, because only positive and significant results are of interest. Moreover, multiple comparisons increase significantly the likelihood of spurious findings as an effect of chance.

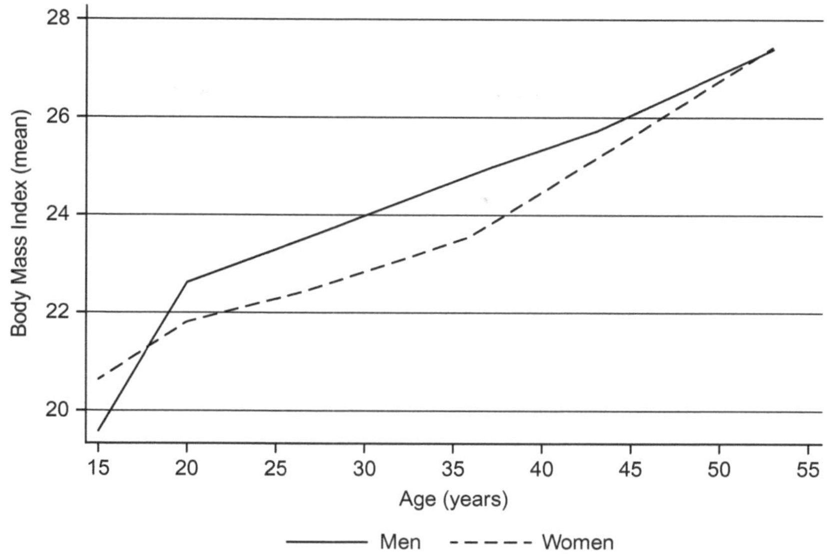

Fig. 3.6. Mean body mass index at different ages in the British Birth Cohort study, by sex (Albanese et al., 2012).

The formulation of a biologically plausible hypothesis is usually informed by a thorough knowledge of the existing scientific literature that may indicate gaps or inconsistencies worth being investigated. The choice of specific research questions within a given hypothesis will determine the outcome(s), exposure(s), and potential confounders and relevant covariates to consider. For each of these, the most appropriate available measure can be determined and the variables to be used in the secondary analysis of existing data can be selected. For instance, the association between obesity and dementia is biologically plausible because adiposity may cause brain damage through vascular, endocrine, and dysmetabolic mechanisms (Gustafson, 2006), and may indirectly influence cognitive reserve (Corley et al., 2010). Because obesity is increasing in all world regions, if it increases the risk of dementia, the public health implications would be significant, given the expected steep increases in the number of cases due to the graying of societies (Prince et al., 2013). However, despite several studies in the last decade which have investigated the association between obesity and dementia, the evidence is markedly inconsistent. A number of reasons can explain the heterogeneity of results (Luchsinger and Gustafson, 2009). In particular, because dementia causes weight loss that starts several years before and accelerates after the clinical onset (Stewart et al., 2005), a clear distinction between exposures to excess body weight in mid and late life allows us to disentangle issues of directionality that arise when associations may be disease- and age-confounded. A life-course approach of adiposity and dementia has been proposed on the basis of these and several other considerations (Gustafson, 2008), and a few hypotheses can be considered. Weight gain during "critical periods" may have irreversible consequences on the structure and function of the brain. Alternatively, varying effects of adiposity on dementia risk may exist (i.e., sensitive periods), not least as a consequence of reverse causality. Finally, rather than the timing of exposure during the life course, the duration of exposure and cumulative effects of obesity may be more relevant.

Steps of statistical analysis in life-course epidemiology

The choice of statistical analysis depends on the hypothesis under study and on whether the outcome is a continuous (i.e., blood pressure, body weight, etc.), dichotomic (presence or absence of disease), ordinal (i.e., disease severity, mild, moderate, or severe), or categoric variable (i.e. dementia subtype, AD, vascular dementia, Lewy body, etc.).

The first step of statistical analysis consists in a graphic inspection of the data, with a focus on the time-varying exposure. The aim is to visually appreciate whether and how the exposure changes over time by plotting the distribution of the mean values (y-axis) of the exposure by the follow-ups years after baseline (or by age, in the case of birth cohort studies, in which data are "balanced") (x-axis). For the example above, this graph would display the study sample's mean body mass index (BMI: body weight in kilograms divided by squared height in meters) at the different follow-ups of the study (Fig. 3.6).

Figure 3.6 plots repeated measures of BMI over time, which may have a number of limitations. These measures

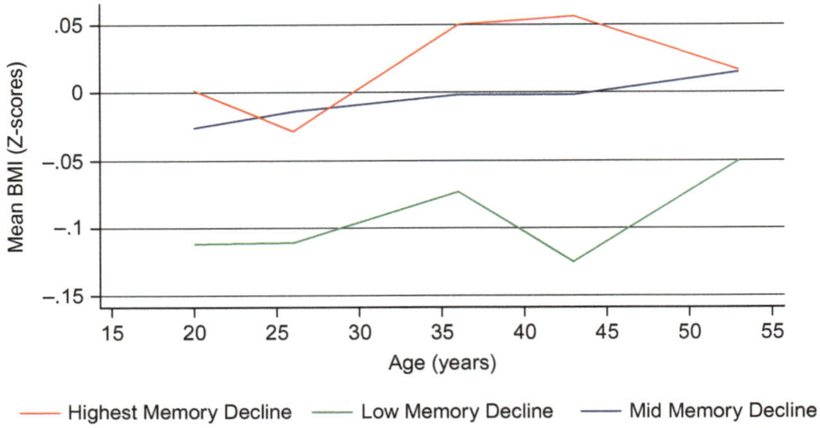

Fig. 3.7. *Z*-scores of body mass index (BMI) at different ages in adulthood by tertiles of memory impairment (decline) in late life. Those with lower decline of memory in late life have lower maintained body weight throughout adulthood compared to those with mid and high memory decline (Albanese et al., 2012).

have high interdependency and greater correlations between them the shorter the interval between follow-ups. Moreover, varying measurement errors may exist as a function of age at the time of measurement and attrition of participants due to death and nonparticipation may introduce bias. Finally, missing may not be considered at random if the exposure status at an early time point influences the likelihood of participation in a subsequent follow-up.

The standardization of raw values of the exposure (i.e., BMI) allows the direct comparison of the relationship between the exposure and the outcome of interest across periods. *Z*-scores are obtained subtracting the overall mean value of the exposure from each observation and dividing it by the standard deviation of the mean (SD) at each age. The *Z*-scores of the exposure can be plotted by outcome status to give a simple picture of risk profiles or paths, which are easy to interpret (Fig. 3.7). Plotting *Z*-scores by outcome status is straightforward in the case of binary (e.g., diagnosis, yes/no) or categoric outcomes (e.g., mild, moderate, severe levels of symptoms) and it requires the categorization of continuous variables (e.g., cognitive test scores), for example, using centiles. *Z*-score plots are extremely important to explore data, and can inform the best approach to conduct the main analysis. However, they have limited applicability because of the need to categorize continuous outcomes, and do not allow any statistical inference or adjustment for potential confounders. Moreover, missing data may cause samples to vary at each follow-up, limiting the interpretation of the plots themselves.

The association between exposure at different ages and the outcome of interest can be investigated conducting repeated univariate regression models (i.e., including one exposure at a time only). The regression coefficients and 95% confidence intervals of the associations between exposure *Z*-scores at each follow-up and the outcome can be plotted to visualize any difference in direction or magnitude (Fig. 3.8).

The multiple regressions approach allows one to formally test whether the associations of a late outcome with exposure status measured at different time points (ages) vary over time. The differences between the regression coefficients can be formally tested using postestimation techniques, including Wald's tests. Moreover, there is no need to categorize continuous outcomes, and models can be adjusted for potential confounders. However, neither sensitive periods nor cumulative effects can be formally investigated.

The first step to investigate sensitive periods is to model change in exposure status between intervals. This

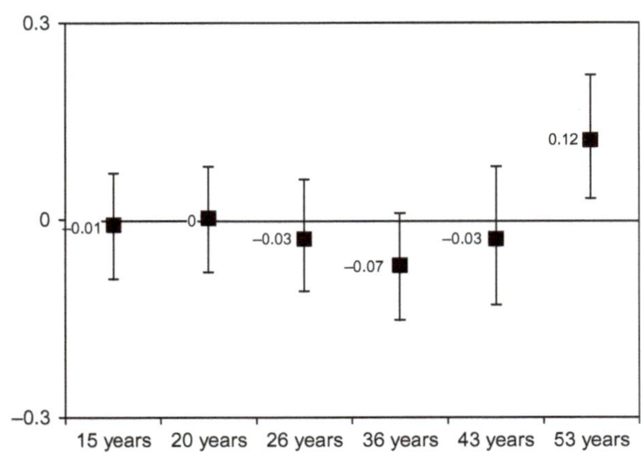

Fig. 3.8. Regression coefficients (95% confidence interval) of the association between cognitive score in late life with *Z*-scores of body mass index (BMI) at different ages. Higher BMI at age 53 was significantly associated with better cognitive performance; associations are not significant at all other ages (Albanese et al., 2012).

change may be computed in various ways. Growth trajectory models can be used to model the changes in a characteristic (the exposure but also covariates) over time. Individuals' growth patterns can be captured fitting random-effects, multilevel models. The major strength of these models is their great efficiency due to the nonexclusion of participants with incomplete data. However, the missing at random and equal pattern of growth (or change) assumptions limit their use. These assumptions may not hold, for instance, when differential attrition due to exposure status at an early time point is not excluded, and when the average trajectory predicted by the model originates from two opposite and distinct patterns of change rather than from the hypothesized variation in rate of change between individuals. Latent class growth models (Østbye et al., 2010) have been proposed as an alternative approach in which the same pattern of change assumption is not needed. However, the existence of different classes characterized by different change patterns should be hypothesized (and justified) *a priori*, and the classes (or groups) should map to the probabilistic identification of subgroups in the sample under study.

The first and simplest alternative to mixed models and latent class growth models consists in subtracting exposure status at t_1 by exposure status at t_2, and so on. For instance, BMI at baseline is subtracted from BMI at the first follow-up to obtain the change in BMI during the first interval, and all other measures of BMI change are calculated similarly for the following intervals. This change may be, and is probably better expressed as, a percentage, to account for the level of exposure status at the previous time point, and because absolute variations may be less informative than relative variations (5 kg weight loss corresponds to 5% of an individual's weight if he/she is 100 kg but to 10% if he/she is 50 kg of weight). Nevertheless, neither approaches to obtain changes in exposure status over time account for the dependencies and high correlations between repeated observations.

A possible solution is to compute change using residuals. Each exposure measure is regressed on the earlier measure(s) to calculate and save residuals that represent the exposure "velocity." This is the change in exposure status above or below that expected given the earlier measure. The *Z*-scores of these residuals are then used to allow comparisons of the relative associations between the different periods. Sensitive periods are captured exploring the existence of periods when changes in exposure status have stronger associations with outcomes compared to other periods. Postestimation tests are used to formally compare the coefficients for each period.

The commonest and most intuitive way to test the accumulation hypothesis is to use exposure scores. These scores are the individual trajectories that compute the number of times an individual falls into a category (i.e., obese, yes/no) at various ages or follow-ups of the study. This approach has important limitations though and, most importantly, when the associations with a later outcome become stronger over time, whether the findings are best explained by the sensitive periods hypothesis or by cumulative effects of exposure over time cannot be disentangled. This distinction is important and can be done using nonadvanced statistical techniques, but only when the exposure measure is binary (or can be dichotomized into subgroups that have clinical utility, as it is for BMI-defined obesity). Likelihood ratio tests are used with generalized linear models (for noncontinuous outcomes) and partial *F*-test for linear regressions to compare nested parsimonious models with a saturated model that contains all parameters of the exposure at each time point and in all combinations. The nested models of the accumulation of risk hypothesis and the critical period hypothesis have fewer parameters because they may use the trajectories and the "velocity" parameters only, respectively. When the partial *F*-statistic (in the case of the partial *F*-test) is not significant (i.e., $p > 0.05$), the observed variations in the outcome of interest are best explained by the parsimonious model because its goodness of fit is equal to that of the saturated model that contains it (Mishra et al., 2009), and would confirm or dispute the merit of sensitive periods and/or cumulative effects hypotheses to explain the association between exposures measured across the life course and a late outcome.

FURTHER EXAMPLES OF SOURCES OF ERRORS AND BIAS IN EPIDEMIOLOGIC STUDIES

Two-phase designs

The use of multiphase design is very common to measure prevalence and incidence of chronic diseases in large samples of community-dwelling individuals, but it is often erroneously applied. A screening assessment is used in all the sample and the final diagnosis is adjudicated with a more intense diagnostic assessment. However, because screening tools are unlikely to be 100% sensitive, the lack of assessment of those who screen negative may lead to a systematic underestimation of the true prevalence. Moreover, bias may also be introduced if there are differences in the sensitivity of the screening tool between individuals. These biases are rarely considered in both descriptive and analytic epidemiology. For instance, neuropsychological tests like the Mini Mental State Examination (MMSE) have important ceiling effects at younger ages and especially in those with higher educational level, amongst whom cognitive impairment may not be apparent. When these measures are used to screen those who are subsequently thoroughly evaluated to adjudicate the final diagnosis, dementia may be systematically underestimated in those with higher educational level. To reduce this potential bias it is recommended that a random sample of those who screen negative in the first step of the multiphase diagnostic procedure receives the full clinical assessment too. This would determine the proportion of dementia cases amongst this group, and dementia prevalence (and incidence) should be "weighed back" accordingly. Nevertheless, this does not reduce a second source of potential bias due to differential loss at follow-up between the screening and the diagnostic phase. This is extremely common, particularly when screening tools are used in the field (i.e., at the participants' homes) and the indepth diagnostic assessment requires long, clinic-based, or even invasive procedures. One-phase design, in which the same comprehensive diagnostic assessment is submitted to all individuals in the study sample, is not prone to this bias, is more efficient, and should be preferred to two-phase designs (Prince et al., 2013).

Use of medical records to ascertain outcome status and information/measurement bias

A particular type of information/measurement bias (described in the previous chapter) can arise when the ascertainment of the outcome status (i.e., presence or absence of disease) is influenced by health service use and access, which in turn may depend on exposure status. This is potentially very relevant in neuroepidemiology because diagnoses may be difficult to make in the field and in population-based samples, and observational, routinely collected health data, including medical records, are often used instead. However, compared to healthy individuals, those who have worse risk exposure profiles during adulthood are more likely to access and use health services. This increases their probability of being diagnosed "incidentally" with other diseases. While "incidentalomas" may lead to overdiagnosis (Moynihan et al., 2012), they can also produce biased estimates of association between disease and exposures that may have influenced the use and access to health services in the first place. This bias may not be removed and should be carefully considered in the interpretation of results of studies conducted using observational routinely collected health data. Reporting guidelines of these specific types of observational studies have been recently published (i.e. the RECORD statement) (Chen et al., 2016).

Competing risks and informative censoring

A number of risk factors for neurologic diseases may also significantly increase all-cause mortality. When analyzing longitudinal data, the possibility that death may occur in some individuals before and in others after the possible onset of the disease must be accounted for in the statistical analysis of the effect of exposure and covariates on disease incidence. This is done by focusing on the cumulative incidence function instead of the traditional survival function (Fine and Gray, 1999). The former correctly indicates the probability of observing the outcome of interest before death occurs (the competing event), whereas the latter inadequately treats events that are no longer possible (after death) as if these were merely unobservable. The nonapplication of competing-risk models is particularly problematic when the main exposure of interest has no effect on the outcome but it increases mortality risk, because the association between the exposure and the outcome would be spuriously significant or overestimated.

Participants' withdrawal from a study may be influenced by exposure status. This differential attrition leads to selection bias because those who are included in the analytic sample are systematically different from those who are excluded. These censoring mechanisms can and should be considered, for instance, computing inverse probability of censoring weights (IPCWs) that model the effects on censoring probability of the exposure status at baseline and relevant covariates using a logistic model (Robins et al., 2000). In sensitivity analyses the main models are re-run assigning weights to participants in the analytic sample based on the characteristics that predict withdrawal and the two sets of findings are compared to determine the robustness of the main model. IPCW is an efficient alternative to missing-data imputation for, which the missing at random assumption may not hold.

Mediation

Mediators and confounders are covariates associated with both the exposure and the outcome. Because a mediator lies on the causal pathway between a putative cause (the exposure) and the outcome event of interest (the disease), it does not provide an alternative explanation for the observed association; rather, it provides further insights to causal inference and should not be controlled for in the analysis. Mediation should be hypothesized based on previous knowledge and evidence and should be biologically plausible. A mediator must temporally follow a putative cause, while a confounder may preexist. Mediation tests should be based on a prespecified model embedded into the main research hypotheses. Some of the quantitative approaches available to test mediation are relatively simple to implement, although boot strapping is required to calculate the 95% confidence intervals (Fritz and MacKinnon, 2007). A series of regression models are conducted to generate the estimates of all the associations between the exposure, the outcome, and the putative mediator, and the proportion of variance explained by the potential mediator in the association between the main exposure and the outcome is formally tested.

References

Albanese E, Hardy R, Wills A et al. (2012). No association between gain in body mass index across the life course and midlife cognitive function and cognitive reserve – the 1946 British Birth Cohort study. Alzheimers Dement 8: 470–482.

Albanese E, Matthews KA, Zhang J et al. (2016). Hostile attitudes and effortful coping in young adulthood predict cognition 25 years later. Neurology 86: 1227–1234.

Barker DJ (2007). The origins of the developmental origins theory. J Intern Med 261: 412–417.

Barker DJ, Osmond C, Winter P et al. (1989). Weight in infancy and death from ischaemic heart disease. The Lancet 334: 577–580.

Barker DJ, Osmond C, Forsén TJ et al. (2005). Trajectories of growth among children who have coronary events as adults. N Engl J Med 353: 1802–1809.

Chen M, Langan S, Benchimol E (2016). Routinely collected electronic health data and STI research: RECORD extension to the STROBE guidelines. Sex Transm Infect 92: 2–3.

Corley J, Gow AJ, Starr JM et al. (2010). Is body mass index in old age related to cognitive abilities? The Lothian Birth Cohort 1936 Study. Psychol Aging 25: 867–875.

Fine JP, Gray RJ (1999). A proportional hazards model for the subdistribution of a competing risk. J Am Stat Assoc 94: 496–509.

Fritz MS, MacKinnon DP (2007). Required sample size to detect the mediated effect. Psychol Sci 18: 233–239.

Gluckman PD, Hanson MA, Bateson P et al. (2009). Towards a new developmental synthesis: adaptive developmental plasticity and human disease. The Lancet 373: 1654–1657.

Gustafson D (2006). Adiposity indices and dementia. Lancet Neurol 5: 713–720.

Gustafson D (2008). A life course of adiposity and dementia. Eur J Pharmacol 585: 163–175.

Harris TB, Launer LJ, Eiriksdottir G et al. (2007). Age, gene/environment susceptibility – Reykjavik Study: multidisciplinary applied phenomics. Am J Epidemiol 165: 1076–1087.

Heneka MT, Carson MJ, El Khoury J et al. (2015). Neuroinflammation in Alzheimer's disease. Lancet Neurol 14: 388–405.

Jack Jr CR, Knopman DS, Jagust WJ et al. (2010). Hypothetical model of dynamic biomarkers of the Alzheimer's pathological cascade. Lancet Neurol 9: 119–128.

Jellinger KA (2008). The pathology of "vascular dementia": a critical update. J Alzheimers Dis 14: 107–123.

Kuh D, Ben-Shlomo Y, Lynch J et al. (2003). Life course epidemiology. J Epidemiol Community Health 57: 778.

Kuh D, Pierce M, Adams J et al. (2011). Cohort profile: updating the cohort profile for the MRC National Survey of Health and Development: a new clinic-based data collection for ageing research. Int J Epidemiol 40: e1–e9.

Launer L (2005). The epidemiologic study of dementia: a life-long quest? Neurobiol Aging 26: 335–340.

Launer LJ (2007). Next steps in Alzheimer's disease research: interaction between epidemiology and basic science. Curr Alzheimer Res 4: 141–143.

Luchsinger JA, Gustafson DR (2009). Adiposity and Alzheimer's disease. Curr Opin Clin Nutr Metab Care 12: 15–21.

Lupien SJ, de Leon M, de Santi S et al. (1998). Cortisol levels during human aging predict hippocampal atrophy and memory deficits. Nat Neurosci 1: 69–73.

McEwen BS (1998). Protective and damaging effects of stress mediators. N Engl J Med 338: 171–179.

Mishra G, Nitsch D, Black S et al. (2009). A structured approach to modelling the effects of binary exposure variables over the life course. Int J Epidemiol 38: 528–537.

Moynihan R, Doust J, Henry D (2012). Preventing overdiagnosis: how to stop harming the healthy. BMJ 344e3502.

Muller M, Sigurdsson S, Kjartansson O et al. (2014a). Joint effect of mid-and late-life blood pressure on the brain The AGES-Reykjavik Study. Neurology 82: 2187–2195.

Muller M, Sigurdsson S, Kjartansson O et al. (2014b). Birth size and brain function 75 years later. Pediatrics 134: 761–770.

Muller M, Sigurdsson S, Kjartansson O et al. (2016). Late-life brain volume: a life-course approach. The AGES-Reykjavik study. Neurobiol Aging 41: 86–92.

Østbye T, Malhotra R, Landerman LR (2010). Body mass trajectories through adulthood: results from the National Longitudinal Survey of Youth 1979 Cohort (1981–2006). Int J Epidemiol. dyq142.

Popkin BM (2001). The nutrition transition and obesity in the developing world. J Nutr 131: 871S–873S.

Prince M, Bryce R, Albanese E et al. (2013). The global prevalence of dementia: a systematic review and metaanalysis. Alzheimers Dement 9 (63–75): e62.

Prince M, Albanese E, Guerchet M et al. (2014). World Alzheimer Report 2014. Dementia and risk reduction: an analysis of protective and modifiable factors, Alzheimers Disease International, London.

Rees S, Inder T (2005). Fetal and neonatal origins of altered brain development. Early Hum Dev 81: 753–761.

Robins JM, Hernan MA, Brumback B (2000). Marginal structural models and causal inference in epidemiology. Epidemiology. 550–560.

Savva GM, Wharton SB, Ince PG et al. (2009). Age, neuropathology, and dementia. N Engl J Med 360: 2302–2309.

Siervo M, Arnold R, Wells JC et al. (2011). Intentional weight loss in overweight and obese individuals and cognitive function: a systematic review and meta-analysis. Obes Rev 12: 968–983.

Somers, Fund (1971). The Kaiser-Permanente medical care program, Commonwealth Fund, New York.

Stern Y (2012). Cognitive reserve in ageing and Alzheimer's disease. Lancet Neurol 11: 1006–1012.

Stewart R, Masaki K, Xue QL et al. (2005). A 32-year prospective study of change in body weight and incident dementia: the Honolulu-Asia Aging Study. Arch Neurol 62: 55–60.

Stewart R, Xue Q-L, Masaki K et al. (2009). Change in blood pressure and incident dementia a 32-year prospective study. Hypertension 54: 233–240.

Wills AK, Hardy RJ, Black S et al. (2010). Trajectories of overweight and body mass index in adulthood and blood pressure at age 53: the 1946 British birth cohort study. J Hypertens 28: 679–686.

Yusuf S, Reddy S, Ôunpuu S et al. (2001). Global burden of cardiovascular diseases part I: general considerations, the epidemiologic transition, risk factors, and impact of urbanization. Circulation 104: 2746–2753.

Zlokovic BV (2005). Neurovascular mechanisms of Alzheimer's neurodegeneration. Trends Neurosci 28: 202–208.

Handbook of Clinical Neurology, Vol. 138 (3rd series)
Neuroepidemiology
C. Rosano, M.A. Ikram, and M. Ganguli, Editors
http://dx.doi.org/10.1016/B978-0-12-802973-2.00004-5

Chapter 4

Basics of neuroanatomy and neurophysiology

C.K. BARHA[1], L.S. NAGAMATSU[2], AND T. LIU-AMBROSE[1,3]*

[1]*Aging, Mobility, and Cognitive Neuroscience Laboratory, Department of Physical Therapy, Faculty of Medicine and Djavad Mowafaghian Centre for Brain Health, University of British Columbia Vancouver, BC, Canada*

[2]*Exercise, Mobility and Brain Health Laboratory, School of Kinesiology, Faculty of Health Sciences, Western University, London, Ontario, Canada*

[3]*Brain Research Centre, University of British Columbia Vancouver, BC, Canada*

Abstract

This chapter presents an overview of the anatomy and functioning of the central nervous system. We begin the discussion by first examining the cellular basis of neural transmission. Then we present a brief description of the brain's white and gray matter and associated diseases, including a discussion of white-matter lesions. Finally, we place this information into context by discussing how the central nervous system integrates complex information to guide key functional systems, including the visual, auditory, chemosensory, somatic, limbic, motor, and autonomic systems. Where appropriate, we have supplied information pertaining to pathologic and functional outcomes of damage to the central nervous system. Also included is a brief description of important tools and methods used in the study of neuroanatomy and neurophysiology. Overall, this chapter provides a basic review of the concepts required to understand and interpret the clinical disorders and related material presented in the subsequent chapters of this book.

In this chapter, we will present an overview of the central nervous system (CNS) in three main sections. First, we will broadly examine the CNS at the cellular level. Second, we will cover its gross anatomy, focusing on white and gray matter. Lastly, we will place this information into context by discussing key systems that integrate multiple brain structures and processes, including the visual, auditory, chemosensory, somatic, limbic, motor, and autonomic systems. We intend this chapter to serve as a broad review of the neuroanatomy of the CNS, with the purpose of facilitating for non-neurologists the understanding and interpretation of the information discussed in subsequent chapters of this book.

CELLULAR STRUCTURE AND FUNCTION

Neurons

In humans, there are an estimated 100 billion neurons in the brain. Although there are over 200 types of neurons of varying size and complexity with different functions, all neurons are comprised of a cell body (soma) containing the nucleus, dendrites with a tree branch-like structure, and a single axon covered with myelin sheath (Fig. 4.1).

Dendrites extend from the cell body in complex branching patterns and receive incoming synaptic inputs from various sources. Dendritic shafts are studded with

*Correspondence to: Teresa Liu-Ambrose, The University of British Columbia, Djavad Mowafaghian Centre for Brain Health, c/o Liu-Ambrose Laboratory, 2215 Wesbrook Mall, Vancouver, BC Canada V6T 2B5. Tel: +1-604-875-4111 ext. 69056, E-mail: teresa.ambrose@ubc.ca

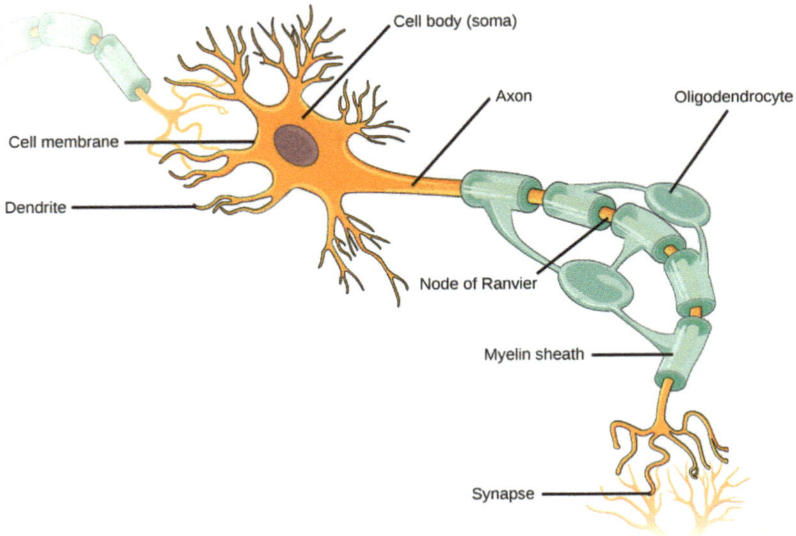

Fig. 4.1. Illustration of a prototypical neuron with important structures labeled. (Courtesy of https://courses.candelalearning.com/biologymajors/chapter/chapter35-the-nervous-system/. Creative Commons Attribution.)

tiny actin-rich membrane protrusions called spines that further increase the receptive surface area of the dendrite. A large degree of heterogeneity exists in dendritic spine morphology, with spines categorized into mushroom-shaped, long, thin, stubby, and filopodia, based on their structure. Functionally, spine morphology is related to synaptic strength, with larger spines, like mushroom spines, indicative of stronger connections. Dendritic spines can be stable or dynamic, changing their morphology in response to incoming stimuli, and are thus a major aspect of neuroplasticity. Importantly, abnormalities in spine number and morphology are seen in numerous psychiatric and neurologic diseases, including schizophrenia, major depressive disorder, and Alzheimer's disease (AD) (Lin and Koleske, 2010).

In contrast to dendrites, axons travel over long distances and carry the output of the neurons. The axon is a specialized cylindric tube structure containing a cytoskeleton consisting of neurofilaments and microtubules, which conducts electric signals from the cell body at the axon hillock to the axon terminal. Many axons are covered by a myelin sheath that increases the speed at which a nerve impulse is propagated along an axon through saltatory conduction.

Neurons connect at the synapse, which is comprised of the axon propagating the signal away from the cell (presynaptic membrane), the dendrite of the receiving neuron (postsynaptic membrane), and the synaptic cleft, approximately 20–30 nm wide. Some synapses occur between an axon and a dendrite or dendritic spine, and are termed axodendritic synapses and are typically excitatory (see below), whereas synapses between an axon and a cell body, called an axosomatic synapse, are typically inhibitory. Synapses between an axon terminal and another axon, axoaxonic synapses, and synapses between dendrites, dendrodendritic synapses, are modulatory (Fig. 4.2).

Neurotransmission

Communication between neurons occurs via electric-chemical processes within and between cells (Fig. 4.3). When a cell is at rest, potassium ions (K^+) are more concentrated on the inside of the cell membrane, while sodium (Na^+) and calcium (Ca^{2+}) ions are more concentrated on the outside. The inside surface of the cell membrane has a negative electric charge compared to the outside surface. The action potential is a very brief reversal (a thousandth of a second) of this, with the inside of the membrane becoming depolarized (more positively charged) relative to the outside. Depolarization of the cell is caused by the rapid influx of Na^+ ions into the cell. When the depolarization of the membrane reaches a critical threshold level, an action potential is generated. The membrane potential then repolarizes and returns to a negative charge through an efflux of K^+ ions. Gradually, the ion concentrations are reset and resting membrane potential is restored in that segment of the neuron.

The depolarization of the presynaptic membrane by action potentials leads to the influx of Ca^{2+} through activated presynaptic voltage-gated Ca^{2+} channels. The rise in presynaptic Ca^{2+} concentration triggers the release of neurotransmitters by causing neurotransmitter-containing synaptic vesicles to move down and fuse with the presynaptic membrane through phosphorylation of proteins called synapsins as well as through the interaction

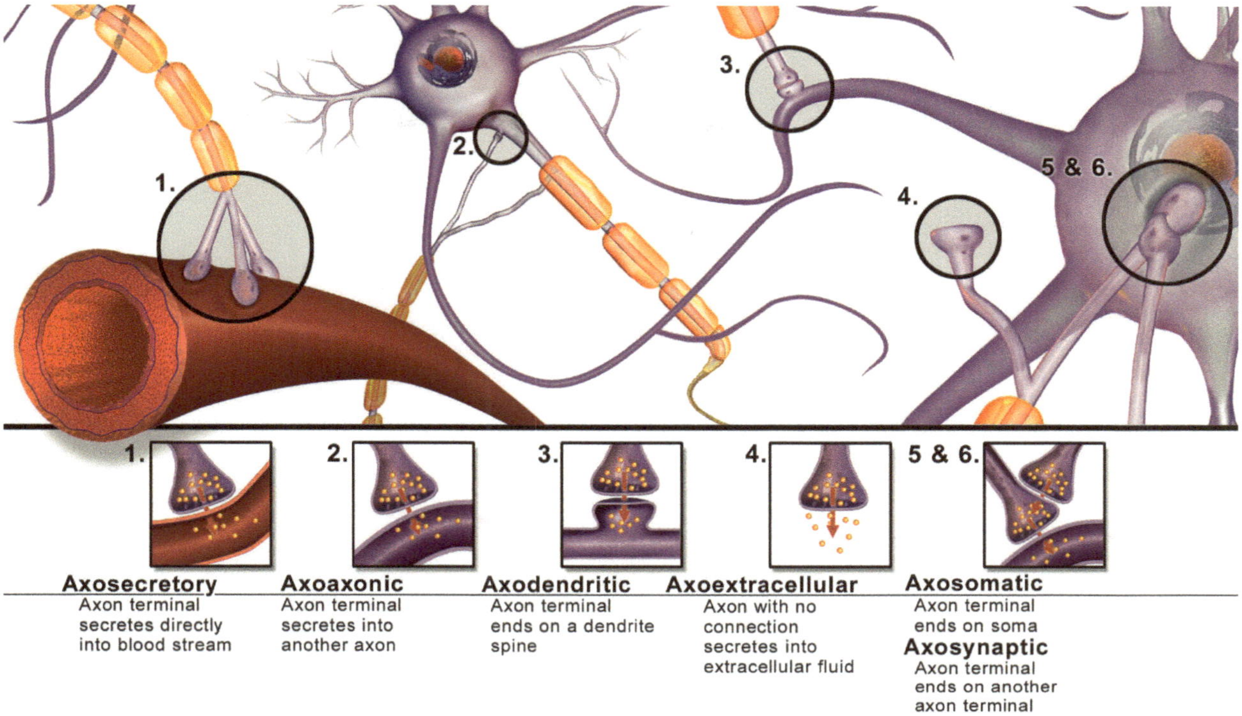

Fig. 4.2. Depiction of the different types of synapses. (Courtesy of Blausen.com staff. Blausen gallery 2014. Wikiversity Journal of Medicine. http://dx.doi.org/10.15347/wjm/2014.010.)

of several other synaptic proteins, each with distinct roles in the process. Importantly, the amount of neurotransmitter released into the synaptic cleft is not fixed and is related to the amount of Ca^{2+} that enters the presynaptic terminal, which is itself related to the frequency of stimulation of the presynaptic membrane. Impairments in synaptic transmission related to defects in the functioning of synaptic vesicles are seen in several neurologic diseases, including myasthenic syndromes, botulism and tetanus, and autism (Humeau et al., 2000; Poot, 2013).

Neurotransmitters diffuse across the synaptic cleft and bind to receptor proteins embedded in the postsynaptic membrane (Fig. 4.4). There are two main families of postsynaptic receptors that differ in how they transduce neurochemical information (neurotransmitter binding) into postsynaptic electric responses. Both families have some receptors that when activated either open ion channels or close ion channels. Ionotropic receptors, also known as ligand-gated ion channels, bind neurotransmitters triggering the receptor's channel to either open or close depending on the specific receptor. Metabotropic receptors, also known as G-protein-coupled receptors, bind neurotransmitters, which activate G-proteins within the cell that then directly or indirectly through secondary messengers trigger ion channels to either open or close. The opening or closing of ion channels and the subsequent movement of ions lead to the generation of postsynaptic potentials that are summated by the neuron to determine whether it will fire or not (see below). Ionotropic receptors mediate rapid, immediate but brief postsynaptic effects, whereas metabotropic receptors produce slower postsynaptic responses that last hundreds of milliseconds or minutes or longer.

The postsynaptic potentials elicited by the binding of neurotransmitters with postsynaptic receptors alter the probability that an action potential will be produced. Depending on the specific neurotransmitter and the receptor, the postsynaptic potential can either be excitatory (excitatory postsynaptic potential, EPSP) or inhibitory (inhibitory postsynaptic potential, IPSP). EPSPs increase the probability that an action potential will be generated by depolarizing the membrane, whereas IPSPs decrease the probability by hyperpolarizing the membrane.

Most neurons receive thousands of inputs, both from excitatory and inhibitory synapses, the majority of which are below the threshold for generating action potentials. Thus a neuron must integrate and summate EPSPs and IPSPs, in both time and space, to determine its own response. If the sum of all the EPSPs and IPSPs results in a depolarization of sufficient amplitude (above threshold) then an action potential will be produced. If the summation does not reach threshold (i.e., inhibition occurs), then an action potential will not be produced.

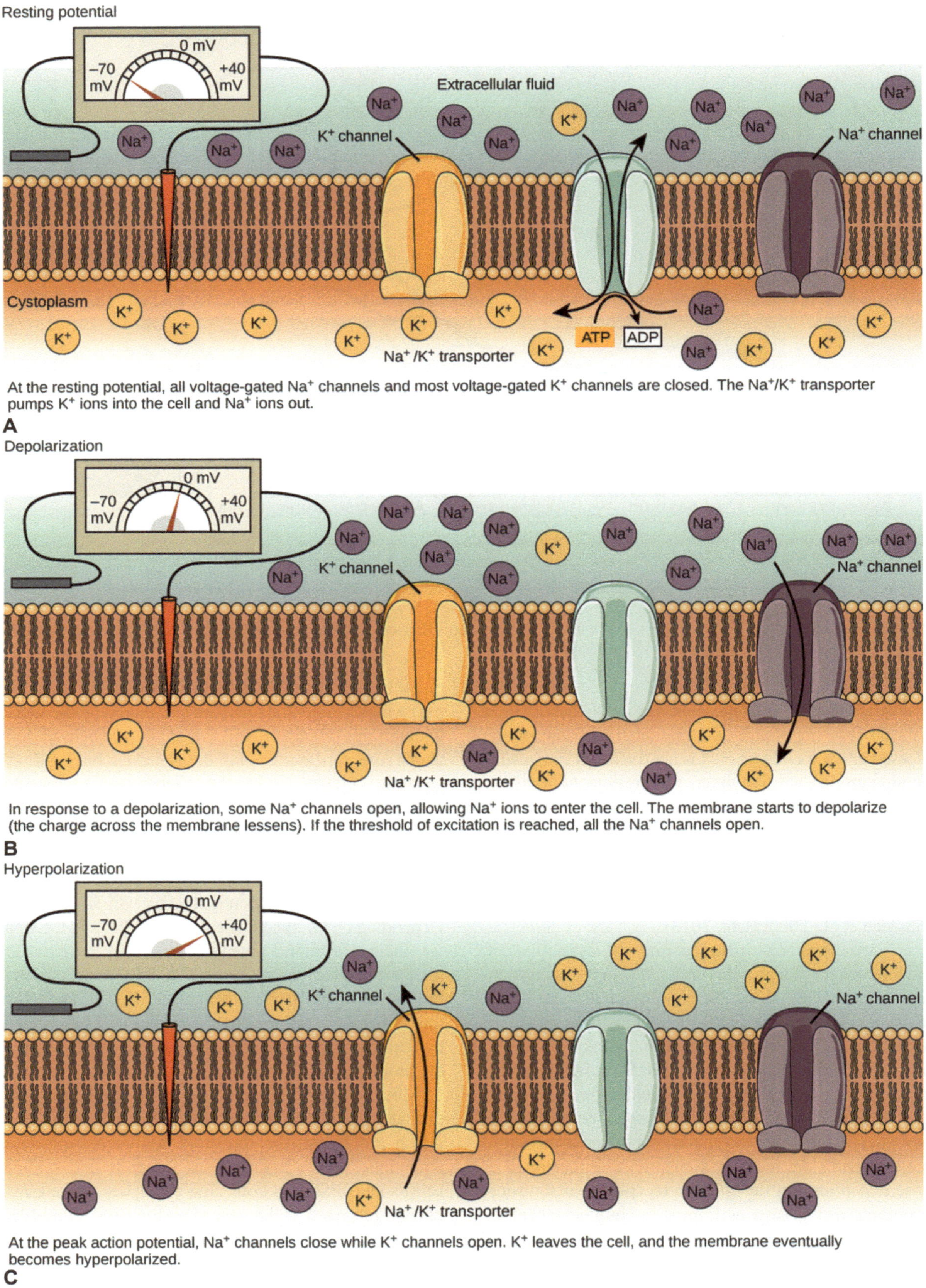

At the resting potential, all voltage-gated Na^+ channels and most voltage-gated K^+ channels are closed. The Na^+/K^+ transporter pumps K^+ ions into the cell and Na^+ ions out.

A

In response to a depolarization, some Na^+ channels open, allowing Na^+ ions to enter the cell. The membrane starts to depolarize (the charge across the membrane lessens). If the threshold of excitation is reached, all the Na^+ channels open.

B

At the peak action potential, Na^+ channels close while K^+ channels open. K^+ leaves the cell, and the membrane eventually becomes hyperpolarized.

C

Fig. 4.3. The action potential of a neuron. (**A**) The resting potential of a neuron. (**B**) Depolarization occurs when the nerve impulse causes Na^+ ions to enter. (**C**) Hyperpolarization occurs when K^+ channels in the membrane open and K^+ ions exit the cell. (Courtesy of https://courses.candelalearning.com/biologymajors/chapter/chapter35-the-nervous-system/. Creative Commons Attribution.)

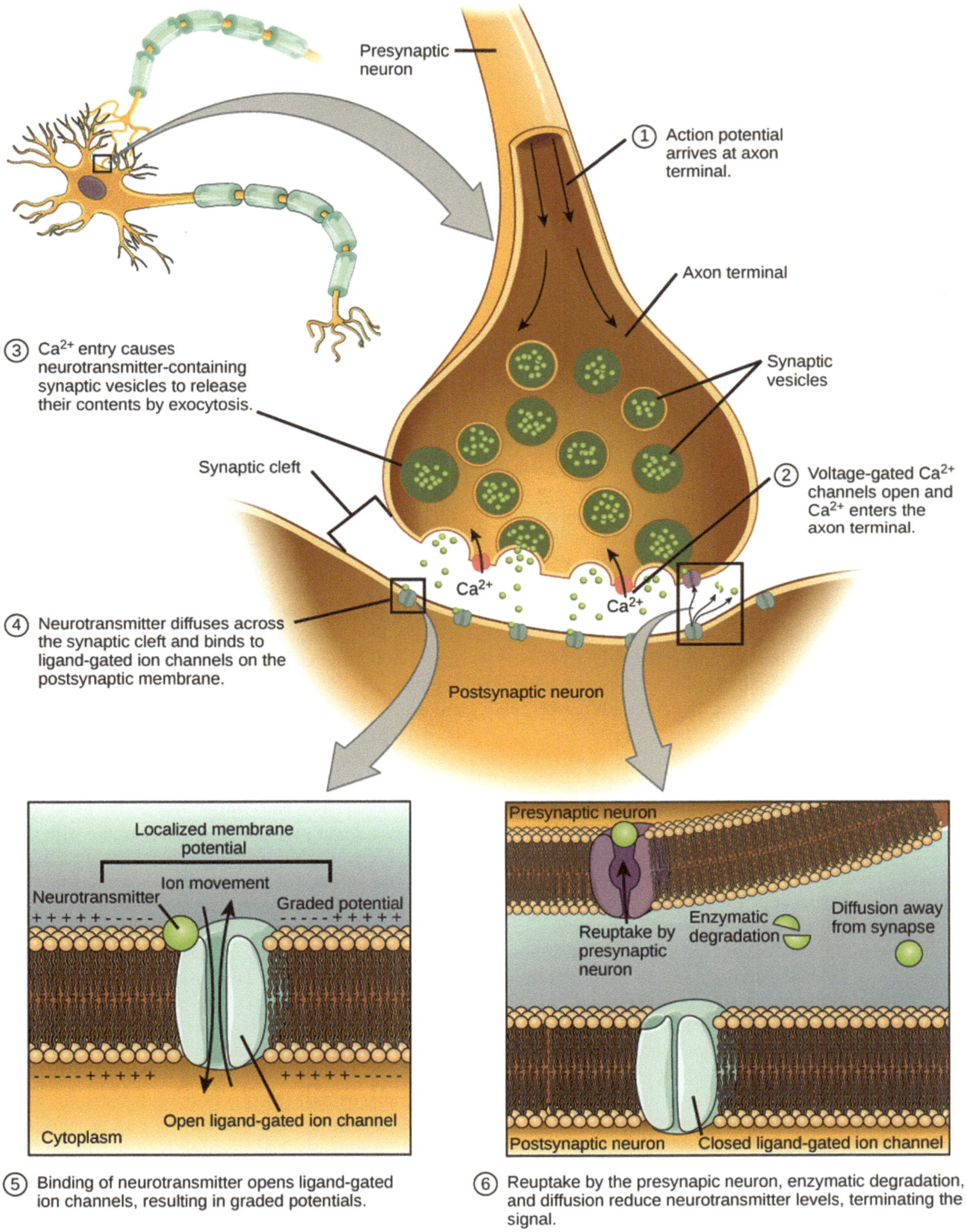

Fig. 4.4. Communication between neurons requires the release of neurotransmitters from the presynaptic membrane of the synapse. The neurotransmitters cross the synaptic cleft and bind to receptors on the postsynaptic membrane. (Courtesy of https://courses.candelalearning.com/biologymajors/chapter/chapter35-the-nervous-system/. Creative Commons Attribution.)

Neurotransmitters

Most neurotransmitters can activate several different receptors, both ionotropic and metabotropic, and some presynaptic terminals can release more than one neurotransmitter. Over 100 neurotransmitters have been identified, and they can be broadly separated into two categories: neuropeptides, which are relatively large molecules of 3–36 amino acids, and small-molecule neurotransmitters.

Glutamate is the most prevalent excitatory neurotransmitter in the brain, with more than half of all brain synapses releasing it. It is synthesized within neurons from the precursor glutamine. Glutamate has many types of ionotropic receptors, including α-amino-3-hydroxy-5-methyl-4-isoxazolepropionic acid (AMPA) receptors, *N*-methyl-D-aspartate (NMDA) receptors, and kainite receptors, and all of these receptors, when activated, produce EPSPs. Once glutamate is released, it is removed from the synaptic cleft by transporters that shuttle it either back into the presynaptic terminal or into nearby glial cells for conversion back to the precursor glutamine, which is then transported back into the presynaptic terminal. Rapid and efficient removal of glutamate is important because abnormally high levels of glutamate within the synaptic cleft can lead to excessive and prolonged excitation of the neuron that leads to neuronal excitotoxicity. Excitotoxicity is thought to cause neuronal damage and death in acute neurologic disorders and possibly in certain chronic diseases such as AD (Lau and Tymianski, 2010).

Gamma-aminobutyric acid (GABA) is the main inhibitory neurotransmitter in the brain and spinal cord. GABA is synthesized within neurons from glutamate by the enzyme glutamic acid decarboxylase (GAD) and a derivative of vitamin B_6. Vitamin B_6 deficiency can dramatically reduce GABA levels in the brain leading to seizures (Kurlemann et al., 1991). GABA is removed from the synaptic cleft in a similar fashion as glutamate. There are two ionotropic receptors for this neurotransmitter, $GABA_A$ and $GABA_C$, and one metabotropic receptor, $GABA_B$. Benzodiazepines that are used to treat anxiety disorders enhance GABA functioning by binding to $GABA_A$ and $GABA_C$ receptors. Barbiturates that are used for anesthesia and treatment of epilepsy also bind to these receptors. Alcohol can also alter the activity of GABA by binding to these receptors.

Acetylcholine, the first chemical substance identified as a neurotransmitter, is the transmitter at all neuromuscular junctions (Karczmar, 1996). It is also found in several brain areas and is involved in cognition and motor control. It is synthesized in the presynaptic terminal from acetyl coenzyme A and choline by choline acetyltransferase. Once released, acetylcholine is not taken back up like many other small-molecule neurotransmitters; instead, it is inactivated by the enzyme acetylcholinesterase within the synaptic cleft. Acetylcholine has two main receptors: the ionotropic nicotinic receptor and the metabotropic muscarinic receptor. These receptors were named after natural compounds that also activate them. Nicotinic receptors are activated by nicotine and muscarinic receptors are activated by muscarine extracted from a poisonous mushroom. Importantly, neurons within the basal forebrain that release acetylcholine into the cortex and hippocampus are lost in AD (Whitehouse et al., 1982). Furthermore, reductions in the number of nicotinic and muscarinic acetylcholine receptors are seen in the cortex of patients with AD (Shimohama et al., 1986). Thus, current medications for AD treatment revolve around enhancing acetylcholine signaling in the brain.

The group of neurotransmitters referred to as the biogenic amines is critically involved in a wide range of functions, including homeostasis, cognition, and mood. Dysfunctions in this group of neurotransmitters are implicated in most psychiatric disorders and also in drug addiction. The catecholamines, dopamine, norepinephrine, and epinephrine, derive from the same precursor, the amino acid tyrosine, within the presynaptic terminal. Epinephrine and norepinephrine are involved in the autonomic nervous system's fight-or-flight response (Chandler et al., 2014). Epinephrine increases attention and concentration and norepinephrine mediates arousal. Dopamine is generally inhibitory and dopamine-producing neurons in the substantia nigra project to the caudate nucleus and putamen along the nigrostriatal pathway, as well as from the ventral tegmental area to the limbic system and cortex along the mesolimbic and mesocortical projections. Dopamine is involved in the coordination of voluntary movement and is also implicated in motivation, reward, and reinforcement (Schultz, 2015). Importantly, many addictive drugs influence dopamine signaling. For example, cocaine and amphetamine work by increasing levels of dopamine within the synaptic cleft. Dopamine has several metabotropic receptor subtypes (D_1–D_5) through which it exerts its influence. Once dopamine activates its receptor, it is degraded within the synaptic cleft by the enzyme catechol-*O*-methyltransferase and within the neuron by monoamine oxidase (MAO). Some antidepressant drugs work by inhibiting these two enzymes. Degeneration of neurons within the substantia nigra is seen in Parkinson's disease, leading to motor dysfunction. Overactivation of the mesocortical projections is thought to be involved in some features of schizophrenia, whereas the mesolimbic projections are thought to be involved in drug addiction.

In addition to the catecholamines, serotonin is a biogenic amine. Serotonin regulates mood, sleep, appetite, and pain (Carlsson, 1987). It is primarily found in neurons in the raphe nuclei that project to the cortex, hippocampus, basal ganglia, thalamus, cerebellum, and spinal cord. Serotonin is synthesized from tryptophan that is found in our diets, and is transported back into the presynaptic terminal via the serotonin transporter for degradation by MAO. Several of the most prescribed antidepressant drugs work by blocking the reuptake of serotonin and are called selective serotonin reuptake inhibitors (SSRI). Many receptors for serotonin have been found, with the majority being metabotropic and one ($5\text{-}HT_3$) being ionotropic. Depending on which receptor is involved, serotonin can be either excitatory or inhibitory. Dysfunction in serotonin signaling at different levels is implicated in depression, bipolar disorder, and anxiety. However, increasing levels of serotonin must be done with care, as too high levels can lead to a condition called serotonin syndrome that can be fatal (Volpi-Abadie et al., 2013).

Glia

Glial cells are the other main cell type found in the CNS. Glia outnumber neurons by about 3:1. There are three types of differentiated glial cells in the CNS: astrocytes, microglia, and oligodendrocytes. Astrocytes are the most numerous glial cells in the brain, forming 20–50% of the volume of several brain areas, and are also the most versatile. They are highly diverse in terms of morphology, including some astrocytes which are star-shaped. Astrocytes have a wide range of functions, including maintaining brain homeostasis, providing physical and metabolic support for neurons, and regulating the content of the extracellular space. More recently, astrocytes have been shown to modulate synaptic transmission through secretion of transmitters and other compounds (Halassa et al., 2007). Microglia comprise about 5–20% of the glial cell population in the brain, and are the main phagocytic cell in the brain. At rest, microglia have a ramified morphology and, once activated by inflammatory signals, they take on ameboid morphology. Together, astrocytes and microglia function within the neuroimmune system to defend against pathologic insults by upregulating neuroinflammatory processes and also by detecting peripheral immune responses. Chronic neuroinflammation is associated with pathology in several brain disorders, ranging from autoimmune disorders and traumatic brain injury to neurodegenerative diseases and neuropsychiatric disorders, including schizophrenia (Skaper et al., 2014; Wang et al., 2015). Oligodendrocytes, the last type of glial cell we will discuss, synthesize and form the myelin sheath that wraps around axonal segments, increasing nerve impulse conduction by 10–100-fold. Myelin in the peripheral nervous system is generated by Schwann cells.

BRAIN STRUCTURE

Gray matter and white matter

The CNS consists of two types of tissue: gray matter and white matter. Gray matter is composed of neuronal cell bodies, dendrites, and synapses, whereas white matter is composed of axons and myelin. White-matter fiber tracts connect different brain regions together to form functional networks. Association fibers connect the various cortical regions within each hemisphere, commissural fibers connect the corresponding regions of the two hemispheres, and projection fibers connect the cortex and lower parts of the brain. White and gray matters follow different patterns of change across the lifespan. For example, white matter develops slower than gray matter but also shows more rapid loss in older age (Pakkenberg and Gundersen, 1997). Innovations in neuroimaging techniques have led to the realization that disorders primarily of white matter also show changes in gray matter, and vice versa. AD has long been thought of as a gray-matter disease with severe gray-matter atrophy seen first in the entorhinal cortex and hippocampus (Janke et al., 2001). However, recently extensive white-matter damage was seen in patients with early-onset and atypical forms of AD (Caso et al., 2015). Thus, white-matter degeneration has been proposed as an early marker of this disease. Conversely, multiple sclerosis, a white-matter disease, has also been linked to early gray-matter damage, which may have consequences for detection and treatment of this disease (Geurts and Barkhof, 2008).

White-matter lesions

Cerebral small-vessel disease is a vascular disorder resulting from an abnormality in the wall or lumen of the small vessels of the brain. This involves vascular lesions such as arteriolosclerosis, lipohyalinosis, and fibrinoid necrosis in the perforating cerebral arterioles and capillaries (Sachdev et al., 2014). These changes may lead to a general disruption of the neurovascular unit of the brain – the neurovascular unit refers to the interactions among glial, neuronal, and vascular elements (Iadecola, 2010). The maintenance of these perforating vessels is essential for maintaining healthy functioning of the brain's nuclei and white-matter networks. Damage to the small vessels of the brain typically results in more chronic, diffuse, and less severe ischemia resulting in white-matter lesions (WML), lacunar infarcts in the white matter and deep gray matter nuclei, dilation of perivascular spaces, microbleeds, and brain atrophy.

Blood supply to cerebral white matter comes from many small arteries arising from larger arteries supplying the cortex. The U fibers, situated just below the cortex, are richly perfused by many short and interdigitated cortical arterioles and are typically spared from ischemic damage (Schmahmann et al., 2008). In contrast, perfusion to deep white matter relies on lenticulostriate and long medullary arterioles which arise from the border zone between the middle cerebral artery and the anterior cerebral artery and are particularly vulnerable to chronic hypoperfusion (Moody et al., 1990). The white matter adjacent to the wall of the lateral ventricles is positioned at the distal end-zone territory of blood supply from the choroidal arteries – this area has been shown to be a low-perfusion region and is very susceptible to ischemic injury, as evidenced by the large volumes of WML in the periventricular regions (Schmahmann et al., 2008).

WML can be imaged using conventional magnetic resonance imaging (MRI). The term white-matter hyperintensities (WMH) collectively refers to lesions that appear hyperintense on T2-weighted, proton density-weighted, and fluid-attenuated inversion recovery MR sequences. These lesions have not cavitated and display a signal that is different from cerebrospinal fluid. With conventional MRI methods, WMH can only be described qualitatively or quantified as volume and are thus limited in their description of lesion morphology. Typically, WMH are described as either punctate lesions in the deep white matter (deep WMH) or confluent lesions appearing as "caps" on the frontal and/or occipital horns and fine "rims" along the walls of the lateral ventricles (periventricular WMH) (Wardlaw et al., 2013). Pathology studies in WMH suggest that several histologic processes are involved in these radiologic lesions. Punctate lesions are associated with dilated perivascular spaces and perivascular gliosis; caps are associated with myelin pallor, gliosis, and arteriosclerosis; and rims are associated with subependymal gliosis and loss of the ependymal lining. Generally, pathology reports have predominantly associated WMH with demyelination and axonal loss (Wardlaw et al., 2015). Structural imaging methods have provided a foundation for understating cerebrovascular lesions; however, in moving forward it is important to employ more specific imaging techniques to gain a better understanding of the microstructural mechanisms that contribute to white-matter damage.

Furthermore, it is important to recognize that ischemic brain injury may contribute to secondary pathologic changes, most notably amyloid-beta (Aβ) plaque accumulation. This is supported by postmortem studies which report vascular comorbidity in 30–60% of AD patients (Jellinger, 2007) and AD pathology in 15–43% of vascular dementia patients (patients with macroinfarctions – interestingly, people with lower volumes of macroinfarctions had more AD pathologic lesions and WML compared with people with higher volumes of macroinfarctions) (Kalaria and Ballard, 1999). In addition, studies suggest that there is a positive-feedback loop effect between Aβ and cerebrovascular dysfunction. Aβ may cause vascular dysregulation by disturbing cerebrovascular regulation (Niwa et al., 2002) by inducing oxidative stress and inflammatory mediators (Aliev et al., 2002). Aβ may also threaten cerebrovascular function by compromising cerebral perfusion, reducing vascular reserves, and increasing the propensity for ischemic damage (Iadecola, 2010). In return, hypoxia and/or ischemia may upregulate β- and γ-secretase activity to promote the cleavage of the Aβ peptide from amyloid precursor protein (APP), resulting in greater Aβ accumulation (Li et al., 2009). Furthermore, studies have found that focal cerebral ischemia produces larger infarcts in mice overexpressing APP (Zhang et al., 1997; Koistinaho et al., 2002). Human studies have found that Aβ is elevated in patients with vascular dementia (Lewis et al., 2006) and patients with AD have a heavier burden of cerebrovascular lesions compared to normal controls (Jellinger and Mitter-Ferstl, 2003).

FUNCTIONAL SYSTEMS

Sensory systems: visual, auditory, chemosensory, somatic systems

Although the five senses are very different from one another, there exist some general fundamental steps that the nervous system takes when processing all sensory information. First, specialized sensory receptors transduce sensory inputs into neural signals (action potentials) that transmit information about the sensory stimulus to the CNS. Second, the CNS has specialized sensory tracts that conduct the neural signals to specific regions of the cerebral cortex and other areas of the brain for sensory processing (Fig. 4.5). In the following sections, we will briefly discuss how these fundamental steps are applied to each sensory modality.

Visual system

When light enters our eye, it is refracted by the cornea and lens, which work in consort to project an inverted image onto the retina. The retina contains photoreceptor cells, the most common of which are rods (sensitive in low-light conditions, distinguishing shape and movement) and cones (sensitive to color). Visual information from the retina travels along the optic nerve via ganglion cells, where it is relayed at the lateral geniculate nucleus in the thalamus before it is sent to the primary visual cortex in the occipital lobe (Fig. 4.5).

The primary visual cortex receives visual inputs from the contralateral visual field (i.e., the left hemisphere receives input from the right visual field), where neurons are retinotopically organized. From here, visual information is passed to the visual association area (extrastriate

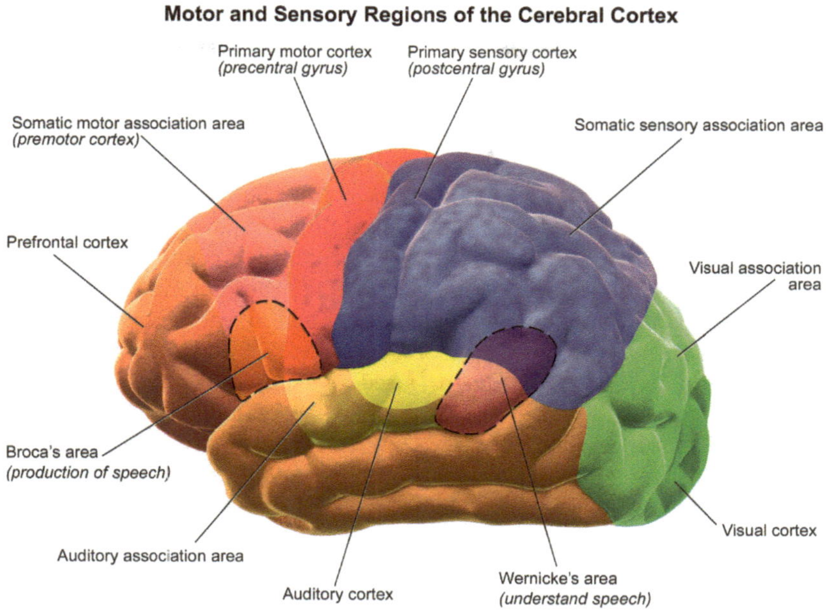

Fig. 4.5. Motor and sensory regions of the cerebral cortex. (Courtesy of Blausen.com staff. Blausen gallery 2014. Wikiversity Journal of Medicine. http://dx.doi.org/10.15347/wjm/2014.010.)

cortex) for further processing and interpretation before being passed to either the ventral or dorsal visual stream. As visual information flows in the posterior–anterior direction through the brain in either visual stream, information is processed and interpreted in a more complex and integrated fashion.

Information processed by the ventral, or "what," visual stream flows from the primary visual cortex to the temporal lobe, with connections to the medial temporal lobe and limbic system. It is responsible for identifying, recognizing, and categorizing objects based on their shape and form. Information passes through the visual stream in a hierarchic manner, with more complex features of the stimuli being identified as the information moves in the posterior to anterior direction across the cortex. Damage that occurs along this path results in very specific impairments. For example, patients with visual agnosia are unable to recognize objects, whereas patients with prosopagnosia ("face blindness") are unable to recognize familiar faces. In contrast, the dorsal, or "how," visual stream flows from the primary visual cortex to the parietal lobe. The dorsal visual stream is critical for spatial attention and guiding motor behaviors. Damage to the posterior parietal cortex can result in optic ataxia, an inability to use visual-spatial information to guide arm movements, such as reaching and grasping.

Visual attention functions to assist us with selectively processing the most relevant stimuli in the complex visual environments we encounter. Information is selectively filtered out in both a top-down and bottom-up manner, according to both relevance and salience. One specific form of selective attention, visual-spatial attention, refers to the selection of information in a particular spatial location. Visual-spatial attention is commonly described as a "spotlight," where attention is focused in one area of visual space, with enhanced perceptual processing of stimuli falling within the spotlight (e.g., Posner, 1980). Visual-spatial attention is typically assessed behaviorally and electrophysiologically, where both reaction times and neural activity in visual cortex are facilitated at attended locations.

Auditory system

The auditory system transforms sound stimuli, in the form of sound waves, into distinct patterns of neural activity that are integrated with information gathered from other sensory systems to help determine behaviors. The external and middle ears collect sound waves and amplify them.

The sound waves are converted into vibrations in the cochlea, a fluid-filled coiled tube in the inner ear. The fluid vibrations move the basilar membrane, which in turn activates the auditory receptor cells, the hair cells. The basilar membrane and hair cells are arranged in a tonotopic map, with certain sections responding optimally to sound of a given frequency. This tonotopic organization is maintained at each level of the auditory system, including the brain auditory areas. The hair cells mediate the mechanoelectric transduction of auditory information. Once displaced by the movement of the basilar membrane, the hair cells release neurotransmitters from their basal end on to neurons just outside the cochlea. The axons of these neurons form the auditory nerve, which carries signals into the cochlear nucleus in the brainstem. The signal-encoding sound is dissected by this structure into its different features, such as frequency and intensity, which are then carried separately but in parallel to higher brain centers. Importantly, these signals are now sent to structures on both sides of the brain, and include the superior olivary nucleus, inferior colliculus in the midbrain, medial geniculate nucleus in the thalamus, and the auditory cortex in the temporal lobe. The tonotopic organization, the mapping of frequencies along a structure, is maintained and carried throughout these areas.

The auditory cortex can be broadly divided into the primary auditory cortex (A1) and the secondary auditory cortex, also known as the belt areas (Fig. 4.5). The A1, located on the superior temporal gyrus, is organized tonographically and supports basic auditory functions, such as frequency discrimination, sound localization, and processing of species-specific sounds. The belt areas, where the tonographic organization is less maintained, receives inputs from lower auditory centers as well as from the A1, and also processes complex sounds. Importantly, Wernicke's area, a region important in comprehending speech, is immediately adjacent to the auditory cortex (Fig. 4.5).

Hearing loss can be divided into two categories: conductive and sensorineural. Middle or external ear disease leads to conduction deafness, which is caused by impaired conduction of sound through the external ear canal. In sensorineural (nerve) deafness there is an inability to transduce sound stimuli into neural signals, and is related to disorders of the inner ear. The gradual loss of auditory functioning that occurs as people age is referred to as presbycusis, and is of the sensorineural form. Typically, with increasing age there is bilateral loss of high-frequency hearing and this involves changes in central auditory processing of stimuli.

Chemosensory systems: olfactory and gustatory systems

The olfactory (smell) and gustatory (taste) systems detect molecules in the environment through specialized receptors in the nose and on the tongue, respectively. The olfactory system processes information about chemical stimuli called odorants found in the environment, and uses this information to influence a wide range of behaviors, including social interactions, reproduction, and feeding. The receptors for the olfactory system are located in the olfactory epithelium, a sheet of neurons that covers about 50% of the surface of the nasal cavity, and are called the olfactory receptor neurons (ORNs). ORNs are bipolar cells with unmyelinated axons extending from their basal region and a single dendritic process on the apical region that has several tiny hair-like protrusions called olfactory cilia. The cilia are covered with olfactory receptors; it seems there may be a particular receptor for a particular set of odorant molecules, and each ORN only expresses a particular receptor. Importantly, the division of neural stem cells found in the olfactory epithelium continuously replenishes the population of ORNs. The airborne odorant molecules enter the nasal cavity and interact with the olfactory receptors on the ORN and the chemical signal is transduced into an electric signal in the form of action potentials that are transmitted along the axon of the ORN to the olfactory bulb. Within the olfactory bulb, the axons make glutamatergic (glutamate-releasing) synapses with mitral cells within spherical structures called glomerulus. The axons of the mitral cells form a bundle called the lateral olfactory tract. This tract transmits the olfactory signal to the higher brain centers involved in olfaction, with the pyriform cortex located in the temporal lobe being the major target. The other targets of the lateral olfactory tract are the accessory olfactory nuclei, olfactory tubercle, entorhinal cortex, and the amygdala. The pyriform cortex sends olfactory inputs to many other brain targets to influence a large number of behaviors. For example, pyriform projects to the mediodorsal nucleus of the thalamus and to the hippocampus are believed to influence memory related to olfaction. And projections from the pyriform cortex to the hypothalamus and amygdala regulate appetite and reproductive behaviors.

Interestingly, there is a naturally occurring decrease in olfactory functioning with increasing chronologic age in humans (Doty and Kamath, 2014). There is also evidence to suggest even greater distorted olfactory functioning in age-associated neurodegenerative disorders such as AD (Nordin and Murphy, 1998). Furthermore, changes in olfaction are also often seen in the early stages of Parkinson's disease (Doty, 2012). Olfactory dysfunction is also seen in many patients with schizophrenia, and has been proposed as an early warning sign of the disease (Turetsky et al., 2009). Although the causes of olfactory deficits in these diseases are still under investigation, many possibilities have been proposed. For example, there may be an age-related reduction in the neural stem cell population that normally replenishes ORNs. The neurogenic environment within the epithelium may be altered such that newly generated ORNs are no longer supported. Furthermore, a genetic contribution to the olfactory dysfunction seen in many of these diseases has also been proposed.

The gustatory system processes chemical and physical information of ingested substances, principally of food. The chemosensory receptors that transduce chemical information into electric signal are found in taste buds, which are specialized epithelial structures located on the surface of the tongue, soft palate, pharynx, and the upper esophagus. Each taste bud has a tiny opening at its surface, around which taste cells amass, and it is here that chemical taste molecules interact with the taste cells. Taste cells are short-lived, epithelial cells with lifespans ranging from 2 days to 3 weeks (Hamamichi et al., 2006). Specific taste molecules activate specific taste cells. Taste molecules are typically classified into five categories: salty, sour, sweet, bitter, and umami (savory) (Kurihara, 2015). Different regions of the tongue respond maximally to specific taste categories. The tip of the tongue is most responsive to salty, sweet, and umami, whereas the sides and back of the tongue are most responsive to sour and bitter. The activation of a taste cell by a taste molecule initiates an intracellular signal cascade that leads to the release of the neurotransmitter serotonin onto the cranial sensory ganglion neurons, which relay the information from taste cells to the nucleus of the solitary tract in the medulla in the brainstem. From here, information is passed to the ventral posterior medial nucleus of the thalamus, which in turn relays the information to neurons in the insular taste cortex in the temporal lobe and areas in the frontal lobe. These neurons also respond to the visual, somatic, and olfactory inputs. The nucleus of the solitary tract in the medulla also sends reciprocal projections to the hypothalamus and amygdala to mediate the emotional aspects of appetite and eating, as well as to influence homeostatic behaviors associated with gustation.

Gustatory disorders can have a large, negative impact on a patient's health and quality of life. The most common gustatory disorder is phantom taste perception, or phantogeusia, in which the patient experiences a continuous unpleasant taste in the mouth in the absence of any stimuli (Maheswaran et al., 2014). Patients with reduced ability to taste suffer from a condition called hypogeusia, and ageusia is the rare case where a patient has no ability to taste (Maheswaran et al., 2014). Gustatory disorders can be congenital or acquired through injury or illness.

Somatic sensory system: proprioception and touch

The somatic sensory system mediates a number of sensations that are transduced by receptors in the skin or muscle. One important subsystem uses skin receptors to mediate the sensation of touch, and the other subsystem uses receptors in muscles, tendons, and joints to mediate proprioception – the ability to sense the position of our own limbs and body parts in space. For both of these subsystems, somatic sensations begin from activity detected by specific receptors on afferent nerve fibers whose processes branch within skin or muscle. These afferent nerve fibers conduct the signal to cell bodies that reside within the dorsal root ganglia for the body or within the cranial nerve ganglia for the head. These afferent fibers conduct the action potentials past the cell bodies in the ganglia until they reach their synaptic terminals in the target structures within the CNS.

Within the skin and muscle, specialized sensory receptor cells called mechanoreceptors typically encapsulate the afferent fibers used for touch and proprioception. For touch, the skin contains four morphologically different mechanoreceptor cells (Zimmerman et al., 2014). Meissner corpuscles are found in the tips of the dermal papillae, close to the skin surface. Merkel cells are found in the epidermis. Ruffini corpuscles are found in the next layer of skin, the dermis, and pacinian corpuscles are found deep in the dermis or even the subcutaneous layer. For proprioception, other types of mechanoreceptors provide information about the position of limbs and other body parts in space. The three receptors and their functions are: (1) muscle spindles that are found in striated muscles and provide information about changes in muscle length; (2) Golgi tendon organs that are found in tendons and provide information about changes in muscle tension; and (3) joint receptors that give signals when joint movements come close to range limits (Proske and Gandevia, 2012).

Tactile sensory information gathered by the mechanoreceptors in the skin of the body enters the spinal cord through the dorsal roots and ascends ipsilaterally through the dorsal column to the lower medulla. In the lower medulla, axons carrying information originating from the upper body synapse onto neurons in the cuneate nucleus subdivision of the dorsal column nuclei, while axons from the lower body synapse onto neurons in the gracile nucleus. The axons exiting from the dorsal column nuclei, called the internal arcuate fibers, decussate (cross over) and form the medial lemniscus which project to the ventral posterior lateral nucleus of the thalamus, which in turn send their axons to the primary somatosensory cortex (SI) as well as to the secondary somatosensory cortex (SII) (Fig. 4.5). Cutaneous mechanoreceptor information from afferents originating in the face enters the spinal cord by a separate set of neurons that are located in the trigeminal ganglion. The central processes of trigeminal ganglion cells enter the brainstem at the pons and synapse with neurons in the trigeminal brainstem complex. From here information is sent to the ventral posterior medial nucleus of the thalamus, where it is then sent to the SI and SII regions of the cerebral cortex.

Proprioceptive sensory information travels along a very similar central pathway as tactile information. However, a few important differences exist that underlie the importance of proprioceptive information in motor reflexes. When proprioceptive afferents originating from the lower body first enter the spinal cord, they separate into two branches. One branch synapses onto neurons in the dorsal and ventral horns of the sacral region of the spinal cord to mediate reflexes such as the knee-jerk reflex (see below). The other branch synapses with neurons in Clarke's nucleus in the lumbar region of the spinal cord. These neurons send axons that travel up to the cerebellum, with collaterals branching off to synapse with neurons in the dorsal column nuclei. The cerebellum requires this proprioceptive information to properly regulate voluntary movements. The proprioceptive neurons in the dorsal column nuclei then send axons along a similar route as tactile information, with neural signals traveling to the ventral posterior lateral nucleus of the thalamus and from there to the SI and SII regions of the cerebral cortex. Proprioceptive afferents originating from the upper body enter the spinal cord at higher regions than afferents from the lower body, but follow a similar route to the dorsal column nuclei and cerebellum.

SI processes both tactile and proprioceptive sensory information. SI is located in the postcentral gyrus of the cerebral cortex and contains four distinct regions, called Brodmann's areas 3a, 3b, 1, and 2. Each of these distinct regions contains its own complete representation of the body using the same topographic organization that is shown in the homunculus. Tactile sensory information is processed mainly in areas 3b and 1, whereas proprioceptive information is processed in area 3a. Neurons in area 2 process both types of sensory information. However, many interconnections exist between the four areas of the SI. All regions of SI send projections to SII, where sensory information is further processed before being sent to limbic structures such as the amygdala and hippocampus. Area 2 in the SI also sends projections to the parietal cortex, specifically areas 5 and 7, which in turn send projections to the primary motor and premotor areas of the frontal lobe. This pathway is critical for integrating sensory and motor information for the execution of voluntary movements.

Limbic system

The limbic system has a central role in many basic functions required for survival, including memory, reproduction, and nutrition. Importantly, it is involved in the experience and expression of emotion. The limbic system is comprised of several functionally and anatomically interconnected cortical and subcortical structures located in the frontal and temporal lobes as well as several multimodal association areas that integrate different sensory inputs (Fig. 4.6). These structures include the orbital and medial prefrontal cortex, insular cortex, cingulate gyrus, parahippocampal gyrus, ventral portions of the basal ganglia, the mediodorsal nucleus of the thalamus, mammillary bodies of the hypothalamus, the hippocampus, and the amygdala. It has been thought of as the "feeling and reacting brain", in contrast to the "thinking brain", the cerebral cortex.

In terms of memory formation, the entorhinal cortex of the parahippocampal gyrus receives cognitive and

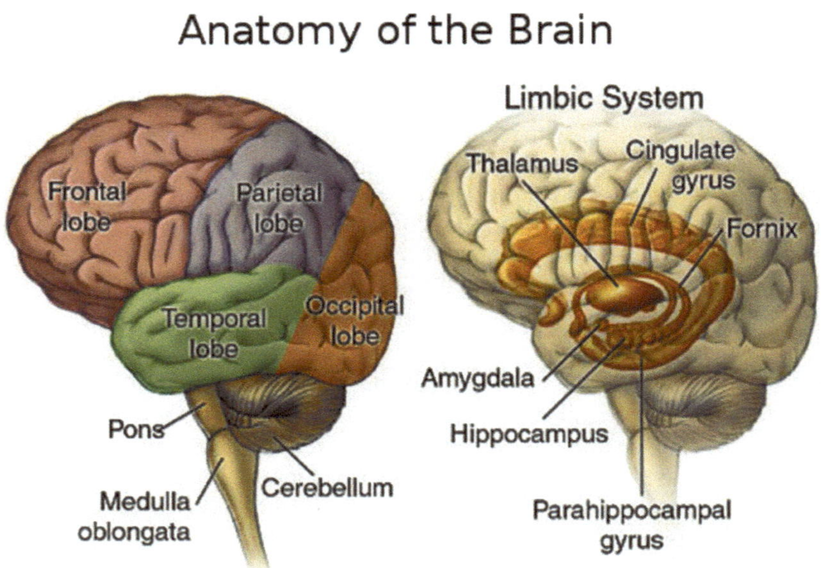

Fig. 4.6. Anatomy of the brain and the limbic system. The brain is divided into the brainstem, cerebellum, and cerebrum. The outer layer of the cerebrum, called the cerebral cortex, is divided into four lobes: frontal, parietal, temporal, and occipital. The limbic system is composed of many interconnected structures, several of which are depicted in the right-hand illustration. (From http://www.wpclipart.com/medical/anatomy/brain/brain_2/brain_anatomy.png.)

sensory information from the association areas of the cortex, which it transmits to the hippocampus for consolidation. Long-term potential of granule and pyramidal cells in the hippocampus is thought to be the cellular mechanism for memory formation. Consolidated information is retrieved by the entorhinal cortex and sent to the association areas for encoding into long-term memory.

The limbic system is involved in addiction (Orsini et al., 2015), as many drugs of abuse act on elements of this system, which can lead to the dysregulation of the emotional processing abilities of the limbic system altering reward and punishment signaling in the brain. The neuropsychiatric disorder schizophrenia is related to altered limbic system functioning. Studies have shown reduced volumes of the frontal lobes, medial temporal lobes, and thalamus related to loss of neuropil and reduced neuronal size in patients suffering from this disorder. Schizophrenia seems to be also related to dopamine, as antipsychotic medications used to treat schizophrenia block dopamine transmission at the D_2 receptor. Damage to the limbic system can also lead to the Klüver–Bucy syndrome, a rare neurobehavioral disorder. The syndrome causes individuals to put objects into their mouths, become hypersexual, show a lack of normal fear and emotional responses, and have memory loss (Lilly et al., 1983). Another disorder associated with the limbic system is Korsakoff's syndrome, also known as amnestic confabulatory syndrome. The disorder typically results from thiamine deficiency from chronic alcoholism, and leads to loss of recent memory (Kril and Harper, 2012).

Motor system

The motor system responds to the sensory information gathered and processed by the sensory system by generating movements and other behaviors. Movement is controlled at numerous neural levels. Ultimately, muscles are innervated by the lower motor neurons located in the brainstem and spinal cord. Descending projections from upper motor neurons in the brainstem and the cerebral cortex influence the activity of the lower motor neurons. The brainstem is required for regulating basic movements, maintaining balance, orientating gaze, and controlling posture. Within the cerebral cortex, the primary motor cortex and the premotor cortex, two highly interconnected and adjacent areas, are essential for planning, initiating, and directing sequences of voluntary movements (Fig. 4.5).

Lesions of upper motor neurons lead to disruptions in the initiation of voluntary muscle activity, weakening or paralysis of the muscles involved, increased muscle tone, withdrawal spasms, and abnormal reflexes (e.g., Babinski's response: see below). Neuronal inputs reach muscles via the lower motor neurons. Lesions of lower motor neurons can lead to weakness or flaccid paralysis of the muscles involved, decreased muscle tone, muscle atrophy, and decreased reflexes of the muscles involved. Amyotrophic lateral sclerosis involves degeneration of both lower and upper motor neurons, eventually leading to the loss of control of voluntary movement.

Complex circuits involving other brain structures, including the cerebellum and the basal ganglia, regulate the activity of the upper motor neurons in the primary motor cortex and premotor cortex. The cerebellum monitors discrepancies between intended movements and movements actually done. It then utilizes the difference or the error it detects to fix the ongoing movement. The cerebellum also partakes in motor learning when it stores the error information to influence future movements. The neurons in the deep cerebellar nuclei, located deep within the cerebellum, output a corrective signal to the upper motor neurons that have been modified by the descending inputs from the cerebellar cortex, which is located on the surface of the cerebellum. Patients with damage to the cerebellum exhibit persistent errors in movement. Depending on the extent and location of the damage, there will be disrupted coordination of ongoing movements using some muscle groups, but not others. Regardless of where the damage is, lesions to the cerebellum lead to an inability to perform smooth, precisely directed movements (cerebellar ataxia) (Morton and Bastian, 2007).

The basal ganglia also influences movements by altering the activity of upper motor neurons. Several basal ganglia nuclei make a subcortical loop that links many parts of the cerebral cortex to the upper motor neurons to suppress or stop unwanted movements, as well as control the initiation of voluntary movements. Medium spiny neurons in the dorsal striatum (caudate nucleus and putamen), the main input areas of the basal ganglia, receive inputs from the cerebral cortex along the corticostriatal pathway, and then send projections to the globus pallidus and substantia nigra pars reticulata, the two main output areas of the basal ganglia. The medium spiny neurons in the dorsal striatum also receive inputs from neurons in the substantia nigra pars compacta that release dopamine that modulate effectiveness of the inputs from the cerebral cortex. Damage to any one of the components of the basal ganglia circuit disrupts the structure's regulatory control over upper motor neurons, resulting in disordered movements. In Parkinson's disease loss of dopamine-producing neurons in the substantia nigra pars compacta leads to an increase in the inhibitory output from the basal ganglia; movements are difficult to initiate and slowed, called bradykinesia, and a tremor is typically seen; and rigidity of limbs caused by increased muscle tone occurs (Berardelli et al., 2001). Disease onset typically occurs between 50 and 65 years. Huntington's disease, an autosomal disorder caused by a mutation in the *Huntingtin* gene on chromosome 4, involves a loss of GABA-releasing inhibitory medium spiny neurons in the dorsal striatum. This pathophysiology leads to jerky, random, uncontrollable movements called chorea (Estrada-Sanchez and Rebec, 2013). Disease onset typically occurs between 35 and 45 years of age.

A reflex is a stereotyped, automatic motor response to a defined stimulus produced by a reflex arc. Typically, a sensory receptor detects a stimulus and action potentials are initiated and travel along the axons of the sensory neurons to the CNS. The axons synapse with interneurons in the gray matter of the spinal cord, which in turn synapse with motor neurons that send axons out of the spinal cord to the muscles or glands in the peripheral nervous system to elicit a response to the stimulus. There are two main types of reflexes in the body: autonomic or visceral reflexes and somatic reflexes. Autonomic reflexes function to maintain homeostasis in the body by controlling and regulating smooth-muscle cells, cardiac muscle cells, and glands. Somatic reflexes control skeletal muscles and typically function to protect the physical body. Monosynaptic reflexes involve simple pathways in which sensory neurons directly interact with motor neurons. Polysynaptic reflexes are more complex and involve one or more interneurons. Many reflexes are integrated within the spinal cord, while others are integrated within the brain. Importantly, higher brain centers,

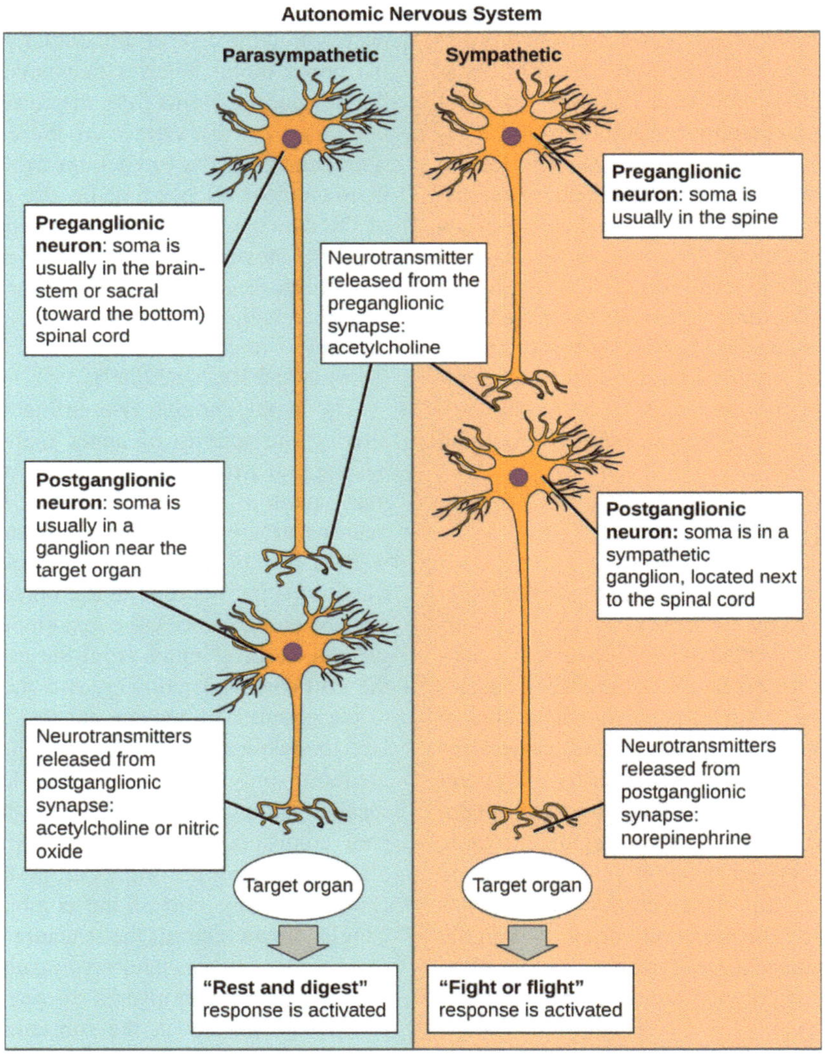

Fig. 4.7. Within the autonomic nervous system the sympathetic division mediates the "flight or flight" response, while the parasympathetic division mediates the "rest and digest" response. Within both divisions, preganglionic neurons in the central nervous system release the neurotransmitter acetylcholine on to the postganglionic neurons in the peripheral nervous system. The postganglionic neurons act on target organs, via the neurotransmitter acetylcholine in the parasympathetic division and the neurotransmitter norepinephrine in the sympathetic division. (Courtesy of https://courses.candelalearning.com/biologymajors/chapter/chapter35-the-nervous-system/. Creative Commons Attribution.)

such as the prefrontal cortex, can exert an influence over reflexes by suppressing or exaggerating them.

The simplest reflex is the stretch reflex, in which the sudden stretch of a muscle results in its contraction. The reflex arc begins when the muscle spindles, the sensitivity of which is regulated by gamma motor neurons, detect the stretching of the muscle. Sensory neurons conduct the action potentials from the muscle spindles to the alpha motor neurons in the spinal cord. The alpha motor neurons in turn innervate the muscle and cause it to contract. This reflex is a monosynaptic reflex. On the other hand, the withdrawal reflex is a polysynaptic reflex. The withdrawal, or flexor, reflex removes a body part from a pain-eliciting stimulus. The reflex arc begins when the pain receptors are stimulated and the axons of the sensory neurons conduct the action potentials to the spinal cord. In the spinal cord the axons branch, and one branch synapses with ascending fibers that travel to the brain and the other branch synapses with excitatory interneurons, which in turn synapse with alpha motor neurons that stimulate muscles that lead to the withdrawal of the body part.

Autonomic

The CNS has two neural outputs: the somatic motor system, which innervates and commands skeletal muscles through motor neurons in the ventral horn of the spinal cord, and the autonomic nervous system, which regulates the functions of the body's internal organs through the sympathetic, parasympathetic, and enteric nervous systems. The sympathetic and parasympathetic divisions operate in parallel, with the former involved in the "fight or flight" response and the latter involved in "rest and digest" functions. The sympathetic nervous system originates in the spinal cord, with the cell bodies of the preganglionic neurons located in the thoracic and lumbar regions. The axons of these neurons, via the ventral roots, synapse onto postganglionic neurons in the sympathetic chain that are located next to the spinal column. The parasympathetic nervous system originates in the sacral region of the spinal cord and the medulla in the brainstem. The preganglionic neurons synapse onto postganglionic neurons in the parasympathetic ganglion that are located next to, or in, the target organs.

The preganglionic neurons of both the sympathetic and parasympathetic divisions release the neurotransmitter acetylcholine. The postganglionic neurons of the parasympathetic system also release acetylcholine; however, postganglionic sympathetic neurons release norepinephrine. (Fig. 4.7). The enteric nervous system is the extensive system of neurons located in the walls of the gastrointestinal tract, gallbladder, and pancreas that control the functioning of the gut independently of the sympathetic and parasympathetic systems (Furness et al., 2014). The hypothalamus, in particular the periventricular zone, is the main regulator of the autonomic nervous system. This brain area receives and integrates diverse information from many other brain structures and controls the preganglionic neurons in the medulla and spinal cord. Another important brain structure in the control of the autonomic nervous system is the nucleus of the solitary tract of the medulla in the brainstem.

References

Aliev G, Smith MA, Seyidova D et al. (2002). The role of oxidative stress in the pathophysiology of cerebrovascular lesions in Alzheimer's disease. Brain Pathol 12: 21–35.

Berardelli A, Rothwell JC, Thompson PD et al. (2001). Pathophysiology of bradykinesia in Parkinson's disease. Brain 124: 2131–2146.

Carlsson A (1987). Perspectives on the discovery of central monoaminergic neurotransmission. Annu Rev Neurosci 10: 19–40.

Caso F, Agosta F, Mattavelli D et al. (2015). White matter degeneration in atypical Alzheimer disease. Radiology 277: 162–172.

Chandler DJ, Waterhouse BD, Gao WJ (2014). New perspectives on catecholaminergic regulation of executive circuits: evidence for independent modulation of prefrontal functions by midbrain dopaminergic and noradrenergic neurons. Front Neural Circ 8: 53.

Doty RL (2012). Olfaction in Parkinson's disease and related disorders. Neurobiol Dis 46: 527–552.

Doty RL, Kamath V (2014). The influences of age on olfaction: a review. Front Psychol 5: 20.

Estrada-Sanchez AM, Rebec GV (2013). Role of cerebral cortex in the neuropathology of Huntington's disease. Front Neural Circ 7: 19.

Furness JB, Callaghan BP, Rivera LR et al. (2014). The enteric nervous system and gastrointestinal innervation: integrated local and central control. Adv Exp Med Biol 817: 39–71.

Geurts JJ, Barkhof F (2008). Grey matter pathology in multiple sclerosis. Lancet Neurol 7: 841–851.

Halassa MM, Fellin T, Haydon PG (2007). The tripartite synapse: roles for gliotransmission in health and disease. Trends Mol Med 13: 54–63.

Hamamichi R, Asano-Miyoshi M, Emori Y (2006). Taste bud contains both short-lived and long-lived cell populations. Neuroscience 141: 2129–2138.

Humeau Y, Doussau F, Grant NJ et al. (2000). How botulinum and tetanus neurotoxins block neurotransmitter release. Biochimie 82: 427–446.

Iadecola C (2010). The overlap between neurodegenerative and vascular factors in the pathogenesis of dementia. Acta Neuropathol 120: 287–296.

Janke AL, de Zubicaray G, Rose SE et al. (2001). 4D deformation modeling of cortical disease progression in Alzheimer's dementia. Magn Reson Med 46: 661–666.

Jellinger KA (2007). The enigma of vascular cognitive disorder and vascular dementia. Acta Neuropathol 113: 349–388.

Jellinger KMD, Mitter-Ferstl EP (2003). The impact of cerebrovascular lesions in Alzheimer disease. J Neurol 250: 1050–1055.

Kalaria RN, Ballard C (1999). Overlap between pathology of Alzheimer disease and vascular dementia. Alzheimer Dis Assoc Disord 13 (Suppl 3): S115–S123.

Karczmar AG (1996). Otto Loewi lecture. Loewi's discovery and the XXI century. Prog Brain Res 109: 1–27. xvii.

Koistinaho M, Kettunen MI, Goldsteins G et al. (2002). Beta-amyloid precursor protein transgenic mice that harbor diffuse A beta deposits but do not form plaques show increased ischemic vulnerability: role of inflammation. Proc Natl Acad Sci U S A 99: 1610–1615.

Kril JJ, Harper CG (2012). Neuroanatomy and neuropathology associated with Korsakoff's syndrome. Neuropsychol Rev 22: 72–80.

Kurihara K (2015). Umami the fifth basic taste: history of studies on receptor mechanisms and role as a food flavor. Biomed Res Int 2015: 189402.

Kurlemann G, Menges EM, Palm DG (1991). Low level of GABA in CSF in vitamin B6-dependent seizures. Dev Med Child Neurol 33: 749–750.

Lau A, Tymianski M (2010). Glutamate receptors, neurotoxicity and neurodegeneration. Pflugers Arch 460: 525–542.

Lewis H, Beher D, Cookson N et al. (2006). Quantification of Alzheimer pathology in ageing and dementia: age-related accumulation of amyloid-beta(42) peptide in vascular dementia. Neuropathol Appl Neurobiol 32: 103–118.

Li L, Zhang X, Yang D et al. (2009). Hypoxia increases Abeta generation by altering beta- and gamma-cleavage of APP. Neurobiol Aging 30: 1091–1098.

Lilly R, Cummings JL, Benson DF et al. (1983). The human Kluver–Bucy syndrome. Neurology 33: 1141–1145.

Lin YC, Koleske AJ (2010). Mechanisms of synapse and dendrite maintenance and their disruption in psychiatric and neurodegenerative disorders. Annu Rev Neurosci 33: 349–378.

Maheswaran T, Abikshyeet P, Sitra G et al. (2014). Gustatory dysfunction. J Pharm Bioallied Sci 6: S30–S33.

Moody D, Bell M, Challa V (1990). Features of the cerebral vascular pattern that predict vulnerability to perfusion or oxygenation deficiency: an anatomic study. Am J Neuroradiol 11: 431–439.

Morton SM, Bastian AJ (2007). Mechanisms of cerebellar gait ataxia. Cerebellum 6: 79–86.

Niwa K, Kazama K, Younkin L et al. (2002). Cerebrovascular autoregulation is profoundly impaired in mice overexpressing amyloid precursor protein. Am J Physiol Heart Circ Physiol 283: H315–H323.

Nordin S, Murphy C (1998). Odor memory in normal aging and Alzheimer's disease. Ann N Y Acad Sci 855: 686–693.

Orsini CA, Moorman DE, Young JW et al. (2015). Neural mechanisms regulating different forms of risk-related decision-making: insights from animal models. Neurosci Biobehav Rev 58: 147–167.

Pakkenberg B, Gundersen HJ (1997). Neocortical neuron number in humans: effect of sex and age. J Comp Neurol 384: 312–320.

Poot M (2013). Towards identification of individual etiologies by resolving genomic and biological conundrums in patients with autism spectrum disorders. Mol Syndromol 4: 213–226.

Posner MI (1980). Orienting of attention. Q J Exp Psychol 32: 3–25.

Proske U, Gandevia SC (2012). The proprioceptive senses: their roles in signaling body shape, body position and movement, and muscle force. Physiol Rev 92: 1651–1697.

Sachdev P, Kalaria R, O'Brien J et al. (2014). Diagnostic criteria for vascular cognitive disorders: a VASCOG statement. Alzheimer Dis Assoc Disord 28: 206–218.

Schmahmann JD, Smith EE, Eichler FS et al. (2008). Cerebral white matter. Ann N Y Acad Sci 1142: 266–309.

Schultz W (2015). Neuronal reward and decision signals: from theories to data. Physiol Rev 95: 853–951.

Shimohama S, Taniguchi T, Fujiwara M et al. (1986). Changes in nicotinic and muscarinic cholinergic receptors in Alzheimer-type dementia. J Neurochem 46: 288–293.

Skaper SD, Facci L, Giusti P (2014). Neuroinflammation, microglia and mast cells in the pathophysiology of neurocognitive disorders: a review. CNS Neurol Disord Drug Targets 13: 1654–1666.

Turetsky BI, Hahn CG, Borgmann-Winter K et al. (2009). Scents and nonsense: olfactory dysfunction in schizophrenia. Schizophr Bull 35: 1117–1131.

Volpi-Abadie J, Kaye AM, Kaye AD (2013). Serotonin syndrome. Ochsner J 13: 533–540.

Wang Q, Liu Y, Zhou J (2015). Neuroinflammation in Parkinson's disease and its potential as therapeutic target. Transl Neurodegener 4: 19.

Wardlaw JM, Smith EE, Biessels GJ et al. (2013). Neuroimaging standards for research into small vessel disease and its contribution to ageing and neurodegeneration. Lancet Neurol 12: 822–838.

Wardlaw JM, Valdes Hernandez MC, Munoz-Maniega S (2015). What are white matter hyperintensities made of? Relevance to vascular cognitive impairment. J Am Heart Assoc 4: 001140.

Whitehouse PJ, Price DL, Struble RG et al. (1982). Alzheimer's disease and senile dementia: loss of neurons in the basal forebrain. Science 215: 1237–1239.

Zhang F, Eckman C, Younkin S et al. (1997). Increased susceptibility to ischemic brain damage in transgenic mice overexpressing the amyloid precursor protein. J Neurosci 17: 7655–7661.

Zimmerman A, Bai L, Ginty DD (2014). The gentle touch receptors of mammalian skin. Science 346: 950–954.

Handbook of Clinical Neurology, Vol. 138 (3rd series)
Neuroepidemiology
C. Rosano, M.A. Ikram, and M. Ganguli, Editors
http://dx.doi.org/10.1016/B978-0-12-802973-2.00005-7

Chapter 5

Population imaging in neuroepidemiology

M.W. VERNOOIJ[1]*, M. DE GROOT[2], AND D. BOS[1]

[1]*Departments of Radiology and Epidemiology, Erasmus MC, University Medical Center Rotterdam, Rotterdam, The Netherlands*

[2]*Departments of Radiology and Medical Informatics, Erasmus MC, University Medical Center Rotterdam, Rotterdam, The Netherlands*

Abstract

Neuroepidemiologic studies have traditionally focused on studying associations between determinants and neurologic outcomes, while treating the pathway in between both as a "black box." With the rise of noninvasive, advanced neuroimaging techniques, it has become possible to directly study brain changes occurring in this "black box." This importantly aids to unravel disease pathways, find new markers of disease, or identify subjects at risk of disease. Imaging in neuroepidemiologic studies is also called population neuroimaging. This chapter discusses the rationale of population neuroimaging, the different imaging modalities that can be applied, and the various ways to extract visual or quantitative information from these images. Population neuroimaging is a fast-progressing field, partly due to new techniques and partly due to the growing need for collaboration, harmonization, and standardization among studies. Considerations for future applications of imaging in neuroepidemiology are discussed against this background.

POPULATION IMAGING: DEFINITION AND RATIONALE

Many neurologic diseases, especially those occurring in older age, have a long subclinical phase during which a person is asymptomatic and does not seek medical attention. As a consequence, once symptoms manifest, in many instances the pathologic changes caused by the disease process are already advanced and mostly irreversible. The paradigm shift in clinical practice and in research from focus on a cure towards focus on prevention and prediction of disease also leads to a shift in attention towards understanding disease pathophysiology in its earliest stage. This includes finding (modifiable) determinants in the subclinical phase, and predicting which individuals are most at risk or would benefit most from preventive actions.

To study disease in the asymptomatic stage, population-based studies are of great importance. Typically, in this type of study, rich phenotyping is performed and a multitude of potential disease determinants are assessed at baseline. Another key property of population-based studies is that individuals are followed over time for occurrence of disease. This allows associations to be studied between genetic and nongenetic determinants and a range of disease outcomes (Fig. 5.1). This study design is very suitable for research on etiology and disease prediction, especially for multifactorial diseases, though not efficient for studying rare diseases.

A drawback of these types of association analyses is that the exact pathway of disease usually remains a "black box." Little is known on intermediate processes or on the exact changes that occur in the tissue at risk. In the present era, in which knowledge of disease pathophysiology and of molecular pathways on which to intervene is becoming increasingly important, there is a growing need to open this black box. This is exactly where radiologic imaging can fill a void (Fig. 5.1). Imaging can, noninvasively or minimally invasively, show the changes that occur in the human body that may reflect early disease, intermediate factors, or risk indicators of disease.

*Correspondence to: Meike W. Vernooij, MD, PhD, Departments of Radiology and Epidemiology, Erasmus MC, University Medical Center Rotterdam, PO Box 2040, 3000 CA Rotterdam, The Netherlands. E-mail: m.vernooij@erasmusmc.nl

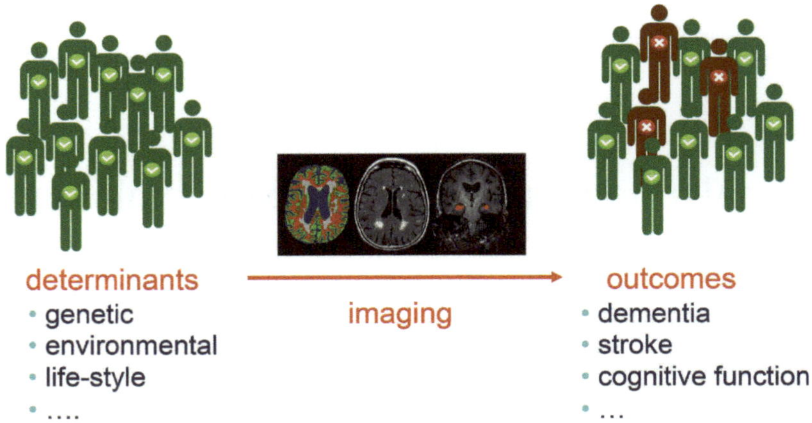

Fig. 5.1. The role of population imaging in epidemiologic studies.

Whereas presence of disease is often a dichotomous state, with little possibility of capturing variation across subjects, imaging offers more objectivity, reproducibility, the ability to monitor change over time and to express changes in a quantitative way, yielding more power to detect small changes and effects (Table 5.1). During the past decade this has led to the introduction of the term population imaging, by which is meant, the large-scale application and analysis of medical images in controlled population cohorts.

Population imaging: not similar to clinical imaging

It should be understood that population imaging differs from medical imaging as applied in clinical practice in several ways. Whereas in a clinical setting a single imaging examination in itself may give the ultimate clue to a certain diagnosis, images acquired in population imaging should always be regarded in a much broader context. The strength of population imaging studies lies in the combination of imaging data with a variety of other information collected on the participants, be it derived from questionnaires, physiologic examinations, or laboratory measurements. Many population imaging studies comprise large datasets, challenging the researchers how to handle these not only in terms of data transfer and storage, but also regarding extraction of information (visual rating versus automated processing). Second, due to the nature of these studies, usually conducted in nonclinical populations of (presumed) healthy individuals, population imaging has certain restrictions that pertain to the burden put on persons, e.g., exposure to radiation, invasive procedures, or the duration of the examination. Third, there are drawbacks related to population imaging on a societal and ethical level, e.g., detection on the images of medical problems that are not related to the research examination, so-called incidental findings.

Table 5.1

Advantages of population imaging
Direct examination of tissue at risk
Noninvasive or minimally invasive
Presymptomatic/asymptomatic phase
Objective
Reproducible
Quantification
Monitoring of change over time
Identification of individuals at risk
Understanding causal pathways
Finding new biomarkers

POPULATION IMAGING MODALITIES: COSTS, RISKS, AND ACCESSIBILITY

As already noted, there are various imaging modalities which can nowadays be applied at the level of population-based research (Table 5.2). In the current section these imaging modalities will be discussed in the context of the major research areas in neuroepidemiology: vascular imaging and brain imaging.

Vascular imaging

For already over a century, the role of vascular disease in neurodegenerative and vascular brain pathology has been an important topic of research. In 1907, it was Alois Alzheimer who first described "arteriosclerotic changes" as a brain autopsy finding in one of his patients with an "unusual illness of the cerebral cortex" (Alzheimer, 1907) – the disease which we now know as Alzheimer's disease. Yet, it has only been during the last decades that research on the role of atherosclerosis in brain pathology has been taken to the next level, largely as a result of the rise of medical imaging. Due to the rapid advances in imaging techniques in combination with the improved

Table 5.2

Comparison of the imaging modalities mainly used in population-based studies

Imaging modality	Advantages	Disadvantages
Ultrasound	Safe, noninvasive Inexpensive Widely available Quick Qualitative vascular imaging No absolute contraindications	Highly operator-dependent Few possibilities for quantification No brain imaging
Computed tomography (CT)	Relatively inexpensive Widely available Relatively quick Quantitative vascular imaging No absolute contraindications	Radiation exposure Contrast administration for brain imaging
Magnetic resonance imaging (MRI)	Relatively safe, noninvasive Structural brain imaging Functional brain imaging Quantitative vascular imaging	Expensive Time consuming Contraindicated when metal in body
Nuclear imaging	Information on brain function Information on brain metabolism	Radiation exposure Invasive Expensive Low spatial resolution (when not combined with CT or MRI)
Retinal imaging	Safe, quick Direct visualization of small blood vessels	No direct cerebral measurements

access to various imaging modalities, we are now able to directly and noninvasively visualize the pathophysiologic process of atherosclerosis *in vivo*. This has facilitated imaging of large populations of asymptomatic individuals to detect atherosclerosis in the subclinical phase and further elucidate its involvement in the development of neurodegenerative and vascular brain pathology.

In the following section different imaging modalities, which are specifically suitable for noninvasive vascular imaging, are highlighted.

Carotid intima-media thickness

Ultrasound examinations of the carotid artery provided the first measures of atherosclerosis that have been used as markers of systemic vascular disease (Owen et al., 2011). By far the most popular of these markers is the carotid intima-media thickness (cIMT). The rationale behind cIMT measurement is that thickening of the intimal layer of the blood vessel is one of the earliest stages of atherosclerosis. This thickening is hypothesized to be mainly due to the accumulation of lipid and atherosclerotic plaque formation.

Although cIMT is still a relatively popular measure in current-day research, owing to the fact that it is noninvasive, safe, and relatively inexpensive, there are certain major drawbacks to consider when using this measure as a proxy for atherosclerosis. First, it is known that the intimal layers of blood vessels thicken with age, regardless of the presence of atherosclerosis (Finn et al., 2010). This inevitably means that in some instances vessels will be marked as diseased, while in fact the thickening is not due to atherosclerotic disease, but is an age-related, physiologic change, and likely does not influence health outcomes. Second, the measurement is generally performed on the common carotid artery, because this is relatively easy and provides more reproducible results. However, clinically meaningful atherosclerosis typically occurs at the carotid bifurcation and is less prevalent in the common carotid artery (Howard et al., 1993). Following this, it is not unlikely that the presence of atherosclerosis will be underestimated in a considerable amount of individuals.

Nonetheless, cIMT has contributed considerably to our insight into the role of atherosclerosis in neurodegenerative and vascular brain disease. For example, a thorough meta-analysis of various epidemiologic studies demonstrated that a larger cIMT is associated with a higher risk of stroke (van den Oord et al., 2013). Moreover, it is associated with the presence and severity of cerebral small-vessel disease, cerebral atrophy, and worse cognitive performance (Wendell et al., 2009; Owen et al., 2011; van den Oord et al., 2013).

In sum, carotid ultrasound was among the very first imaging modalities that could be applied to a broad public, and has provided important first insights into the relationship between atherosclerosis and subsequent brain pathology. Yet, given its considerable drawbacks, cIMT has now begun to be overpowered by imaging modalities that provide more accurate information on both the presence and amount of atherosclerosis.

Transcranial ultrasound

Transcranial ultrasound is an interesting imaging method to acquire indirect measures of cerebral vascular disease, which is based on the "Doppler principle." At first purely based on Doppler, but later also in combination with B-mode imaging (transcranial color-coded duplex), transcranial ultrasound has become a frequently used imaging method. Using transcranial ultrasound, one can assess the flow in the intracranial vasculature and detect potential threats in the blood supply of the brain, such as emboli or stenosis (Aaslid et al., 1982). One of the most frequently used measures is the pulsatility index, which is calculated from the difference between the peak systolic velocity and end diastolic velocity within the vessels (Kidwell et al., 2001).

Although transcranial ultrasound is completely noninvasive and inexpensive, its popularity from a population-based perspective is only modest. Although few in number, some population-based studies applied transcranial ultrasound, and found that transcranial ultrasound-based measures were associated with an increased risk of stroke and mortality (Bos et al., 2007; Lopez-Cancio et al., 2012; Portegies et al., 2014). The reasons for the limited use of transcranial ultrasound are likely due to certain issues specific to the application of this technique. The first, and probably most difficult, step in the examination is to find a suitable acoustic window through which the investigation can be performed (Sarkar et al., 2007). The three main windows to get access to the intracranial arteries are the transtemporal window, the transorbital window, and the transforaminal window, for the circle of Willis, the intracranial internal carotid artery, and the basilar artery, respectively. In up to 20% of individuals, there will be difficulties in finding such an acoustic window to perform the examination (Sarkar et al., 2007). This directly demonstrates the high operator dependency of transcranial ultrasound and the need to acquire considerable skills and expertise in order to provide meaningful results.

Vascular calcification

Nonenhanced computed tomography (CT) has become one of the major players in vascular imaging, and has acquired a firm position in the assessment of vascular disease given its unique and unprecedented ability to detect calcium (Agatston et al., 1990; Sangiorgi et al., 1998; Owen et al., 2011). Calcification is a frequent characteristic of the atherosclerotic plaque and has been established as a reliable marker of the total underlying atherosclerotic plaque burden (Rumberger et al., 1995; Sangiorgi et al., 1998). Until recently, the main focus of population-based studies using CT for the assessment of atherosclerosis was restricted to the coronary arteries (Owen et al., 2011). Yet, additional vessel beds can easily be assessed within that same examination, because of the speed of CT. Indeed, the assessment of the amount of vascular calcification in vessels other than the coronary arteries, especially those supplying the brain, is rapidly gaining interest (Allison et al., 2004; Arenillas, 2011; Bos et al., 2014). Visualization and quantification of calcification in the aortic arch, the carotid artery bifurcation, or even the intracranial part of the internal carotid artery provide unique information on the local burden of atherosclerosis. Subsequently, the amount of calcification in these vessels is strongly related to vascular brain pathology (Nandalur et al., 2006; Elias-Smale et al., 2010; Bos et al., 2014), such as white-matter hyperintensities (WMH) or silent infarcts. Moreover, intracranial carotid artery calcification (Fig. 5.2) was demonstrated to be the strongest risk factor for clinical stroke,

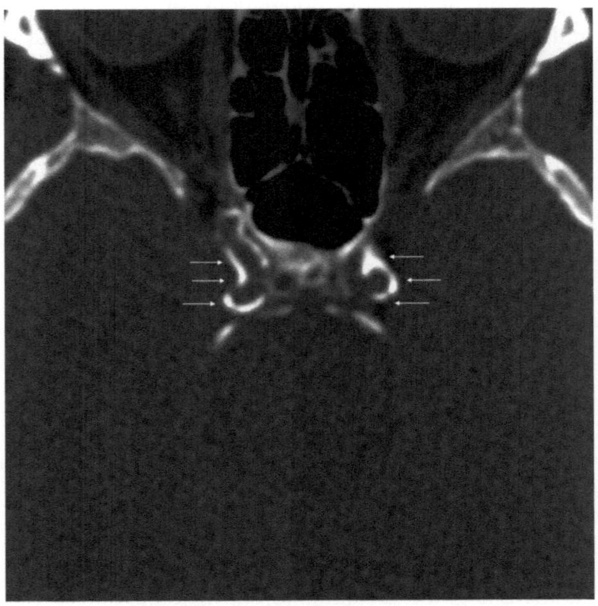

Fig. 5.2. Example of calcification in the intracranial part of the internal carotid artery (arrows).

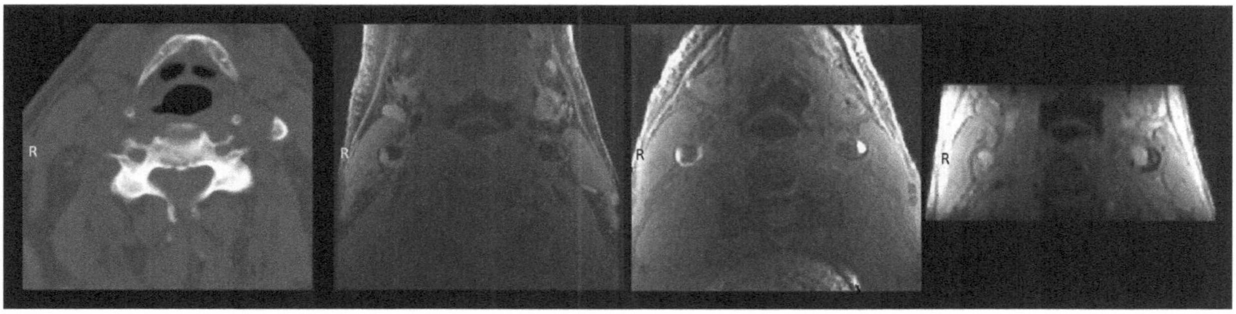

Fig. 5.3. Example of atherosclerotic plaque components on magnetic resonance imaging (MRI) and computed tomography (CT). From left to right: CT image showing a considerable amount of calcification in the left carotid artery; three MRI sequences showing different plaque components, such as lipid (first image), intraplaque hemorrhage (second image), and calcification (third image).

independent of the presence or amount of atherosclerosis at other locations such as the carotid artery bifurcation (Bos et al., 2014). Besides contributing to vascular brain pathology, larger CT-based calcification volumes in these various vessel beds contribute to cerebral atrophy, cognitive decline, and even dementia (Vidal et al., 2010; Bos et al., 2015).

Despite the great strengths of nonenhanced CT in vascular imaging, it is noteworthy that the costs of one CT examination are considerably higher than those of the above-mentioned ultrasound techniques. Moreover, there is the challenge of radiation exposure which may hamper its applicability on a larger scale (Smith-Bindman et al., 2009). Yet, the newer generations of CT scanners, especially dual-source and dual-energy scanners, deposit significantly lower radiation doses (Lambert et al., 2014), and the expectation is that this will decrease even further in the near future.

Noncalcified plaque components: intraplaque hemorrhage and lipid

As depicted above, with nonenhanced CT it is not possible to visualize the complete atherosclerotic plaque, but only the calcified part. Although the assessment of atherosclerotic calcification is an accurate and robust estimate of the underlying plaque burden, information on the so-called noncalcified plaque components may provide additional information on plaque vulnerability (Alsheikh-Ali et al., 2010). This is where the use of magnetic resonance imaging (MRI) comes into play. The use of different contrast weightings such as T1, T2, or proton-density weighting, and the use of optimized sequences allow for a great potential for plaque characterization, mainly with regard to the noncalcified plaque components such as lipid or intraplaque hemorrhage (Cai et al., 2002; Owen et al., 2011). On the contrary, the ability of MRI to detect and quantify calcification is inferior to that of CT, which directly demonstrates the complementary value of these imaging modalities with regard to vascular imaging (Fig. 5.3).

To date, most research on MRI-based atherosclerosis has been on the carotid arteries, and mainly in clinical studies (Ota et al., 2009; Underhill et al., 2010). Population-based studies applying carotid artery MRI examinations are rapidly increasing and correlations between several cardiovascular risk factors and noncalcified plaque components have already been demonstrated (Underhill et al., 2010; van den Bouwhuijsen et al., 2012; Selwaness et al., 2014). Yet, the exact relation between these plaque components and clinical stroke is still to be elucidated.

Besides its great potential with regard to vascular imaging, there are also certain limitations to the use of MRI in the population-based setting. First, compared to the previously discussed imaging modalities, MRI is by far the most expensive examination. Second, there are absolute contraindications for MRI, which include the presence of ferromagnetic material in the body and claustrophobia. Third, MRI examinations are far more time consuming than any other of the above-mentioned imaging modalities.

Brain imaging

Brain imaging has become one of the leading imaging applications in current-day population-based neuroimaging research, and centers around MRI (Mueller et al., 2005; Morris et al., 2009; Ikram et al., 2011). Besides its excellent capacity to visualize brain structure, MRI allows visualization of brain function using the properties of the cerebral circulation. Moreover, this can all be done without exposing individuals to harmful radiation or the administration of contrast material, two important drawbacks which currently restrict broad use of CT in this field (Smith-Bindman et al., 2009). Yet, it should also be noted that there are certain restrictions to the use of MRI. Most important are the absolute contraindications (i.e., pacemakers or certain metal objects in the body) and claustrophobia.

Both structural and functional MRI will be discussed in more detail in the following paragraphs. Other, less commonly used methods of brain imaging in the population-based setting include nuclear imaging techniques. Both positron emission tomography (PET) and single-photon emission computed tomography (SPECT) will also shortly be highlighted. Finally, we will highlight the role of retinal imaging.

Structural brain imaging

An MRI examination of the brain provides a tremendous amount of information on brain structure. Making use of the magnetization principle on the hydrogen atoms in different types of tissue, different-weighted image contrasts can be obtained (Weishaupt et al., 2008), without administering contrast material. Besides these basic sequences, there are many more advanced sequences, such as diffusion-weighted sequences, perfusion sequences, or susceptibility-weighted sequences, which all contribute to our understanding of the brain.

The huge value of the various measures that can be extracted from all of these MRI-based sequences (which are discussed in detail in the next section) becomes evident from the current body of literature. Especially with regard to common clinically relevant brain disorders such as strokes, cognitive disorders, and dementia, brain imaging has provided unprecedented insight into the etiology and pathophysiology of these diseases (Jack, 2012; Jack and Holtzman, 2013). Hippocampal volume, for example, is established as a major marker for cognitive deterioration and dementia (den Heijer et al., 2010; Jack, 2012). Another example includes the presence and amount of WMH and lacunae, which proved to be important markers of subsequent clinical stroke and dementia (Vermeer et al., 2003; DeCarli, 2006; Debette and Markus, 2010; O'Sullivan, 2010). There is also increasing interest in the use of MRI-derived brain markers in clinical prediction models for both cerebrovascular and neurodegenerative diseases. Early and subtle changes in the brain as detected by MRI might identify persons at high risk of such diseases at an early stage. Against the background of the current paradigm shift from cure to prevention in medicine, this is a key element in the development of effective preventive strategies.

Functional brain imaging

A relatively novel feature of MRI is the possibility to perform noninvasive functional imaging (fMRI). The magnetic properties of oxygenated and deoxygenated hemoglobin allow fMRI imaging to produce an overview of the hemodynamics in the brain, which is directly linked to brain activity (Buxton, 2002). Although this technique has many other applications, one main field of research in fMRI centers around the functionality of the brain in neurodegenerative disease. Its use in larger population-based studies is still limited but will likely rapidly increase in the near future given the fact that this technique provides information on brain function in a completely noninvasive way.

The use of fMRI is already very popular in neuroscience studies on clinical patients with dementia. Within this field, fMRI was primarily used to compare brain activity patterns in individuals during the performance of various simple tasks in the MRI scanner, such as visual tasks on face or word recognition (Lustig et al., 2003; Sperling et al., 2003). Later, it also became apparent that insights could be derived from the synchronicity of localized activation patterns at rest (i.e., when the subject in the scanner does not perform any specific tasks), so-called resting-state networks (Fig. 5.4).

One of the most prominent and investigated of these networks is the default-mode network (Raichle and Snyder, 2007). Although very intriguing, its exact function has not been fully elucidated. The general hypothesis is that this network represents underlying physiologic processes in the brain that are unrelated to any particular thought or idea. Various clinical studies have demonstrated a decrease in default-mode network activity and connectivity in persons with dementia, and cognitive decline (Simic et al., 2014). Data from population-based cohorts will provide a wealth of information on the normal variation of these resting-state networks and are expected to provide in the near future information on specific changes in these networks that may represent a predisposition to subsequent neurodegenerative disease. Yet, several important challenges with regard to the large-scale application of resting-state- and task-related fMRI in population-based

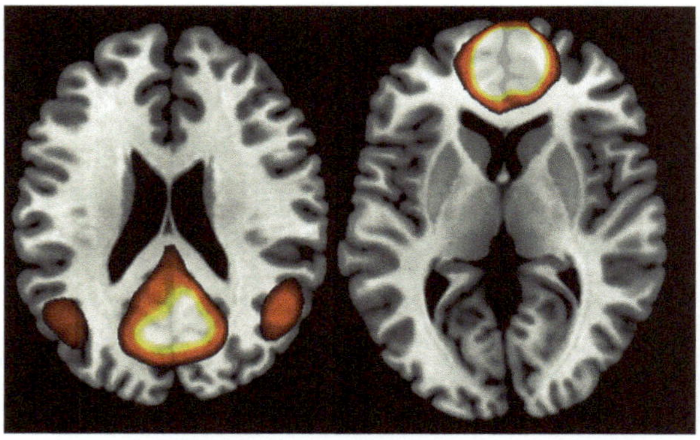

Fig. 5.4. Example of the default-mode network as identified with resting-state functional magnetic resonance imaging.

studies need to be addressed. First, in general, the addition of MRI sequences to an existing imaging protocol is restricted by the duration of the session, especially in population-based studies that involve generally healthy participants. Specifically, for resting-state fMRI, at least 6 minutes of data acquisition are necessary to obtain meaningful data (van Dijk et al., 2011), which may exceed the time that such studies allow for healthy participants to spend in the scanner. Second, more specific to the mechanism underlying fMRI is the assumption that signal change measured is related to changes in blood flow and blood oxygenation. This directly highlights the direct dependency of the signal change on changes in the vascular system. Examples include heart rate variability, stenosis in important afferent vessels to the brain, but also respiration (Kotsoni et al., 2006). Third, and specific to task-related fMRI, concerns the differences in the performance of tasks between subjects. It is very important that these differences are not due to reasons that are unrelated to the underlying research question. In other words, the tasks should be carefully chosen to be understood and similarly challenging for all study participants. It is, for example, known that more difficult or challenging tasks may alter the blood flow and pattern of activation (Bookheimer et al., 1996). Especially, in the population-based setting this might be challenging given the broad variation across the study subjects.

Nuclear brain imaging

Especially in studies on dementia and cognitive impairment there is a role for nuclear brain imaging, specifically PET or SPECT (McConathy and Sheline, 2015). The general idea behind these methods is to obtain information on brain function and metabolism by using radioactive tracers whose *in vivo* uptake patterns indicate and quantify the presence or activity of specific biochemical processes (McConathy and Sheline, 2015). A novel approach in nuclear imaging is the combination of PET with MRI (Catana et al., 2012). Using the excellent capacity of MRI to characterize brain tissue in combination with the sensitivity of PET for brain metabolism seems promising to further improve on nuclear brain imaging (Catana et al., 2012).

From the perspective of neurodegenerative brain disease, PET techniques have mainly been applied in clinical studies on dementia patients, and provided important insights into the etiology of dementia and cognitive decline. Specifically, the direct visualization of amyloid-beta deposits using a radioactive substance called Pittsburgh compound B (PIB) has been a major milestone in nuclear brain imaging (Fig. 5.5) (Zhang et al., 2014). Using PIB it is possible to differentiate between different types of dementia, specifically Alzheimer's disease and frontotemporal dementia (Svedberg et al., 2012; Zhang et al., 2014), something that was only possible before using histopathology of postmortem brains.

Although providing unique information with regard to brain metabolism, from a population-based perspective, it is unlikely that these techniques will be applied in the near future, not in the last place because of the fact that a radioactive tracer is required which is injected directly into the vascular system of the subject. This, in combination with the high costs and the radiation exposure, hampers the application on a large scale.

Retinal imaging

Another imaging technique which has increasingly been used to derive biomarkers for brain pathology is retinal imaging. Using the eye as "window" to the brain, various proxies for brain pathology may be investigated by studying the retina and its vasculature. The two most commonly used retinal imaging techniques are fundus photography and optical coherence tomography (OCT). Both are potentially interesting with regard to providing novel markers for brain pathology.

The concept of the fundus photography (Fig. 5.6) originates from 1910 (Gullstrand, 1910), and is to date still the most frequently used tool to image the retina for the detection of various diseases, such as diabetic retinopathy, glaucoma, or age-related macular degeneration. Yet, fundus photography may also be used to obtain information on the brain. It was found that the blood vessels of the retina share various characteristics with the cerebral vessels, on the level of anatomy and physiology. This follows from the fact that the retina is an extension of the diencephalon and has a blood–retinal barrier which serves a comparable function to that of the blood–brain barrier (Baker et al., 2008). In this light, it has been argued that changes in the vessels of the retina reflect similar changes in the cerebral vessels. Indeed, measures derived from the retinal vessels, such as arteriolar or venular diameters, or the presence of retinopathy, have been related to a higher prevalence of stroke and cerebral white-matter lesions (Baker et al., 2008). In addition, increasing evidence supports an association between retinal vascular changes and dementia, especially Alzheimer's disease (Cheung et al., 2014), but further research on this topic is needed.

Another, relatively novel, retinal imaging technique is OCT (Huang et al., 1991). OCT is a noninvasive medical imaging method that creates *in vivo* cross-sectional images of the biologic tissue of the eye (Fig. 5.7). The technique was developed in 1991 (Huang et al., 1991),

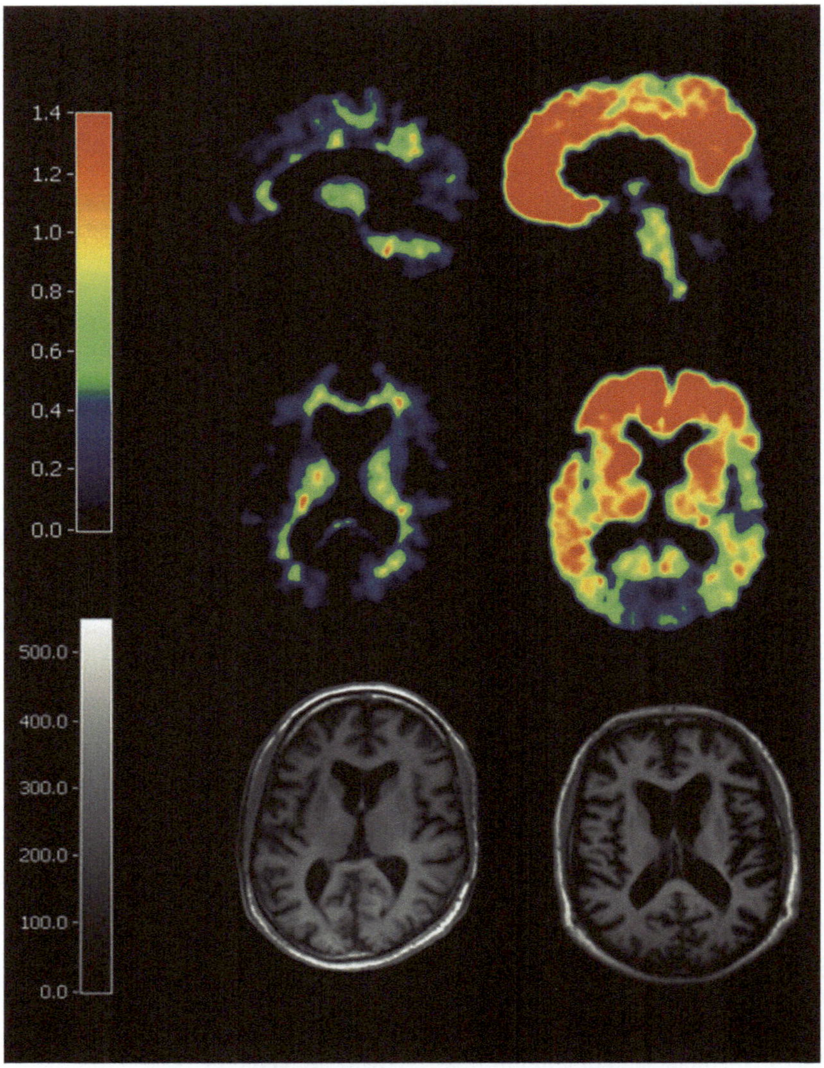

Fig. 5.5. Example of a positron emission tomography examination allowing direct visualization of amyloid-beta deposits using a radioactive substance called Pittsburgh compound B (PIB).

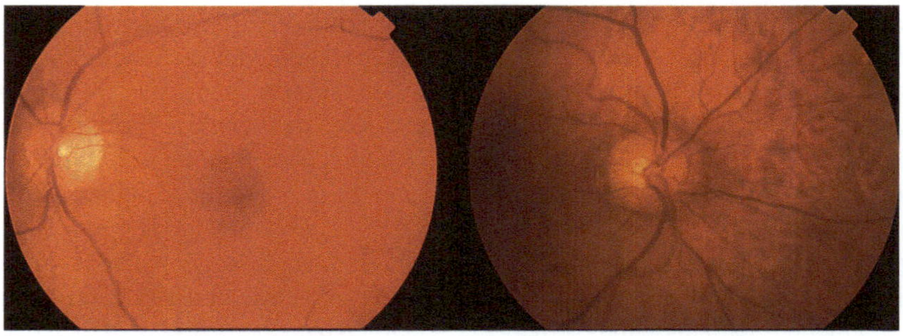

Fig. 5.6. Example of fundus photography.

and is based on interferometry of light to detect structures at the micrometer scale. It is widely used in ophthalmology to capture corneal and retinal abnormalities. Its use in research on brain pathology to date is largely limited to studies on multiple sclerosis (Frohman et al., 2008), but may have the potential to provide us with novel markers of vascular, and possibly also neurodegenerative, brain disease.

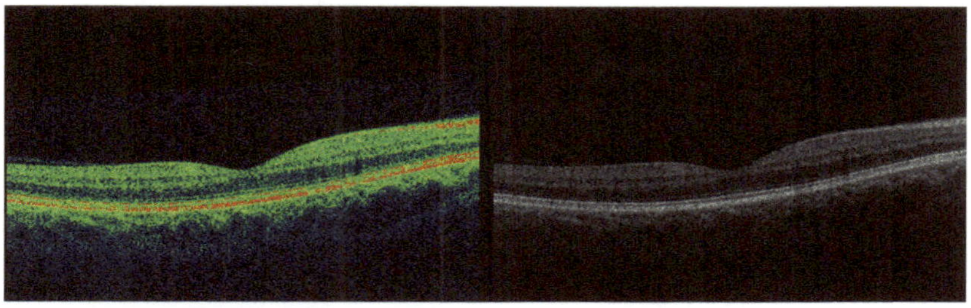

Fig. 5.7. Optical coherence tomography examination.

IMAGE INTERPRETATION

When we acquire medical images, we do so with the aim of increasing understanding of the anatomy, physiology, or pathophysiology of the person under investigation. However this understanding does not come automatically with the acquisition of the image itself. Rather, it requires interpretation of the image in one way or another. This interpretation can be made by an expert observer or by a dedicated computer algorithm, and can be qualitative or quantitative. The same holds for imaging in neuroepidemiology, and the following sections will describe the types of interpretation most commonly applied.

Visual interpretation

Visual interpretation can be an advantageous approach to extracting relevant characteristics from neurologic images for multiple reasons. First and foremost, it is most closely related to clinical routine, meaning that clinical expertise and visual rating scales can directly be used to interpret images, but also that possible findings can easily be applied in clinical practice. Second, visual interpretation does not require advanced computing knowledge or software. Third, it may be more robust in complex visual traits or lesions. However, visual interpretation suffers from inter- and intrarater variability, and these factors therefore need to be controlled or at least evaluated. More importantly, typical cohort sizes in neuroepidemiologic studies which are in the order of hundreds or thousands may quickly escalate the time required for visual interpretation beyond the point of feasibility. In this case, automated or semiautomated algorithms need to be developed, tested, and applied.

When a discernable image feature becomes the focus of research, this potential imaging biomarker is usually first interpreted visually. Over time, when the feature becomes more established, algorithms to automatically interpret the feature will be proposed. The delay by which algorithms can become available is caused in part by the long trajectories associated with their development and validation, but also by the dependence on established definitions for visual interpretation. Ongoing algorithm refinement and increasing ease of use may then eventually lead the automated interpretations to become established imaging biomarkers.

Visual interpretation is or has been used in a number of research settings, a number of which will be discussed.

White-matter hyperintensities

These lesions are typically identified as hyperintense regions on T2-weighted or T2 fluid-attenuated inversion recovery (T2-FLAIR) MRI sequences. Leukoariosis lesions, also called white-matter lesions, are generally thought to have an ischemic origin (Pantoni and Garcia, 1997) and have a high prevalence in the elderly. While they generally manifest without clinical symptoms, they have been associated with an increased risk of stroke, dementia, and death (Debette and Markus, 2010). Several visual rating scales have been developed to characterize the burden of disease (Fazekas et al., 1987; Scheltens et al., 1993; Wahlund et al., 2001), as well as progression scales for longitudinal investigations (Schmidt et al., 1999; Prins et al., 2004a). Use of these rating scales in population-based settings has been instrumental in the past decades in establishing an important role of vascular lesions in stroke, cognitive decline, and dementia (Prins et al., 2004b, 2005). Yet, the interobserver variability of these visual scales and their limited ability to capture biologic variation and progression over time have led to automated algorithms winning favor (see further below).

Brain atrophy

Atrophy in specific brain regions has been attributed to multiple diseases, including medial-temporal atrophy in Alzheimer's disease, and frontotemporal atrophy in frontotemporal dementia. Visual interpretation of specific regional atrophy patterns has proven useful (Scheltens et al., 1992; Golomb et al., 1993; Koedam et al., 2011), but the ability of the human eye to accurately identify volume change is limited and visual rating of atrophy therefore remains crude. We will see that in order to overcome this limitation, a quantification step is needed.

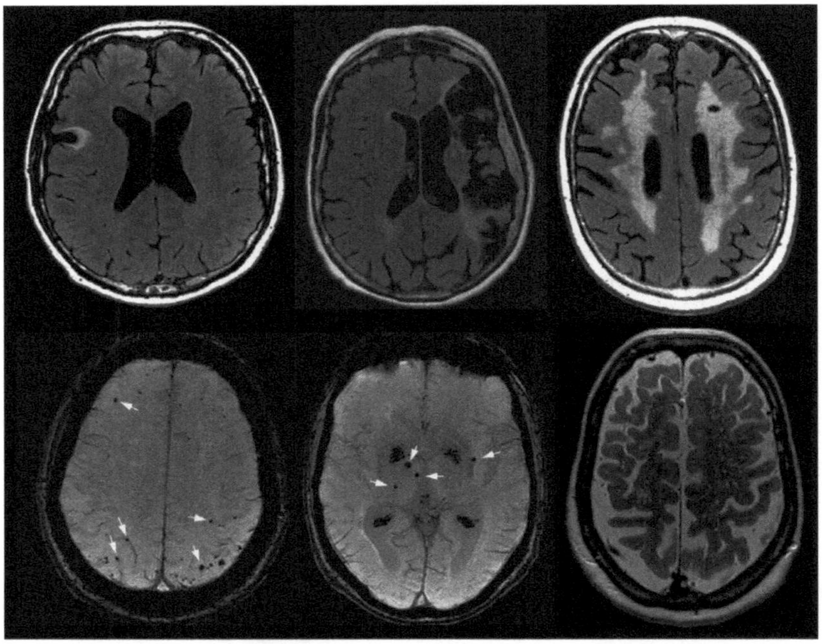

Fig. 5.8. Examples of visually identified lesions. From left to right, top to bottom: small cortical infarct, large cortical infarct, lacunar infarct, lobar microbleeds (arrows), deep microbleeds (arrows), dilated Virchow–Robin spaces (enlarged perivascular spaces).

Other visual findings

Multiple other discrete findings in the brain have been investigated. Isolated and less prevalent findings, such as cerebral microbleeds, cortical macro- and microinfarcts and lacunar infarcts, shown in Figure 5.8, have been interpreted visually (Greenberg et al., 2009; van Veluw et al., 2015a). These findings are less prevalent, and may often be represented numerically, which obviates the need for a visual rating scale. Yet, there is increasing evidence that anatomic distribution of these lesions contains important information, which would call for either manual labeling or automated detection.

Quantitative imaging

Central to quantitative imaging is the ability to extract quantitative measures from (medical) images. For the majority of imaging modalities, including the examples discussed above, additional steps are required in order to extract quantitative measures. Some imaging modalities, however, are intrinsically quantitative. This section will also discuss approaches for the statistical analysis of neuroepidemiologic imaging studies.

Quantitative measurements on nonquantitative imaging modalities

In order to extract quantitative measurements from imaging modalities that are not intrinsically quantitative there is a wide range of technologies available. These techniques can focus on volumetric estimates, but also on other geometric properties, such as lengths and surface areas, and also on temporal geometric aspects, including the quantification of movement. Approaches can rely on human annotation, full automation, or a mixture of these. Figure 5.9 gives some examples of automated quantification techniques.

Brain atrophy: structure segmentation

To sensitively detect minute volumetric brain changes, an explicit volumetric segmentation is needed. For smaller brain regions such as the hippocampus, this might be feasible to do by hand. Automated brain structure segmentation approaches have also been developed.

One of the most straightforward ways to segment an anatomic structure is by using an annotated brain atlas. Anatomic information, encoded in a manually segmented subject image or template, is transferred to the subject image using image registration (Maintz and Viergever, 1998), an approach graphically depicted in Figure 5.10. Image registration is used to find an affine or more complicated nonlinear transformation that relates the structural images with a transformation or a deformation field. After the transformation has been computed on the structural images, it can be used to transform the label mask image (of the annotated brain structure) to the subject under investigation. The position

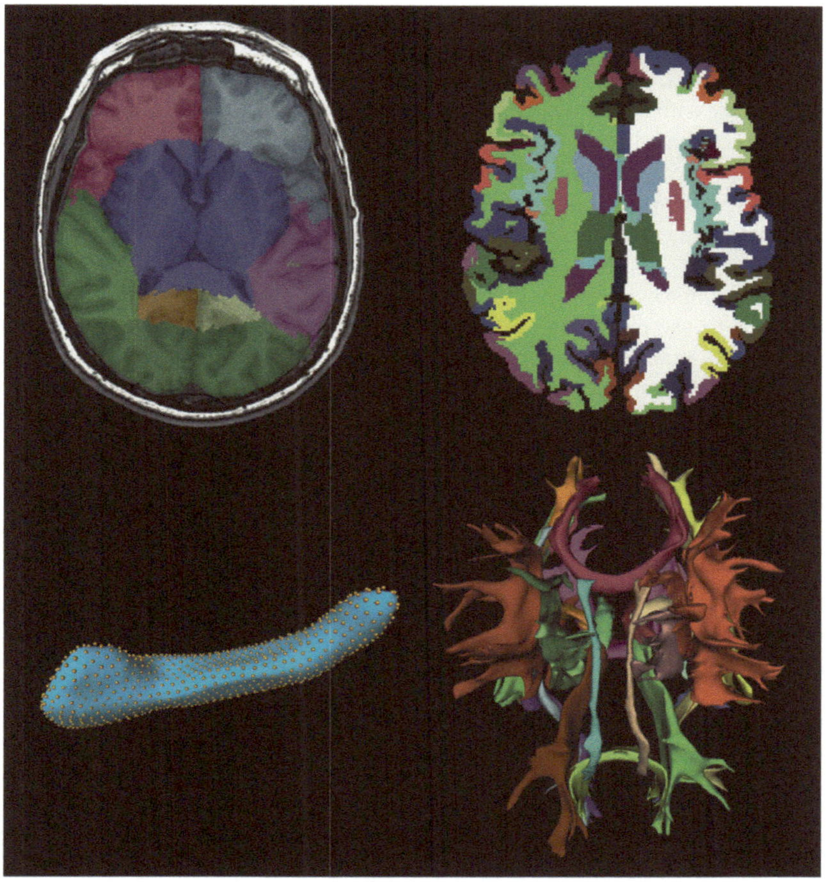

Fig. 5.9. Examples of automated quantification techniques. From left to right, top to bottom: atlas-based lobar segmentation, FreeSurfer-based cortical parcelation and subcortical structure segmentation, hippocampal shape model, diffusion tractography-based white-matter structure segmentation.

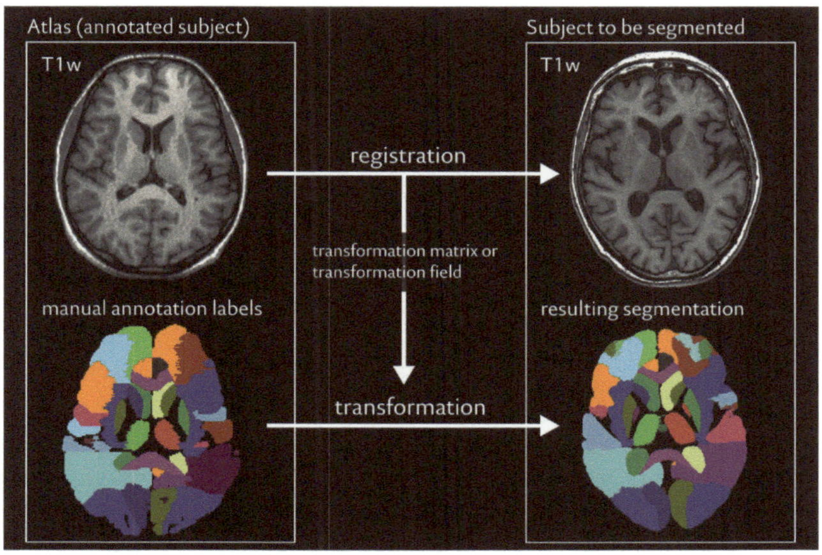

Fig. 5.10. Atlas-based image segmentation. Image annotation labels are transformed from atlas space to the individual subject space with the use of image registration, which establishes anatomic correspondence between both subjects.

of the atlas-based structure label is then our estimate for the position of the subject-specific structure. While easy to implement, the atlas-based image segmentation approach has difficulty coping with anatomic variability, as it relies on the anatomy of the atlas image correctly representing the subject anatomy. This variability can be accounted for by using a probabilistic atlas, as is used in the subcortical structure segmentation in FreeSurfer (Fischl et al., 2002).

An important strategy to improve the robustness of atlas-based segmentation is to use more than one atlas image. By registering multiple annotated atlases to each subject, failed registrations or nonrepresentative atlases have less influence on the final segmentation result. Multiatlas segmentation approaches can use a simple averaging of the transformed label images, a fusion approach referred to as majority voting, but the development of more refined label fusion strategies is an active area of research. More evolved approaches attempt to take the uncertainty of each of the atlases into account, such as the simultaneous truth and performance level estimation approach (Warfield et al., 2004), as well as selection of the best-matching atlases (Aljabar et al., 2009), and specific attention to the atlas accuracy at the local level (Jorge Cardoso et al., 2013). Knowledge based on one or more atlases can also be used as an anatomic prior when fused with other (e.g., appearance) features (van der Lijn et al., 2012).

Atlas-based segmentation approaches suffer in the presence of lesions, as these lesions will lead to misrepresentation of the subject anatomy, and are weak in segmenting lesions themselves for the same reason. But overall, the use of atlas-based information for brain structure segmentation has produced highly accurate segmentation approaches (Mendrik et al., 2015).

Brain atrophy: tissue segmentation

Segmenting gray matter, white matter, cerebrospinal fluid, and nonbrain tissues is pertinent to the investigation of general brain atrophy, and is often a prerequisite for tissue-specific studies. The effort required to manually perform a full-brain segmentation of high-resolution structural imaging is however almost unreasonable. Approaches for fully automated tissue segmentation, each with its own merit, have therefore been developed.

Many structural MR images are prone to intensity inhomogeneities over the field of view. In order to perform automated tissue segmentation, these inhomogeneities need to either be compensated for or the segmentation approach needs to be robust against them. Compensation of inhomogeneities is often performed as a preprocessing step for the tissue segmentation (Sled et al., 1998), but may also form an integral part of the segmentation process. As the bias field and the tissue composition together drive the appearance of the structural images, the latter approach enables both unknowns to be estimated in one process (Zhang et al., 2001; Ashburner and Friston, 2005).

To address specific concerns, multiple tissue segmentation approaches have been introduced. Approaches that operate solely on image intensities make few assumptions and are therefore relatively robust against lesions (Cocosco et al., 2003; Vrooman et al., 2007). Prior anatomic information can be incorporated in the form of tissue atlases to initialize the segmentation (Van Leemput et al., 1999) or to enhance robustness, as in the unified segmentation approach (Ashburner and Friston, 2005). Information on tissue topology can also inform the segmentation, either locally, e.g., via the use of Markov random fields (Van Leemput et al., 1999; Zhang et al., 2001) or by modeling the complete tissue boundary, as is done in FreeSurfer (Fischl and Dale, 2000).

An example image of multiple segmentation results on the same dataset is shown in Figure 5.11.

Lesion segmentation

Automatic identification of lesions is a very different task from the previously described segmentations. Lesions may or may not be present, while tissues and brain structures can be assumed to be present. Furthermore, lesions are subject to large variability, both in type and thereby in appearance, as well as in location. This variability is hard to deal with in automated detection. Existing approaches are therefore limited to relatively prevalent lesion types that are easy to distinguish and that have a consistent appearance.

Automated approaches have been developed for the segmentation of WMH. Many of these rely on FLAIR image sequences (Anbeek et al., 2004; de Boer et al., 2009), but with the use of regularization and prior knowledge on spatial distribution of WMH, approaches including FreeSurfer can do without (Fischl et al., 2002; Schwarz et al., 2009).

Approaches for the automated detection of cerebral microbleeds have also been proposed, although methods still require more or less involved manual input to achieve satisfactory results (Seghier et al., 2011; Kuijf et al., 2013).

Quantitative interpretation of nonquantitative images

While many MR sequences probe (quantifiable) physical tissue properties, most of the data acquired is only weighted by the physical property, and as such cannot directly be interpreted quantitatively. However, even with nonquantitative images, it is possible to perform a normalization to transform the image in a quantitative scale that allows direct comparison across subjects. This

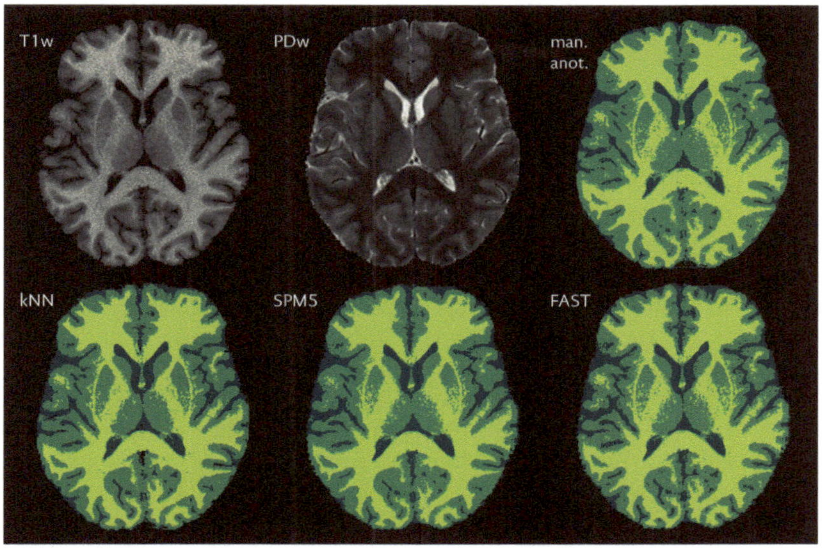

Fig. 5.11. Examples of automated tissue segmentation algorithms. From left to right, top to bottom: T1-weighted imaging, proton density-weighted imaging, manual annotation of tissue classes (cerebrospinal fluid, gray matter, and white matter), k-nearest neighbor classifier (Cocosco et al., 2003), SPM5 segmentation (Ashburner and Friston, 2005), FAST segmentation (Zhang et al., 2001). (Data courtesy of Dr. Renske de Boer.)

approach has been used for the analysis of tissue in the vicinity of WMH (Maillard et al., 2013) and in tissue later converting to WMH (de Groot et al., 2013).

Quantitative imaging modalities

Imaging modalities that are intrinsically quantitative permit direct quantitative analysis of tissue properties. In the scope of neuroepidemiology, a number of quantitative MRI contrasts are especially relevant. CT, while intrinsically quantitative (by measuring Hounsfield units), is less relevant as the radiation exposure is often hard to justify for observational studies. Multiple MR contrasts do however provide useful information in the context of neuroepidemiologic analysis. Diffusion tensor imaging (DTI), for example, provides quantitative information on microstructural tissue properties (Beaulieu, 2002). Metrics provided by DTI include the mean diffusivity of the tissue, and the fractional anisotropy, which measures the orientational dependence of the diffusion signal and encodes the degree of microstructural organization of the tissue. Epidemiologic studies have used DTI, for example, to investigate tissue changes with aging (Sexton et al., 2014), cognitive performance (Kantarci et al., 2011), and vascular risk factors such as smoking (Gons et al., 2011). Relevant information may also be encoded in the tissue T1 and T2 relaxation times. Prolongation of the T1 relaxation time has been identified in temporal white matter (Raz et al., 1990). Quantitative MR contrasts are actively researched, and probes to biophysical quantities such as blood flow will in time become relevant in neuroepidemiology.

Population statistics

The general paradigm for imaging analyses in cohorts is to acquire quantitative measurements for each participant, and then to aggregate these measurements into population statistics or group analyses. The aggregation of measurements depends on the type of image data available and on the research question posed. In general, though, quantitative images may be analyzed in a voxelwise fashion, projected on a parameterization, or aggregated over a volume. In these group analyses, statistical power is directly influenced by the measurement precision, vesting great significance in the ability to extract precise and accurate quantitative measurements. The next sections deal with two main analysis paradigms for neuroepidemiologic studies: voxel-based analysis and region-of-interest analysis. Analysis approaches that are more specific to dedicated hypotheses are not touched upon.

Voxel-based analysis

In order to investigate effects on the brain that are thought to occur locally, voxel-based analyses offer fast and automated insight. The general paradigm in voxel-based analyses is to establish anatomic correspondence, and then to compute voxelwise statistics in a reference space. To establish anatomic correspondence, image registration is commonly used. Often-used examples include voxel-based morphometry for gray matter, and voxel-based analysis, which includes the tract-based spatial statistics framework for white-matter diffusion analysis. Voxel-based analyses are especially relevant for exploratory analyses, as they offer insight into the localization of effects and thereby generate new hypotheses for further investigations.

Optimized voxel-based morphometry (Good et al., 2001) brings tissue segmentations to standard space using image registration, and takes into account the local compression and expansion needed to align each subject to standard space. After this preprocessing, for every gray-matter voxel in standard space, a general hypothesis can be tested, e.g., does gray-matter density reduce with age? Clustering of statistics can be adopted to increase the sensitivity for finding regions expressing the same behavior in relation to the hypothesis being tested.

Voxel-based analyses in diffusion MRI similarly transform the diffusion metric images (e.g., fractional anisotropy) to standard space using image registration. As these metric images are quantitative, however, they allow for direct statistical inference. Voxel-based analysis of diffusion data is fast and straightforward to implement, but a number of concerns have been identified with the approach. One of the concerns is the dependence of the approach on an *a priori* hypothesized extent of the finding (Jones et al., 2005). Another concern arises in the periphery of white-matter tracts, where atrophy and microstructural change is difficult to disambiguate. The tract-based spatial statistics approach (Smith et al., 2006) was introduced to mitigate these concerns by constraining the analysis to the central voxels in the white-matter tracts. This approach has found widespread adoption, owing to the ease of operation and the sensitivity to microstructural change.

Region-of-interest analysis

Anatomic knowledge can be incorporated in the analysis when a more specific regional effect is hypothesized. Sensitivity to detect change can be increased by aggregating measurements over a specific volume: firstly, by providing more robust measurements, and secondly, by reducing the number of tests performed and thereby reducing the penalty incurred in the correction for multiple testing.

Region-of-interest analyses can be used to compare volumes of anatomic structures obtained by either manual or automated delineations. Quantitative imaging modalities on the other hand offer direct analysis on the measurements aggregated over the region of interest. The region of interest can then be either a segmented neuroanatomic structure, or a (manually) placed volume such as a pillbox inside a larger homogeneous area, e.g. the corpus callosum.

CHALLENGES

Incidental findings

Though population imaging is usually not aimed at discovering underlying medical conditions (contrary to medical screening), the images acquired may reveal an abnormality unknown to the participating subject but with potential clinical relevance. These incidental findings have been described with varying frequencies in neuroimaging studies, depending on study population and imaging parameters (Illes et al., 2006; Vernooij et al., 2007; Lehman et al., 2009; Morris et al., 2009; Hartwigsen et al., 2010; Gur et al., 2013; Sandeman et al., 2013; Bos et al., 2016), but in the adult population the vast majority comprise benign brain tumors like meningiomas and unruptured intracranial aneurysms (Vernooij et al., 2007; Bos et al., 2016). A wide range of findings has been described, and depending on the imaging technique used, could encompass everything from (treatable) pituitary macroadenomas to (untreatable) high-grade malignant tumors (Fig. 5.12). This poses several challenges that are still largely unresolved, but should be the focus of research in the coming decades and should also be a point of consideration for researchers starting a population imaging study.

Firstly, there are no clear guidelines for the management of incidental findings on brain imaging research. To start, there is no clear consensus on whether images should be read, if at all, by technicians, students, (trained) researchers, or relevant medical specialists. Associated costs and logistical difficulties, e.g., no access to specialist readings, with these various scenarios may put a great burden on researchers. Further, protocols for disclosure of findings to research participants are not uniform, not only regarding what findings are judged clinically relevant, but also how and by whom participants should be informed and what further medical management should be. In large part, this is hampered by a lack of information on the natural course of these findings. Also, scarce information is available on the psychologic harm that is inflicted on presumed healthy persons when confronted with an unexpected finding (Schmidt et al., 2012). There is a growing awareness among researchers that incidental findings are an unavoidable byproduct of population imaging research and that proper management regarding informed consent, protocols for image handling, and management of detected findings should be part of any study protocol. This is paralleled by the development of multiple ethical frameworks, all of which have in common that a contingency plan should be in place for the management of incidental findings (Illes et al., 2006; Wolf et al., 2008). Also several population-based imaging studies are currently addressing participant expectations, informed consent procedures,

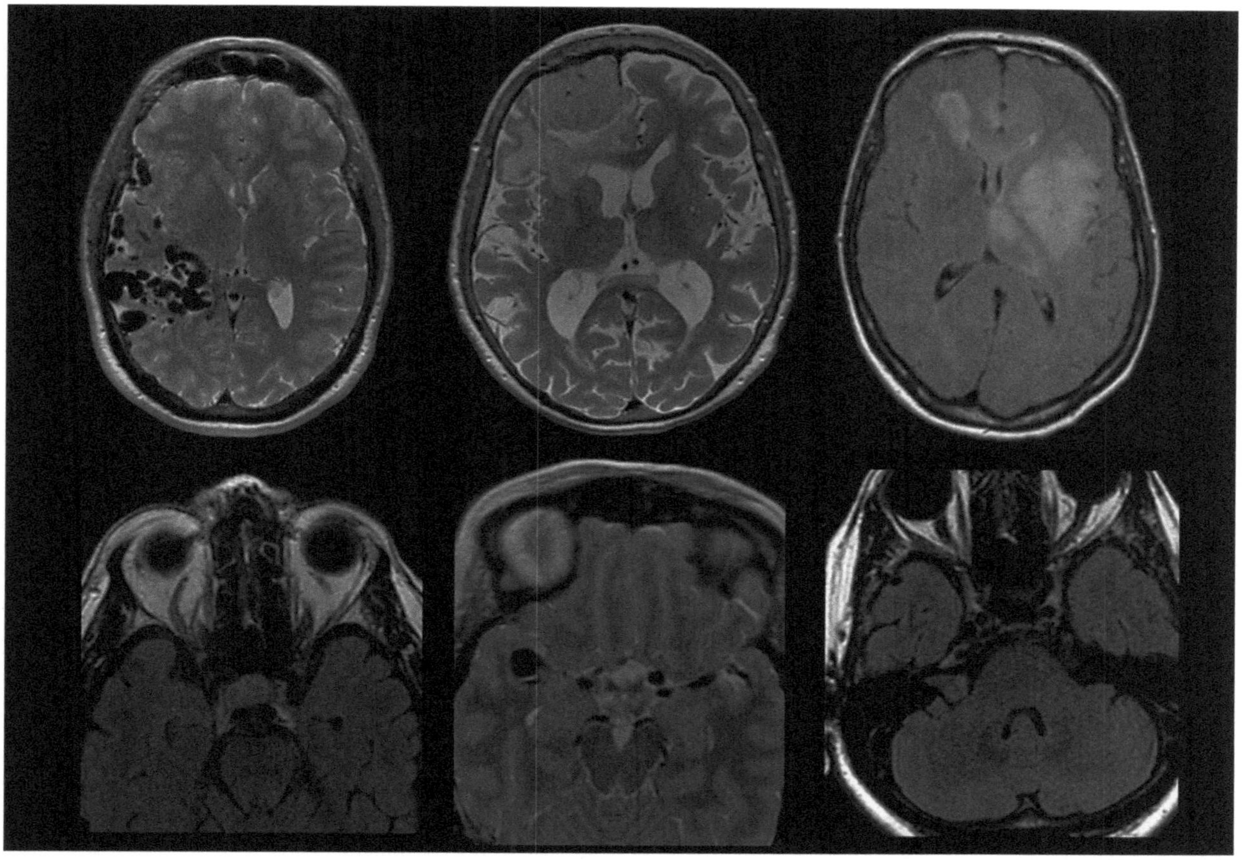

Fig. 5.12. Pictorial examples of incidental findings discovered in a population-based setting of middle-aged and elderly subjects. From left to right, top to bottom: arteriovenous malformations, meningioma, glioma, pituitary tumor, aneurysm, vestibular schwannoma.

and psychologic stress caused by incidental findings (Hegenscheid et al., 2009; Petersen et al., 2013; Bamberg et al., 2015; Hofman et al., 2015), which will hopefully yield new evidence that can bring important progress in this area.

Challenges of longitudinal data acquisition

Costs of image acquisition and image analyses

An important topic, already briefly highlighted, is the costs associated with the application of imaging on a population level.

Overall, the costs associated with population-based imaging generally comprise the following components: acquisition of a dedicated research scanner or costs of the use of a hospital-based scanner, development and continuation of a well-functioning research infrastructure, including personnel, storage facilities of large-scale imaging data, and the analyses of the imaging (i.e., image processing). Still, these overall costs can vary considerably, depending on the imaging modality (e.g., MRI versus CT versus ultrasound), and on the planned number of study participants, which directly influence the costs for infrastructure, data storage, and data analyses.

To date, several long-standing prospective population-based studies, for example, the Framingham Study, the Rotterdam Study, or the AGES-Reykjavik Study (Dawber et al., 1951; Harris et al., 2007; Hofman et al., 2015), and some newer initiatives such as the UK biobank (Petersen et al., 2013) and German National Cohort (Bamberg et al., 2015), have firmly implemented imaging in the core study protocol. Yet, as mentioned above, setting up such an imaging infrastructure is expensive and poses a great limitation for smaller studies or for the development of population-based imaging studies in developing countries.

Reproducibility of acquisition

Given the increase in population-based imaging studies, with the accompanying huge increase in imaging data and computer algorithms to analyze these data, there has been a movement towards the development of standards for the acquisition of imaging data and computer algorithms. These standards or guidelines may be used to monitor and guard reproducibility issues (Kessler et al., 2015; Raunig et al., 2015).

Reproducibility is influenced by multiple issues that may be divided into two main categories, namely factors related to the acquisition of the imaging examinations, and factors related to the processing of the imaging data.

When considering image acquisition, there are various factors that should be taken into account. First, stability of imaging hardware is an important and difficult point to consider. The rapid advances in imaging techniques, especially in MRI, are astonishing, but also provide difficulties for population-based research with respect to interscanner reproducibility. Especially from a longitudinal point of view, studies in which imaging is applied at multiple time points over a relatively long period of time should consider that upgrading to the latest technologic advances will reduce the comparability of images over time. Yet, by not keeping up with the technologic advances at some point, the imaging techniques used may become obsolete, which may hamper the credibility of the research. The above-mentioned standards may provide guidance in this difficult dilemma (Kessler et al., 2015; Raunig et al., 2015). It is important to note that many of these considerations apply both to the imaging hardware, as well as to the scanner software.

Second, the design of the scan protocol is vital for adequate and reproducible data acquisition. One of the most important issues to keep in mind is to restrict the length of the scan protocol. Longer scan protocols will lead to an increase in motion artefacts due to the individual getting uncomfortable inside the scanner, directly affecting the scan quality. Moreover, as for changes in hardware, as discussed above, changes in the scan protocol may affect the reproducibility and comparability of different scans.

Provenance of image-processing algorithms

Advances in image-processing algorithms are rapidly made. Yet, in general, processing should consist of standardized pipelines. The imaging processing tools in these pipelines can consist of commercially available software packages (see Chapter 3 for various examples), but also of validated, custom-made software packages, developed by individual study centers. It is however important to make sure that the results of image analysis approaches remain reproducible over time. This is important both to warrant the reproducibility of published findings, but also to make sure that future follow-up examinations can be processed in the same way as the examinations already acquired. Organizational and infrastructural conditions need to be met in order to provide continued reproducibility of analysis approaches.

Image analysis research continues to provide more, and more accurate biomarkers that are relevant in population imaging studies, as discussed above. Software algorithms, just like protocols, are updated from time to time, which introduces incompatibilities. To maintain reproducibility, algorithms need to be versioned ("frozen") and for each dataset it should be recorded what the exact version of each algorithm was that was applied. Unfortunately, this goes much further than recording just the algorithm version as most software algorithms depend on many packages, each of which in turn is also versioned. It is an important organizational task to have version control for each biomarker (algorithm) and to record, for each biomarker and each dataset, the complete version information of all relevant software used in extracting that biomarker, i.e., the full data provenance.

The computational infrastructure also needs to support future reproducibility. Mainly this means support for algorithms with multiple versions, and support in turn for multiple versions of the software dependencies used by these algorithms. This can be accomplished in multiple ways, ranging from modular systems where consecutive versions coexist (creating an ever-growing library) to fully isolated virtual computers dedicated to producing a single biomarker. It is worth noting that this is still an area of active development. But as these approaches all require some degree of architectural choices upfront, it is important to consider this aspect when designing a study.

Big data challenges

The bulk of neuroimaging data in neuroepidemiologic studies present "big data" challenges in multiple domains such as in data handling, data management, and data analysis.

Data management deals with the many different aspects of data quality, security, and availability. Focusing on security and availability, data management procedures need to allow authorized-only access, while providing safeguard from alterations. Procedures need to cope with the diverse nature of the various sources of data in large investigations, which besides imaging may include genetic or proteomic data. Access control is most easily facilitated in a centralized storage with access control, but could also use fine-grained encryption in combination with a key management system.

In the design of data management procedures, it is important to keep data handling in mind. As such, the two concerns require an integrated approach. When investigators access imaging data for analysis, multiple participants' images are often analyzed simultaneously using compute clusters. This approach is necessary to make computationally intensive methods feasible, but it does require high-bandwidth data storage and network. At the same time, the number of individual files acquired as raw, two-dimensional imaging data may easily run into the tens of millions for a large imaging study. For these reasons, imaging studies put considerable demands on the storage infrastructure, which, combined with demands for data security and availability, requires careful upfront planning with specialized parties in order for demands to be met. Data management policies may however allow for *ad hoc* solutions such as securely sending anonymized data to existing compute facilities or cloud services via either (temporary) high-speed connections or via physical drives or tapes, which may alleviate concerns for specific computationally intensive investigations.

Large neuroimaging studies also pose severe challenges for appropriate statistical analysis. These challenges mostly originate in the dependency of resource requirements on the number of participants in a single analysis. Specifically, analyses that include twice the number of participants may require twice the amount of memory or compute power. In some applications, this scaling may even be quadratic, effectively limiting the number of participants that may be included in an analysis. However, with this limitation becoming more and more urgent as a result of increasingly large studies, methodologic developments are being introduced that aim to circumvent these limitations. Examples include approximations for independent and principal component analyses (Hyvärinen, 2013; Smith et al., 2014).

Challenges to open the "black box"

Many population imaging initiatives are focused on increasing our sensitivity to detect brain changes in the earliest stage of disease by means of imaging. Studies from the past decade have shown that this approach has added value in understanding disease development and potentially in disease prediction. A good example is the quantification of white-matter microstructural integrity, through application of DTI (Le Bihan et al., 2001). This advanced imaging marker has been shown to contain information beyond that present in macrostructural markers, for example, in explaining cognitive decline (Vernooij et al., 2009). Also, recent work suggests that alteration of tissue microstructure is truly an early marker of disease, as microstructural changes were shown to precede macrostructural pathology in the brain (de Groot et al., 2013; Maillard et al., 2013). Yet, a better sensitivity for detecting underlying pathology will not suffice for translation to clinical practice. Specificity also needs to be high, in order to distinguish those who are likely to suffer from disease from those who are not at risk. Advances in this field are expected from several areas. First, studies correlating new imaging markers to findings on pathology are crucial to understand the specificity of many new imaging markers. Recent work on cortical microinfarcts shows how instrumental these correlation studies are in establishing new imaging markers for disease (van Veluw et al., 2015b). Second, rather than evaluating imaging markers as isolated findings, the field should move towards integration of imaging findings in a multimodal approach into models that help classify subjects better according to their likelihood of disease and/or aid in better disease prediction. Advanced data-handling techniques such as machine learning, using either *a priori* defined markers or data-driven approaches (Young et al., 2014), will be important in this respect. Third, there is an urgent demand for more longitudinal data to better understand progression of imaging markers over time. This is crucial to be able to factor out the variation caused by normal aging, disentangle different trajectories, and unravel the temporal order with which certain markers appear in the course of disease (Jack et al., 2010). The Alzheimer's Disease Neuroimaging Initiative study, a longitudinal study focused on investigation of cognitive, clinical, and biomarker trajectories across the entire spectrum of Alzheimer's disease, is a good example of an observational study that through rigorous assessment of a large array of biomarkers aims to inform the design of clinical trials and accelerate the development of drugs (Hua et al., 2016). Yet, substantial challenges remain in unraveling the temporal order of (imaging) markers in disease progression, with the most important ones being needed for extensive follow-up (as dementia development, for example, has an average course of 15 years), and the need to study individuals in the earliest disease states, necessitating the inclusion of younger subjects (who thus have to be followed for a longer time until occurrence of disease).

FUTURE PERSPECTIVES

Considering all the above, there are a number of future perspectives and recommendations for imaging in neuroepidemiologic studies.

Firstly, these consider the need for collaborative efforts in a number of areas of population imaging. As the field is striving towards more detailed phenotyping of more subtle subclinical changes, in order to find earlier markers of disease, a single epidemiologic study is nowadays often underpowered to show meaningful effects and associations. This is especially true with the increasing number of hypothesis-free, whole-brain studies (voxel-based morphometry), in combination with other multidimensional data, e.g., genomics, metabolomics, or proteomics (Petrella et al., 2008; Fiandaca et al., 2014). There is a growing number of collaborative efforts from international consortia that bring together imaging data from cohort studies, e.g. the Cohorts for Heart and Aging Research in Genomic Epidemiology (CHARGE) consortium (Psaty et al., 2009) and the Enhancing Neuro Imaging Genetics through Meta Analysis (ENIGMA) consortium (Thompson et al., 2014). For these collaborative efforts to be successful, standardization of imaging acquisition and image processing/rating as well as terminology is important in order to pool imaging data and jointly interpret results. Though there have been some efforts to define minimum imaging standards and common terminology for neuroimaging studies (Wardlaw et al., 2013), more work in this area is needed to meet the future needs for collaborations.

Medical imaging is a rapidly progressing field, in which new technologies become available at a high rate. These consist of both advances in hardware (e.g., high magnetic field strength, dual-source CT, PET–MRI fusion) and software (e.g., arterial spin labeling for brain perfusion, T1 and T2 mapping for new MRI contrasts (Dieringer et al., 2014), as well as new contrast media and ligands used for imaging (e.g., tau-PET imaging: Villemagne et al., 2015)). Such developments on the imaging side will unlock new doors towards studying brain changes in a variety of disease processes, increasing the spatial and temporal resolution, revealing so far unknown parameters of disease and aiding in measuring brain function as well as structure.

Besides advances in imaging hardware and software, important future influence is particularly expected for developments in image processing. Automated quantification of brain tissue, structure, and function and the ability to measure (microscopic) changes in the brain invisible to the naked eye, or too complicated to grasp by a visual observer, greatly expand our palette of imaging biomarkers in neuroepidemiology. Examples of these from the past decade comprise shape-based analyses, connectivity matrices, and brain networks (Ferreira and Busatto, 2013; Yang et al., 2014; Gutman et al., 2015).

This new array of imaging markers that become available, either through new acquisition schemes or advance processing, bring about the need to develop new analytic approaches that combine all the information derived from the images. Rather than evaluating a single marker (or single imaging modality), the field should move towards integrating all information derived from the various imaging modalities and processing algorithms into models that combine all for a better understanding of disease etiology and improved disease prediction.

In summary, the above-mentioned and future advances in imaging (analysis) techniques will increase our sensitivity to early brain changes, help identify new pathways towards disease, and aid in understand disease etiology better. Incorporation of these techniques in neuroepidemiologic studies will depend on availability, and associated risks and costs.

References

Aaslid R, Markwalder TM, Nornes H (1982). Noninvasive transcranial Doppler ultrasound recording of flow velocity in basal cerebral arteries. J Neurosurg 57: 769–774.

Agatston AS, Janowitz WR, Hildner FJ et al. (1990). Quantification of coronary artery calcium using ultrafast computed tomography. J Am Coll Cardiol 15: 827–832.

Aljabar P, Heckemann RA, Hammers A et al. (2009). Multi-atlas based segmentation of brain images: atlas selection and its effect on accuracy. Neuroimage 46: 726–738.

Allison MA, Criqui MH, Wright CM (2004). Patterns and risk factors for systemic calcified atherosclerosis. Arterioscler Thromb Vasc Biol 24: 331–336.

Alsheikh-Ali AA, Kitsios GD, Balk EM et al. (2010). The vulnerable atherosclerotic plaque: scope of the literature. Ann Intern Med 153: 387–395.

Alzheimer A (1907). Über eine eigenartige Erkrankung der Hirnrinde. Allgemeine Zeitschrift für Psychiatrie und psychisch-gerichtliche Medizin 64: 146–148.

Anbeek P, Vincken KL, van Osch MJ et al. (2004). Probabilistic segmentation of white matter lesions in MR imaging. Neuroimage 21: 1037–1044.

Arenillas JF (2011). Intracranial atherosclerosis: current concepts. Stroke 42: S20–S23.

Ashburner J, Friston KJ (2005). Unified segmentation. Neuroimage 26: 839–851.

Baker ML, Hand PJ, Wang JJ et al. (2008). Retinal signs and stroke: revisiting the link between the eye and brain. Stroke 39: 1371–1379.

Bamberg F, Kauczor HU, Weckbach S et al. (2015). Whole-body MR imaging in the German national cohort: rationale, design, and technical background. Radiology 277: 206–220.

Beaulieu C (2002). The basis of anisotropic water diffusion in the nervous system – a technical review. NMR Biomed 15: 435–455.

Bookheimer SY, Dapretto M, Blaxton T et al. (1996). Functional MRI of the hippocampus during short term memory tasks: parametric responses to task difficulty and stimulus novelty. Neuroimage 3: S531.

Bos MJ, Koudstaal PJ, Hofman A et al. (2007). Transcranial Doppler hemodynamic parameters and risk of stroke: the Rotterdam study. Stroke 38: 2453–2458.

Bos D, Portegies ML, van der Lugt A et al. (2014). Intracranial carotid artery atherosclerosis and the risk of stroke in whites: the Rotterdam study. JAMA Neurol 71: 405–411.

Bos D, Vernooij MW, de Bruijn RF et al. (2015). Atherosclerotic calcification is related to a higher risk of dementia and cognitive decline. Alzheimers Dement 11: 639–647.

Bos D, Poels MMF, Adams HHH et al. (2016). Prevalence, clinical management, and natural course of incidental findings on brain MR images: the population-based Rotterdam scan study. Radiology. http://dx.doi.org/10.1148/radiol.2016160218.

Buxton RB (2002). Introduction to functional magnetic resonance imaging: Principles and techniques, Cambridge University Press, Cambridge.

Cai JM, Hatsukami TS, Ferguson MS et al. (2002). Classification of human carotid atherosclerotic lesions with in vivo multicontrast magnetic resonance imaging. Circulation 106: 1368–1373.

Catana C, Drzezga A, Heiss WD et al. (2012). PET/MRI for neurologic applications. J Nucl Med 53: 1916–1925.

Cheung CY, Ong YT, Ikram MK et al. (2014). Retinal microvasculature in Alzheimer's disease. J Alzheimers Dis 42: S339–S352.

Cocosco CA, Zijdenbos AP, Evans AC (2003). A fully automatic and robust brain MRI tissue classification method. Med Image Anal 7: 513–527.

Dawber TR, Meadors GF, Moore FE (1951). Epidemiological approaches to heart disease: the Framingham Study. Am J Public Health Nations Health 41: 279–286.

de Boer R, Vrooman HA, van der Lijn F et al. (2009). White matter lesion extension to automatic brain tissue segmentation on MRI. Neuroimage 45: 1151–1161.

de Groot M, Verhaaren BF, de Boer R et al. (2013). Changes in normal-appearing white matter precede development of white matter lesions. Stroke 44: 1037–1042.

Debette S, Markus HS (2010). The clinical importance of white matter hyperintensities on brain magnetic resonance imaging: systematic review and meta-analysis. BMJ 341: c3666.

DeCarli CS (2006). When two are worse than one: stroke and Alzheimer disease. Neurology 67: 1326–1327.

den Heijer T, van der Lijn F, Koudstaal PJ et al. (2010). A 10-year follow-up of hippocampal volume on magnetic resonance imaging in early dementia and cognitive decline. Brain 133: 1163–1172.

Dieringer MA, Deimling M, Santoro D et al. (2014). Rapid parametric mapping of the longitudinal relaxation time T1 using two-dimensional variable flip angle magnetic resonance imaging at 1.5 Tesla, 3 Tesla, and 7 Tesla. PLoS One 9: e91318.

Elias-Smale SE, Odink AE, Wieberdink RG et al. (2010). Carotid, aortic arch and coronary calcification are related to history of stroke: the Rotterdam Study. Atherosclerosis 212: 656–660.

Fazekas F, Chawluk JB, Alavi A et al. (1987). MR signal abnormalities at 1.5 T in Alzheimer's dementia and normal aging. AJR Am J Roentgenol 149: 351–356.

Ferreira LK, Busatto GF (2013). Resting-state functional connectivity in normal brain aging. Neurosci Biobehav Rev 37: 384–400.

Fiandaca MS, Mapstone ME, Cheema AK et al. (2014). The critical need for defining preclinical biomarkers in Alzheimer's disease. Alzheimers Dement 10: S196–S212.

Finn AV, Kolodgie FD, Virmani R (2010). Correlation between carotid intimal/medial thickness and atherosclerosis: a point of view from pathology. Arterioscler Thromb Vasc Biol 30: 177–181.

Fischl B, Dale AM (2000). Measuring the thickness of the human cerebral cortex from magnetic resonance images. Proc Natl Acad Sci U S A 97: 11050–11055.

Fischl B, Salat DH, Busa E et al. (2002). Whole brain segmentation: automated labeling of neuroanatomical structures in the human brain. Neuron 33: 341–355.

Frohman EM, Fujimoto JG, Frohman TC et al. (2008). Optical coherence tomography: a window into the mechanisms of multiple sclerosis. Nat Clin Pract Neurol 4: 664–675.

Golomb J, de Leon MJ, Kluger A et al. (1993). Hippocampal atrophy in normal aging. An association with recent memory impairment. Arch Neurol 50: 967–973.

Gons RA, van Norden AG, de Laat KF et al. (2011). Cigarette smoking is associated with reduced microstructural integrity of cerebral white matter. Brain 134: 2116–2124.

Good CD, Johnsrude IS, Ashburner J et al. (2001). A voxel-based morphometric study of ageing in 465 normal adult human brains. Neuroimage 14: 21–36.

Greenberg SM, Vernooij MW, Cordonnier C et al. (2009). Cerebral microbleeds: a guide to detection and interpretation. Lancet Neurol 8: 165–174.

Gullstrand A (1910). Neue Methoden der Reflexlosen Ophtalmoskopie. Berichte Deutsche Ophtalmologische Gesellschaft 36.

Gur RE, Kaltman D, Melhem ER et al. (2013). Incidental findings in youths volunteering for brain MRI research. AJNR Am J Neuroradiol 34: 2021–2025.

Gutman BA, Wang Y, Yanovsky I et al. (2015). Empowering imaging biomarkers of Alzheimer's disease. Neurobiol Aging 36 (Suppl 1): S69–S80.

Harris TB, Launer LJ, Eiriksdottir G et al. (2007). Age, gene/environment susceptibility-Reykjavik study: multidisciplinary applied phenomics. Am J Epidemiol 165: 1076–1087.

Hartwigsen G, Siebner HR, Deuschl G et al. (2010). Incidental findings are frequent in young healthy individuals undergoing magnetic resonance imaging in brain research imaging studies: a prospective single-center study. J Comput Assist Tomogr 34: 596–600.

Hegenscheid K, Kuhn JP, Volzke H et al. (2009). Whole-body magnetic resonance imaging of healthy volunteers: pilot study results from the population-based SHIP study. Rofo 181: 748–759.

Hofman A, Brusselle GG, Darwish Murad S et al. (2015). The Rotterdam Study: 2016 objectives and design update. Eur J Epidemiol 30: 661–708.

Howard G, Sharrett AR, Heiss G et al. (1993). Carotid artery intimal-medial thickness distribution in general populations as evaluated by B-mode ultrasound. ARIC Investigators. Stroke 24: 1297–1304.

Hua X, Ching CR, Mezher A et al. (2016). MRI-based brain atrophy rates in ADNI phase 2: acceleration and enrichment considerations for clinical trials. Neurobiol Aging 37: 26–37.

Huang D, Swanson EA, Lin CP et al. (1991). Optical coherence tomography. Science 254: 1178–1181.

Hyvärinen A (2013). Independent component analysis: recent advances. Philos Trans A Math Phys Eng Sci 371.

Ikram MA, van der Lugt A, Niessen WJ et al. (2011). The Rotterdam Scan Study: design and update up to 2012. Eur J Epidemiol 26: 811–824.

Illes J, Kirschen MP, Edwards E et al. (2006). Ethics. Incidental findings in brain imaging research. Science 311: 783–784.

Jack Jr CR (2012). Alzheimer disease: new concepts on its neurobiology and the clinical role imaging will play. Radiology 263: 344–361.

Jack Jr CR, Holtzman DM (2013). Biomarker modeling of Alzheimer's disease. Neuron 80: 1347–1358.

Jack CR, Knopman Jr DS, Jaqust WJ et al. (2010). Hypothetical model of dynamic biomarkers of the Alzheimer's pathological cascade. Lancet Neurol 9: 119–128.

Jones DK, Symms MR, Cercignani M et al. (2005). The effect of filter size on VBM analyses of DT-MRI data. Neuroimage 26: 546–554.

Jorge Cardoso M, Leung K, Modat M et al. (2013). STEPS: Similarity and Truth Estimation for Propagated Segmentations and its application to hippocampal segmentation and brain parcelation. Med Image Anal 17: 671–684.

Kantarci K, Senjem ML, Avula R et al. (2011). Diffusion tensor imaging and cognitive function in older adults with no dementia. Neurology 77: 26–34.

Kessler LG, Barnhart HX, Buckler AJ et al. (2015). The emerging science of quantitative imaging biomarkers terminology and definitions for scientific studies and regulatory submissions. Stat Methods Med Res 24: 9–26.

Kidwell CS, el-Saden S, Livshits Z et al. (2001). Transcranial Doppler pulsatility indices as a measure of diffuse small-vessel disease. J Neuroimaging 11: 229–235.

Koedam EL, Lehmann M, van der Flier WM et al. (2011). Visual assessment of posterior atrophy development of a MRI rating scale. Eur Radiol 21: 2618–2625.

Kotsoni E, Byrd D, Casey BJ (2006). Special considerations for functional magnetic resonance imaging of pediatric populations. J Magn Reson Imaging 23: 877–886.

Kuijf HJ, Brundel M, de Bresser J et al. (2013). Semi-automated detection of cerebral microbleeds on 3.0 T MR images. PLoS One 8: e66610.

Lambert J, MacKenzie JD, Cody DD et al. (2014). Techniques and tactics for optimizing CT dose in adults and children: state of the art and future advances. J Am Coll Radiol 11: 262–266.

Le Bihan D, Mangin JF, Poupon C et al. (2001). Diffusion tensor imaging: concepts and applications. J Magn Reson Imaging 13: 534–546.

Lehman SJ, Abbara S, Cury RC et al. (2009). Significance of cardiac computed tomography incidental findings in acute chest pain. Am J Med 122: 543–549.

Lopez-Cancio E, Galan A, Dorado L et al. (2012). Biological signatures of asymptomatic extra- and intracranial atherosclerosis: the Barcelona-AsIA (Asymptomatic Intracranial Atherosclerosis) study. Stroke 43: 2712–2719.

Lustig C, Snyder AZ, Bhakta M et al. (2003). Functional deactivations: change with age and dementia of the Alzheimer type. Proc Natl Acad Sci U S A 100: 14504–14509.

Maillard P, Carmichael O, Harvey D et al. (2013). FLAIR and diffusion MRI signals are independent predictors of white matter hyperintensities. AJNR Am J Neuroradiol 34: 54–61.

Maintz JB, Viergever MA (1998). A survey of medical image registration. Med Image Anal 2: 1–36.

McConathy J, Sheline YI (2015). Imaging biomarkers associated with cognitive decline: a review. Biol Psychiatry 77: 685–692.

Mendrik AM, Vincken KL, Kuijf HJ et al. (2015). MRBrainS challenge: online evaluation framework for brain image segmentation in 3T MRI scans. Comput Intell Neurosci 2015: 813696.

Morris Z, Whiteley WN, Longstreth Jr WT et al. (2009). Incidental findings on brain magnetic resonance imaging: systematic review and meta-analysis. BMJ 339: b3016.

Mueller SG, Weiner MW, Thal LJ et al. (2005). The Alzheimer's disease neuroimaging initiative. Neuroimaging Clin N Am 15: 869–877. xi-xii.

Nandalur KR, Baskurt E, Hagspiel KD et al. (2006). Carotid artery calcification on CT may independently predict stroke risk. AJR Am J Roentgenol 186: 547–552.

O'Sullivan M (2010). Imaging small vessel disease: lesion topography, networks, and cognitive deficits investigated with MRI. Stroke 41: S154–S158.

Ota H, Yu W, Underhill HR et al. (2009). Hemorrhage and large lipid-rich necrotic cores are independently associated with thin or ruptured fibrous caps: an in vivo 3 T MRI study. Arterioscler Thromb Vasc Biol 29: 1696–1701.

Owen DR, Lindsay AC, Choudhury RP et al. (2011). Imaging of atherosclerosis. Annu Rev Med 62: 25–40.

Pantoni L, Garcia JH (1997). Pathogenesis of leukoaraiosis: a review. Stroke 28: 652–659.

Petersen SE, Matthews PM, Bamberg F et al. (2013). Imaging in population science: cardiovascular magnetic resonance in 100,000 participants of UK Biobank – rationale, challenges and approaches. J Cardiovasc Magn Reson 15: 46.

Petrella JR, Mattay VS, Doraiswamy PM (2008). Imaging genetics of brain longevity and mental wellness: the next frontier? Radiology 246: 20–32.

Portegies ML, de Bruijn RF, Hofman A et al. (2014). Cerebral vasomotor reactivity and risk of mortality: the Rotterdam Study. Stroke 45: 42–47.

Prins ND, van Dijk EJ, den Heijer T et al. (2004a). Cerebral white matter lesions and the risk of dementia. Arch Neurol 61: 1531–1534.

Prins ND, van Straaten EC, van Dijk EJ et al. (2004b). Measuring progression of cerebral white matter lesions on MRI: visual rating and volumetrics. Neurology 62: 1533–1539.

Prins ND, van Dijk EJ, den Heijer T et al. (2005). Cerebral small-vessel disease and decline in information processing speed, executive function and memory. Brain 128: 2034–2041.

Psaty BM, O'Donnell CJ, Gudnason V et al. (2009). Cohorts for Heart and Aging Research in Genomic Epidemiology

(CHARGE) Consortium: design of prospective meta-analyses of genome-wide association studies from 5 cohorts. Circ Cardiovasc Genet 2: 73–80.

Raichle ME, Snyder AZ (2007). A default mode of brain function: a brief history of an evolving idea. Neuroimage 37: 1083–1090. discussion 1097–1089.

Raunig DL, McShane LM, Pennello G et al. (2015). Quantitative imaging biomarkers: a review of statistical methods for technical performance assessment. Stat Methods Med Res 24: 27–67.

Raz N, Millman D, Sarpel G (1990). Cerebral correlates of cognitive aging: gray-white-matter differentiation in the medial temporal lobes, and fluid versus crystallized abilities. Psychobiology 18: 475–481.

Rumberger JA, Simons DB, Fitzpatrick LA et al. (1995). Coronary artery calcium area by electron-beam computed tomography and coronary atherosclerotic plaque area. A histopathologic correlative study. Circulation 92: 2157–2162.

Sandeman EM, Hernandez Mdel C, Morris Z et al. (2013). Incidental findings on brain MR imaging in older community-dwelling subjects are common but serious medical consequences are rare: a cohort study. PLoS One 8: e71467.

Sangiorgi G, Rumberger JA, Severson A et al. (1998). Arterial calcification and not lumen stenosis is highly correlated with atherosclerotic plaque burden in humans: a histologic study of 723 coronary artery segments using nondecalcifying methodology. J Am Coll Cardiol 31: 126–133.

Sarkar S, Ghosh S, Ghosh SK et al. (2007). Role of transcranial Doppler ultrasonography in stroke. Postgrad Med J 83: 683–689.

Scheltens P, Leys D, Barkhof F et al. (1992). Atrophy of medial temporal lobes on MRI in "probable" Alzheimer's disease and normal ageing: diagnostic value and neuropsychological correlates. J Neurol Neurosurg Psychiatry 55: 967–972.

Scheltens P, Barkhof F, Leys D et al. (1993). A semiquantative rating scale for the assessment of signal hyperintensities on magnetic resonance imaging. J Neurol Sci 114: 7–12.

Schmidt R, Fazekas F, Kapeller P et al. (1999). MRI white matter hyperintensities: three-year follow-up of the Austrian Stroke Prevention Study. Neurology 53: 132–139.

Schmidt S, Duun-Henriksen AK, Norgaard K (2012). Psychosocial factors and adherence to continuous glucose monitoring in type 1 diabetes. J Diabetes Sci Technol 6: 986–987.

Schwarz C, Fletcher E, DeCarli C et al. (2009). Fully-automated white matter hyperintensity detection with anatomical prior knowledge and without FLAIR. Inf Process Med Imaging 21: 239–251.

Seghier ML, Kolanko MA, Leff AP et al. (2011). Microbleed detection using automated segmentation (MIDAS): a new method applicable to standard clinical MR images. PLoS One 6: e17547.

Selwaness M, van den Bouwhuijsen Q, van Onkelen RS et al. (2014). Atherosclerotic plaque in the left carotid artery is more vulnerable than in the right. Stroke 45: 3226–3230.

Sexton CE, Walhovd KB, Storsve AB et al. (2014). Accelerated changes in white matter microstructure during aging: a longitudinal diffusion tensor imaging study. J Neurosci 34: 15425–15436.

Simic G, Babic M, Borovecki F et al. (2014). Early failure of the default-mode network and the pathogenesis of Alzheimer's disease. CNS Neurosci Ther 20: 692–698.

Sled JG, Zijdenbos AP, Evans AC (1998). A nonparametric method for automatic correction of intensity nonuniformity in MRI data. IEEE Trans Med Imaging 17: 87–97.

Smith SM, Jenkinson M, Johansen-Berg H et al. (2006). Tract-based spatial statistics: voxelwise analysis of multi-subject diffusion data. Neuroimage 31: 1487–1505.

Smith SM, Hyvärinen A, Varoquaux G et al. (2014). Group-PCA for very large fMRI datasets. Neuroimage 101: 738–749.

Smith-Bindman R, Lipson J, Marcus R et al. (2009). Radiation dose associated with common computed tomography examinations and the associated lifetime attributable risk of cancer. Arch Intern Med 169: 2078–2086.

Sperling RA, Bates JF, Chua EF et al. (2003). fMRI studies of associative encoding in young and elderly controls and mild Alzheimer's disease. J Neurol Neurosurg Psychiatry 74: 44–50.

Svedberg MM, Rahman O, Hall H (2012). Preclinical studies of potential amyloid binding PET/SPECT ligands in Alzheimer's disease. Nucl Med Biol 39: 484–501.

Thompson PM, Stein JL, Medland SE et al. (2014). The ENIGMA Consortium: large-scale collaborative analyses of neuroimaging and genetic data. Brain Imaging Behav 8: 153–182.

Underhill HR, Yuan C, Yarnykh VL et al. (2010). Predictors of surface disruption with MR imaging in asymptomatic carotid artery stenosis. AJNR Am J Neuroradiol 31: 487–493.

van den Bouwhuijsen QJ, Vernooij MW, Hofman A et al. (2012). Determinants of magnetic resonance imaging detected carotid plaque components: the Rotterdam Study. Eur Heart J 33: 221–229.

van den Oord SC, Sijbrands EJ, Ten Kate GL et al. (2013). Carotid intima-media thickness for cardiovascular risk assessment: systematic review and meta-analysis. Atherosclerosis 228: 1–11.

van der Lijn F, de Bruijne M, Klein S et al. (2012). Automated brain structure segmentation based on atlas registration and appearance models. IEEE Trans Med Imaging 31: 276–286.

Van Dijk KR, Hedden T, Venkataraman A et al. (2011). Intrinsic functional connectivity as a tool for human connectomics: theory, properties, and optimization. J Neurophysiol 103: 297–301.

Van Leemput K, Maes F, Vandermeulen D et al. (1999). Automated model-based tissue classification of MR images of the brain. IEEE Trans Med Imaging 18: 897–908.

van Veluw SJ, Hilal S, Kuijf HJ et al. (2015a). Cortical microinfarcts on 3T MRI: clinical correlates in memory-clinic patients. Alzheimers Dement 11: 1500–1509.

van Veluw SJ, Zwanenburg JJ, Rozemuller AJ et al. (2015b). The spectrum of MR detectable cortical microinfarcts: a classification study with 7-tesla postmortem MRI and histopathology. J Cereb Blood Flow Metab 35: 676–683.

Vermeer SE, Prins ND, den Heijer T et al. (2003). Silent brain infarcts and the risk of dementia and cognitive decline. N Engl J Med 348: 1215–1222.

Vernooij MW, Ikram MA, Tanghe HL et al. (2007). Incidental findings on brain MRI in the general population. N Engl J Med 357: 1821–1828.

Vernooij MW, Ikram MA, Vrooman HA et al. (2009). White matter microstructural integrity and cognitive function in a general elderly population. Arch Gen Psychiatry 66: 545–553.

Vidal JS, Sigurdsson S, Jonsdottir MK et al. (2010). Coronary artery calcium, brain function and structure: the AGES-Reykjavik Study. Stroke 41: 891–897.

Villemagne VL, Fodero-Tavoletti MT, Masters CL et al. (2015). Tau imaging: early progress and future directions. Lancet Neurol 14: 114–124.

Vrooman HA, Cocosco CA, van der Lijn F et al. (2007). Multi-spectral brain tissue segmentation using automatically trained k-Nearest-Neighbor classification. Neuroimage 37: 71–81.

Wahlund LO, Barkhof F, Fazekas F et al. (2001). A new rating scale for age-related white matter changes applicable to MRI and CT. Stroke 32: 1318–1322.

Wardlaw JM, Smith EE, Biessels GJ et al. (2013). Neuroimaging standards for research into small vessel disease and its contribution to ageing and neurodegeneration. Lancet Neurol 12: 822–838.

Warfield SK, Zou KH, Wells WM (2004). Simultaneous truth and performance level estimation (STAPLE): an algorithm for the validation of image segmentation. IEEE Trans Med Imaging 23: 903–921.

Weishaupt D, Köchli VD, Marincek B (2008). How Does MRI Work? Springer, Berlin.

Wendell CR, Zonderman AB, Metter EJ et al. (2009). Carotid intimal medial thickness predicts cognitive decline among adults without clinical vascular disease. Stroke 40: 3180–3185.

Wolf SM, Lawrenz FP, Nelson CA et al. (2008). Managing incidental findings in human subjects research: analysis and recommendations. J Law Med Ethics 36: 219–248. 211.

Yang Z, Chang C, Xu T et al. (2014). Connectivity trajectory across lifespan differentiates the precuneus from the default network. Neuroimage 89: 45–56.

Young AL, Oxtoby NP et al. (2014). A data-driven model of biomarker changes in sporadic Alzheimer's disease. Brain 137: 2564–2577.

Zhang Y, Brady M, Smith S (2001). Segmentation of brain MR images through a hidden Markov random field model and the expectation-maximization algorithm. IEEE Trans Med Imaging 20: 45–57.

Zhang S, Smailagic N, Hyde C et al. (2014). (11)C-PIB-PET for the early diagnosis of Alzheimer's disease dementia and other dementias in people with mild cognitive impairment (MCI). Cochrane Database Syst Rev 7: CD010386.

Handbook of Clinical Neurology, Vol. 138 (3rd series)
Neuroepidemiology
C. Rosano, M.A. Ikram, and M. Ganguli, Editors
http://dx.doi.org/10.1016/B978-0-12-802973-2.00006-9

Chapter 6

Use of "omics" technologies to dissect neurologic disease

G. TOSTO[1,2] AND C. REITZ[1,2,3,4]*

[1]*Taub Institute for Research on Alzheimer's Disease and the Aging Brain, Columbia University, New York, NY, USA*

[2]*Gertrude H. Sergievsky Center, Columbia University, New York, NY, USA*

[3]*Department of Neurology, Columbia University, New York, NY, USA*

[4]*Department of Epidemiology, Columbia University, New York, NY, USA*

Abstract

Over the past 5 years, the advent of massively parallel technologies for understanding disease at the molecular level accompanied by simultaneous rapid development of the computational tools needed to analyze and filter such data has revolutionized medical science. These "next-generation" "omics" technologies include next-generation sequencing technology for detection of disease-associated DNA sequence variants, RNA sequencing for transcriptome and noncoding RNA analysis, quantitative detection of epigenomic dynamics, and chromatin immunoprecipitation sequencing analysis for DNA–protein interactions, interactome analysis for networks formed by protein–protein interactions, and metabolome analysis for metabolic systems. The analysis and integration of data derived from massively parallel technologies will significantly deepen our understanding of human disease, will inform functional studies, *in vitro* and *in vivo* model generation, and will advance the development of improved, personalized diagnostic tools and more effective therapeutic targets. In this chapter we review the classic genomic approaches for identifying mechanisms underlying human disease, and summarize the emerging "omics" technologies allowing massively parallel interrogation of biologic systems.

BACKGROUND

The ultimate aim of biomedical research is to acquire a complete understanding of the molecular mechanisms underlying a disease and to make personalized medicine feasible. Over the past two decades, specific genetic risk factors underlying various neurodegenerative disorders have been identified, capitalizing on linkage studies in large families with multiple affected individuals through demonstration of co-segregation with disease status, and candidate gene and whole-genome association studies capitalizing on customized or commercially available genotyping arrays in samples of related individuals or independent cases and controls.

While these methodologic approaches resulted in considerable success and significantly advanced the understanding of the biologic mechanisms underlying numerous neurologic disorders and syndromes, these methods have several limitations (discussed in detail below). In brief, linkage studies require large, multigenerational pedigrees with both affected and unaffected individuals, and this approach yields only approximate genomic regions linked to disease but not specific causative variants. Similarly, genotype array candidate gene or genomewide association studies (GWASs) are designed to screen the genome for the identification of common variants (maf$>$1%), potentially tagging a true underlying disease-associated rather than mapping the true disease-associated mutation. The gold standard for DNA sequencing for the past 25 years, which allows DNA analysis of read-lengths of up to ~1000 bp at the nucleotide level, has been Sanger sequencing (also

*Correspondence to: Christiane Reitz, MD, PhD, Sergievsky Center/Taub Institute, Department of Neurology/Department of Epidemiology, Columbia University, 630 W 168th Street, Rm 19-308, New York NY 10032, USA. Tel: +1-212-305-0865, Fax: +1-212-305-2391, E-mail: cr2101@cumc.columbia.edu

referred to as dideoxy or chain termination sequencing) based on the selective incorporation of chain-terminating dideoxynucleotides (ddNPs) by DNA polymerase during *in vitro* DNA replication (described in detail below).

Over the past 5 years, the advent of massively parallel technologies (named by appending the suffix "-omics") for understanding disease at the molecular level accompanied by simultaneous rapid development of the computational tools needed to analyze and filter such data has revolutionized medical science. These "next-generation" "omics" technologies include applications such as whole-exome (WES) or whole-genome (WGS) sequencing for detection of disease-associated DNA sequence variants ("genomics"), RNA sequencing (RNA-seq) for transcriptome and noncoding RNA analysis ("transcriptomics"), quantitative detection of epigenomic dynamics and chromatin immunoprecipitation sequencing (ChIP-seq) analysis for DNA–protein interactions ("epigenomics"), interactome analysis for networks formed by protein–protein interactions ("proteinomics"), and metabolome analysis for metabolic systems ("metabolomics"). Sophisticated bioinformatics means are essential to process the massive amount of data generated by these "-omics"-based systems biology technologies and extract the critical information. In addition, bioinformatics allows to integrate data derived from the various "omics" fields, acknowledging that each individual field is beginning to reach its inherent limit, making integration with data derived from other "omics" fields crucial for further disentangling of disease etiologies and normal physiologic processes.

The analysis and integration of data derived from massively parallel technologies will significantly deepen our understanding of human disease, inform functional studies and *in vitro* and *in vivo* model generation, and will advance the development of improved, personalized diagnostic tools and more effective therapeutic targets. In this chapter we review the classic genomic approaches for identifying mechanisms underlying human disease, and summarize the emerging "omics" technologies allowing massively parallel interrogation of biologic systems.

OVERVIEW OF -OMICS TECHNOLOGIES

The best understood of all the "omics" disciplines is probably the "genomics" field, which analyzes the function and structure of genomes (i.e., the complete set of DNA within a single cell of an organism), including gene expression and copying of DNA to RNA. "Transcriptomics" is the study of the transcriptome, the complete set of RNA transcripts produced by the genome, including mRNAs, noncoding RNAs, and small RNAs. It allows to determine the transcriptional structure of genes in terms of their start sites, 5′ and 3′ ends, splicing patterns and other posttranscriptional modifications, and to quantify the changing expression levels of each transcript during development and under different conditions. "Proteomics" determines the structure and function of proteins. "Metabolomics" aims to identify and quantify cellular metabolites, i.e., the chemical end result of upstream genetic activity. Metabolomics is therefore perceived as the "chemical fingerprint" and the bridging gap between genotype and phenotype. Figure 6.1 summarizes the interaction between the various "omics" systems.

INTEGRATION OF OMICS DATA

Biologic pathways are often highly interconnected, with regulating, feedforward and feedback mechanisms occurring at various levels. Consequently, physiologic processes and the etiology of a disease and effectiveness of therapeutics can only be fully understood when the complete set of interconnected biologic pathways is studied. This holistic computational and mathematic modeling of complex biologic systems through integration of high-throughput omics data is called systems biology. Over the past years, besides the need for massively parallel technology generating the biologic data, the bioinformatics software able to process, integrate, analyze, and visualize the massive amounts of data in the context of previous knowledge has been the rate-limiting step. Figure 6.2 shows the integration of biology, next-generation sequencing (NGS) technology, and bioinformatics in systems biology.

GENOMICS

Genetic linkage studies

Traditionally, the search for genetic causes of a disease gene begins with genetic linkage analysis as a screen of the genome in order to identify the rough genomic location of disease-associated genes. A genetic linkage study is a family-based method used to map a disease or trait to a genomic location by demonstrating co-segregation of the disease or trait with genetic markers with known chromosomal location throughout a pedigree. This method is based on the principle that genomic regions showing co-segregation are more likely to contain a causal genetic variant than regions not demonstrating co-segregation. Linkage analysis methods can be applied to both major gene disorders (analyzed by parametric linkage methods) and complex diseases (analyzed by model-free or nonparametric linkage methods) and are particularly useful for the identification of genes that are inherited in a Mendelian fashion.

Genomic loci that are close enough together on the same chromosome segregate together more often than do loci on different chromosomes. Loci on different chromosomes segregate together purely by chance. Each

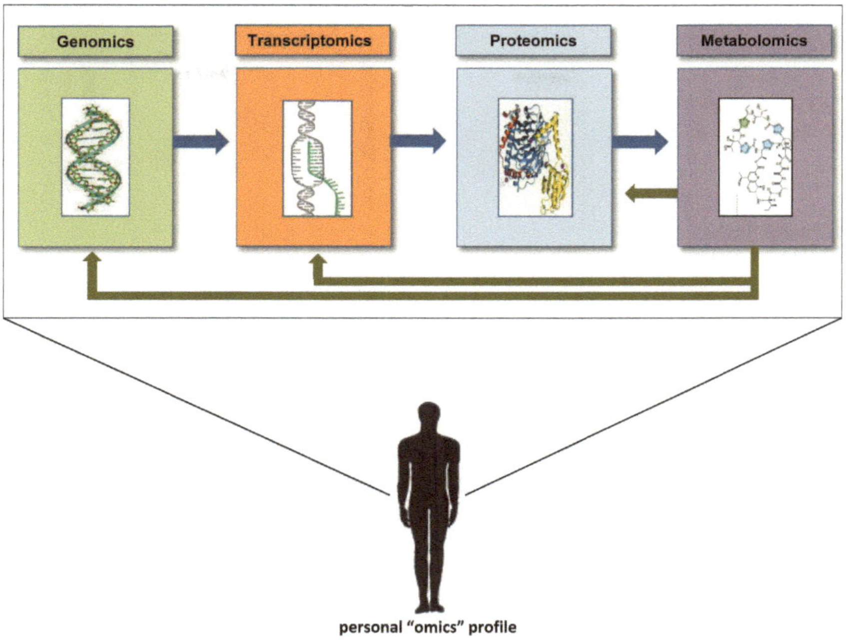

Fig. 6.1. Interaction between genomics, transcriptomics, proteomics, and metabolomics. Blue arrows indicate production; brown arrows indicate regulation.

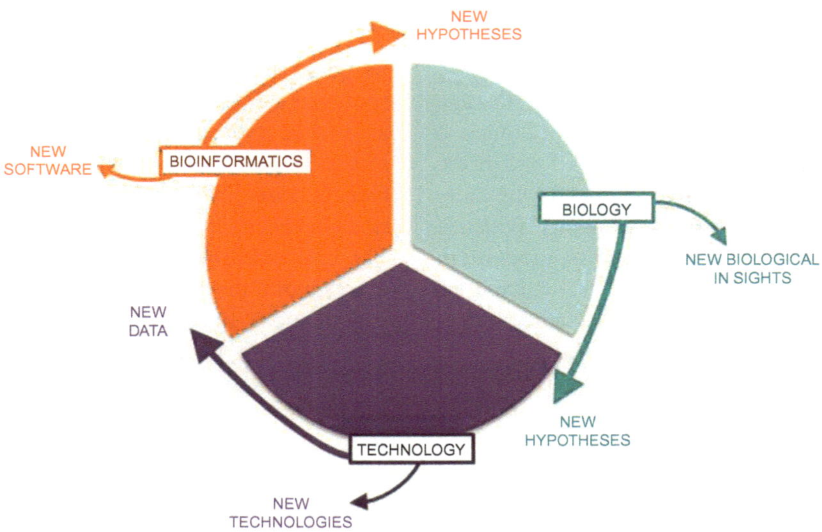

Fig. 6.2. Integration of biology, next-generation sequencing technology, and bioinformatics in systems biology.

genotype for one genetic marker or locus is made up of two alleles, one inherited from each parent. Specific alleles are in gametic phase when they are co-inherited from the same parent – i.e., they were present together in the single transmitted gamete originating from that parent. The further apart two loci are on the same chromosome, the more likely it is that a recombination event at meiosis will break up the co-segregation (Fig. 6.3A). The main quantity of interest in parametric linkage analysis is the recombination fraction θ (the probability of recombination between two loci at meiosis). By genotyping genetic markers and studying their segregation through pedigrees, it is possible to infer their position relative to each other on the genome. Similarly, disease-associated alleles can be mapped by measuring recombination against a panel of genotyped markers. In most cases, recombination will occur frequently, indicating that the disease allele and the marker are far apart. Markers in close proximity, however, will tend not to recombine with the disease gene; in this case, the marker

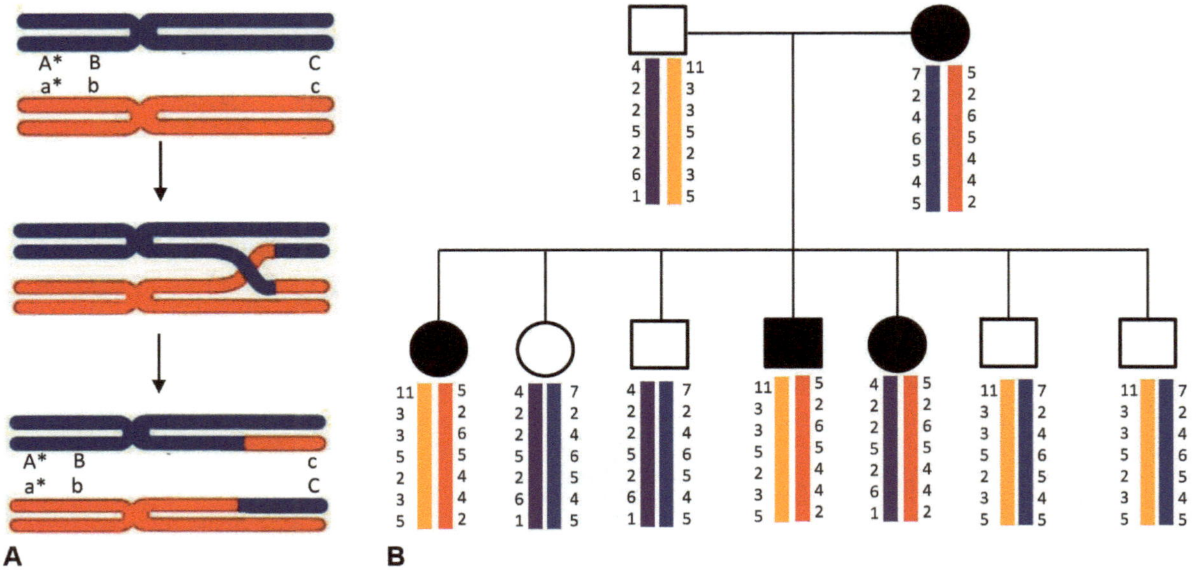

Fig. 6.3. Principle of genetic linkage analysis. (**A**) The diagram shows paternal (blue) and maternal (red) chromosomes. Three alleles are shown, labeled A*, B, and C. The capital letters represent the paternal alleles; the lower-case letters represent the maternal alleles. During recombination DNA strands cross over between the paired chromosomes. When the cross-over is resolved, the maternal and paternal alleles are mixed (recombined). These mixed alleles are then passed on to the sperm or eggs. If A* is the disease gene and B and C are genetic markers, recombination is likely to occur much more frequently between A and C than it is between A and B. This allows the disease gene to be mapped relative to the markers B and C. (**B**) The haplotype marked in red segregates with disease throughout a pedigree.

and the disease allele are said to be in "linkage disequilibrium" (LD) (Slatkin, 2008). Ideally, linkage analysis identifies close markers that flank the disease gene and define a candidate region of <10 cM in the genome, which then can be further screened for the causative mutations by finemapping or sequencing. Evidence for linkage is most commonly expressed as a logarithm (base 10) of the odds (LOD) score, calculated as LOD = Z = log 10 (probability of birth sequence with a given linkage value/probability of birth sequence with no linkage). Traditionally, an LOD score of 3 or higher, indicating 1000:1 odds that the linkage being observed did not occur by chance, is considered as evidence for a high probability of genetic linkage. However, also lower LOD scores can contain disease-associated alleles and should be considered when screening for disease-associated genomic regions.

Candidate gene and genomewide association studies

In contrast to highly heritable monogenic diseases, where linkage analyses remain an important analytic technique for the identification of disease-associated genomic regions, common neurologic and nonneurologic disorders such as cardiovascular disease or Alzheimer's disease are caused by a combination of multiple environmental risk factors and common and rare genetic variants in genes belonging to various pathogenic pathways (polygenic disease, complex disorder). In this scenario, a genetic association study can demonstrate higher statistical power if the sample size is sufficiently large. Genetic association studies typically study the associations between single-nucleotide polymorphisms (SNPs: Fig. 6.4) and diverse outcomes, such as the presence or absence of a disease (i.e., cases vs. healthy controls) or a continuous trait (e.g., memory score, hippocampus volume). The main predictor in the analyses is the SNP, i.e., a variation in the DNA sequence present in a specific population and in which a single nucleotide – A, T, C, or G – in the genome differs between members of a biologic species or paired chromosomes. The polymorphism is simply part of the natural genetic variation within a population that creates diversity.

Subject 1

AAGCAATTGTGTGCCCATATAGT

Subject 2

AAGCAATTGTATGCCCATATAGT

Fig. 6.4. Single-nucleotide polymorphism. The two individuals represented here differ for a single nucleotide in this specific trait of DNA.

The majority of SNPs have only two alleles (in the case illustrated in Fig. 6.4, the variation is identified by a guanine and an adenine). The genomic distribution of SNPs is not homogeneous; they occur in noncoding regions more frequently than in coding regions. In most circumstances, natural selection tends to eliminate deleterious variants aiming for the most favorable genetic adaptation. There are several exceptions to this rule. A classic example is the allele for sickle-cell hemoglobin in Mediterranean and African countries: individuals homozygous for the allele die young from sickle-cell disease but those homozygous for the "normal" beta chain are more likely to die from falciparum malaria. Ultimately subjects heterozygous for the "ill" allele are more likely to survive the infection. Another common scenario is what is called founder effect. A population might have begun with a small number of individuals carrying a particular allele which end up being overrepresented in the set of descendants. For example, the couple Ariaantje and Gerrit Jansz emigrated from Holland to South Africa, one of them bringing along an allele for porphyria. Today, the prevalence of porphyria in South Africa is of more than 30 000 individuals.

It has to be noted that SNPs commonly detected by genotyping chips (i.e., DNA microarray used to detect polymorphisms) are mostly located in the introns, i.e., they do not usually lie in coding regions of the genome. Therefore, these SNPs generally do not represent the true disease-associated markers, but merely "flag" the region where the true disease-associated SNP is located (Fig. 6.5). The genotyped SNP and the true disease-associated SNP are said to be in LD. The concept of LD is discussed in further detail below.

Genetic association analyses are usually performed with an additive genetic model where the SNP is coded as 0, 1, or 2 based on the number of reference alleles. With the SNP as main predictors, analyses are adjusted for common confounders, such as sex, age at evaluation or disease onset. If one type of the variant (one allele) is statistically more frequent in people with the disease, the SNP is said to be associated with the disease.

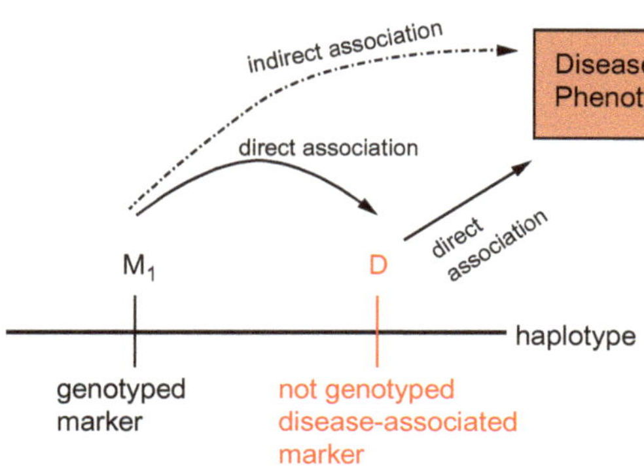

Fig. 6.5. The picture here describes a common scenario when a genotyped single-nucleotide polymorphism (M1, part of commonly used commercial genotyping arrays) is found associated with the disease only with indirect causal link; the true variant (D), not captured by the array, is instead the true cause of the disease but also in linkage disequilibrium with M1.

In contrast to methods which specifically test SNPs lying within one or a few genetic regions (candidate-gene approach), a GWAS screens for risk variants throughout the entire genome ("noncandidate-driven" or "hypothesis-free" approach). A freely available resource that lists most of the GWASs conducted in recent years can be consulted online (http:/www.genome.gov/GWAStudies).

One of the main confounders in genetic association analyses is population stratification, i.e., the possibility that the study sample comprises individuals derived from a population containing subgroups that are genetically distinct. This ultimately produces an increased type I error. The latter scenario is common because datasets that are large enough to achieve sufficient statistical power are often assembled by combining several individual samples. Particularly in studies of independent cases and controls, population substructure can yield false-positive associations: a common scenario is when the allele frequency differs between cases and control samples regardless of the disease outcome, but due to sampling from different populations (Fig. 6.6).

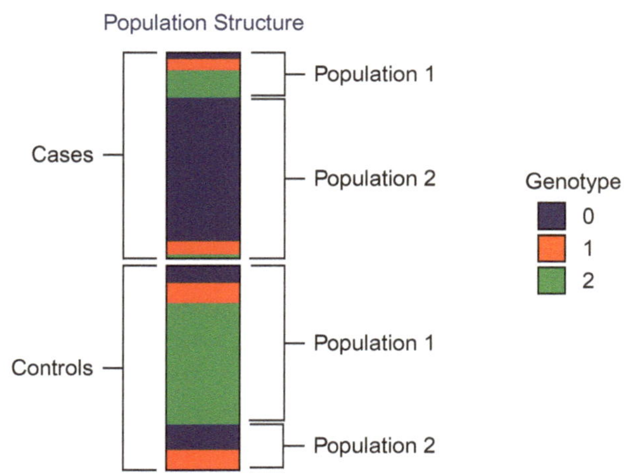

Fig. 6.6. Population substructure is a common confounder in genetic studies, particularly when multiple datasets are combined in one joint sample or when admixed populations are being explored. Here, a common scenario is depicted: cases and controls (the most used design in association studies) are not homogeneous with respect to the specific marker investigated. Clearly, "cases" are disproportionally sampled from subjects homozygous for the minor allele (0) as compared to the "controls," who are mostly homozygous for the major allele 2.

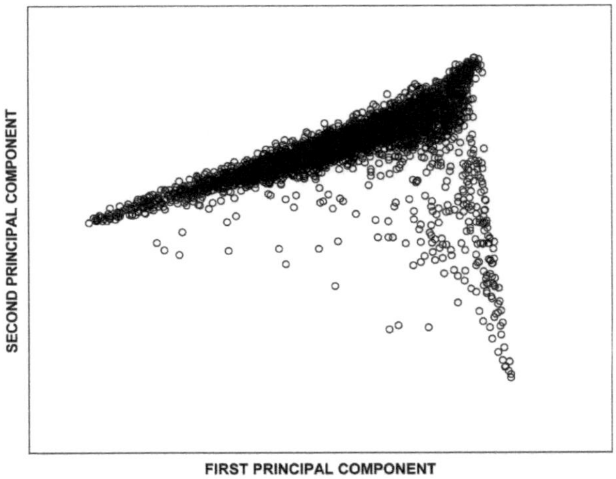

Fig. 6.7. Principal component analysis. Here the first and the second axis of variation are represented for the selected population; each dot represents a single individual.

The principal tool to correct for this bias and to investigate the structure underlying a population is principal components analysis, which was introduced in genetic research by pioneering studies of Cavalli-Sforza (Menozzi et al., 1978). Using axes of variability (i.e., principal components, PCs) as covariates is considered the best approach to correct for stratification and several algorithms have been proposed in the last decade (e.g. EIGENSTRAT; Price et al., 2006). Figure 6.7 shows a graphic example of population structure as measured on the first two PCs: most of the variability is captured by PC1, represented on the *x*-axis, with the second component accounting for additional variability.

BOX 1. Linkage disequilibrium (LD) is an index that tell us whether any two given SNPs are inherited together in an offspring. Two to hundreds of SNPs can be in (strong) LD as to constitute short or long LD blocks.

When genomewide data are not available (for example, in replication studies focusing on top hits from the discovery phase only), using ancestry-informative markers (i.e., a set of SNPs which shows substantially different frequencies between different ancestral populations) can be useful, although the best ancestry estimates are obtained using a very large number of random markers.

Replication across different studies is traditionally considered the necessary step to confirm that the observed association is real. Nevertheless several aspects need to be considered whenever different populations are assessed: it is common to observe different loci associated with the same disease across independent samples, especially if derived from different ethnic groups. This phenomenon (allelic heterogeneity) is expected in complex diseases because of the occurrence of pathogenic mutations across multiple domains of disease genes. Also common is the absence of a specific variant in some datasets or ethnic groups (locus heterogeneity). An additional scenario that adds complexity in the field of genetic epidemiology is differences in LD patterns. LD patterns depend on the ancestral background, as populations that have experienced bottlenecks (such as Caucasians) tend to show longer LD blocks and less evidence of recombination events when compared to populations having experienced less severe bottlenecks (such as Africans). As variants observed in GWASs commonly tag the true functional variant potentially not covered by the genotyping, tagging SNPs may not be replicable across studies if differences in LD in this genomic region exist, even when the causal variant is effectively shared.

Sometimes it is not uncommon to find an association with the disease for the same variant but in the opposite direction (e.g., in one sample the SNP is a risk factor for the onset of the disease, whereas in another one it is protective). Recombination events produce chromosomes that are mosaic of chromosomal regions originating from distinct ancestry, and the so-called flip-flop phenomenon is a well-known issue when interpreting transethnic discordant results.

The most common design for genetic association studies is the case-control design, employing samples of unrelated individuals. Nevertheless, recent advances in statistical and bioinformatics tools allow researchers to include more complex relationships between study individuals. Related individuals belonging to trios or larger pedigrees can be jointly analyzed by applying models that take into account the same family as a cluster (e.g., general estimating equation) or relationship between subjects independently from the reported pedigree but estimated directly from the genome (e.g., mixed models). The latter can model population structure, family structure, and cryptic relatedness. The basic approach is to model phenotypes using a mixture of fixed effects and random effects. Fixed effects include the candidate SNP and optional covariates such as sex or age, while random effects are based on a phenotypic covariance matrix, which is modeled as a sum of heritable and nonheritable random variation.

Some of the missing heritability of common complex disorders is likely carried by rare variants that may be identified in ongoing NGS projects (discussed in the next section), as opposed to the more common variants discovered by GWAS.

DNA sequencing

The gold standard for DNA sequencing for the past 25 years has been the chain termination method develop by Frederick Sanger (Sanger sequencing) in 1977. In automated high-throughput environments, DNA to be sequenced is either prepared and amplified by polymerase chain reaction (PCR) amplification using primers flanking the target or by cloning into a bacterial plasmid. Subsequently, sequencing is performed in an automated cycle-sequencing reaction in a thermal cycler, i.e., successive rounds of template denaturation, primer annealing, and primer extension (chain elongation). The primer is complementary to the known sequence immediately flanking the region of interest. Chain elongation occurs through incorporation of deoxynucleotides at the 3' end of the primer by basepair matching to the template. The extension product grows by the formation of a phosphodiester bridge between the 3′-hydroxyl group on the primer and the 5′-phosphate group of the incoming deoxynucleotide. Growth occurs in the $5' \rightarrow 3'$ direction. The growing sequence chain is randomly terminated by the incorporation of fluorescently labeled ddNTPs (chain termination) lacking the 3′-hydroxyl group, preventing further elongation of the chain and creating DNA fragments of varying lengths.

During subsequent capillary electrophoresis, the products of the cycle-sequencing reaction are injected electrokinetically into capillaries filled with polymer gel. High voltage causes the negatively charged DNA fragments to move towards the positive electrode through the path of a laser beam which causes the dyes on the fragments to fluoresce in four colors corresponding to the nucleotides A, C, G, or T. An optical detection device on the genetic analyzer detects the fluorescence signal. Software translates these traces into DNA sequence and generates error probabilities for each base call. Simultaneous electrophoresis in 96 or 384 independent capillaries provides a limited level of parallelization. To date, Sanger biochemistry can achieve read-lengths of up to ~1000 bp, and yields per-base “raw” accuracies as high as 99.999%.

Next-generation DNA sequencing

In 2003, the first major DNA sequencing effort, a $3 billion, 13-year-long international research project to determine the complete sequence of the human genome (Human Genome Project), was completed using Sanger technology. Since then, the demand for low-cost, high-throughput sequencing methods during which millions of fragments of DNA from a single sample can be sequenced simultaneously has continuously increased, acknowledging the value of mapping of disease-associated or protective variants for medical science and public health systems. As a result, several novel NGS methods have been developed which enable sequencing of an entire genome in less than 1 day. The four major NGS platforms currently on the market, each applying distinctive methods, are the Ion Torrent's PGM, Pacific Biosciences (PacBio)'s RS II, the Illumina HiSeq2000, and the Illumina MiSeq.

Template preparation

The first step in the sequencing process is the building and amplification of a library of nucleic acids (DNA or complementary DNA (cDNA)). Libraries are constructed by fragmenting the DNA (or cDNA) sample and ligating synthetic oligonucleotides of a known sequence (“adaptor”) on to the ends of the DNA fragments. Once constructed, libraries are clonally amplified in preparation for sequencing. The Illumina platforms capitalize on bridge amplification to form template clusters on a flow cell in Illumina's “cluster station,” an automated flow cell processor (Fig. 6.8). In this technology, the flow cell surface is coated with single-stranded oligonucleotides that correspond to the sequences of the adaptors ligated during the sample preparation stage. Single-stranded, adaptor-ligated fragments are bound to the surface of the flow cell exposed to reagents for polymerase-based extension. Priming occurs as the free/distal end of a ligated fragment “bridges” to a complementary oligo on the surface. Repeated denaturation and extension result in localized amplification of single molecules in millions of unique locations across the flow cell surface. This process occurs in what is referred to as Illumina's “cluster station,” an automated flow cell processor.

The PGM utilizes emulsion PCR to amplify single library fragments on to microbeads (ion sphere particles).

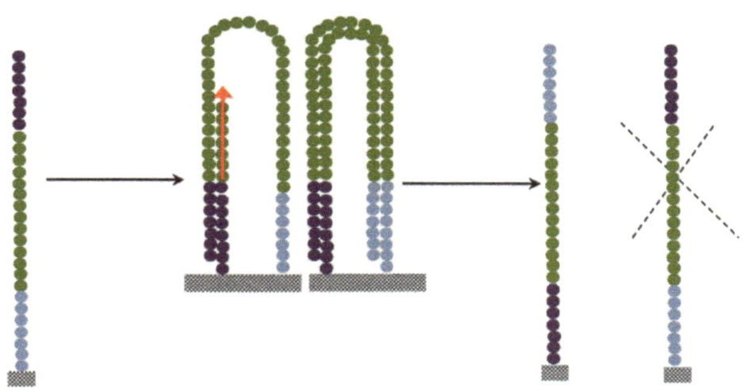

Fig. 6.8. Bridge amplification. DNA is randomly fragmented and adaptors are ligated to both ends of the fragments. Single-stranded fragments are bound randomly to the inside surface of the flow cells. Solid-phase bridge amplification incorporates deoxynucleoside triphosphate nucleotides, building double-stranded bridges on the solid-phase substrates. Denaturation leaves in each flow cell a cluster of cloned single-stranded amplified DNA molecules.

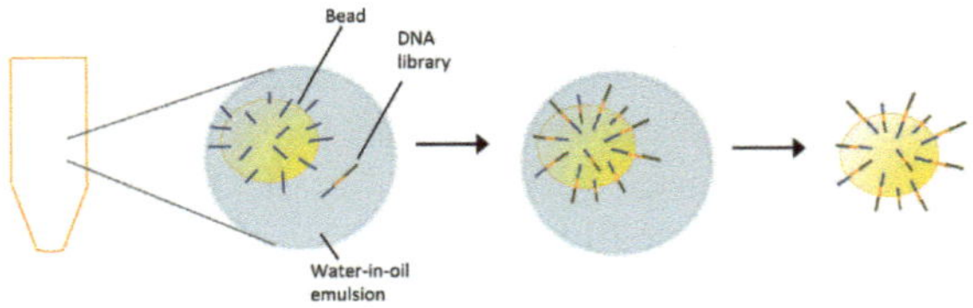

Fig. 6.9. Emulsion polymerase chain reaction (PCR). Beads, on which library DNA is captured via the adaptor sequence, are encapsulated in a water-in-oil emulsion. Within the emulsion PCR amplifies the library DNA. After amplification, the emulsion is disseminated and the beads are collected.

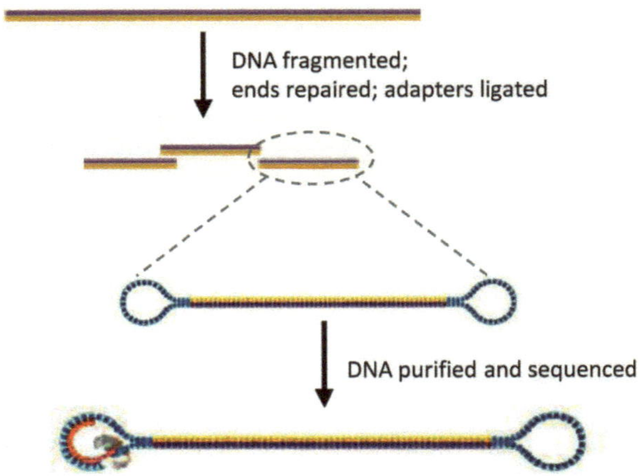

Fig. 6.10. Single-molecule, real-time (SMRT) technology workflow. The input DNA is first fragmented. Then, the ends are repaired and hairpin adaptors are ligated to both ends of each fragment. After purification, the SMRT bell template undergoes the sequencing reaction in which a polymerase opens the SMRT bell into a circular template and generates independent reads.

Emulsion PCR isolates individual DNA molecules along with primer-coated beads in aqueous droplets within an oil phase. PCR then coats each bead with clonal copies of the DNA molecule, followed by immobilization for the subsequent sequencing phase (Fig. 6.9).

The PacBio system utilizes single-molecule, real-time (SMRT) technology without the need for amplification (Fig. 6.10). In this technology, the submitted DNA serves as a direct template for the sequencing reaction via the ligation of hairpin adaptors that create a circular template. Real-time sequencing occurs continuously across this circular template (both forward and reverse of the same DNA molecule), generating extra-long read-lengths and high consensus accuracy. The quality score increases linearly with the number of times the molecule is sequenced. However, as the input DNA quality will be directly reflected in the sequencing results, any DNA damage or contaminants in the input material will affect the performance of the system.

Sequencing and imaging

The Illumina platforms employ a "sequencing-by-synthesis" approach derived from the Sanger sequencing methodology (Bentley et al., 2008). During each sequencing cycle, a single labeled deoxynucleoside triphosphate (dNTP) is added to the nucleic acid chain. The nucleotide label serves as a terminator for polymerization, so after each dNTP incorporation, the fluorescent dye is imaged to identify the base and then enzymatically cleaved to allow incorporation of the next nucleotide. Since all four reversible terminator-bound dNTPs (A, C, T, and G) are present as single, separate molecules, natural competition minimizes incorporation bias. Tens of millions of clusters are sequenced in parallel on the flow cell surface. Base calls are made directly from signal intensity measurements during each cycle.

The Ion Torrent PGM is based on semiconductor technology and capitalizes on the principle that, when a nucleotide is incorporated by the polymerase in the growing DNA chain, a proton is released, resulting in a local pH change (Rothberg et al., 2011). Each proton-sensing microwell of the Ion Torrent semiconductor sequencing chip contains $\sim$1 million copies of a DNA molecule. The sequencer sequentially floods the chip with each of the four bases. If a nucleotide complements the sequence of the DNA molecule in a particular microwell, it will be incorporated and hydrogen ions are released. The pH of the solution in that well changes and is detected by the ion sensor. If there are two identical bases on the DNA strand, the voltage is double, and the chip records two identical bases. If a nucleotide that floods the chip is not a match, no voltage change is recorded and no base is called.

As mentioned above, the PacBio RS employs SMRT sequencing technology, a parallelized single-molecule DNA sequencing by synthesis technology (Eid et al., 2009). For sequencing, a single DNA polymerase enzyme is affixed at the bottom of a zero-mode waveguide (ZMW), with the single molecule of DNA as the template. The ZMW creates an illuminated observation volume that is small enough to observe only a single nucleotide of DNA being incorporated by DNA polymerase. Each of the four DNA bases is attached to one of four different fluorescent dyes. When a nucleotide is incorporated by the DNA polymerase, the fluorescent tag is cleaved off and diffuses out of the observation area of the ZMW, where its fluorescence is no longer observable. A detector detects the fluorescent signal of the nucleotide incorporation, and the base call is made according to the corresponding fluorescence of the dye.

Additional, commonly employed sequencing/imaging methods are pyrosequencing and sequencing by ligation. Pyrosequencing, which is utilized by the Roche 454

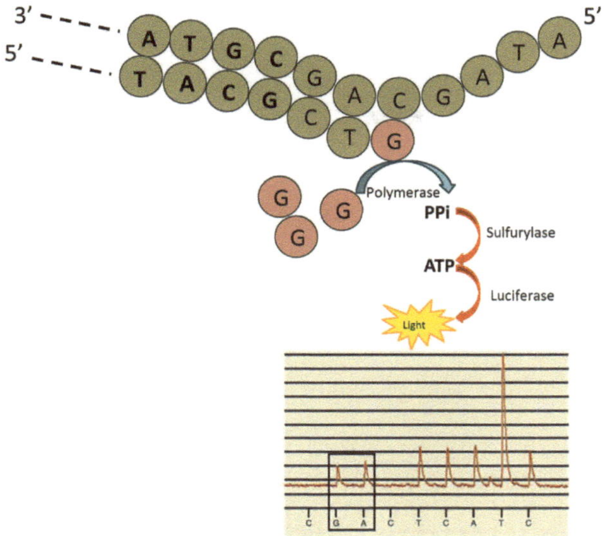

Fig. 6.11. Principles of pyrosequencing. Catalyzed by polymerase, nucleotides (in beige) are sequentially added to form the complementary strand of the single-stranded template, releasing one molecule of pyrophosphate (PPi) for every incorporated nucleotide. Sulfurylase converts PPi to adenosine triphosphate (ATP), which is subsequently used to convert luciferin to visible light. The amount of generated light is proportional to the amount of incorporated nucleotides at each nucleotide dispensation, and can be measured by a pyrogram.

platform and is based on the "sequencing by synthesis" principle, capitalizes on the detection of released pyrophosphate (PPi) during DNA synthesis (Fig. 6.11). After a nucleic acid polymerization reaction in which inorganic PPi is released as a result of nucleotide incorporation by polymerase, released PPi is converted to adenosine triphosphate (ATP) by ATP sulfurylase, providing the energy to luciferase to oxidize luciferin and generate visible light. This light is proportional to the number of incorporated nucleotides. Because the added nucleotide is known, the sequence of the template can be determined.

Sequencing by ligation is employed by the SOLiD technology (Fig. 6.12). After emulsion PCR, beads are deposited on to a glass slide and primers hybridize to the adaptor sequence on the template beads. Four fluorescently labeled probes compete for ligation to the sequencing primer, catalyzed by the enzyme DNA ligase. Multiple cycles of ligation, fluorescence detection, and cleavage are performed, with the number of cycles determining the eventual read-length. Following a series of ligation cycles, the extension product is removed and the template is reset with a primer complementary to the n-1 position for additional rounds of ligation cycles.

RNA sequencing/transcriptomics

The term "transcriptome" describes the full range of messenger RNA (mRNA) molecules expressed by an organism, a population of cells or a single cell, both in terms of type and quantity. In contrast to the genome which is characterized by its stability, the transcriptome can vary with external environmental conditions and, with the exception of mRNA degradation phenomena such as transcriptional attenuation, reflects the genes that are being actively expressed at any given time.

Since the development of DNA microarray technology in the 1990s, hybridization-based microarrays have been the technology of choice for large-scale studies of gene expression. The ability of these arrays to simultaneously analyze thousands of transcripts has led to significant insights into molecular mechanisms underlying a wide range of physiologic conditions, developmental processes, disease states, pharmacogenomic responses, and the evolution of gene regulation in different species. However, array technology has several limitations, including low accuracy of expression measurements for transcripts present in low or very high abundance, and the limitation of transcript interrogation on probes present on the array.

In the past 5 years, DNA microarray has been replaced by RNA-seq as the method of choice for whole-genome transcriptome profiling. RNA-seq is the direct sequencing of transcripts by high-throughput sequencing technologies. In a typical RNA-seq experiment, a population of long total or poly(A)+RNAs is converted through RNA or DNA fragmentation into a library of cDNA fragments. After attachment of sequencing adaptors to one or both ends, short sequences from each cDNA are obtained using high-throughput single-end or pair-end sequencing by synthesis (Fig. 6.13). The resulting sequence reads are aligned with a reference transcriptome or genome. The number of reads mapping to a scaled region of genome/transcriptome space reflects the expression level of the gene.

RNA-seq has several advantages compared to other transcriptome analysis technologies. It is the first sequencing-based method that allows the entire transcriptome to be surveyed in a high-throughput, quantitative fashion. It provides both full-genome-scale gene expression levels, as well as single-base resolution, making this technique also highly valuable for organisms lacking reference genomes. It can provide precise location of transcription boundaries, information on exon connectivity and sequence variation, and the ability to distinguish different isoforms and allelic expression in the transcribed regions. In contrast to DNA microarrays, which have significant background noise and lack sensitivity for genes expressed at either very low or very high

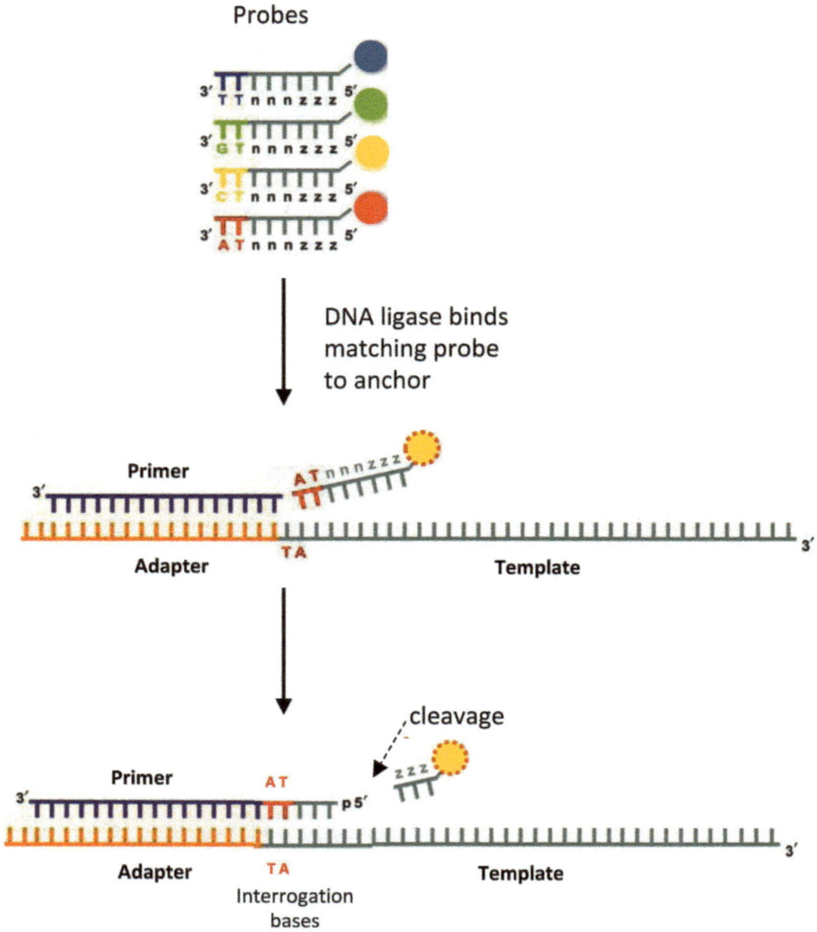

Fig. 6.12. Sequencing by ligation. The template with annealed primer is interrogated by probes representing 16 possible two-base combinations. After annealing and ligation of a matching probe, fluorescence is recorded before cleavage of the last three degenerate probe bases. Before the next sequencing step, the 5′ end of the cleaved probe is phosphorylated (not shown).

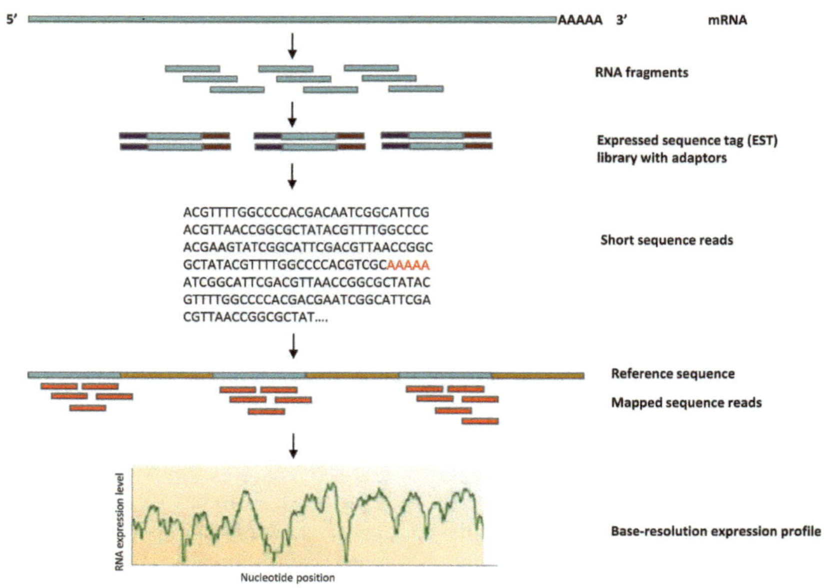

Fig. 6.13. Typical RNA-sequencing experiment.

levels, RNA-seq has low background signal and does not have an upper limit for quantification, allowing transcript detection over a large dynamic range of expression levels. For both technical and biologic replicates, RNA-seq findings show high reproducibility. Finally, RNA-seq requires a lower amount of RNA as the cloning and amplification steps can be circumvented.

Chromatin immunoprecipitation sequencing/microarray

Essential for understanding biologic processes during normal development and disease is also the assessment of protein–DNA interactions at a genomewide level to understand how proteins interact with DNA to regulate gene expression. A commonly used method to analyze protein interactions with DNA or RNA and to identify binding sites of DNA-associated proteins is chromatin immunoprecipitation with massively parallel DNA sequencing (ChIP-seq) or microarray (ChIP-chip). In a typical experiment, tissues or cells are treated with a chemical agent, usually formaldehyde, to cross-link proteins to DNA. Then, chromatin is sheared to target size and the protein of interest with its bound DNA fragment is enriched by purification with an antibody. After enrichment, cross-links are reversed, and the enriched DNA is purified and prepared for analysis. In ChIP-seq, the DNA is analyzed by high-throughput DNA sequencing. In ChIP-chip, the DNA is fluorescently labeled and hybridized to a DNA microarray, along with differentially labeled reference DNA.

Sequencing errors

A major limitation of the described high-throughput NGS technologies, which applies to WES, WGS, targeted sequencing, RNA-seq and ChIP-seq, is the high rate of erroneous base calls generated during sample preparation and sequencing, which results in miscalling of bases, causing misalignment of short reads and erroneous genome assembly (Gundry and Vijg, 2012). Experimental errors can occur in each step of the sequencing process (i.e., sample preparation, library preparation, sequencing, and imaging; Table 6.1). The error rate lies on the order of one incorrect base call per 100–1000 sequenced nucleotides. Given that there are billions of nucleotides in a human genome, even low rates of experimental error result in high error rates in variant calls.

Sample preparation

Sample degradation or contamination during sample isolation and preservation can lead to sequencing errors. Insufficient quantities of high-quality genomic material can increase amplification errors, resulting in decreased read depth (Akbari et al., 2005). Introduction of contaminating microbial DNA or exogenous human DNA during sample preparation, for example, through contaminated molecular biology-grade water, PCR reagents, or DNA extraction kits, can also generate erroneous results. The presence of contaminating DNA is in particular a challenge when only a low amount of sequencing material is available. In these cases, the low amount of starting material may be effectively swamped by the contaminating DNA and generate misleading results.

Table 6.1

Sources of experimental errors in next-generation sequencing

Batch effects
Sample preparation
User errors (for example, contamination, mislabeling)
Low DNA amount
DNA/RNA degradation
Library preparation
User errors (for example, contamination)
Polymerase chain reaction amplification errors
Primer bias
3′ end capture bias (RNA sequencing)
Split (chimeric) reads
Barcode/adaptor errors
Machine failure
Sequencing and imaging
User errors (for example, cluster cross-talk)
Incorrect incorporation of nucleotides
Nonfluorescing/damaged nucleotides, overlapping signals
T fluorophore accumulation
Strand bias
Machine failure

Library preparation

Errors occurring during library preparation can lead to uneven coverage, sequence changes, and interruption of sequence tags. Differential denaturation leading to significant GC% differences between alleles, DNA fragmentation, differential length of PCR products, or stochastic fluctuation in the number of copies of each allele when the initial number of genomes sampled is small, can cause preferential amplification of one allele relative to another (Walsh et al., 1992). Library amplification is subject to unmeasured primer biases, such as primer bias in multiple displacement amplification (Hutchison et al., 2005), mispriming in PCR target enrichment (Hodges et al., 2007), and incorporation of sequence errors during both clonal amplification and PCR cycling (Aird et al., 2011). Inadequately designed barcodes or adaptors can lead to erroneously assigned sequences with read loss and decreased read quality, and cross-contamination of datasets (Bystrykh, 2012). Also split (chimeric) reads (sequences formed from two or more accidentally joined clones), which can arise from long-insert paired-end libraries, can have serious ramifications for variant calling, *de novo* assembly, and structural variant (SV) prediction (Koboldt et al., 2010).

Sequencing and imaging

The current NGS platforms have sequencing and imaging error types that are specific to the platforms (Table 6.1) (Xuan et al., 2013). These include introduction of incorrect bases during clonal amplification of templates leading to substitution errors; errors arising from single-strand DNA folding or sequence-specific alterations in enzyme preference; reads that are subject to false insertions and deletions (indels) from nonfluorescing nucleotides (Fuller et al., 2009; Roberts et al., 2013); and erroneous counting of homopolymer stretches on pyrosequencing or semiconductor-sequencing platforms, resulting in carry-forward insertion and deletion errors (Yang et al., 2013). Cross-talk occurs when there is an overlap in signals of the dye emission frequencies. This overlap can lead to confusion of the nucleotide G with nucleotide T, and of A with C. Dephasing occurs with stepwise addition methods when growing primers move out of synchronicity for any given cycle. Lagging strands (for example, $n-1$ from the expected cycle) result from incomplete extension, and leading strands (for example, $n+1$) result from the addition of multiple nucleotides or probes in a population of identical templates. T fluorophore accumulation results in more Ts being incorrectly attributed towards the ends of reads.

Batch effects

Because measurements are affected by laboratory conditions, reagent lots and personnel difference batch effects are widespread, posing a common issue in NGS analysis.

Validation and replication

Current efforts in sequencing error mitigation mainly rely on postprocessing filtering strategies, including filtering for sequencing read depth, base call quality, short-read alignment quality, variant call quality, known variants, strand bias, allelic imbalance, and sequence context (Lynch, 2010; Lawrence et al., 2013; Roberts et al., 2013). These filtering techniques can further be optimized by increasing read depth, and by technical, biologic, and cross-platform replication. By increasing the number of short reads, variant calling on easily sequenced regions can be improved, and the number of false negatives (i.e., missed true variants) and sometimes the number of false positives (i.e., nonvariants that are incorrectly detected as variants) can be reduced. Technical replication (i.e., the repeat analysis of the same sample), biologic replication (analysis of multiple biologic samples from the same host under the same conditions), and cross-platform replication (i.e., integration of sequencing data from different technologies) can further improve the specificity and sensitivity of bioinformatics filtering approaches (Reumers et al., 2012). However, even with the use of replicates, some types of errors cannot be identified and addressed without further technologic advances and improvements in bioinformatic processing. For example, indels (Ye et al., 2009), paralogues, and other repetitive sequences (Jurka et al., 2005) often confound NGS short-read alignment (Li and Durbin, 2010), which results in mismapped reads and variant call errors. Other sources of errors, which cannot be addressed with replicates alone, can result from limitations in software performance and configuration during analysis, such as read trimming/clipping and filtering (i.e., the removal of low-quality strings of bases, sections of reads, and reads containing user-provided sequences) (Lindgreen, 2012), measurement of allelic bias (bias toward higher mapping rates of one allele compared with the alternative allele) (Degner et al., 2009) and calculation of variant call confidence (McKenna et al., 2010). Finally, incorrect variant calls can also occur when reads are mapped to uncomplete or misassembled reference genomes or transcriptomes (Genovese et al., 2013). These errors will decrease in frequency as reference genome assemblies and annotations are updated.

Duplex sequencing

As described above, all NGS approaches have a relatively high error rate, on the order of one incorrect base call per 100–1000 sequenced nucleotides (Shendure and Ji, 2008; Glenn, 2011). While this degree of error rate might be acceptable when studying inherited genetic variants, it limits the analysis of genetic alterations occurring in only a fraction of cells (subclonal mutations). Duplex sequencing is an NGS method capable of detecting a single mutation among $>1\times10^7$ wild-type nucleotides, enabling the identification of ultrarare genetic alterations in highly heterogeneous cell populations. It capitalizes on the fact that DNA is double-strand, with one strand reciprocally encoding the sequence information of its complement. Sequencing adaptors harboring random complementary double-stranded nucleotide sequences are ligated to the sample DNA of interest. Individually labeled strands are then PCR-amplified, creating sequence "families" that share a common tag sequence derived from the two original complementary strands (single-strand consensus sequence). Mutations are only considered as "true" when the variant is present in the consensus sequence of both DNA strands. Mutations that are only detected in the consensus sequence of one strand are regarded as technical errors.

Duplex sequencing yields unprecedented accuracy in sequencing of double-stranded DNA, with a >10 000-fold improvement compared with conventional NGS. In theory, it can be applied to any double-stranded DNA sample, but it is ideal for small genomic regions of <1 Mb in size.

PROTEOMICS

The study of DNA, described up to this paragraph, does not fully disentangle the human and other organisms' complexity: in fact, it is well known that DNA–RNA (and subsequential steps through the protein formation and its functions; Fig. 6.14) are not fully correlated (Li et al., 2011). Proteomics has emerged as a promising field of research in the context of complex human diseases and it refers to the analysis of the "proteome" (i.e., the whole set of proteins). The latter is a dynamic set, depending on the state of the cell and the organ to which it belongs, in contrast to the genome, which is fixed throughout the life of an organism. The human genome encodes 26 000–31 000 proteins. Several factors, including alternative splicing, protein modifications and degradations, assembly in complexes and protein–protein interaction, and their signal roles in complex pathways, make the study of proteomics challenging.

Proteomics is a complex field of study consisting of three main areas of research: (1) structural proteomics, which investigates the 3D structure of a protein; (2) functional proteomics, which focuses on protein–protein(s) and protein–other molecule(s) interactions; and (3) clinical proteomics, dedicated to the study of protein as potential biomarkers of various diseases (Drabik et al., 2007).

Several methods have been developed in the last decades to study the entire profile or specific proteins.

Two-dimensional electrophoresis

The traditional and still widely used method of protein separation is two-dimensional gel electrophoresis (2DGE) (O'Farrell, 1975). 2DGE relies on the properties

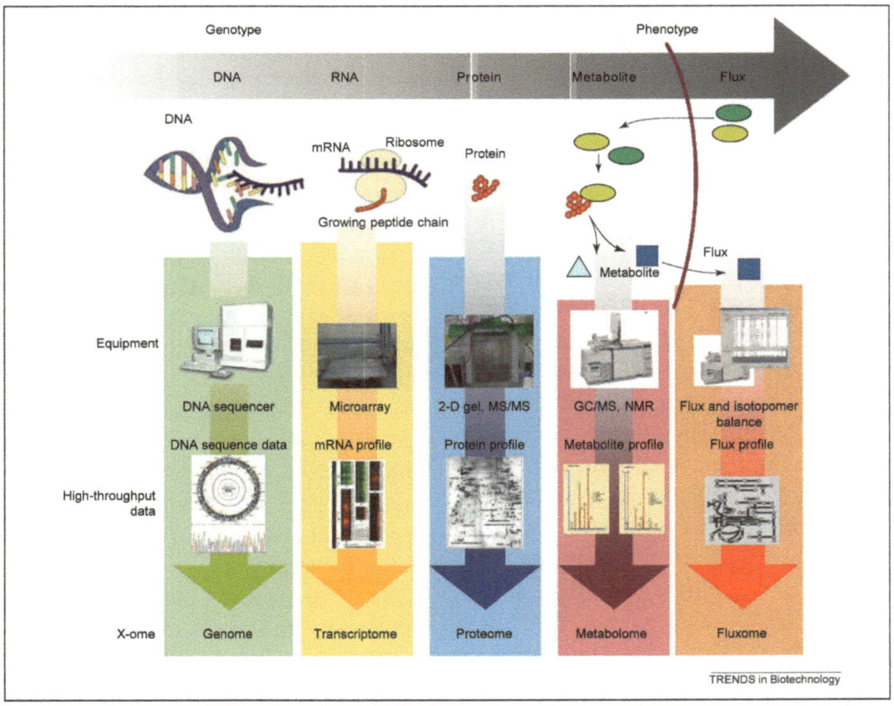

Fig. 6.14. Genomics advanced by the development of DNA sequencing is now accompanied by transcriptome profiling using DNA microarrays and RNA sequencing. Proteome profiling is joining the high-throughput race as two-dimensional (2D)-gel electrophoresis combined with mass spectrography is advancing. Metabolome profiling is also rapidly advancing with the development of better gas chromatography–mass spectrometry (GC/MS), liquid chromatography–mass spectrometry (LC/MS), and nuclear magnetic resonance (NMR) technologies. Isotopomer profiling followed by challenging with isotopically labeled substrate allows determination of flux profiles in the cell (fluxome). MS/MS, tandem mass spectrometry. (Adapted from Lee et al., 2005.)

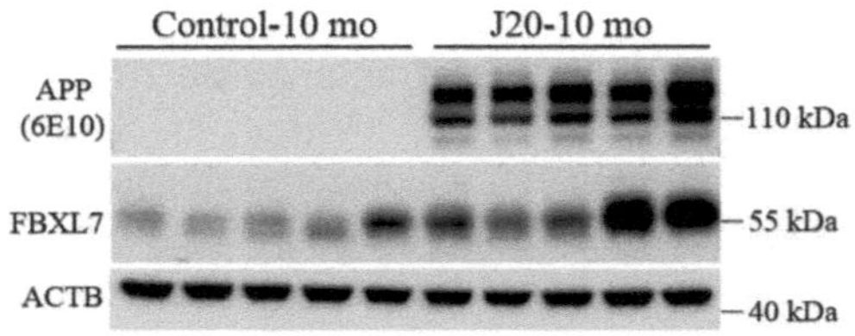

Fig. 6.15. Two-dimensional gel electrophoresis (2DGE): a common scenario where 2DGE is employed. Here, two mouse models are employed to study the expression of the *FBXL7* gene. The disease-transgenic mice ("J20-10 mo", an Alzheimer's-type mouse model) shows overexpression of the protein encoded by the gene, as compared to wild-type littermates ("Control-10 mo"). (Reproduced from Tosto et al., 2015.)

of charged molecules to migrate through a matrix when an electric field is applied: this allows the separation of complex protein mixtures by molecular charge (pI) and by molecular mass (Mr) (in the first and in the second dimension respectively) and each protein migrates to a specific location in a polyacrylamide gel. Proteins separated by this technique can then be visualized, for example, by staining with silver. The final result consists of a protein map in which different proteins are separated into spots, which provide numerous information, including molecular weight, charge, and quantity, as well as possible posttranslational modifications: this is particularly useful when two samples with different characteristics (disease vs. healthy controls, treated vs. nontreated) are to be compared (Fig. 6.15). 2DGE is extensively used, although predominantly for qualitative experiments because of its many limitations (i.e., reproducibility, inability to detect low abundant and hydrophobic proteins, low sensitivity in identifying proteins with very low or very high pH values ($pH < 3$ or > 10) and molecular masses too small ($Mr < 10$ kDa) or too large ($Mr > 150$ kDa)). Posttranslational modifications that alter the Mr and/or the pI of a protein are identified by the shift in location of the corresponding protein spot on the proteomic pattern. Further, in combination with Western blotting, antibodies specific for posttranslational modifications can reveal spots on 2DGE containing proteins with these modifications. Proteins can also be fluorescently labeled with fluorescent dyes prior to 2DGE.

Gel-free proteomics techniques are emerging as the methods of choice for quantitatively comparing protein levels among biologic proteomes, since they are more sensitive and reproducible than two-dimensional gel-based methods. In addition, samples can be directly analyzed by mass spectrometry (MS: see below), whereas samples processed through gel-based techniques need further processing.

Mass spectrometry

MS is considered one of the main tools in the proteomic field (Wright et al., 2012). It detects and quantifies proteins and their mass/charge ratio shifts due to post-translational modifications, of either intact proteins or peptides that derived from the original proteins. This method is very precise, being able to distinguish proteins that differ in composition even by a single hydrogen atom.

Analysis of proteins by MS occurs in three major steps: (1) protein ionization and generation of gas-phase ions; (2) separation of ions according to their mass/charge ratio; and (3) detection of ions. Nevertheless, MS is not able to handle complex protein mixtures from unprocessed human biospecimens. In order to carry out an experiment, the proteins derived from a specific tissue or fluid need to be first digested into peptides; additional technologies such as organelle or protein fractionation or affinity capture have been developed to reduce the complexity of proteins. Depending on the aim of the study, the resulting samples can be either prefractionated (i.e., the peptides are fractioned according to their physico-chemical properties) or enriched (i.e., selection based on modified peptides such as phosphorylation, demethylation, etc.). Subsequently, the complexity of this pre-processed sample is reduced by ultra-high-performance liquid chromatography before the mass spectrometer records their mass/charge ratio.

Protein microarrays

Protein microarrays are powerful tools for capturing and measuring proteins from biologic samples in a high-throughput fashion. A protein microarray typically consists of a small piece of glass or plastic coated with thousands of capture reagents (molecules that match specific proteins). This technology allows scientists to isolate and study many potential biomarker proteins. Protein microarrays can be miniaturized to contain tens of thousands of capture features arranged in a grid, each specific for a given protein; therefore, they are considered a multiplexed device – for example, they can test for multiple biomarkers simultaneously, which is essential for clinical use.

Functional protein microarrays are composed of arrays containing full-length functional proteins or protein domains in order to study, for example, protein–protein, protein–DNA, protein–RNA, protein–phospholipid, and protein–small-molecule interactions.

Clinical application of proteomics

Proteomics has been used to study two key aspects of the condition: (1) identification of biomarkers and (2)

Table 6.2

Types of biomarkers and related example of common biomarkers used in neuroscience

Type of biomarkers	Example
DNA-based biomarkers	DNA methylation MicroRNA
Blood-based biomarkers	Protein fragments (e.g., Abeta) Interleukins
Cerebrospinal fluid-based biomarkers	Protein fragments (e.g., Aβ42) Tau Albumin

understanding of the basic mechanisms of the disease. Several types of biomarkers can be detected (Table 6.2). An emerging field for proteomics is the accumulation of misfolded peptide/protein aggregation in key organs (Ren et al., 2014). Protein homeostasis and folding capacity (proteostasis) is, for example, a key pathway investigated in aging and neurodegenerative disorders in general. Posttranslationally, modification of protein residues is a key event for phosphorylation of tau in Alzheimer's disease neurofibrillary tangles, or cleavage and posttranslational modifications in the amyloid-beta peptides.

METABOLOMICS

With the term "metabolomics" is intended the study of the metabolites and/or low-molecular-weight intermediates (generally <2 kDa) profile at a cell or organ (Wood, 2014). Metabolomics studies the changes in the metabolome, which ultimately reflects all previous steps described in this chapter, i.e., genomic, transcriptomic, and proteomic changes. Therefore it represents an accurate biochemical phenotype of the organism in health and disease. The metabolome can be defined as the ultimate product of gene, mRNA, and protein activities. Several biofluids are the target of this field of research, albeit, for neuroscience, it has to be considered that the blood–brain barrier limits the passage of many metabolites, and, thus, it is questionable whether a biomarker produced within the brain may be detected within easily accessible biofluids, such as blood or urine. The optimal biofluid for metabolomics studies of neurologic or psychiatric disorders is the cerebrospinal fluid (CSF), which is in intimate contact with brain structures (Fiandaca et al., 2014). However, CSF studies are complicated in that they generally involve sampling of the spinal CSF, which is far removed from rostral brain structures where it is formed, and also by the dynamics of CSF export to the blood stream. Brain microdialysis is also used to perfuse local brain areas and obtain a profile of localized metabolism. However, CSF collection and brain microdialysis are both invasive procedures. Plasma is the alternative biofluid of choice, as it is more easily sampled and interacts with all tissues, providing an integrated output of changes in the homeostasis of this central compartment. Saliva metabolomics is also an area of increasing investigation because of the ease of obtaining this biofluid. Investigation of patient plasma cell populations (erythrocytes, platelets, and lymphocytes) also provides useful information of cellular metabolism in relation to plasma measurements.

Techniques for metabolomics are shared with the proteomics approach and mainly rely on MS (see above) along with nuclear magnetic resonance spectroscopy. The latter, albeit less sensitive as compared to MS, is highly reproducible, noninvasive, cheaper, and requires little preparation for the sample to study.

The high number of metabolites found in relation with physiologic or pathologic brain activity can be consulted online (http:/www.hmdb.ca), where thousands of metabolite entries can be freely found.

CONCLUSION

The recent advances in systems biology have significantly transformed the field of modern healthcare from symptom-based disease diagnosis and treatment to the possibility of precision medicine, in which patients are optimally treated based on their individual molecular characteristics. The development of high-throughput technologies such as NGS and MS, in combination with sophisticated bioinformatics software for data analysis and visualization, has enabled scientists and clinicians to examine genomes, transcriptomes, proteomes, metabolomes, and other "omics" information in unprecedented detail. This combined "omics" information provides scientists with the ability to gain a more comprehensive view of the way in which cellular processes work and the interactions influencing the steps from genotype to phenotype, allows a global profiling of health and disease, and opens avenues for precision medicine, enabling personalized health monitoring, prevention, and therapy. While in certain fields – in particular, oncologic diseases – precision medicine is already routinely applied in clinical settings, in neurologic diseases "omics" technologies have in recent years significantly advanced the understanding of the molecular underpinnings but have not yet reached the level of knowledge needed for translation into clinical practice. Nevertheless, we are confident that, with the increased application of "omics" technologies in neurologic research and ongoing advances in "omics" technologies, this will be achieved in the near future and will significantly improve patient care.

References

Aird D, Ross MG, Chen WS et al. (2011). Analyzing and minimizing PCR amplification bias in Illumina sequencing libraries. Genome Biol 12: R18.

Akbari M, Hansen MD, Halgunset J et al. (2005). Low copy number DNA template can render polymerase chain reaction error prone in a sequence-dependent manner. J Mol Diagn 7: 36–39.

Bentley DR, Balasubramanian S, Swerdlow HP et al. (2008). Accurate whole human genome sequencing using reversible terminator chemistry. Nature 456: 53–59.

Bystrykh LV (2012). Generalized DNA barcode design based on Hamming codes. PLoS One 7: e36852.

Degner JF, Marioni JC, Pai AA et al. (2009). Effect of read-mapping biases on detecting allele-specific expression from RNA-sequencing data. Bioinformatics 25: 3207–3212.

Drabik A, Bierczynska-Krzysik A, Bodzon-Kulakowska A et al. (2007). Proteomics in neurosciences. Mass Spectrom Rev 26: 432–450.

Eid J, Fehr A, Gray J et al. (2009). Real-time DNA sequencing from single polymerase molecules. Science 323: 133–138.

Fiandaca MS, Mapstone ME, Cheema AK et al. (2014). The critical need for defining preclinical biomarkers in Alzheimer's disease. Alzheimers Dement 10: S196–S212.

Fuller CW, Middendorf LR, Benner SA et al. (2009). The challenges of sequencing by synthesis. Nat Biotechnol 27: 1013–1023.

Genovese G, Handsaker RE, Li H et al. (2013). Using population admixture to help complete maps of the human genome. Nat Genet 45 (406–414): 414e401–414e402.

Glenn TC (2011). Field guide to next-generation DNA sequencers. Mol Ecol Resour 11: 759–769.

Gundry M, Vijg J (2012). Direct mutation analysis by high-throughput sequencing: from germline to low-abundant, somatic variants. Mutat Res 729: 1–15.

Hodges E, Xuan Z, Balija V et al. (2007). Genome-wide in situ exon capture for selective resequencing. Nat Genet 39: 1522–1527.

Hutchison 3rd CA, Smith HO, Pfannkoch C et al. (2005). Cell-free cloning using phi29 DNA polymerase. Proc Natl Acad Sci U S A 102: 17332–17336.

Jurka J, Kapitonov VV, Pavlicek A et al. (2005). Repbase Update, a database of eukaryotic repetitive elements. Cytogenet Genome Res 110: 462–467.

Koboldt DC, Ding L, Mardis ER et al. (2010). Challenges of sequencing human genomes. Brief Bioinform 11: 484–498.

Lawrence MS, Stojanov P, Polak P et al. (2013). Mutational heterogeneity in cancer and the search for new cancer-associated genes. Nature 499: 214–218.

Lee SY, Lee DY, Kim TY (2005). Systems biotechnology for strain improvement. Trends Biotechnol 23: 349–358.

Li H, Durbin R (2010). Fast and accurate long-read alignment with Burrows-Wheeler transform. Bioinformatics 26: 589–595.

Li M, Wang IX, Li Y et al. (2011). Widespread RNA and DNA sequence differences in the human transcriptome. Science 333: 53–58.

Lindgreen S (2012). AdapterRemoval: easy cleaning of next-generation sequencing reads. BMC Res Notes 5: 337.

Lynch M (2010). Rate, molecular spectrum, and consequences of human mutation. Proc Natl Acad Sci U S A 107: 961–968.

McKenna A, Hanna M, Banks E et al. (2010). The Genome Analysis Toolkit: a MapReduce framework for analyzing next-generation DNA sequencing data. Genome Res 20: 1297–1303.

Menozzi P, Piazza A, Cavalli-Sforza L (1978). Synthetic maps of human gene frequencies in Europeans. Science 201: 786–792.

O'Farrell PH (1975). High resolution two-dimensional electrophoresis of proteins. J Biol Chem 250: 4007–4021.

Price AL, Patterson NJ, Plenge RM et al. (2006). Principal components analysis corrects for stratification in genome-wide association studies. Nat Genet 38: 904–909.

Ren RJ, Dammer EB, Wang G et al. (2014). Proteomics of protein post-translational modifications implicated in neurodegeneration. Transl Neurodegener 3: 23.

Reumers J, De Rijk P, Zhao H et al. (2012). Optimized filtering reduces the error rate in detecting genomic variants by short-read sequencing. Nat Biotechnol 30: 61–68.

Roberts RJ, Carneiro MO, Schatz MC (2013). The advantages of SMRT sequencing. Genome Biol 14: 405.

Rothberg JM, Hinz W, Rearick TM et al. (2011). An integrated semiconductor device enabling non-optical genome sequencing. Nature 475: 348–352.

Shendure J, Ji H (2008). Next-generation DNA sequencing. Nat Biotechnol 26: 1135–1145.

Slatkin M (2008). Linkage disequilibrium – understanding the evolutionary past and mapping the medical future. Nat Rev Genet 9: 477–485.

Tosto G, Fu H, Vardarajan BN et al. (2015). F-box/LRR-repeat protein 7 is genetically associated with Alzheimer's disease. Ann Clin Transl Neurol 2: 810–820.

Walsh PS, Erlich HA, Higuchi R (1992). Preferential PCR amplification of alleles: mechanisms and solutions. PCR Methods Appl 1: 241–250.

Wood PL (2014). Mass spectrometry strategies for clinical metabolomics and lipidomics in psychiatry, neurology, and neuro-oncology. Neuropsychopharmacology 39: 24–33.

Wright PC, Noirel J, Ow SY et al. (2012). A review of current proteomics technologies with a survey on their widespread use in reproductive biology investigations. Theriogenology 77 (738–765): e752.

Xuan J, Yu Y, Qing T et al. (2013). Next-generation sequencing in the clinic: promises and challenges. Cancer Lett 340: 284–295.

Yang X, Chockalingam SP, Aluru S (2013). A survey of error-correction methods for next-generation sequencing. Brief Bioinform 14: 56–66.

Ye K, Schulz MH, Long Q et al. (2009). Pindel: a pattern growth approach to detect break points of large deletions and medium sized insertions from paired-end short reads. Bioinformatics 25: 2865–2871.

Handbook of Clinical Neurology, Vol. 138 (3rd series)
Neuroepidemiology
C. Rosano, M.A. Ikram, and M. Ganguli, Editors
http://dx.doi.org/10.1016/B978-0-12-802973-2.00007-0

Chapter 7

Neuropsychologic assessment

P. PALTA[1], B. SNITZ[2], AND M.C. CARLSON[3]*

[1]*Department of Epidemiology, University of North Carolina at Chapel Hill, Chapel Hill, NC, USA*

[2]*Department of Neurology, University of Pittsburgh, Pittsburgh, PA, USA*

[3]*Department of Mental Health, Johns Hopkins Bloomberg School of Public Health, Baltimore, MD, USA*

Abstract

In this chapter, we review the use of neuropsychologic assessment in epidemiologic studies. First, we provide a brief introduction to the history of clinical neuropsychology and neuropsychologic assessment. We expand on the principal components of a neuropsychologic assessment and cognitive domains most commonly examined. This chapter also seeks to highlight specific domains and tests with validated psychometric properties that are widely accepted in clinical practice, as well as how data from a neuropsychologic test should be interpreted. Additionally, the important roles that neuropsychologic assessments play in tracking normative changes, patient diagnoses, care, and research will be discussed. Factors to consider when deciding on the inclusion of test instruments for a research study will also be reviewed. Lastly, we shed light on the contributions that neuropsychology has played in epidemiologic studies, as well as some challenges frequently faced when participating in this field of research.

INTRODUCTION

The link between behavioral expressions and underlying brain structure and function was the driving force for the emergence of the field of applied clinical neuropsychology. Efforts to explain the connection between our brain and behavior extend as far back as the early 17th century (Castro-Caldas and Grafman, 2000). In an era when neuroimaging was not yet on our technologic horizon, Franz Joseph Gall was the first to hypothesize explanations for observed behavioral differences between his schoolmates. Gall first linked cerebral lesions with aphasia or language difficulties (Gall, 1809), establishing a foundation for the enormous efforts made by Bouillaud and Broca on functional localization, yielding a greater breadth of knowledge on the specialized functions of certain areas of the brain. The field of clinical neuropsychology became a practical necessity during the World War I era, when the co-occurrence of brain injuries and behavioral issues among war veterans became more apparent (Lishman, 1968, 1973, 1988), therefore increasing the demand for rehabilitation programs and prompting initiatives to establish neuropsychology training programs. Neuropsychologic assessment forms the basis of the field of clinical neuropsychology.

Neuropsychologic assessment is the use of standardized behavioral or cognitive tasks to detect impairments in cognition to make inferences about brain function. The utility of information from a neuropsychologic assessment can be comprehensive, from screening to diagnoses and, most importantly, treatment. The principal component of a complete neuropsychologic assessment involves the examination of several domains of cognition, including memory, executive function, psychomotor speed/attention, visuospatial construction, and language. Each of these cognitive domains is predominantly mapped or localized to specific regions or networks of the brain to infer brain functions. Imaging modalities have complemented the role of traditional neuropsychologic testing in the localization of overt brain pathologies. However, neuropsychologic assessment is unique in its measurement of behavioral function

*Correspondence to: Michelle C. Carlson, Department of Mental Health, Johns Hopkins Bloomberg School of Public Health, 624 N. Broadway, Hampton House 805, Baltimore MD 21205, USA. Tel: +1-410-614-4887, E-mail: mcarlso2@jhu.edu

closely linked to, yet independent from, brain injury or disease.

In this chapter, we expand on the contents of a neuropsychologic assessment, and also discuss more broadly the role that neuropsychology plays in tracking normative changes, patient care, and in epidemiologic research. In order to understand whether a patient's or participant's behavior is atypical, researchers often rely on normative data from within a sample or from other well-defined samples that take into account factors known to impact an individual's abilities, including demographic and lifestyle factors (e.g., education, age, medical status, history of chronic conditions, emotional functioning, activity level, sleep patterns, and medication intake), among many other potentially relevant factors. Advances have been made over recent years regarding assessment and interpretation of neuropsychologic data; however, at the same time that we shed light on opportunities, we also identify the challenges frequently faced in this field of research. Most importantly, it is imperative that we place an individual's neuropsychologic test scores in context, both in terms of broader epidemiology and sociodemographic patterns and in relation to one's own prior (premorbid) function, which may or may not be known.

ASSESSMENT OF COGNITIVE DOMAINS AND RELATED BRAIN FUNCTIONS

Assessing cognitive functions is multifaceted and combines information from neuropsychologic testing, neuroimaging, neurologic and psychiatric examinations, demographics, and medical history. Neuropsychologic testing does not inform whether lesions are present, as does neuroimaging; rather, the goal is to use standardized behavioral/cognitive tests to make inferences about normative and nonnormative brain function. Particularly, neuropsychologic test performance has been linked to disease-typical patterns and specific brain functions, thereby aiding the diagnoses of central nervous system disorders. Disease-typical patterns are often identified based on performance in cognitive domains. For example, cerebrovascular disease-related cognitive decline and traumatic brain injuries tend to affect areas of the brain responsible for executive functioning (Selnes and Vinters, 2006; Gorelick et al., 2011; Cristofori et al., 2015). We now discuss in greater depth the most commonly studied and widely affected domains of cognition: memory, visual-spatial construction, language, attention/psychomotor speed, and executive functioning. We will highlight specific domains and tests with validated psychometric properties that are widely accepted in clinical practice.

It is important to note that cognitive tests, although often times categorized as domain-specific, are typically multifactorial with respect to task demands. For example, a test of executive function will first establish whether an individual is able to attend to the simpler components of the test, including visual search (e.g., Trail Making Test (TMT), Part A) or expressive language (e.g., Digit Span Forward). Furthermore, no single test can evaluate all aspects of a cognitive domain; rather it is intended to evaluate a primary component of a given domain. Table 7.1 summarizes cognitive domains, their

Table 7.1

Principal cognitive domains of a neuropsychologic assessment and commonly used neurocognitive tests

Domain	Brain-related biologic relevance	Example tests*
Memory (working and long-term)	Prefrontal, parietal, cingulate cortex, related thalamic regions, hippocampus, associated medial temporal structures	California Verbal Learning Test (Delis et al., 2000); CERAD Word List Learning; WMS-IV Logical Memory (Wechsler, 2009); Digit Span Forward (Wechsler, 1981)
Visual-spatial construction	Parietal lobe	Rey–Osterrieth Figure Copy (Rey, 1941; Osterrieth, 1944); Benton Visual Retention Test (Benton, 1946)
Language	Left-hemisphere language network (Broca and Wernicke areas)	Boston Naming Test (Kaplan et al., 1983); Token Test (De Renzi and Vignolo, 1962)
Attention/psychomotor speed	Dorsolateral prefrontal cortex, posterior parietal, anterior cingulate cortex, subcortical connections	Digit Span Forward (Wechsler, 1981); Trail Making Test, Part A (Reitan, 1944)
Executive functions	Frontal network (dorsolateral, prefrontal, orbitofrontal, posterior parietal, cingulate cortex, basal ganglia)	Digit Span Backward (Wechsler, 1981); Trail Making Test, Part B (Reitan, 1944); Stroop Color-Word Test (Stroop, 1935); Clock Drawing (Agrell and Dehlin, 1998)

*A sample of tests included is recommended for cognition assessment using the National Institutes of Health Toolbox (Weintraub et al., 2013). WMS, Wechsler Memory Scale; CERAD, Consortium to Establish a Registry for Alzheimer's Disease.

associated targeted areas of brain function, and validated neurocognitive tests commonly used to ascertain domain performance. It is beyond the scope of this chapter to describe all relevant neuropsychologic tests currently in use. Test descriptions and normative data are available for a number of tests and can be found in other books devoted solely to neuropsychologic assessment (Strauss et al., 2006; Lezak et al., 2012).

Memory

Memory involves the ability to encode and store stimuli for retrieval under different conditions. The Atkinson–Shiffrin memory model first operationalized memory into two important subcomponents: working memory/short-term memory and long-term memory (Atkinson and Shiffrin, 1968). Working memory is the temporary storage of verbal and/or visual information for brief periods and is subject to limited capacity. A familiar example is the storage of a nine-digit phone number as one prepares to dial the number (Miller, 1956). Once the phone number is dialed, it is no longer needed in memory. Baddeley and Hitch operationalized and mainstreamed the concept of working memory in the 1970s using a multicomponent model (Baddeley and Hitch, 1974; Baddeley, 2007). In this model, Baddeley and Hitch described three ordered components: the attentional processing unit (the central executive), which determines whether the information is stored, and two temporary storage units (phonologic loop and visuospatial sketchpad) to determine how the information is stored.

Working memory is often measured with digit span tests, such as the Wechsler Adult Intelligence Scale (WAIS) Digit Span Forward or Backward test (Wechsler, 1981). In this test, participants are orally presented with a string of digits that they are required to repeat back in the same order in which they were presented to them. To increase the difficulty of the tasks, participants may be asked to repeat back the string of digits in reverse order. Past research suggests that adults are, on average, able to process at most chunks of seven digits, plus or minus two digits (Miller, 1956).

The greater the length of time information is kept in short-term memory and rehearsed repeatedly will yield information entered into long-term memory. Long-term memory involves the storage of information over prolonged periods of time. Long-term memory is further separated into declarative memory and procedural memory. Declarative or episodic memory involves experiential memories that are either consciously encoded memories related to personal events, or facts and events that are autobiographic in nature (e.g., George Washington was the first President of the USA), but become semantic memories. Procedural memory involves those memories that are not activated by conscious control, such as riding a bike or using a pencil to write.

Several validated tests are available to assess long-term memory, particularly verbal episodic memory. The California Verbal Learning Test (CVLT) (Delis et al., 2000) and Hopkins Verbal Learning Test (Rasmusson et al., 1995; Shapiro et al., 1999) are examples of list-learning tests which include immediate and delayed word recall. For CVLT immediate recall, participants are orally presented with a 16-word list to recall in a series of five trials to assess immediate word recall. Following a 20-minute delay, participants are asked to recall the list to assess delayed word recall.

Visual-spatial construction

Visual-spatial construction involves the ability to manipulate and arrange objects in positional relation to each other and in space. Two classes of activity are typically used to assess visual-spatial construction: drawing and building/assembling. Figure copying of an abstract, complex figure, such as that in the Rey–Osterrieth Complex Figure Test (Rey, 1941; Osterrieth, 1944), is a classic test to measure impairments in visuospatial construction. The specifics of how a picture is copied shed light on areas of the brain affected by lesions. For example, studies of patients with unilateral lesions show that those with left-sided lesions tend to draw small pictures, whereas participants with right-sided lesions tend to draw large pictures, both of which are not drawn to scale of the original figure (Larrabee and Kane, 1983). Data suggest modest age effects on quantitative scores (Strauss et al., 2006), but differences in the quality of the figures have been noted (Ska and Nespoulous, 1987).

Tests that draw on the competencies of building or assembling quantify individuals' spatial abilities. In WAIS Block Design (Wechsler, 1981), participants are presented with two, four, or nine red and white blocks depending on item difficulty. Each block has two red and two white sides and two half-red, half-white sides colored diagonally on the block. The participant is asked to construct a model based on a drawing, using only the blocks provided. As the participant moves through the items, the number of blocks and designs increases in difficulty. Significant age effects are seen, with declines beginning in midlife (Ryan et al., 2000). Older participants are observed to take a longer time to complete the task than their younger counterparts (Ogden, 1990; Salthouse et al., 1996).

Language

Language is the expression of written and/or oral words. It not only involves the fluency and expression of speech, but also the comprehension of speech. Classic cases of language impairments in expression and comprehension are known as aphasias in the Broca and Wernicke regions in the inferior frontal and posterior temporal lobes, respectively, of the left hemisphere, where language is typically localized.

Fluency tests are widely used to test abilities in production of language. The Controlled Oral Word Association Test (Lezak et al., 2012) from the Halstead–Reitan Neuropsychological Battery is a verbal fluency test in which participants are asked to say as many words as possible from a given category and in a specified timeframe (typically 60 seconds). Categories of fluency tests are often initial letters of words (e.g. "f," "s," or "a") or semantic categories, such as animals or fruits. Findings are mixed regarding age effects in verbal fluency tests (Benton and Hamsher, 1994; Hughes and Bryan, 2002). The Boston Naming Test (Kaplan et al., 1983) is a language test of noun retrieval, primarily to diagnosis aphasia, where participants are presented with 60 ink drawings of objects and asked to name them. Age effects are small, with almost no decline seen until the 70s, and thereafter only small declines occur. Considerable variability in test scores has been noted in older adult populations (Van Gorp et al., 1986; Ross et al., 1995; Welch et al., 1996). The Token Test is administered to detect impairments in language comprehension. Circle and square tokens of two sizes and five colors are presented to the participant in random order. Participants are then asked to respond to commands of varying complexity, such as, "Touch the large red circle" or "Touch the small yellow circle and the large green square."

Attention/psychomotor speed

Psychomotor speed and attention involve an individual's ability to rapidly process information and execute a response, in addition to selectively focusing on target stimuli while ignoring irrelevant stimuli. Tests in this domain ascertain an individual's vigilance, reaction time (speed), or ability to divide and shift attention. Set shifting corresponding with divided attention often overlaps with the executive function domain.

The TMT, Part A (Reitan, 1944; Lezak et al., 2012) is a commonly used psychomotor test to assess speed of information processing. Participants are presented with circled numbers 1–25 positioned randomly on a paper, and are asked to connect the numbers in ascending order as fast as possible. Time (in seconds) to task completion is the common metric used. Studies have shown that the time to task completion increases steadily with each decade of age (Ernst et al., 1987; Stuss et al., 1987).

Executive function

Executive function represents the ability to abstract and involves shifting, inhibitory, and working-memory processes that often overlap with several of the other domains, most often memory and attention/psychomotor speed (Miyake et al., 2000). Working memory, specifically, is important for executive function because it involves the temporary storage and processing of new or existing information and is central to the coordination of all processes of executive function. Overall, executive function is the cognitive system responsible for planning, organizing, problem solving, goal setting, and cognitive flexibility. It also underlies concept formation and abstract thinking. Cognitive flexibility, as noted above, is important for set shifting and the process of mentally switching cognitive processes required for a given task.

A prototypical test of executive function that highlights cognitive flexibility is the TMT, Part B, which combines executive, memory, and psychomotor speed/attention processes (Reitan, 1944; Lezak et al., 2012). Participants are presented with numbers and letters positioned randomly on one side of a paper, and are asked to connect the numbers and letters in numeric-alpha ascending order (1-A-2-B-3-C) as fast as possible. Scoring for the TMT, Part B is based on time (in seconds) to completion of task. Studies have shown that the time to task completion increases steadily with each decade of age (Ernst et al., 1987; Stuss et al., 1987).

Response inhibition involves the ability to suppress unnecessary information while problem solving. A commonly used task is the Stroop Color/Word Interference Test (Stroop, 1935), where participants are challenged to state the ink color of the word presented and not the name color-word itself, which is conflicting with the ink color. Declining test performance with increasing age has been well documented (Cohn et al., 1984; Daigneault et al., 1992).

INCORPORATION OF COGNITIVE ASSESSMENTS IN RESEARCH STUDIES

Cognitive tests, although often times categorized as domain-specific, are broad and multifactorial. For example, cognitive demands for TMT, Part B involve working memory, visual-spatial skills, and motor and executive functions. When considering what test instruments to include in a research study, several factors should be considered, including length and ease of administration, participant burden, psychometric properties, generalizability to population under study, and availability of norms to infer appropriate comparisons (Mitrushina et al., 2005). In some cases, a predetermined comprehensive neuropsychologic test battery, such as the Halstead–Reitan Battery, which includes a set of well-validated and reliable cognitive test instruments with available norms, may be used. These batteries are typically administered in clinical vs. epidemiologic settings given the time and resources required to implement, score, and interpret the patterns of performance. Established norms, although important for the interpretation of cognitive test scores, are limited in their generalizability beyond the clinic samples for which the norms were established. Cost constraints and availability of resources (e.g., test administrators) often limit the use of such a full neuropsychologic test battery, therefore encouraging the use of a select battery of tests.

To aid the inclusion of cognitive test assessments into research practice, the National Institutes of Health (NIH) charged an expert panel of neuropsychologists, neurologists, and practitioners to identify a brief and convenient set of validated cognitive tests for use in epidemiologic and longitudinal research, as well as clinical trials (Table 7.2) (Weintraub et al., 2013). This test battery, NIH Toolbox, can be administered in a computerized format and has been nationally standardized to provide a "common currency" among researchers, therefore facilitating the appropriate use of cognitive tasks into epidemiologic studies and clinical trials.

Several other computerized neurocognitive test batteries have been proposed to aid these research and clinical efforts, including the CANTAB (Cambridge Neuropsychologic Test Automated Battery) and COGSTATE.

Table 7.2

Psychometric properties of neuropsychologic tests

Characteristic	Definition
Validity	Effectiveness of the test to measure what it is purposed to measure (e.g., distinguishing individuals with and without cognitive impairment)
Sensitivity	Probability that the test correctly classifies cases or individuals with impaired cognitive functioning (or disease)
Specificity	Probability that the test correctly classifies individuals with normal cognition (or without disease)
Reliability	Repeatability of a test and its ability to produce consistent results
Standard deviation (SD)	Measure of dispersion for test scores, indicating on how far from the mean, on average, are the scores in a distribution

These standardized batteries are simple, fast, and reliable and have several additional desirable characteristics, including: (1) minimization of participant burden; (2) providing comprehensive information on key domains of cognition (e.g., executive control function, memory); (3) translatable across ages (e.g., could be used for children and adults so that one could track trajectories of cognitive decline from childhood to adulthood); (4) previously published and validated, hence comparable across other studies; and (5) ease of administration and reliability of scoring.

APPLICATION OF NEUROPSYCHOLOGIC TESTING FOR PATIENT CARE AND RESEARCH

The utility of information obtained from a neuropsychologic assessment is vast, from screening for the presence of abnormalities to tracking normative changes to diagnoses of conditions and, most importantly, monitoring responses to treatment. Lezak and colleagues (2012) describe the utility of neuropsychologic examinations in several facets of patient care: diagnosis, management and planning of patient care, and treatment (treatment needs and treatment efficacy), in addition to its important contributions to research.

Patient care

Screening

Tests developed to screen for cognitive impairments can be critical for the prevention of further decline in function. In the clinical setting, a simple screener can guide healthcare practitioners on the global cognitive functioning of a patient or participant to be enrolled in a study, thereby allowing them to make important decisions and referrals on care for the patient's mental health status or suitability for inclusion in a study or intervention. Several screening measures for dementia have been developed, including the Mini Mental State Examination (MMSE: Folstein et al., 1975) and the Montreal Cognitive Assessment (MoCA: Nasreddine et al., 2005). Although useful in detecting suspected cognitive impairment, these measures do not provide information on impairments in specific domains of cognition. These tests are also not sensitive for detecting early or subtle cognitive impairments. Of critical need is the development of screening tests that are sensitive to early and/or mild impairments related to underlying brain pathology vs. normal fluctuations in test performance. The use of such tests will help to prevent further cognitive and functional declines.

Diagnosis

In the advent of neuroimaging methods that are reliable and sensitive for diagnostic purposes, neuropsychologic assessments are still important for the diagnosis of neurologic conditions. Neuropsychologic assessment is useful for discriminating between psychiatric and neurologic symptoms as well as distinguishing between different neurologic conditions based on the behavioral attributes that are related to localized brain lesions or atrophy. In some patients, neuroimaging alone may fail to discriminate between vastly different behavioral outcomes, thereby increasing the utility of neuropsychologic assessments to distinguish differences across such patients. For example, Bigler (2001) identified 3 patients with focal left frontal traumatic brain injury of similar size and localization, but who had different behavioral symptoms on neuropsychologic examination, thereby lending relevant information to the diagnosis of the overall neurobehavioral syndrome. The patients experienced different neurobehavioral symptoms, ranging from word fluency, motor, and executive impairments in 1 patient to the absence of impairments in verbal or motor functions in another patient. Therefore, combined information from both neuroimaging and neuropsychologic assessments is useful for highlighting the variability in brain–behavior relationships.

Planning/management of patient care and treatment

Self-management and patient care often depend on an individual's behavioral capabilities in daily life and, moreover, the extent of the patient's cognitive limitations. Neuropsychologic assessment is useful for determining an individual's levels of cognitive impairment and independent function in order to inform treatment planning and management. Neuropsychologic examinations also provide information on the progression of a neurologic condition and treatment effects aiding in the design and adjustment of treatment regimens. For example, repeat cognitive testing in brain-injured patients can provide valuable information as to the patient's recovery as well as the patient's responses to rehabilitation therapy and treatments postinjury, such as during the critical first year following recovery from stroke (Desmond et al., 1996) and traumatic brain injury (Christensen et al., 2008).

Utility of neuropsychologic assessment in the treatment of chronic conditions

Neuropsychologic examination can be informative beyond diagnosis of neurologic conditions for the planning and management of patient care. Other disease conditions associated with comorbid cognitive impairment warrant neuropsychologic examination to aid in treatment planning and management. For example, in the setting of diabetes, self-management is important for the prevention of diabetes-related complications and other adverse outcomes. Poor self-management in diabetes can lead to an increased incidence of diabetic complications, such as micro- and macrovascular disease, and death. Impairments in executive function and memory among individuals with diabetes have been associated with diabetes self-management behaviors. One study found an inverse correlation between cognitive impairment and diabetes self-management behaviors (Thabit et al., 2009). In this study, executive function was found to have a significant impact on diabetes self-management. Individuals with a greater dysexecutive function score, as measured by the Executive Interview 25 (Royall et al., 1992), performed lower on the Summary of Diabetes Self-Care Activities Scale (SDSCA) (Thabit et al., 2009). Global cognitive dysfunction (Munshi et al., 2006) and executive dysfunction (Nguyen et al., 2012) have also been associated with higher hemoglobin A1c levels, a risk factor for many diabetes-related complications and other adverse outcomes. Considering the impact that cognitive dysfunction may have on metabolic measures and future complication risk, specific cognitive screening tools may be tremendously useful if implemented in the diabetes clinical care setting.

Community-based aging cohort studies

Neuropsychologic testing in the theoretic and applied fields of research adds a wealth of information when used to identify modifiable exposures related to healthy and accelerated patterns of cognitive aging. Most commonly, neuropsychologic testing or the cognitive test scores themselves are an outcome in research studies. For example, cardiovascular disease and many cardiovascular disease risk factors (e.g., diabetes (McCrimmon et al., 2012), metabolic syndrome (Panza et al., 2010), and stroke (Tatemichi et al., 1994)) have been studied in relation to cognitive test performance and cognitive impairment. Using a comprehensive neuropsychologic test battery, the Ginkgo Evaluation of Memory Study (GEMS) examined changes in cognitive domain scores to test the effectiveness of ginkgo biloba for the treatment of cognitive decline and dementia (DeKosky et al., 2008; Snitz et al., 2009a).

Neuropsychologic testing is also often used in epidemiologic studies in the definitions or diagnosis of prevalence or incidence of mild cognitive impairment (MCI) or dementia (Derrer et al., 2001). In addition to detecting impairments, repeated neuropsychologic testing is useful for characterizing patterns of age-related cognitive declines that predict risk for adverse health outcomes. In the Asset and Health Dynamics Among the Oldest Old (AHEAD) study, investigators examined whether comorbid cognitive function and depressive symptoms affected the risk of mortality in community-dwelling older adults (Mehta et al., 2003). Another example from the Monongahela Valley Independent Elders' Survey (MoVIES) study was in the evaluation of cognitive tests that best discriminate between presymptomatic Alzheimer's disease and those individuals who remain nondemented (Chen et al., 2001). In the Women's Health and Aging Study II (WHAS II), executive attentional abilities were linked to performance on complex instrumental activities of daily living (Carlson et al., 1999) and observed to decline earlier than, and as rapidly as, memory over a 9-year period (Carlson et al., 2009).

MEASUREMENT AND INTERPRETATION OF NEUROPSYCHOLOGIC DATA

Several important psychometric properties should be considered when selecting a neuropsychologic test for use in research or clinical practice, including and validity and reliability.

Validity

An example of validity is the ability of a test to distinguish between participants with Alzheimer's dementia and depression, where cognitive deficits, such as episodic memory (Goodwin, 1997; Backman et al., 2001), may appear similar across the two conditions. This diagnostic accuracy of a test is based on the sensitivity and specificity (Gordis, 2013), and dependent on specific score cutoffs. The accuracy of sensitivity and specificity estimates also depends on the gold standard to which comparisons are being made. This requires that the comparison gold standard be culturally appropriate and similar to the population being evaluated with a given test. Although the diagnostic accuracy of a test is important, in a clinical setting neuropsychologic testing should be integrated with a neurologic evaluation and neuroimaging to inform a final diagnosis. Finally, test validity, sensitivity, and/or specificity may be affected when ceiling (high-performance) and floor (low-performance) effects are apparent (Duff, 2012), or when normative test score distributions are skewed (i.e., large proportions of normal participants performing on one side of the normal distribution).

Reliability

If a test that is administered repeatedly provides the same results, it has high reliability and therefore provides confidence that the test is consistently measuring what it intends to measure. Issues with reliability come into effect when repeated testing results in practice effects. Practice effects are most noticeable with memory tests where participants are asked to recall the same word list over repeated visits, initiating a learning curve of the information (Benedict and Zgaljardic, 1998). Practice effects may also be evident in other neuropsychologic tests because of a generalized improved test-taking ability with repeated measurements (Beglinger et al., 2005).

Standardization of tests

Behavioral data measured from a neuropsychologic test are often scored, but with the scales varying across tests. These raw scores themselves are not interpretable without either normative reference data or transforming raw scores to a standard scale. Standardization can account for confounding factors, such as age, sex, race, education, and premorbid ability. In an effort to equate units across different tests, common distribution-based scales are used based on the standard deviation (SD) unit (e.g., Z-scores and T-scores). Once a raw test score is standardized to this metric, test scores can be compared to a normal distribution bell curve, allowing for interpretation of test results. This normal distribution bell curve can be used to scale all normally distributed neuropsychologic test data, thereby allowing for comparisons across units and to percentile equivalents. For most commonly used neuropsychologic tests, normative datasets are available in test manuals and/or textbooks (Mitrushina et al., 2005; Strauss et al., 2006; Lezak et al., 2012). Normative datasets contain samples of individuals, typically screened for cognitive impairment, often stratified by demographic factors such as age, gender, and education level. The goal is to provide the expected performance distribution on a test within a sample of healthy individuals.

The National Institute on Aging–Alzheimer's Association workgroup on diagnostic guidelines evidences a practical application of this standardization for diagnostic criteria for Alzheimer's disease and MCI. Using culturally appropriate age-normative data, the guidelines provide recommendations on the cognitive characteristics of MCI as cognitive test scores that are 1–1.5 SD units below the mean for age- and education-matched peers (Albert et al., 2011). This SD unit equates to between the third and 16th percentile rank of an age-matched normal distribution. Diagnostic guidelines for dementia in relation to cognitive characteristics include performance on cognitive tests that are approximately 2 SD units below the mean, equivalent to the second and third percentile rank (McKhann et al., 2011). Of note, objective performances in neuropsychologic tests do not solely form the basis for a diagnosis of all-cause dementia.

Additional criteria for a dementia diagnosis include interference with the ability to function at work or at usual activities, decline from prior levels of cognitive functioning, no apparent delirium or other major psychiatric disorder, cognitive impairment as diagnosed from self-report and informant interviews, and impairments observed in multiple domains of cognition.

Although normative datasets are available, often times a demographically or culturally appropriate normative dataset is not available for a given study population. In this case, it is common to generate internal norms within a given study population in order to make comparisons about the relative performance across a specific group of participants. For example, in the Atherosclerosis Risk in Communities (ARIC) study, normative data were recently estimated for eight neuropsychologic tests in older African American and Caucasian participants because race-specific norms that accounted for subclinical/latent neurologic disease and risk factors were not previously available (Schneider et al., 2015). This is valuable for reducing the misclassification of results; however, the external validity or generalizability of the findings may be significantly reduced because of the selectiveness of the ARIC population, which came from only two geographic regions (Forsyth, NC and Jackson, MS) in the USA.

OPPORTUNITIES, CHALLENGES, AND NEXT STEPS

In our aging population, we face an escalating epidemic of chronic conditions with increased survival. Without effective interventions, progression of MCI to dementia may become more widespread. Moreover, an increase in coexisting comorbidities that can precipitate cognitive decline (e.g., earlier onset of type 2 diabetes and longer survival of patients with type 1 diabetes) is resulting in an earlier onset (e.g., middle-aged) of cognitive complications. Reducing the incidence of neurologic disorders is a major public health challenge that demands greater attention and necessitates a well-trained scientific community who can work at the interfaces of epidemiology and clinical neuropsychology.

Incorporating neuropsychology into epidemiologic studies

Neuropsychologic testing offers many opportunities for research and clinical practice. In clinical practice, it is essential for tracking normative changes, diagnoses, treatment, and overall planning of care for patients with cognitive disorders. In the research setting, we rely on neuropsychologic testing to provide data on exposure and outcome ascertainment, namely for the detection of preclinical and clinical cognitive disorders, and in some cases for a diagnosis (e.g., dementia, traumatic brain injury, or stroke). Neuroepidemiologic studies provide an opportunity to identify novel risk factors and modifiers of dementia and other cognitive disorders at the population level, which may serve as possible intervention targets earlier in the course of symptom or disease manifestations.

We turn to dementia as a first model of the evolution of applied neuropsychology at population level of research and epidemiology. At the advent of many historic population-based longitudinal studies in the 1980s and 1990s, the risk determinants of cognitive decline and dementia were not of primary interest. Large cohort studies of older adults, such as the Cardiovascular Health Study (CHS) (Fried et al., 1991) and Women's Health Initiative (WHI, 1998), did not incorporate extensive neuropsychologic testing in the initial cohort exam visits. As dementia became more salient as a cognitive outcome due to our aging population, cognitive screening tests became of interest. CHS began to prospectively administer its first cognitive tests, the MMSE and the Digit Symbol Substitution Test (DSST) in the second year of follow-up. With decades of follow-up data now available, CHS investigators have used these neuropsychologic assessments to assist in the classification of and trends in: (1) cognitive function; (2) MCI (Lopez et al., 2003b); and (3) dementia (Lopez et al., 2003a, c, 2005, 2012; Podewils et al., 2005; Kuller et al., 2016). Furthermore, both risk and protective factors of these salient outcomes have also been identified, and were recently grouped within the Institute of Medicine's (IOM) *Cognitive Aging* report (IOM, 2015) as (1) lifestyle and physical environment (e.g., physical activity and exercise, diet, smoking) (Huang et al., 2005; Podewils et al., 2005; Barnes et al., 2010; Erickson et al., 2010); and (2) health and medical factors (e.g., cerebrovascular and cardiovascular disease risk factors, adiposity, inflammation) (Johnston et al., 2004; Rea et al., 2005; Luchsinger et al., 2013).

As the need for understanding patterns and predictors of dementia has increased, several other epidemiologic studies have also incorporated neuropsychologic assessments into their follow-up visits. Among these cohort studies are the Framingham Heart Study (Elias et al., 2000; Seshadri et al., 2011; Wolf, 2012), the ARIC study cohort (Cerhan et al., 1998; Knopman et al., 2009, 2016), the Women's Health and Aging Study II (Carlson et al., 1999), the Rotterdam Study (Ott et al., 1998; Hoogendam et al., 2014), and the Baltimore Longitudinal Study of Aging (Kawas et al., 2000). Additionally, epidemiologic studies specifically designed to examine the prevalence of, and risk factors for, dementia include the Honolulu-Asia Aging Study (Gelber et al., 2012), the Women's Health Initiative Memory Study

(Shumaker et al., 1998), the Cache County Study on Memory Health and Aging (Hayden and Welsh-Bohmer, 2012), and the Monongahela-Youghiogheny Healthy Aging Team (Ganguli et al., 2010).

Neuropsychologic testing has also been effectively incorporated in population studies of other neurologic diseases. A large cohort study of multiple sclerosis (MS) in Israel used comprehensive computerized testing in 1500 MS patients from a centralized registry (Achiron et al., 2013). This cross-sectional study examined patients with a wide range of disease durations (up to 55 years). Cognitive deficits were associated with disease duration but were apparent only at 5 years or later after onset, suggestive of an important therapeutic window in which to intervene on cognitive declines. By 10 years postonset, 38% of the cohort had documented mild to severe cognitive impairment. Applications of neuropsychologic assessment in the epidemiologic stroke literature include major cohort studies, such as the REasons for Geographic and Racial Differences in Stroke (REGARDS) study and the Framingham Heart Studies. For example, in the REGARDS cohort, investigators reported increased incident cognitive impairment over 4 years among stroke-belt residents compared to other parts of the country, in over 30 000 participants aged 45 and older (Wadley et al., 2011). This regional disparity in a cognitive outcome mirrors the regional disparities in stroke mortality, suggestive of shared risk factors. In the Framingham Offspring Study, magnetic resonance imaging measures of vascular brain damage (i.e., white-matter hyperintensities) predicted incident MCI in over 2000 participants over 6 years, with the outcome assessed by a multidomain neuropsychologic test battery (Debette et al., 2010). All of these studies are using neuropsychologic assessment to help address key questions about age-associated neurodegenerative syndromes in an aging society.

Challenges to the application of neuropsychology in epidemiologic studies

There are also several challenges to neuropsychologic assessment to consider, particularly in a research setting. Cost constraints associated with in-person vs. telephone test administration, study personnel, and time have limited the incorporation of neuropsychologic testing into many population-based cohort studies. This limitation can be addressed, in part, through the incorporation of brief, computerized tests that can sensitively measure age-related declines and impairments with greater precision and less time (Carlson et al., 2008; Kasper et al., 2011; Duchek et al., 2013). The use of global cognitive tests, such as the MMSE (Folstein et al., 1975) and the expanded, modified MMSE (Teng and Chui, 1987), became the vanguard of neuropsychologic testing in cohort studies and were often the sole measurement to ascertain cognitive function. Although useful as a global cognitive measure to ascertain dementia-associated impairment, the MMSE is limited on specific domains of cognition, lacks sensitivity for detecting early cognitive impairment or MCI, and has a ceiling effect in young and educated populations (Hoops et al., 2009). As the use of the MMSE has begun to lose traction due to these limitations, the administration of the MoCA (Nasreddine et al., 2005) has become more widespread given that it measures various cognitive abilities beyond just global cognitive function. Overall, the use of neuropsychologic testing in large population-based research is increasing, as interest grows in clinically relevant questions on primary and secondary prevention of dementia and its clinical sequelae.

Although the incorporation of neuropsychologic testing has increased in the last decade, the use of a multidomain, comprehensive assessment that is administered longitudinally at multiple time points is still limited. Several cohort studies began with the administration of a subset of cognitive tests. The ARIC study (ARIC, 1989), a large biracial cohort whose primary aim was to examine risk factors for atherosclerosis and cardiovascular disease, administered three tests of language, memory, and executive function (Word Fluency, Delayed Word Recall, and the DSST, respectively) at two visits separated by 6 years and a third visit 15 years later. The study has contributed novel findings on the effects of midlife vascular risk factors on cognitive decline (Knopman et al., 2009; Rawlings et al., 2014). The Women's Health and Aging Study II (Guralnik et al., 1995), designed to examine the causes and courses of disability among older adults, administered tests of memory, psychomotor speed, and executive function (Hopkins Verbal Learning Test-Revised (Brandt, 1991), TMT, Parts A and B (Reitan, 1944) respectively) at 18-month intervals over 9 years (Carlson et al., 2009). The study has contributed novel findings on the role of physical function (Krall et al., 2014), lifestyle (Carlson et al., 2012) and physical activity (Clark et al., 2015), depression (Rosenberg et al., 2010), cholesterol (Mielke et al., 2008), and inflammation (Palta et al., 2014) on cognitive declines and impairment.

GEMS (Snitz et al., 2009b) is one of the few studies with the strength of having longitudinal data on cognitive tests for several domains of cognition. With the primary goal of examining the preventive effect of ginkgo biloba on dementia in community-dwelling, cognitively unimpaired or mildly impaired adults, GEMS administered several tests of cognition to ascertain multidimensional cognition over a median follow-up of 6.1 years. GEMS has contributed findings on cognition, neuroimaging outcomes, and dementia, including the effects of antihypertensive medication (Yasar et al., 2013) and statin (Bettermann et al., 2012) use on dementia and multidimensional cognitive function, and the associations between cognitive trajectories and beta-amyloid deposition (Snitz et al., 2013). The availability of such longitudinal cognitive test data is also informative for the ascertainment of dementia, MCI, and its subtypes. For example, the incorporation of longitudinal data on the MMSE and the DSST in CHS coupled with a detailed neuropsychologic assessment allowed for the identification of nonamnestic, as well as amnestic, forms of MCI in this population-based cohort (Lopez et al., 2005, 2006).

The major challenge that still remains is the question of whether to incorporate less rather than more cognitive tests in population-based studies. This will be dependent on the study design, research question and exposures, outcomes of interest, ease of administration, and study resources. As we think about next steps for the integration of neuropsychologic assessment into epidemiologic studies, it is important that we recognize the utility and advantages that it can provide, as well as increased efficiencies in the context of limited resources. A greater understanding of these opportunities and challenges will lead to a broader and more widespread integration of neuropsychology into epidemiologic studies so that we may better understand preclinical and clinical precursors related to neuropathologic disorders.

References

Achiron A, Chapman J, Magalashvili D et al. (2013). Modeling of cognitive impairment by disease duration in multiple sclerosis: a cross-sectional study. PLoS One 8. e71058.

Agrell B, Dehlin O (1998). The clock-drawing test. Age Ageing 27: 399–403.

Albert MS, DeKosky ST, Dickson D et al. (2011). The diagnosis of mild cognitive impairment due to Alzheimer's disease: recommendations from the National Institute on Aging-Alzheimer's Association workgroups on diagnostic guidelines for Alzheimer's disease. Alzheimers Dement 7: 270–279.

ARIC (1989). The Atherosclerosis Risk in Communities (ARIC) Study: design and objectives. Am J Epidemiol 129: 687–702.

Atkinson RC, Shiffrin RM (1968). Human memory: a proposed system and its control processes. Psychol Learn Motiv 2: 89–195.

Backman L, Small BJ, Fratiglioni L (2001). Stability of the preclinical episodic memory deficit in Alzheimer's disease. Brain 124: 96–102.

Baddeley AD (2007). Working Memory, Thought, and Action, Oxford University Press, New York.

Baddeley AD, Hitch G (1974). Working Memory. In: G Bower (Ed.), The psychology of learning and motivation, Academic Press, New York.

Barnes DE, Haight TJ, Mehta KM et al. (2010). Secondhand smoke, vascular disease, and dementia incidence: findings from the cardiovascular health cognition study. Am J Epidemiol 171: 292–302.

Beglinger LJ, Gaydos B, Tangphao-Daniels O et al. (2005). Practice effects and the use of alternate forms in serial neuropsychological testing. Arch Clin Neuropsychol 20: 517–529.

Benedict RH, Zgaljardic DJ (1998). Practice effects during repeated administrations of memory tests with and without alternate forms. J Clin Exp Neuropsychol 20: 339–352.

Benton AL (1946). A visual retention test for clinical use, The Psychological Corporation, New York, NY.

Benton AL, Hamsher K (1994). Multilingual Aphasia Examination, AJA, Iowa City, IA.

Bettermann K, Arnold AM, Williamson J et al. (2012). Statins, risk of dementia, and cognitive function: secondary analysis of the ginkgo evaluation of memory study. J Stroke Cerebrovasc Dis 21: 436–444.

Bigler ED (2001). Neuropsychological testing defines the neurobehavioral significance of neuroimaging-identified abnormalities. Arch Clin Neuropsychol 16: 227–236.

Brandt J (1991). The Hopkins Verbal Learning Test: development of a new memory test with six equivalent forms. Clin Neuropsychol 5: 125–142.

Carlson MC, Fried LP, Xue QL et al. (1999). Association between executive attention and physical functional performance in community-dwelling older women. J Gerontol B Psychol Sci Soc Sci 54: S262–S270.

Carlson MC, Saczynski JS, Rebok GW et al. (2008). Exploring the effects of an "everyday" activity program on executive function and memory in older adults: Experience Corps. Gerontologist 48: 793–801.

Carlson MC, Xue QL, Zhou J et al. (2009). Executive decline and dysfunction precedes declines in memory: the Women's Health and Aging Study II. J Gerontol A Biol Sci Med Sci 64: 110–117.

Carlson MC, Parisi JM, Xia J et al. (2012). Lifestyle activities and memory: variety may be the spice of life. The women's health and aging study II. J Int Neuropsychol Soc 18: 286–294.

Castro-Caldas A, Grafman J (2000). Those were the (phrenological) days. Neuroscientist 6: 297–302.

Cerhan JR, Folsom AR, Mortimer JA et al. (1998). Correlates of cognitive function in middle-aged adults. Atherosclerosis Risk in Communities (ARIC) Study Investigators. Gerontology 44: 95–105.

Chen P, Ratcliff R, Belle SH et al. (2001). Patterns of cognitive decline in pre-symptomatic Alzheimer's disease: a prospective community study. Arch Gen Psychiatry 58: 853–858.

Christensen BK, Colella B, Inness E et al. (2008). Recovery of cognitive function after traumatic brain injury: a multilevel modeling analysis of Canadian outcomes. Arch Phys Med Rehabil 89: S3–S15.

Clark S, Parisi J, Kuo J et al. (2015). Physical activity is associated with reduced risk of executive function impairment in older women. J Aging Health 28: 726–739.

Cohn NB, Dustman RE, Bradford DC (1984). Age-related decrements in Stroop Color Test performance. J Clin Psychol 40: 1244–1250.

Cristofori I, Zhong W, Chau A et al. (2015). White and gray matter contributions to executive function recovery after traumatic brain injury. Neurology 84: 1394–1401.

Daigneault S, Braun CM, Whitaker HA (1992). Early effects of normal aging on perseverative and non-perseverative prefrontal measures. Dev Neuropsychol 8: 99–114.

De Renzi E, Vignolo LA (1962). The token test: a sensitive test to detect receptive disturbances in aphasics. Brain 85: 665–678.

Debette S, Beiser A, DeCarli C et al. (2010). Association of MRI markers of vascular brain injury with incident stroke, mild cognitive impairment, dementia, and mortality: the Framingham Offspring Study. Stroke 41: 600–606.

DeKosky ST, Williamson JD, Fitzpatrick AL et al. (2008). Ginkgo biloba for prevention of dementia: a randomized controlled trial. JAMA 300: 2253–2262.

Delis DC, Kramer JH, Kaplan E et al. (2000). California Verbal Learning Test, Psychological Corporation, San Antonio, TX.

Derrer DS, Howieson DB, Mueller EA et al. (2001). Memory testing in dementia: how much is enough? J Geriatr Psychiatry Neurol 14: 1–6.

Desmond DW, Moroney JT, Sano M et al. (1996). Recovery of cognitive function after stroke. Stroke 27: 1798–1803.

Duchek JM, Balota DA, Thomas JB et al. (2013). Relationship between Stroop performance and resting state functional connectivity in cognitively normal older adults. Neuropsychology 27: 516–528.

Duff K (2012). Evidence-based indicators of neuropsychological change in the individual patient: relevant concepts and methods. Arch Clin Neuropsychol 27: 248–261.

Elias MF, Beiser A, Wolf PA et al. (2000). The preclinical phase of Alzheimer disease: a 22-year prospective study of the Framingham Cohort. Arch Neurol 57: 808–813.

Erickson KI, Raji CA, Lopez OL et al. (2010). Physical activity predicts gray matter volume in late adulthood: the Cardiovascular Health Study. Neurology 75: 1415–1422.

Ernst J, Warner MH, Townes BD et al. (1987). Age group differences on neuropsychological battery performance in a neuropsychiatric population: an international descriptive study with replications. Arch Clin Neuropsychol 2: 1–12.

Folstein MF, Folstein SE, McHugh PR (1975). Mini-mental state. A practical method for grading the cognitive state of patients for the clinician. J Psychiatr Res 12: 189–198.

Fried LP, Borhani NO, Enright P et al. (1991). The Cardiovascular Health Study: design and rationale. Ann Epidemiol 1: 263–276.

Gall FJ (1809). Recherches sur le système nerveux en général et sur celui du cerveau en particulier, Schoell, Paris.

Ganguli M, Chang CC, Snitz BE et al. (2010). Prevalence of mild cognitive impairment by multiple classifications: The Monongahela-Youghiogheny Healthy Aging Team (MYHAT) project. Am J Geriatr Psychiatry 18 (8): 674–683. http://dx.doi.org/10.1097/JGP.0b013e3181cdee4f.

Gelber RP, Launer LJ, White LR (2012). The Honolulu-Asia Aging Study: epidemiologic and neuropathologic research on cognitive impairment. Curr Alzheimer Res 9: 664–672.

Goodwin GM (1997). Neuropsychological and neuroimaging evidence for the involvement of the frontal lobes in depression. J Psychopharmacol 11: 115–122.

Gordis L (2013). Epidemiology. PA, Saunders, Philadelphia.

Gorelick PB, Scuteri A, Black SE et al. (2011). Vascular contributions to cognitive impairment and dementia: a statement for healthcare professionals from the American Heart Association/American Stroke Association. Stroke 42: 2672–2713.

Guralnik JM, Fried LP, Simonsick EM et al. (1995). The Women's Health and Aging Study: Health and Social Characteristics of Older Women with Disability, National Institutes of Health, National Institute on Aging, Bethesda, MD.

Hayden KM, Welsh-Bohmer KA (2012). Epidemiology of cognitive aging and Alzheimer's disease: contributions of the Cache County Utah study of memory, health and aging. Curr Top Behav Neurosci 10: 3–31.

Hoogendam YY, Hofman A, van der Geest JN et al. (2014). Patterns of cognitive function in aging: the Rotterdam Study. Eur J Epidemiol 29: 133–140.

Hoops S, Nazem S, Siderowf AD et al. (2009). Validity of the MoCA and MMSE in the detection of MCI and dementia in Parkinson disease. Neurology 73: 1738–1745.

Huang TL, Zandi PP, Tucker KL et al. (2005). Benefits of fatty fish on dementia risk are stronger for those without APOE epsilon4. Neurology 65: 1409–1414.

Hughes DL, Bryan J (2002). Adult age differences in strategy use during verbal fluency performance. J Clin Exp Neuropsychol 24: 642–654.

IOM (2015). Cognitive aging: Progress in understanding and opportunities for action. Washington, DC.

Johnston SC, O'Meara ES, Manolio TA et al. (2004). Cognitive impairment and decline are associated with carotid artery disease in patients without clinically evident cerebrovascular disease. Ann Intern Med 140: 237–247.

Kaplan E, Goodglass H, Weintraub S (1983). Boston Naming Test, Lea & Febiger, Philadelphia, PA.

Kasper JD, Edwards B, Freedman VA et al. (2011). Advances in Survey Assessment of Disability in Older Adults: Measuring Physical and Cognitive Capacity in the National Health and Aging Trends Study (NHATS). In: 10th Conference on Health Survey Research Methods, National Center for Health Statistics, Hyattsville, MD.

Kawas C, Gray S, Brookmeyer R et al. (2000). Age-specific incidence rates of Alzheimer's disease: the Baltimore Longitudinal Study of Aging. Neurology 54: 2072–2077.

Knopman DS, Mosley TH, Catellier DJ et al. (2009). Fourteen-year longitudinal study of vascular risk factors, APOE genotype, and cognition: the ARIC MRI Study. Alzheimers Dement 5: 207–214.

Knopman DS, Gottesman RF, Sharrett AR et al. (2016). Mild Cognitive Impairment and Dementia Prevalence: The Atherosclerosis Risk in Communities Neurocognitive Study (ARIC-NCS). Alzheimers Dement (Amst) 2: 1–11.

Krall JR, Carlson MC, Fried LP et al. (2014). Examining the dynamic, bidirectional associations between cognitive and physical functioning in older adults. Am J Epidemiol 180: 838–846.

Kuller LH, Lopez OL, Becker JT et al. (2016). Risk of dementia and death in the long-term follow-up of the Pittsburgh Cardiovascular Health Study-Cognition Study. Alzheimers Dement 12: 170–183.

Larrabee GJ, Kane RL (1983). Differential drawing size associated with unilateral brain damage. Neuropsychologia 21: 173–177.

Lezak M, Howieson D, Bigler ED et al. (2012). Neuropsychological Assessment, Oxford University Press, New York, NY.

Lishman WA (1968). Brain damage in relation to psychiatric disability after head injury. Br J Psychiatry 114: 373–410.

Lishman WA (1973). The psychiatric sequelae of head injury: a review. Psychol Med 3: 304–318.

Lishman WA (1988). Physiogenesis and psychogenesis in the 'post-concussional syndrome'. Br J Psychiatry 153: 460–469.

Lopez OL, Jagust WJ, DeKosky ST et al. (2003a). Prevalence and classification of mild cognitive impairment in the Cardiovascular Health Study Cognition Study: part 1. Arch Neurol 60: 1385–1389.

Lopez OL, Jagust WJ, Dulberg C et al. (2003b). Risk factors for mild cognitive impairment in the Cardiovascular Health Study Cognition Study: part 2. Arch Neurol 60: 1394–1399.

Lopez OL, Kuller LH, Fitzpatrick A et al. (2003c). Evaluation of dementia in the cardiovascular health cognition study. Neuroepidemiology 22: 1–12.

Lopez OL, Kuller LH, Becker JT et al. (2005). Classification of vascular dementia in the Cardiovascular Health Study Cognition Study. Neurology 64: 1539–1547.

Lopez OL, Becker JT, Jagust WJ et al. (2006). Neuropsychological characteristics of mild cognitive impairment subgroups. J Neurol Neurosurg Psychiatry 77: 159–165.

Lopez OL, Becker JT, Chang YF et al. (2012). Incidence of mild cognitive impairment in the Pittsburgh Cardiovascular Health Study-Cognition Study. Neurology 79: 1599–1606.

Luchsinger JA, Biggs ML, Kizer JR et al. (2013). Adiposity and cognitive decline in the cardiovascular health study. Neuroepidemiology 40: 274–281.

McCrimmon RJ, Ryan CM, Frier BM (2012). Diabetes and cognitive dysfunction. Lancet 379: 2291–2299.

McKhann GM, Knopman DS, Chertkow H et al. (2011). The diagnosis of dementia due to Alzheimer's disease: recommendations from the National Institute on Aging-Alzheimer's Association workgroups on diagnostic guidelines for Alzheimer's disease. Alzheimers Dement 7: 263–269.

Mehta KM, Yaffe K, Langa KM et al. (2003). Additive effects of cognitive function and depressive symptoms on mortality in elderly community-living adults. Journals of Gerontology. Series A. Biol Sci Med Sci 58: M461–M467.

Mielke MM, Xue QL, Zhou J et al. (2008). Baseline serum cholesterol is selectively associated with motor speed and not rates of cognitive decline: the Women's Health and Aging Study II. Journals of Gerontology. Series A. Biol Sci Med Sci 63: 619–624.

Miller GA (1956). The magical number seven plus or minus two: some limits on our capacity for processing information. Psychol Rev 63: 81–97.

Mitrushina M, Boone KB, Razani J et al. (2005). Handbook of Normative Data for Neuropsychological Assessment, 2nd Edition. Oxford University Press, New York, NY, US.

Miyake A, Friedman NP, Emerson MJ et al. (2000). The unity and diversity of executive functions and their contributions to complex "frontal lobe" tasks: a latent variable analysis. Cogn Psychol 41: 49–100.

Munshi M, Grande L, Hayes M et al. (2006). Cognitive dysfunction is associated with poor diabetes control in older adults. Diabetes Care 29: 1794–1799.

Nasreddine ZS, Phillips NA, Bedirian V et al. (2005). The Montreal Cognitive Assessment, MoCA: a brief screening tool for mild cognitive impairment. J Am Geriatr Soc 53: 695–699.

Nguyen HT, Arcury TA, Grzywacz JG et al. (2012). The association of mental conditions with blood glucose levels in older adults with diabetes. Aging Ment Health 16: 950–957.

Ogden JA (1990). Spatial abilities and deficits in aging adn age-related disorders. In: F Boller, J Grafman (Eds.), Handbook of Neuropsychology, Elsevier, Amsterdam.

Osterrieth PA (1944). Filetest de copie d'une figure complex: Contribution a l'etude de la perception et de la memoire. Arch Psychol 30.

Ott A, Breteler MM, van Harskamp F et al. (1998). Incidence and risk of dementia. The Rotterdam Study. Am J Epidemiol 147: 574–580.

Palta P, Xue QL, Deal JA et al. (2014). Interleukin-6 and C-reactive protein levels and 9-year cognitive decline in community-dwelling older women: the Women's Health and Aging Study II. The Journals of Gerontology. Series A. Biol Sci Med Sci 70: 873–878.

Panza F, Frisardi V, Capurso C et al. (2010). Metabolic syndrome and cognitive impairment: current epidemiology and possible underlying mechanisms. J Alzheimers Dis 21: 691–724.

Podewils LJ, Guallar E, Kuller LH et al. (2005). Physical activity, APOE genotype, and dementia risk: findings from the Cardiovascular Health Cognition Study. Am J Epidemiol 161: 639–651.

Rasmusson DX, Bylsma FW, Brandt J (1995). Stability of performance on the Hopkins Verbal Learning Test. Arch Clin Neuropsychol 10: 21–26.

Rawlings AM, Sharrett AR, Schneider AL et al. (2014). Diabetes in midlife and cognitive change over 20 years: a cohort study. Ann Intern Med 161: 785–793.

Rea TD, Breitner JC, Psaty BM et al. (2005). Statin use and the risk of incident dementia: the Cardiovascular Health Study. Arch Neurol 62: 1047–1051.

Reitan RM (1944). Army Individual Test Battery, Adjutant General's Office, War Department, U.S. Army, Washington, DC.

Rey A (1941). L'examen psychologique dans les cas d'encephalopathie traumatique (les problèmes). Arch Psychol 28: 215–285.

Rosenberg PB, Mielke MM, Xue QL et al. (2010). Depressive symptoms predict incident cognitive impairment in cognitive healthy older women. Am J Geriatr Psychiatr 18: 204–211.

Ross TP, Lichtenberg PA, Christensen BK (1995). Normative data on the Boston naming test for elderly adults in a demographically diverse medical sample. Clin Neuropsychol 9: 321–325.

Royall DR, Mahurin RK, Gray KF (1992). Bedside assessment of executive cognitive impairment: the executive interview. J Am Geriatr Soc 40: 1221–1226.

Ryan JJ, Sattler JM, Lopez SJ (2000). Age effects on Wechsler Adult Intelligence Scale-III subtests. Arch Clin Neuropsychol 15: 311–317.

Salthouse TA, Fristoe N, Rhee SH (1996). How localized are age-related effects on neuropsychological measures? Neuropsychology 10: 272–285.

Schneider AL, Sharrett AR, Gottesman RF et al. (2015). Normative data for 8 neuropsychological tests in older blacks and whites from the atherosclerosis risk in communities (ARIC) study. Alzheimer Dis Assoc Disord 29: 32–44.

Selnes OA, Vinters HV (2006). Vascular cognitive impairment. Nat Clin Pract Neurol 2: 538–547.

Seshadri S, Beiser A, Au R et al. (2011). Operationalizing diagnostic criteria for Alzheimer's disease and other age-related cognitive impairment-Part 2. Alzheimers Dement 7: 35–52.

Shapiro AM, Benedict RH, Schretlen D et al. (1999). Construct and concurrent validity of the Hopkins Verbal Learning Test-revised. Clin Neuropsychol 13: 348–358.

Shumaker SA, Reboussin BA, Espeland MA et al. (1998). The Women's Health Initiative Memory Study (WHIMS): a trial of the effect of estrogen therapy in preventing and slowing the progression of dementia. Control Clin Trials 19: 604–621.

Ska B, Nespoulous JL (1987). Pantomimes and aging. J Clin Exp Neuropsychol 9: 754–766.

Snitz BE, O'Meara ES, Carlson MC et al. (2009a). Ginkgo biloba for preventing cognitive decline in older adults: a randomized trial. JAMA 302: 2663–2670.

Snitz BE, Saxton J, Lopez OL et al. (2009b). Identifying mild cognitive impairment at baseline in the Ginkgo Evaluation of Memory (GEM) study. Aging Ment Health 13: 171–182.

Snitz BE, Weissfeld LA, Lopez OL et al. (2013). Cognitive trajectories associated with beta-amyloid deposition in the oldest-old without dementia. Neurology 80: 1378–1384.

Strauss E, Sherman E, Spreen O (2006). A Compendium of Neuropsychological Tests: Administration, Norms and Commentary, Oxford University Press, New York, NY.

Stroop JR (1935). Studies of interference in serial verbal reactions. J Exp Psychol 18: 643–662.

Stuss DT, Stethem LL, Poirier CA (1987). Comparison of three tests of attention and rapid information processing across six age groups. Clin Neuropsychol 1: 139–152.

Tatemichi TK, Desmond DW, Stern Y et al. (1994). Cognitive impairment after stroke: frequency, patterns, and relationship to functional abilities. J Neurol Neurosurg Psychiatry 57: 202–207.

Teng EL, Chui HC (1987). The Modified Mini-Mental State (3MS) examination. J Clin Psychiatry 48: 314–318.

Thabit H, Kennelly SM, Bhagarva A et al. (2009). Utilization of Frontal Assessment Battery and Executive Interview 25 in assessing for dysexecutive syndrome and its association with diabetes self-care in elderly patients with type 2 diabetes mellitus. Diabetes Res Clin Pract 86: 208–212.

Van Gorp WG, Satz P, Kiersch ME et al. (1986). Normative data on the Boston Naming Test for a group of normal older adults. J Clin Exp Neuropsychol 8: 702–705.

Wadley VG, Unverzagt FW, McGuire LC et al. (2011). Incident cognitive impairment is elevated in the stroke belt: the REGARDS study. Ann Neurol 70: 229–236.

Wechsler D (1981). Wechsler Adult Intelligence Scale, Pearson, San Antonio, TX.

Wechsler D (2009). Wechsler Memory Scale-Fourth Edition (WMS-IV) technical and interpretive manual. Pearson, San Antonio, TX.

Weintraub S, Dikmen SS, Heaton RK et al. (2013). Cognition assessment using the NIH Toolbox. Neurology 80: S54–S64.

Welch LW, Doineau D, Johnson S et al. (1996). Educational and gender normative data for the Boston Naming Test in a group of older adults. Brain Lang 53: 260–266.

WHI (1998). Design of the Women's Health Initiative clinical trial and observational study. Control Clin Trials 19: 61–109.

Wolf PA (2012). Contributions of the Framingham Heart Study to stroke and dementia epidemiologic research at 60 years. Arch Neurol 69: 567–571.

Yasar S, Xia J, Yao W et al. (2013). Antihypertensive drugs decrease risk of Alzheimer disease: Ginkgo Evaluation of Memory Study. Neurology 81: 896–903.

Section 2

Epidemiology of disorders primarily affecting the nervous system

Handbook of Clinical Neurology, Vol. 138 (3rd series)
Neuroepidemiology
C. Rosano, M.A. Ikram, and M. Ganguli, Editors
http://dx.doi.org/10.1016/B978-0-12-802973-2.00008-2

Chapter 8

Dementias

S.F. SACUIU*

Department of Neuropsychiatry, Sahlgrenska University Hospital and Department of Psychiatry and Neurochemistry, University of Gothenburg Institute of Neuroscience and Physiology, Gothenburg, Sweden

Abstract

This chapter will focus on the descriptive, analytic, and intervention-oriented epidemiology of dementia and its most frequent etiologic type due to Alzheimer's disease. The chapter opens with a brief presentation of the concept of dementia, followed by the presentation of dementia of the Alzheimer type (DAT), including natural history, clinical manifestation, neuropathology, medical prognosis, and management. Further, the chapter presents the prevalence and incidence of dementia, with special consideration of secular trends in prevalence and incidence of DAT, and prognosis of the socioeconomic impact of dementia. Thereafter the main risk factors for DAT are covered. The chapter also addresses the results of ongoing therapeutic and preventive intervention trials for DAT. Finally, the future challenges of the epidemiology of dementia with a focus on the impact of the new diagnostic criteria for neurocognitive disorders, as well as the development of biomarkers for DAT and other types of dementia, will be briefly discussed.

INTRODUCTION

Large population studies and international epidemiologic consortia have made possible the standardization of research criteria and contributed to the refinement of the clinical diagnosis of dementia. The ultimate goal of continuously ongoing revisions of the diagnosis criteria has been to improve detection of dementia in mild stages. Etiologic classification of different types of dementia has proven to be a difficult task in both clinical and research settings. Currently only dementia of the Alzheimer type (DAT) has well-established diagnosis criteria for epidemiologic research. The standardized criteria have made possible the discovery and implementation of symptomatic treatment against cognitive, behavioral, and functional changes in DAT. Drugs and vaccines with putative disease-modifying effect against DAT are in different phases of clinical trials. The initiation of potentially disease-modifying treatments as early as possible from the onset of DAT is prioritized in clinical trials. Objective and sensitive markers for different types of dementia are avidly sought to improve early detection. In this respect, observational studies help to identify putative biomarkers of dementia in different population groups. Some findings indicate that individuals' age, genetic susceptibility, and lifestyle in early adulthood may modulate the pattern of cognitive decline in old age, the profiles of plasma and cerebrospinal fluid (CSF) proteins, and brain morphologic and functional changes associated with the development of dementia. On the other hand, analytic epidemiology that addresses risk factors of dementia remains a powerful instrument to identify disease mechanisms that may provide new avenues for therapeutic and preventive interventions. There are also indications that the prevalence of dementia may be decreasing in later generations in some parts of the Western world, while worryingly, it may be increasing in Asia. It is estimated that low-income countries will face the largest incidence of dementia in the decades to come. At the same time, the majority of the neuroepidemiologic studies of dementia are set in high-income countries. It is still unclear if the prevalence, incidence,

*Correspondence to: Simona F. Sacuiu, MD, Ph.D, GU-Institute of Neuroscience and Physiology, Department of Psychiatry and Neurochemistry, Wallinsgatan 6, 431 41 Mölndal, Sweden. Tel: +46-031-343-86-63, Fax: +46-031-87-13-79, E-mail: Simona.Sacuiu@neuro.gu.se

and risk factors for dementia reported from countries with high income apply to other parts of the world. The need for worldwide population-based surveys may thus be emphasized as a future research priority, along with the standardization of methods in epidemiologic studies of dementia. The ultimate goal of population surveys, either observational or analytic, is to provide solid grounds for intervention studies to reduce the worldwide burden of dementia.

BRIEF DEFINITION AND HISTORY OF DEMENTIA

Dementia is a syndrome of cognitive impairment or cognitive decline that affects independent living. A new dementia case is being registered worldwide every 4 seconds (Duthey, 2013). Although most dementias in older adults are attributable to Alzheimer's disease (AD), the cause of underlying pathologic processes remains unknown in many cases.

An overview of the concept of dementia through time by Berchtold and Cotman (1998) underscored that already in the early days of medicine practitioners were able to identify a syndrome of severe disturbance of reasoning and behavior that was acquired in late adulthood and resulted in gradual loss of ability to interact with the surroundings and cope with everyday life challenges. The decline in cognitive abilities accompanied by behavioral changes was for a long time, and may still be, erroneously associated by common sense with mere aging. Apparently, the term "senile dementia" relating to pathology of the nervous system was first used in William Cullen's classification of diseases from 1776 (Berchtold and Cotman, 1998). Cullen, an 18th-century Scottish pathologist, defined senile dementia as a nervous disease caused by "decay of perception and memory in old age, *amentia senilis*" (Berchtold and Cotman, 1998). The term "dementia" was incorporated in medical dictionaries and encyclopedias by the 18th century (Boller and Forbes, 1998; Sachdev, 2000). It took, though, a couple of centuries for dedicated medical practitioners and scientists around the world to come closer to a generally accepted definition of dementia and to begin to understand the underpinnings of this complex syndrome. In modern era, dementia with its most common etiologic variant due to AD has imposed its urgency on public health sectors as the epidemic of old age. The most recent estimate of the prevalence of dementia worldwide reported 35.6 million people affected in 2010, a figure that will increase to 115.4 million people with dementia by 2050 (WHO, 2012; Prince et al., 2013). The inherent costs of such a disease burden are deemed insurmountable for any nation in the world.

DEMENTIA OF ALZHEIMERS TYPE

Dementia due to AD typically presents as an insidious decline in cognitive functioning. Initial memory deficits are circumscribed to short-term memory, and learning new information becomes difficult. Although these changes may appear at first as benign forgetfulness, which is a common condition with aging (Schaie and Willis, 1993; Gazzaley et al., 2005), memory decline due to AD is progressive and irreversible (APA, 2013). Progression is slow and gradually affects other areas of cognition, i.e., language, visuospatial, and executive abilities.

DAT diagnosis

The available guidelines to establish an etiology attributable to AD for probable and possible DAT diagnosis in research and clinical settings are based on National Institute of Neurological and Communicable Disorders and Stroke and Alzheimer's Disease and Related Disorders Association (NINCDS-ADRDA) criteria (McKhann et al., 1984). These criteria have been revised with new recommendations in the National Institute on Aging–Alzheimer's Association (NIA-AA) criteria (McKhann et al., 2011). A dementia diagnosis is required as a first step in the diagnosis of DAT. The current core clinical criteria for dementia or "major neurocognitive disorder" diagnosis, as defined by the *Diagnostic and Statistical Manual of Mental Disorders* (DSM-5), require a significant cognitive decline in at least one cognitive domain (learning and memory, complex attention, executive function, language, perceptual-motor ability, including praxis and gnosis, and social cognition) (APA, 2013). The cognitive decline has to substantially affect independence in everyday activities, and not to be due to delirium or other identifiable medical condition (APA, 2013). Further, clinical criteria for DAT require an insidious onset of dementia symptomatology represented by decline in at least two cognitive domains, one of which should be learning and memory, over months to years, and the absence of other identifiable causes of cognitive decline (McKhann et al., 2011; APA, 2013). The novelty in the 2011 McKhann revision was to address the development of the *in vivo* diagnosis of DAT by proposing the use in research settings of peripheral biologic markers, i.e., changes in protein profiles in the CSF and morphologic and functional brain neuroimaging changes that have been shown to reflect the pathologic process of AD. The impact of the newly revised criteria on prevalence and incidence of DAT remains to be evaluated in large representative studies.

Brain changes in AD

The phenotypic brain changes in AD consist of: (1) build-up of nonsoluble amyloid-beta peptide (Aβ) plaques in the brain parenchyma, as standardized in A–C stages according to the Consortium to Establish a Registry for AD (CERAD) (Mirra et al., 1991); accompanied by (2) gradual intracellular accumulation of neurofibrillary tangles (NFT) in brain neurons (Braak I–VI stages) (Braak and Braak, 1995). The definitive diagnosis of DAT requires postmortem neuropathologic confirmation by identification of CERAD amyloid plaque density and Braak NFT stage (Newell et al., 1999). Neuropathology studies have reported a wide range of correlation figures, from 64% to 86% overlap, between clinical and pathologic diagnoses of DAT (Tierney et al., 1988; Mirra et al., 1991; Newell et al., 1999; Zaccai et al., 2006). Figures in the higher end were confirmed by a more recent study that reported 83% overlap between the clinically diagnosed DAT cases and the postmortem neuropathologic diagnosis of DAT (Beach et al., 2012). In this latest study, the neuropathologic diagnosis showed 71% sensitivity to detect clinically probable DAT, but sensitivity increased to 83% with more permissive diagnosis criteria that allowed probable and possible DAT to be taken into account. The reverse was true for specificity of neuropathology diagnosis to detect DAT in the same study, with restrictive criteria increasing the specificity from 44% to 71% (Beach et al., 2012). Cortical lesions in AD are associated with accumulation of NFT in the transentorhinal neurons of the medial temporal lobes. Some studies described NFT aggregates in the brainstem nuclei (Grinberg et al., 2009; Braak et al., 2011). It is, however, uncertain if NFT changes at the brainstem level detected even in young adults (Braak et al., 2013) are associated with later development of DAT. Cognitive changes in DAT are otherwise strongly associated with the accumulation of NFT (Nelson et al., 2012).

Although data from patients who develop DAT have shown retrospectively that these individuals tend to accumulate Aβ in the brain at least 10 years before they were diagnosed with DAT (Jack et al., 2013; Villemagne et al., 2013), the prediction of later development of DAT based solely on Aβ accumulation in the brain parenchyma is highly nonspecific (Reiman et al., 2009). Only the accumulation of both Aβ and NFT in the brain constitutes the definite neuropathologic signs of DAT. However, the postmortem defined pathologic stages are not clearly associated with a characteristic symptomatology during the development of DAT and may even lack clinical expression in individuals with a high cognitive reserve. In accordance with these observations, currently proposed NIA-AA AD criteria recognize a preclinical stage, i.e., cognitively intact individuals who present biomarkers of the disease as above, a mild cognitive impairment (MCI) stage, and, finally, a dementia stage (McKhann et al., 2011). It remains a challenge for clinical practitioners and researchers to distinguish MCI from mild stages of DAT, as will be discussed below.

DAT clinical severity stages and their pathologic correlates

According to DSM-5, neurocognitive disorders may be clinically classified by severity stage according to performance on quantifiable neuropsychologic assessments and level of independence in everyday activities in mild and major neurocognitive disorders, the latter including mild, moderate, and severe stages of dementia (APA, 2013). While there is no golden standard in neuropsychologic examination for the purpose of clinical dementia diagnosis, the Clinical Dementia Rating (CDR) scale is frequently used in the assessment of everyday function in DAT in both clinical and research settings. The CDR is based on a semistructured interview with the patient and a knowledgeable informant and covers six domains likely to be affected in DAT – memory, orientation, judgement and problem solving, community affairs, home and hobbies, and personal care (Morris, 1993). A CDR score of zero indicates a cognitively intact individual, a 0.5 score corresponds to questionable impairment, and scores of 1, 2, and 3 pertain to mild, moderate, and severe impairment, respectively (Morris, 1993). The questionable DAT stage with a CDR score of 0.5 is generally associated with the syndrome of MCI.

MCI is currently accepted as a clinically identifiable correlate of the initial neuropathologic stages of dementia (DeCarli, 2003; Andreasen and Blennow, 2005; Apostolova and Thompson, 2008; Petersen, 2011). Individuals with MCI perform at least 1–1.5 standard deviations (SD) below the norm for age and education on standardized neuropsychologic tests, but they carry out most daily activities satisfactorily. These cases are thus encompassed by the DSM-5 "minor neurocognitive disorder," for which the cognitive decline should not interfere with performing everyday activities (APA, 2013). There are amnestic and nonamnestic types of MCI syndromes, i.e., other cognitive domains than memory may be affected (Winblad et al., 2004). The evidence converges on amnestic MCI, either as isolated memory impairment or with other cognitive domains minimally affected (multiple domains amnestic MCI), representing an initial symptomatic phase in the development of DAT in older segments of the population (Amieva et al., 2004; Aggarwal et al., 2005; Visser et al., 2006; Palmer et al., 2008; Sacuiu et al., 2009). Nonamnestic MCI seems to have a heterogeneous underlying pathology that may not always be related to a neurodegenerative process. This aspect could also explain why more strictly characterized MCI in clinical settings are more likely to progress to DAT (Petersen et al., 1999; Alexopoulos et al., 2006) than MCI in the community, where a mixed brain pathology, including vascular insults or other types of pathology, may be more frequent (Ritchie et al., 2001; Schneider et al., 2003; Artero et al., 2006). In recent years pathologic AD changes related to low Aβ and high tau-protein concentration in the CSF have been evidenced in amnestic MCI, low Aβ as a peripheral marker of Aβ accumulation in the brain, and high tau-protein as a peripheral marker of NFT (Andreasen et al., 2003; Visser et al., 2009). Parallel development of structural neuroimaging techniques with better resolution and the discovery of brain amyloid labeling have made it possible to show in amnestic MCI both pathologic atrophy of the hippocampus regions and amyloid deposits in the cortex (Small et al., 2006; Schuff and Zhu, 2007; Mosconi et al., 2008; Nobili et al., 2008). These changes are thought to reflect ongoing AD pathologic changes in the brain at the MCI stage. By the time a clear pattern of cognitive signs has emerged and a mild stage of DAT can be clinically diagnosed according to current DAT criteria, the NFT have spread throughout the limbic system and Aβ plaques may be found in brain parenchyma. In moderate and severe stages of the disease NFT and Aβ plaques are found across the whole cortex. However, the clinical phenotypic presentation of DAT varies largely, as the disease may also affect initially and to different extents the executive, visuoperceptual, or language domains (Storey et al., 2002).

Familial and sporadic DAT

The predominant phenotype of DAT described above, with memory decline followed by other cognitive impairments and later by functional and behavioral changes, is typically encountered in familial AD due to autosomal-dominant genetic mutations (ADAD). There are three known mutations involved in ADAD: either one of the genes coding for amyloid precursor protein on chromosome 21 (Chartier-Harlin et al., 1991), presenilin 1 (on chromosome 14), and presenilin 2 (on chromosome 1) (Sherrington et al., 1995; Haass and De Strooper, 1999). ADAD starts at early ages of around 30–50 years, with cases after 60 years rarely reported (mostly in those who inherit presenilin 2 mutations) (Bateman et al., 2011). Dominantly inherited forms of AD, also called familial AD, represent about 1% of the total DAT burden (Bateman et al., 2011; Lambert et al., 2014).

Although the sporadic form of DAT, more frequently encountered after the age of 65 years, may present the same symptomatology as the ADAD, comorbidities play a substantial role in the multifaceted clinical aspects of sporadic, late-life DAT (Sadowski et al., 2004; Chui et al., 2012). With some rare exceptions, e.g., the international epidemiologic study Dominantly Inherited Alzheimer's Network (DIAN), epidemiologic studies report on sporadic DAT.

Pathologic mechanism of DAT

The currently dominant pathologic model of DAT is mainly based on findings from ADAD case series and postulates that brain accumulation of Aβ at toxic levels due to an acquired imbalance between its production and clearance triggers neuronal and synaptic dysfunction and ultimately neuronal death (amyloid cascade hypothesis: Hardy and Higgins, 1992; Selkoe, 2000; Tanzi and Bertram, 2005). This model was also supported by findings in patients with Down syndrome known to develop amyloid plaques in the brain early in life. This initial hypothetic pathologic cascade has been amended to reflect new molecular findings related to the complex structure of Aβ proteins and their physiologic role in the brain (Ballard et al., 2011; Herrup, 2015; Puzzo et al., 2015). The initial hypothetic pathologic cascade has also been challenged by proponents of the tau-pathology cascade, who have suggested that intraneuronal tau-protein changes induce the first pathologic lesions that later trigger amyloid plaque building (Braak and Del Tredici, 2015). Neither one of these hypotheses seems to explain by itself the diverse pathologic changes observed in sporadic DAT. This standpoint is also reflected by repeated failure of clinical trials that targeted amyloid, as shown by some recent comprehensive reviews of DAT therapeutic trials (Mangialasche et al., 2010; Karran et al., 2011; Lambracht-Washington and Rosenberg, 2013).

Throughout recent decades other hypotheses have been formulated that involve multifactorial factors, both genetic and environmental. Large efforts have been undertaken lately to detect genotypes and thus mechanisms involved in sporadic DAT. More on the genetic determinants of sporadic DAT follows in the analytic epidemiology part of this chapter. Briefly, let us mention here that recent results from the International Genomics of Alzheimer's Project (IGAP) using genomewide association studies (GWAS) have identified over 20 susceptibility gene loci for sporadic, late-life DAT. These findings converge on the implication of immune mechanisms and regulation of transport of amyloid precursors in DAT (Lambert and Amouyel, 2011).

Prognosis and outcomes

The prognosis is similar for sporadic DAT without comorbidities and the autosomal-dominant form of DAT in terms of functional outcome and survival (Bateman et al., 2011). DAT-affected individuals suffer a gradual cognitive and mental (i.e., emotional and behavioral) decline. At the mild stage of the disease DAT-affected individuals require only minimal help for their more complex daily routines and may continue living in their home. At moderate and severe stages of the disease, the cognitive function is so much reduced that assisted living and care supervision are necessary. Survival after DAT diagnosis converges at around 8 years on average in individuals who received the diagnosis in the mild stage at around 70–75 years, survival being reduced in those cases diagnosed at older ages and in men with DAT (Xie et al., 2008; Kua et al., 2014). The survival period may appear shorter after diagnosis in people with high cognitive function due to a large cognitive reserve (Stern et al., 1995; Stern, 1998). It has been suggested that larger cognitive reserve may mask the initial typical cognitive decline, thus delaying DAT diagnosis. The more severe the stage of the disease at diagnosis the shorter is the survival, with an average duration of 5.5 years in the mild phase and 3 years in moderate and severe phases (Kua et al., 2014). Overall, survival after DAT diagnosis ranges between 3 and 12 years.

Comorbidity in DAT

Multiple comorbidities are a frequent occurrence in the elderly and influence DAT prognosis, hospitalization, and institutionalization. Some of the most frequently encountered comorbidities will be discussed below.

- Cardiovascular diseases: As the leading cause of death worldwide, cardiovascular diseases co-occurring with DAT curb survival (Bunch et al., 2010; Chaudhry et al., 2010). Cerebrovascular conditions impact also on severity and mortality in DAT. For example, in the prospective Cardiovascular Health Cognition Study in four US communities, with an average of 6.5 years of follow-up, the average estimated survival in DAT was 7 years, whereas DAT with vascular pathology (mixed dementia cases) had shorter survival of 5.4 years (Fitzpatrick et al., 2005).
- Cancer: The relation between cancers, the other leading cause of death in the world, and DAT has not been elucidated. Although some reports have suggested a protective effect of cancer in DAT (Roe et al., 2010; Driver et al., 2012), this interpretation has been methodologically arguable. On the one hand both diseases share some similarities regarding risk profiles, and on the other hand both are associated with increased mortality and thus represent a competitive risk to each other in survival analyses (Ganguli, 2015). Another issue that has started to gain more attention lately is cognitive impairment in cancer survivors and the impact on the development of dementia in this group of patients. Interestingly, these observations spurred discussions regarding the mitogenic role of the neurotransmitter acetylcholine in cancer, as well as putative common pathogenesis between neurodegenerative processes and cancer, both largely unexplored avenues (Bennett and Leurgans, 2010).
- Other common chronic diseases that are associated with increased mortality and increased risk of dementia and DAT are diabetes mellitus (DM) (Hassing et al., 2002; Xu et al., 2009b; Huang et al., 2014) and pulmonary diseases that reduce lung capacity (i.e., chronic obstructive lung disease) (Guo et al., 2007; Liao et al., 2015).

New statistic approaches that are amenable to exploring multiple comorbidities and competing risks (Putter et al., 2007) may shed light on the determinants of survival in elderly people with dementia.

The cause of death in DAT without other severe comorbidities may be due to the brain degenerative process in itself or terminal complications in more severe stages of DAT, e.g., pneumonia, dehydration (Ganguli et al., 2005a).

Outcomes in DAT

Institutionalization, or transfer from own residence to a long-term care facility, is especially important to consider in people with dementia for resource planning in healthcare. An overview of the findings in EURODEM concerted studies based on eight European population surveys of individuals aged 65 years and older showed that prevalent and incident dementia cases were more likely than unaffected individuals to already receive or to enter institutionalized care at 2-, 3-, and 4-year follow-up, with a median time to institutionalization of 3.25 years from diagnosis (Jagger et al., 2000). This figure is in line with earlier accounts from the USA showing that dementia cases entered institutionalized care on average 2–3 years after diagnosis (Severson et al., 1994; Heyman et al., 1997). There was a significant interaction between institutionalization and age at baseline in both prevalent and incident cases of dementia in EURODEM, most likely reflecting an increased risk of dementia with aging (Jagger et al., 2000).

Although differences in the national health policies regarding institutionalization have to be taken into account when considering this socioeconomic aspect of dementia, large representative population studies with a case-control design may offer some guidance for health-care planning across nations. The level of care required in institutions is more often qualified and thus more costly. A recent longitudinal population-based study from France, the Three City study, with a 10-year follow-up at regular intervals, showed that earlier referral to dementia specialized units has a significant impact on institutionalization, but no further functional decline in basic activities of daily living was registered in those who were referred early to specialists (Pimouguet et al., 2015). Thus, although individuals could retain some level of basic functioning after diagnosis, they were transferred to qualified care settings. This study points out an important aspect of dementia management, namely that it is imperative not only to be able to diagnose dementia in its incipient stages, but also to tailor the at-home care and treatments appropriately to the actual level of impairment in order to ensure the patients a good quality of life in their own home and avoid institutionalization with incurred higher costs.

Therapeutic management in DAT

Currently available pharmachologic therapy against DAT based on cholinesterase inhibitors (ChEIs) was first approved in some European countries and the USA in 1992–1993 in the form of tacrine, later discontinued due to liver toxicity. The ChEIs that are now available worldwide are represented by donepezil, galantamine, and rivastigmine. These drugs increase acetylcholine concentration at the cholinergic synapse by inhibiting the enzyme that removes acetylcholine from the synaptic cleft. Thus ChEIs act toward maintaining the appropriate substrate for attention and memory functions (Bartus et al., 1982; Sarter and Bruno, 2004). ChEIs are also involved in modulating signal transduction pathways coupled to toxic amyloid production and tau-protein phosphorylation, and regional brain blood flow, which may suggest a disease-modifying effect of these inhibitors (Lane et al., 2004). All ChEIs have been approved for treatment of DAT from mild to moderate stages. The use of donepezil and rivastigmine has been recently extended to moderate and severe stages of DAT. Another medication for DAT in moderate to severe stages, that was approved in Europe in 2002 and in the USA in 2003, is memantine. Memantine is an antagonist of the *N*-methyl-D-aspartate (NMDA) receptor that is thought to reduce glutamate excitotoxic activity in neurodegeneration.

Although some skepticism as to the benefits of both ChEI and memantine may be voiced, practitioners directly involved in the care of patients with dementia have learned to appreciate the qualitative effects of pharmacologic intervention with DAT medication. A short time after the release of ChEIs, numerous randomized controlled trials (RCT) showed the benefits of early-initiated ChEI medication, in that the rate of cognitive decline was slower in DAT patients receiving long-term treatment very close from the onset (Corey-Bloom, 2002), the rate of increasing functional disability was reduced (Potkin, 2002), and behavioral symptoms such as agitation, psychosis, and motor aberrant behavior benefited from ChEIs as well (Grossberg, 2002). Cholinergic stimulation has been associated with weight loss, physical weakness, and syncope as its most notable adverse effects (Buckley and Salpeter, 2015). A Cochrane systematic review from 2006 that included 13 randomized, double-blind, placebo-controlled trials with ChEIs in DAT (mild, moderate, and severe) showed benefits, but the authors underlined that the drugs are mainly effective in mild to moderate DAT (Birks, 2006). In the most recent evaluation of dementia medications, based on a systematic review of 257 RCTs and observational studies on dementia or MCI, the authors report the same benefits as previously, but they underscored that these benefits seem to wane over time and question the DAT medication in patients over 85 years (Buckley and Salpeter, 2015). This latest report also concludes that there are no benefits of ChEIs in the advanced stage of dementia, confirming the results of the Cochrane analysis mentioned above. However, the inclusion of MCI and all types of dementias in the systematic review from 2015 may have contaminated the results toward reports of less benefits, since ChEIs have not been shown to prevent decline to dementia in MCI (Cooper et al., 2013) or to have any effect in other non-DAT-associated neurodegenerative disorders, except in Lewy body and Parkinson-associated dementia (Rolinski et al., 2012). The effect of ChEIs cannot be predicted in current clinical practice before the initiation of the treatment, and the optimal duration of treatment is unknown (Birks, 2006; Cacabelos, 2007).

The other category of medication represented by memantine has also shown clinical benefits in moderate to severe DAT in terms of global status, cognition, activities of daily living, and behavior (especially agitation and aggression), and the medication was well tolerated, although some adverse effects, consisting mostly of somnolence and weight gain, have been mentioned (McShane et al., 2006; Yang et al., 2013). In current clinical practice the DAT medication is continued as long as the patient sustains a meaningful interaction with family members and caregivers, which is used as a proxy for maintaining good quality of life.

The latest addition to the therapeutic arsenal in DAT in the USA in 2014 was the combined memantine extended-release donepezil (Namzaric) with indication for moderate to severe DAT (Howard et al., 2012; Ehret and Chamberlin, 2015). The European Medicines Agency refused the marketing of the combination of memantine and donepezil (Acrescent) in the EU in 2012 due to lack of studies and negative results in those studies in which memantine was added to donepezil treatment (EMA/CHMP, 2012).

Individual strategies in choosing the appropriate dementia medication are highly recommended due to the individual course of the disease and different response to the drugs and their adverse effects. A switch between different ChEIs should be considered if necessary to reduce adverse effects and ensure a better quality of life for the patient. Also, the correct timing of the start of single or mixed drug therapy, a ChEI and memantine, along the progression of DAT to more severe stages needs qualified and continuous assessment of the patient from early MCI stage.

Clinical trials in DAT

Although clinical trials with active and passive antiamyloid vaccines (Wisniewski and Konietzko, 2008; Mangialasche et al., 2010) have not proved yet that this may be an effective therapy for use in humans with DAT, indications that the compounds were modifying the amyloid accumulation have been reported in ancillary analyses (Fox et al., 2005; Holmes et al., 2008). Although the agent bapineuzumab based on monoclonal antibodies targeting Aβ has not been successful in phase III clinical trials in terms of primary outcomes, cognition and function (Vellas et al., 2013; Salloway et al., 2014), a similar agent, solanezumab, made its way through several phase III efficacy trials in both asymptomatic older individuals at risk for AD and in symptomatic AD in prodromal MCI or mild DAT stage, with results expected in the near future.

Other similar agents are undergoing examination: ganterenumab (subcutaneously administered monoclonal antibody against Aβ) is being tested in a multicenter, randomized, double-blind, placebo-controlled, parallel-group study for efficacy and safety in a phase III trial in mild DAT on a 104-week period (trial registered at ClinicalTrials.gov identifier NCT02051608, recruiting) with primary outcomes including changes in brain amyloid load over time using florbetapir positron emission tomography and changes in scores in Alzheimer's Disease Activity Scale-Cognitive Subscale 13 (ADAS-Cog 13) and Alzheimer's Disease Cooperative Study-Activities of Daily Living (ADCS-ADL). Aducanumab is another promising monoclonal antibody against Aβ currently being tested in two phase III trials (trials registered at ClinicalTrials.gov identifiers NCT02484547 recruiting, NCT02477800 recruiting), both testing monthly doses of aducanumab in early-stage DAT against cognitive and functional decline measured using the CDR-sum of boxes as primary outcome. At the same time there are ongoing trials with active compounds, for example, the active antiamyloid vaccine CAD106 that showed encouraging results regarding safety profile in phase I and II trials (Winblad et al., 2012; Farlow et al., 2015). In the latter phase II study, active immunization with CDAT106 was repeatedly administered to 58 patients with mild DAT in a multicenter randomized, double-blind, placebo-controlled trial over 52 weeks, and 45 of these patients entered the open-label extension study of 66 weeks. In this study the serologic response to active immune stimulation was obtained in 64% of patients, the adverse effects profile did not require discontinuation, and there were very few serious adverse effects in the form of different types of hemorrhages, but without signs of brain inflammation due to induced T-cell responses (Farlow et al., 2015). A phase III study with CAD106 in individuals at risk of developing DAT is planned (ClinicalTrials.gov identifier NCT02565511).

Meanwhile, other potentially therapeutic compounds are in the pipeline based on hypotheses generated mainly by epidemiologic observations and wide genomic association studies. Some of the working hypotheses under scrutiny address tau-pathology in DAT (Mitchell et al., 2002; Small and Duff, 2008), whereas others target enzymes involved in the production of toxic beta-amyloid in the DAT brain, for example, Aβ precursor protein-cleaving enzyme 1 (BACE1) (Yan and Vassar, 2014), while still others address oxidative (Honda et al., 2004) and inflammatory mechanisms in the brain (Heneka et al., 2015), and hormone-enhanced cognition (Henderson, 2010), to name the most prominent therapeutic avenues open to investigation. These hypotheses are not mutually exclusive and addressing multiple mechanisms could be a more efficient way of addressing the complex neurodegenerative pathology of DAT.

Another aspect of the individual management of DAT consists of nonpharmachologic strategies, namely adequate control of other health issues, ensuring appropriate safety measures regarding dwelling, driving, and access to firearms, as well as legal counseling and support groups for the patient and caregivers (Olazaran et al., 2010; Cummings et al., 2015).

The realistic objective for the management of DAT patients and their family is to ensure quality of life during the progression of the disease.

DAT management costs

The costs for DAT and other dementias in the USA were estimated at about \$226 billion a year for 2015 (Alzheimer's Association, 2015), with most costs attributed to nursing home or assisted living facilities. Delays in dementia onset or severe symptomatology that hastens institutionalization may bring important benefits not only to those who suffer from the disease and their families, but also to public health costs. Higher awareness in the general population and increased referrals for investigation of memory impairments along with DAT becoming amenable to diagnosis at the primary care level have contributed to increased opportunities to diagnose DAT in milder stages. Therapy directed to alleviate the symptoms should preferably be initiated at the mild stage of the disease when cognitive function is not severely impaired. A 2-year delay in the onset of dementia, using predictions based on incidence studies from the USA, may reduce the prevalence of dementia by 23% by the year 2047 and a 5-year delayed onset may promote a 47% reduction in dementia prevalence (Brookmeyer et al., 1998). Similar figures were derived for Australia, namely that delaying the onset of dementia by 2 years reduces the prevalence of dementia by 21% in 2050, while a 5-year delay in onset will reduce the prevalence by 44% (Jorm et al., 2005, based on demographic figures for Australia in the 1990s). A recent update of these estimates adjusted to 2001 demographics for Australia showed slightly lower figures: 16% reduced prevalence by 2040 for a 2-year delay in the onset of dementia and 37% reduction by 5-year delay of onset (Vickland et al., 2010). This discrepancy may reflect changes in the demographic structure, with more individuals than predicted in the 1990s entering the age groups at increased risk of developing DAT, the effect of previously unaccounted risk factors or increased exposure to known risk factors for the development of dementia (e.g., lifestyle contributing to metabolic syndrome with obesity and hypertension, increased survival after cardio- and cerebrovascular insults) (Vickland et al., 2010). Regardless of the actual point estimates, a reduction in prevalence by only 2% by 2040 compared to current projections may bring a reduction in cost of US$ 153 billion, excluding informal care costs (Hurd et al., 2015).

An even more beneficial and desirable goal is to reduce the impact of the disease by primary prevention. It has been recently estimated that approximately a third of the population attributable risk of DAT is due to seven potentially modifiable risk factors – low education level, low physical activity, depression, smoking, diabetes, midlife hypertension, and obesity – and interventions that aim at 10% relative reduction per decade in any of these factors have the potential to reduce the prevalence of DAT by around 8% by 2050 compared to current projections (Norton et al., 2014). More about primary prevention that addresses potentially modifiable risk factors follows in the section on intervention epidemiology later in this chapter.

DESCRIPTIVE EPIDEMIOLOGY OF DEMENTIA

Healthcare management in modern society uses reports from epidemiologic observations of dementia outcome to adjust healthcare policies and practices to current needs based on prevalence. Future healthcare system challenges are predicted based on incidence of dementia. Prevalence and incidence figures also help to evaluate the effect of preventive interventions on known risk factors on the development of dementia in the population. Moreover, they may indirectly point out the presence of putative risk and protective factors in populations that show differences in either prevalence or incidence of dementia. The use of similar methodologies in epidemiologic studies is thus essential for valid comparisons. Participants in epidemiologic studies of dementia are usually selected based on cognitive screening. Large randomized studies without screening provide ecologically valid results, but may not always be economically feasible. Usually random representative samples are selected. The following step in an epidemiologic study is to diagnose dementia in the whole sample or in those who screen positive and a representative subsample of those who screen negative. Finally an etiologic diagnosis is established based on the diagnosis criteria in use at the time of the study. Two of the most pervasive factors that may affect the outcome in prevalence and incidence studies will be discussed below.

Diagnostic criteria

Although international collaborative efforts have succeeded in standardizing the diagnostic criteria for dementia, different screening methods and different systems of diagnosis may affect the prevalence and incidence of dementia reported in the literature. For example, criteria based on the DSM system used extensively in research perform better at capturing mild forms of dementia than the *International Classification of Diseases* (ICD) system used for coding diseases in hospital records (World Health Organization, 2010), which contributed to the establishment of the DSM system as the criteria of choice in research settings. For many of the etiologic subtypes of dementia, excluding DAT, no established consensus criteria are used in research. Usually more than one set of criteria may be available at the time of the study, which makes it difficult to compare parameters from observational epidemiologic studies of etiologic subtypes of dementia.

Cohort effects

We are currently witnessing unprecedented increases in the prevalence of chronic diseases due to the eradication of most deadly, acute infectious illnesses. Also, as a consequence of demographic changes, with a larger proportion of people reaching older ages worldwide, there is an accumulation of chronic morbidities, specifically in older segments of the population. Among the morbidities of old age, dementia is a major contributor to reduced quality of life, institutionalization, and mortality. Changes in the structure of population and in society between generations, so-called cohort effects, may change the expression of a disease and affect the prevalence and incidence in epidemiologic studies. The descriptive parameters of dementia have been drawn from earlier cohorts born at the beginning of the 20th century and therefore they do not fully reflect the disease status of the postwar baby boomer generation. Furthermore, new challenges imposed by the "digital generations" are soon to be tackled by healthcare systems in terms of diagnosis in mild stages of the disease and long-term management. There have been indications that dementia may present different cognitive phenotypes between generations due to increased brain and cognitive reserve in the later-born cohorts (Schaie et al., 2005; Llewellyn and Matthews, 2009; Sacuiu et al., 2010). Thus, continuous updates of descriptive epidemiologic parameters by following different birth cohorts are essential for healthcare planning.

PREVALENCE OF DEMENTIA

Prevalence informs on the number of individuals in a representative population afflicted by a disease at a specific point in time (snapshot counts). Thus, any changes in the criteria used to define a disease at different periods in time may affect prevalence. Moreover, methods available to operationalize the same criteria may affect prevalence.

The discrepancy in prevalence of dementia due to different diagnostic criteria was exemplified in a representative urban Scandinavian population of 1019 individuals, aged 70 years and older, enrolled without screening (response rate 68.2%), in which the use of DSM-IV criteria resulted in 9.6% prevalent cases of dementia (6.3%, according to DSM-III-R criteria that were in use before 1994). The application of ICD-10 (World Health Organization, 2010) resulted in 3.1% dementia prevalence (1.2% with ICD-9 criteria that were in use before 1992) (Wancata et al., 2007).

A large epidemiologic survey, the Canadian Study of Health and Aging (CSHA), reported on a representative sample of 10 263 people 65 years and older from which a subsample of 1879 individuals, mean age 80.4 years, were selected by using a conventional two-stage sampling method (randomization and screening). This study evidenced the same discrepancy between the DSM and ICD diagnosis systems, with a dementia prevalence of 13.7% when using DSM-IV (17.3% when using DSM-III-R criteria) and 3.1% when using ICD-10 (5% according to ICD-9) (Erkinjuntti et al., 1997). However, the CSHA reported larger prevalence of dementia than the Scandinavian study irrespective of the diagnosis system used. Moreover, in the CSHA the older criteria captured more dementia cases than the newer criteria, in contrast to the results reported by the Scandinavian study. While both studies included institutionalized and community-dwelling participants, the Scandinavian study was age-biased toward younger elderly by including a 70-year-old cohort, an age group at which dementia is less prevalent compared to older ages. Also, detailed cognitive examinations were performed by neuropsychologists and physicians to detect dementia in the CSHA in all institutionalized individuals and in the community sample after screening, whereas in the Scandinavian study the assessments were performed by trained psychiatric research nurses in the whole population and dementia diagnosis was algorithm-based using different diagnostic criteria (Wancata et al., 2007). Also, the ICD-10 explicitly requires that the information on dementia should be corroborated by information from caregivers, a criterion that may not always be fulfilled in elderly patients, and it was therefore excluded from the diagnosis in the Scandinavian report (Wancata et al., 2007). Methodologic differences may, thus, explain the discrepancies in prevalence reported by these two representative studies.

Although the introduction of standardized DSM diagnostic criteria for dementia in research, and specific etiologic criteria for DAT (McKhann et al., 1984) and vascular dementia (VaD) (Gold et al., 1997), have resulted in more consistent figures of prevalence of dementia across nations, a recent review of studies of prevalence of dementia reported between 2000 and 2012 revealed further discrepancies in the instruments used for screening and to standardize diagnostic information, and in strategies used to account for those who refused participation, which all may distort the actual prevalence figures (Misiak et al., 2013). Moreover, it has been reported in the literature that half of prevalent cases may be found in high-income countries, about a third in middle-income countries and around 14% in low-income countries (Banerjee, 2012). However, the 10/66 Dementia Research Group Population-Based Surveys in low and middle income countries in Latin-America, India and China pointed out that the DSM-system of diagnosis for dementia may underestimate the burden of disease in these regions (Prince et al., 2003). By using the so-called 10/66 dementia diagnosis based on an algorithm using information from structured clinical and informant interviews, a structured neurologic examination and results from cognitive tests that allow "culture and education-fair dementia diagnosis" (Prince et al., 2007), it was shown that the prevalence of dementia in a low-income, developing country is similar to that in middle and high income countries (Llibre Rodriguez et al., 2008). There is, though, a fundamental lack of large scale, descriptive epidemiologic studies from low-income countries. Thus, there is a continuous need for well-conducted, multinational, epidemiologic surveys that use similar methods.

According to the most recent estimate of the worldwide burden of dementia at age 60 years and over, 35.6 million people were affected by dementia as of 2010 (Prince et al., 2013). Detailed figures from this meta-analysis were incorporated in the WHO report on dementia from February 2013 (Duthey, 2013). It has also been estimated that dementia prevalence worldwide will rise to 115.4 million people by 2050, an estimate based on extrapolation of prevalence figures obtained from representative epidemiologic studies to the expected world population in 2050 (Prince et al., 2013). Some of the assumptions made by this type of projection consist of constant prevalence and rate of progression as well as constant risk factors over time, independence of risk factors, equal mortality risks between healthy elderly and elderly with dementia, and no regional differences between prevalence and risk factors of dementia. Violations of these assumptions make the extrapolation a quite crude estimation of the expected dementia prevalence, as discussed by the authors (Prince et al., 2013). Although the accuracy of available projections may be somewhat limited, they are necessary to orient the management of the disease in population. Scenarios that take into account behavioral modifications of lifestyle risk factors at different ages (windows of opportunity), as well as stratification by age with regard to prevalence of comorbidities known to increase the risk of DAT and mortality (e.g., traumatic brain injury (TBI), vascular brain insults such as transitory ischemic attacks and stroke, DM, hypertension) as well as correlations between risk factors by age strata, could be more informative (Norton et al., 2013).

DAT is the most prevalent type of dementia at age 60 years and older, representing approximately 60–70% of all dementias, with another 10% of all dementias attributed to mixed AD and VaD (Duthey, 2013). The prevalence of DAT cases starting earlier than age 60 years is estimated at around 6–9% of all dementias, but the global estimate is uncertain due to limited accounts across world regions (Duthey, 2013). DIAN is the only established cohort of familial AD with early debut that gathers data from the USA and European countries. However, inherited forms of DAT represent only 1% of all dementias with debut earlier than age 65 years (Bateman et al., 2011). Other types of dementia are more frequent at younger ages, e.g., VaD and frontotemporal-lobe dementia (Duthey, 2013; Lambert et al., 2014). However, a recent Australian epidemiologic study that reported the prevalence of dementia for the year 2008 in an urban sample at ages below 65 years found a high prevalence of DAT of almost 18% which was more frequent than VaD (13%) and frontotemporal-lobe dementia (11%); DAT prevalence in this sample was only surpassed by that of alcohol-related dementia (Withall et al., 2014). This distribution of different types of dementia in a young urban sample (catchment area represented by four districts in Eastern Sydney) may reflect selection bias, but it may also reflect secular trends in dementia etiology in the newer generations, together with earlier diagnosis of DAT due to increased awareness of DAT in the population and among medical professionals.

It is important to keep in mind that epidemiologic studies without neuropathologic confirmation have a high degree of uncertainty regarding the etiologic diagnosis. Moreover, there is a high degree of overlap between different pathologies of dementia (Storey et al., 2002; Zaccai et al., 2006), especially with increasing age. The majority of epidemiologic studies point though toward an increased prevalence of DAT with age, from 5% to 10% at ages 65–75 years (Plassman et al., 2007; Kalaria et al., 2008; Tola-Arribas et al., 2013; Wu et al., 2013; Ding et al., 2014; Velazquez-Brizuela et al., 2014), to around 15% at age 85 (Skoog et al., 1993; von Strauss et al., 1999) and over 50% at 90 years and older (von Strauss et al., 1999; Borjesson-Hanson et al., 2004). A US prevalence study showed slightly lower prevalence of DAT of 37% in the very old (Plassman et al., 2007). This finding was supported by a neuropathologic study of 532 Finnish elderly 85 years and older followed up for the outcome of dementia between 1991 and 1999 that showed that 33% of individuals had neuropathologically diagnosed DAT (Polvikoski et al., 2001). However, the reports of etiologic types of dementia in the very old should be interpreted cautiously. The etiologic diagnosis of dementia in the very old is difficult due to high prevalence of other morbidities that may affect the presentation of the typical DAT phenotype.

DAT seems to be more frequent in women, especially in the very old (Jorm et al., 1987; Fratiglioni et al., 1997; Borjesson-Hanson et al., 2004). The biased sex distribution has been attributed to longer lifespan in women than in men, a standpoint supported by similar incidence of DAT in men and women up to the 80th decade of life, but it is still unclear if this difference is underpinned by biologic or social factors (Bachman et al., 1993; Evans et al., 1999; Fitzpatrick et al., 2004).

In the context of global warning regarding dementia as an epidemic of the new century with a devastating impact on societal health costs, it is worth mentioning that secular trends of decline in the prevalence of dementia have been indicated by a couple of recent studies from Europe and the USA. In the Zaragoza Study in Spain, a 5.2% dementia prevalence was reported for the period 1988–1989, standardized by age and sex to the structure of European population 1992 (global prevalence 5.5%) and 3.9% for the period 1994–1996 adjusted prevalence (global prevalence 5.9%) (using DSM-IV criteria in institutionalized and noninstitutionalized populations) (Lobo et al., 2007). The authors suggested that the decline over a short period of time in age-adjusted prevalence, but not global prevalence, may be due to unidentified exposure factors in different age strata that may counterbalance the effects of increased prevalence of dementia with age (Lobo et al., 2007).

The decline in age-adjusted prevalence has been confirmed by a large epidemiologic study from the UK, the Medical Research Council (MRC) Cognitive Function and Aging Study (CFAS), that compared two cohorts of elderly, 65 years and older, from the same regions of the UK across a time span of 20 years. The prevalence of dementia decreased by 1.8% between the two waves, 1990–1993 CFAS-I 8.3% versus 2008–2011 CFAS-I 6.5% (sex- and age-specific prevalence standardized to the population in 1991 and 2011, respectively, using DSM-IIIR criteria in both community dwellers and in care homes) (Matthews et al., 2013). However, comparisons of septuagenarian cohorts examined 30 years apart in Gothenburg, Sweden (i.e., 70-year-olds examined 1971–1972 and 2000–2001, and 75-year-olds examined 1976–1977 and 2005–2006), found no differences in the prevalence of dementia across cohorts (Wiberg et al., 2013). A synopsis of secular trends of dementia in Western Europe has been recently published (Wu et al., 2016). Data from the USA are not consistent either, with a national survey indicating the same declining trends in dementia prevalence as most European studies (Langa et al., 2008), but another survey contradicted these findings (Rocca et al., 2011). On the other hand, a review of eight population studies from Japan showed an increased prevalence of dementia between 1992 and 2008 (Dodge et al., 2012). The Japanese authors could not analyze specific prevalences of dementia subtypes due to large variations in the criteria used for diagnoses, but they speculate that increased prevalence of vascular diseases may have contributed to this trend (Dodge et al., 2012). Detailed investigations of worldwide trends of dementia prevalence could help to understand the impact of exposure to known risk factors in different parts of the world, as well as reveal as-yet unknown protective or risk factors.

INCIDENCE OF DEMENTIA

Incidence is calculated as the number of newly diagnosed cases of dementia by person-years at risk, the latter being defined as the time period in years from the start of study (baseline) until dementia diagnosis, dropout (death or refusal to participate at follow-up), or end of the study. According to the most recent EURODEM meta-analysis that used pooled data from eight well-established population studies across Europe, the incidence of dementia increases exponentially with age (Fratiglioni et al., 2000). This report confirmed earlier results based on data from 23 surveys across the USA, Europe, and Asia published between 1966 and 1997 (Jorm and Jolley, 1998). However, in the EURODEM pooled cohorts the incidence of dementia seems to differ by sex in that it enters a plateau in men after age 80 years, while it continues to increase in women (Fratiglioni et al., 2000). Interestingly, the incidence of dementia at age 80 years and over was higher in populations from northern European regions (Finland, Sweden, Denmark, The Netherlands, and the UK) than in populations from southern regions (Spain and France) (Fratiglioni et al., 2000).

An account of the incidence of dementia in South America and China from the 10/66 Dementia Research Group has also shown that older age and female sex, among other factors, were independently associated with dementia incidence (Prince et al., 2012). Cohort studies from North America have also shown increased incidence of dementia with age, but the results regarding sex differences are divergent, with most North American studies finding no sex differences (Bachman et al., 1993; Kokmen et al., 1993; CSHA Working-Group, 2000; Kukull et al., 2002; Fitzpatrick et al., 2004), whereas others reported either a trend of higher incidence of DAT in women (Kawas et al., 2000) or higher incidence of all dementias in men (Ganguli et al., 2000). For example, the Framingham Study cohort showed an increased incidence of dementia from 7.0 per 1000 person-years between ages 65 and 69 years to 118 per 1000 person-years between ages 85 and 89 years, without sex differences (Bachman et al., 1993). Similar findings were reported by the CSHA, although there was a slightly higher incidence of all dementias in men in their 70s in this study, which was thought to be due to a higher incidence of VaD and a higher prevalence of cognitive impairment at baseline (CSHA Working-Group, 2000).

In the MoVIES project there was an increased incidence of all dementias in men (29.4%, compared to 23.1% in women), but only when MCI (CDR $\geq$0.5) was included, a discrepancy explained by socioeconomic disparities between men and women in the catchment area that could have increased the risk for cognitive impairment in men (Ganguli et al., 2000). Of course, as evidenced above in prevalence studies, many of the differences in incidence reports may be attributable to methodologic factors such as operationalization of diagnosis criteria, screening methods, and representativeness of the sample. Other specific factors that may affect incidence studies are the length of follow-up periods. Dementia and especially DAT develop slowly and long periods of follow-up are necessary to capture an ecologically valid outcome in representative samples. The younger the baseline sample, the longer the follow-up needs to be due to lower prevalence of dementia, at least DAT, at younger ages. Therefore incidence studies are more costly than prevalence, snapshot studies, which is reflected in fewer reports available on incidence than on prevalence of dementia.

Another specific issue related to long follow-ups for dementia is the bias due to those lost at follow-up, which could be due to higher mortality in this group or higher comorbidity burden and thus a hindrance to active participation in long epidemiologic surveys. A valid study design must address this issue, for example, by using close-informant interviews, death registers, or hospital registers for missing data in those lost to follow-up. Although this proxy information may not be complete, it will give a better estimate of the incidence of dementia, as has been suggested by some accounts (Aevarsson and Skoog, 1996; Sacuiu et al., 2005; Polvikoski et al., 2006).

DAT represents 60–70% of all incident dementias. DAT rates increase exponentially with age, doubling every 5–6 added years of adult life, with some regional variations (slightly longer doubling time in Europe of 6 years and North America of 5.8 years than in the rest of the world, 5 years) and a nonsignificant trend of sex difference worldwide (5.4 years to double incidence of DAT for women and 6.5 years for men) (Ziegler-Graham et al., 2008).

ANALYTIC EPIDEMIOLOGY: RISK FACTORS FOR DAT

The accepted current view regarding DAT pathogenesis is that this type of dementia is multifactorially determined. The disease state reflects a complex interplay between age and genetic factors, both representing the risk load that cannot be controlled, and lifestyle and comorbidity factors that are amenable to intervention. The main nonmodifiable and modifiable risk factors for which compelling epidemiologic evidence has been accumulated during the last decades will be discussed below.

Nonmodifiable risk factors for DAT

Age

Age is the dominant risk factor for DAT, as indicated by incidence and prevalence studies. Although proponents of the continuum aging–neurodegeneration are inclined to believe that a long lifespan extending to about 130 years may be inherently accompanied by DAT due to brain aging *per se* with dysfunctional repair and apoptosis processes (Terry and Katzman, 2001), it is important to underline that DAT is not the inevitable result of aging (Perls, 2004). The incidence curve in the very old (nonagenarians and centenarians) has not been clearly defined. Although the earlier consensus criteria for DAT that specified dementia debut between 40 and 90 years (McKhann et al., 1984) has been removed from current DAT criteria (McKhann et al., 2011), it is difficult to diagnose DAT in nonagenarians and centenarians due to considerable overlap of neuropathologic findings in the brains of cognitively unaffected very old individuals and in those with cognitive decline (Neuropathology Group of the Medical Research Council Cognitive Function and Ageing Study (MRC-CFAS), 2001; McKhann et al., 2011). This overlap is believed to be due to the confounding effects of a large cognitive reserve that compensates for neuropathologic brain damage. Thus, the neurodegenerative process will induce only a subtle cognitive decline that is difficult to detect with the available screening instruments in those with a large cognitive reserve. Some studies indicate, though, continued increase in the prevalence of dementia and DAT from around 25–30% in 85-year-olds to 50–65% among those 95 years and older (Fratiglioni et al., 1997; Borjesson-Hanson et al., 2004; Lucca et al., 2015), whereas a Finnish study of incidence of dementia in the very old indicated a plateau in dementia incidence after age 85 years (Polvikoski et al., 2006). The latter report confirms the results of an earlier meta-analysis of epidemiologic studies in populations aged 80 years and older (Ritchie and Kildea, 1995). The few centenarian studies available converge on a prevalence of dementia of around 50–65% (Poon et al., 2012; Lucca et al., 2015), with neuropathologic studies showing an increased frequency of mixed pathologies in the very old, especially brain vascular damage together with Alzheimer's pathology (Jellinger and Attems, 2010). Less than 1% of the sample in this neuropathologic study showed negative results (Jellinger and Attems, 2010), a finding that seems to support the continuum brain aging–neurodegeneration theory.

Interestingly, women seem to have an increased risk of developing DAT, but it is not clear if the risk is associated with the female sex *per se* or is merely a consequence of women having a longer lifespan than men (Bachman et al., 1993; Evans et al., 1999; Fitzpatrick et al., 2004), as previously emphasized in the section on the prevalence of dementia, above.

Genetic risk factors for DAT

The major genetic risk factor that has been linked to both inherited and sporadic DAT is the presence of a specific genetic variant (allele) of the APOE gene. Human APOE gene is located on chromosome 19 and has three alleles – ε4, ε3, and ε2 (Mahley and Huang, 2012). Although the role of apolipoproteins E, the end products of these gene variants, is not fully understood, they seem to be involved in the amyloid metabolism at cell level (Nierenberg et al., 2005). There is a dose-related association between the presence of the APOE ε4 allele and the risk of cognitive impairment, with highest risk in homozygote carriers of the ε4 allele (Slooter et al., 1998a; Small et al., 2004; Genin et al., 2011; Mahley and Huang, 2012). Due to the wide spread of the APOE ε4 allele in humans, with approximately 25% of the population being carriers of this genotype, the gene represents a major risk factor for DAT, which is also reflected by an earlier debut by about 8 years in heterozygotes and 15 years in homozygote APOE ε4 carriers compared to noncarriers (Mahley and Huang, 2012).

The APOE ε2 allele has been associated with protective effects against Alzheimer's type of pathology, probably by contributing to the clearance of toxic Aβ from the brain (Chiang et al., 2010; Lambert and Amouyel, 2011; Fouquet et al., 2014). Perhaps the most notable characteristic of the APOE gene variants is that they play a modulating role on the phenotypic expression of the disease, the episodic memory being especially vulnerable to the presence of APOE ε4 (Wilson et al., 2002; Ercoli et al., 2006). Moreover, the APOE ε4 genotype modulates the effect of other risk factors in DAT, e.g., it has been shown that vascular atherosclerosis has larger deleterious effects on cognition in APOE ε4 carriers compared to noncarriers (Slooter et al., 1998b), and alcohol consumption in APOE ε4 carriers increased the risk of dementia in frequent drinkers by seven times (95% confidence interval 1.4–36.6) compared to carriers who never drank (Anttila et al., 2004). In individuals with stroke, APOE ε4 has additive effects on the development of dementia (Zhu et al., 2000). APOE ε4 also confers less favorable outcome in TBI (Mahley and Huang, 2012), which in turn has been related to increased risk for DAT. Moreover, having both APOE ε4 and diabetes increases the risk of DAT (Irie et al., 2008a). Also an increased risk of DAT was found in depressed men who were APOE ε4 carriers (Irie et al., 2008b).

Several other genetic mutations have been identified during the last decade as a result of GWAS in DAT, e.g., CLU, that codes for another apolipoprotein suggested to be involved in the amyloidogenesis, and PICALM, that transcribes into a intracellular protein involved in intracellular trafficking and thus in synaptic neurotransmission (Heyman et al., 1997). The AlzGene website provides open-access summaries of data from genetic studies (after the association found between the candidate genes and DAT was rated according to standardized criteria) and results of meta-analyses of the available AlzGene studies (http://www.alzgene.org/methods.asp). Although the CLU and PICALM associations have been replicated, the AlzGene meta-analyses show a small associated risk of these genes with DAT compared with the risk associated with APOE4 (odds ratios (OR) 3.68), with estimated OR of 0.89 for CLU and 0.88 for PICALM (http://www.alzgene.org/TopResults.asp).

There are currently 22 candidate risk genes associated with DAT (Fox et al., 2005): some are involved in inflammation at cellular level (Pimouguet et al., 2015), whereas others are involved in lipid metabolism and endocytosis (Fox et al., 2005). However, not all of these associations have been replicated (Burns et al., 2011). The AlzGene website currently prepares for inclusion of studies from the IGAP consortium, which represents another step towards gaining more power for genomic studies and perhaps new insights into the complex genetics of sporadic DAT. However, the GWAS relies heavily on statistical power (extremely large samples) and uses a case-control approach (demented versus nondemented) to identify genetic traits. Consequently, GWAS analyses are prone to confounding by yet undetected cognitive decline in the control group. This is especially problematic in DAT since the disease has a long prodromal phase. Therefore new insights are sought from whole-genome and whole-exome trait association studies using next-generation sequencing that has the ability to screen the genome for mutations at basepair resolution. Although these studies have their limitations due to the phenotypic heterogeneity of DAT that generates very small groups of individuals with the same phenotype, e.g., similar profiles of established biomarkers of DAT, the studies are promising in terms of insights related to pathogenetic mechanisms of sporadic DAT (Gilman et al., 2005). By establishing large genetic consortia of DAT, e.g., the Alzheimer's Disease Sequencing Project (DATSP), researchers are trying to overcome these limitations. The DATSP data portal currently offers to approved applicants whole-exome and whole-genome sequencing of 10 939 subjects (https://www.niagads.org/adsp/content/home).

Modifiable risk factors for DAT

Low education level

Low education level is a well-established risk factor for DAT. Other modifiable risk factors are metabolic syndrome *per se* or each of its characteristic comorbid conditions, i.e., obesity, high blood pressure (HBP) and high blood levels of glucose, cholesterol, and triglycerides. Moreover, low blood levels of vitamins B and high levels of plasma homocysteine may increase the risk of DAT. Other factors that have been associated with DAT are smoking, traumatic head injury, and psychiatric disorders such as depression and posttraumatic stress disorder (PTSD). Each of these will be discussed below.

Low education level has been associated with the development of dementia and DAT in particular. The hypothesis that supports this negative association is that schooling may contribute to consolidate the cognitive reserve that makes the brain resilient to neuropathologic changes induced by AD and, thus, delays or prevents dementia later in life (Eriksdotter-Jonhagen et al., 2012). This hypothesis, proposed by Yaakov Stern, is known as the cognitive reserve hypothesis (Stern et al., 2003). On the contrary, higher levels of education have been associated with reduced risk of DAT, especially in APOE ε4 carriers, thus suggesting an interaction effect between these two factors (Wang et al., 2012a). Moreover, education level is closely associated with intellectually stimulating occupations in adult life, which, in turn, have been associated with a reduced risk of DAT (Stern et al., 1994; Karp et al., 2004). The cognitive reserve hypothesis has been supported by *in vivo* findings of more amyloid deposits and lower brain metabolism in individuals with higher education or intellectual occupation compared to individuals with lower education at the same level of cognitive function (Garibotto et al., 2008; Kemppainen et al., 2008).

Some other studies do not support this hypothesis. For example, the Framingham Study reported no relationship between low education and dementia or DAT after age adjustment, but a significant relation was found between low education and non-DAT dementias (Cobb et al., 1995). The authors suggested that vascular risk factors together with smoking, both commonly found in individuals with lower levels of education, were more likely to contribute to the development of the vascular type of dementia. However, the Framingham Study was designed to disentangle the pathogenesis of cardiovascular diseases and therefore was more likely to capture risk factors for VaD and less likely to bring support to previous findings related to education and DAT. On the contrary, a study of 143 twin pairs from the Swedish Twin Registry has shown that low education incurred higher risk of DAT, but not of dementia in general (Gatz et al., 2001). The intellectual involvement throughout life was lower, though, in twins who became demented, supporting the cognitive hypothesis proposed by Stern.

Role of HBP, glucose and lipid dysregulation, and metabolic syndrome in DAT

The metabolic syndrome and each of its characteristic features, HBP, glucose metabolism, and lipid dysregulation with high blood levels of glucose, cholesterol, and triglycerides have been associated with an increased risk for dementia and DAT (Frisardi et al., 2010). All these factors are manageable and thus represent an opportunity to reduce the global burden of dementia. First we will review the role of the independent risk factors named above and after we will address the role of the metabolic syndrome in the development of dementia.

High blood pressure

HBP has been associated with cognitive impairment in longitudinal studies with long follow-up (Elias et al., 1993; Launer et al., 1999) and also with the development of dementia (VaD and DAT) (Skoog et al., 1996; Qiu et al., 2003; Kivipelto et al., 2006; Stewart et al., 2009). In a study by Launer et al. (1999), it was shown that in midlife Japanese-American men, an increase in systolic blood pressure with every 10 mmHg over 110 mmHg increased the risk of poor cognitive function up to 9%, the strength of the relation being somewhat reduced by additional risk factors such as comorbid stroke, coronary heart disease, and atherosclerosis. At the same time, cross-sectional studies (Rockwood et al., 1996) and studies with short follow-up (Guo et al., 1999; Qiu et al., 2004) have shown an association between low blood pressure and dementia. Taken together, these results suggested that HBP in midlife or in late adulthood is common in individuals who will later develop dementia. Additionally, it appears that a significant decline in blood pressure may occur a few years previous to dementia diagnosis.

In a study conducted by Skoog et al. (1996), a representative population sample of 70-year-olds was followed up for the development of dementia (VaD and DAT) over a 15-year period and brain scans were performed in participants at age 85 years. This study showed not only that incident DAT 10 years later was associated with HBP at age 70 years, but also that brain white-matter lesions were more common at age 85 years in individuals with HBP than in those with normal blood pressure at baseline. Based on these results, it was suggested that long exposure to HBP may induce lesions in the deep penetrating brain vessels and thus increase the risk of developing DAT (Skoog et al., 1996).

The role of HBP and small-vessel disease in DAT has later been confirmed in other studies, as reviewed by Qiu et al. (2005). Unfortunately, the majority of randomized

HBP therapy trials initiated in the elderly to reduce the incidence of dementia (main outcome) could not show benefits (McGuinness et al., 2009). Moreover, in the naturalistic longitudinal studies, irrespective of the length of the period of follow-up, the use of antihypertensive medication did not affect the relation between HBP and dementia (Skoog et al., 1996; Guo et al., 1999). These findings suggested that antihypertensive medication, if initiated late in life, may be less effective against vascular changes that have started a long time before. Thus, HBP should be addressed early in midlife before neurodegenerative brain lesions occur. At the same time, insufficient perfusion of the brain due to low blood pressure, especially low diastolic pressure, may play a role in the neurodegenerative process (Verghese et al., 2003). On the other hand, it has been difficult to demonstrate an iatrogenic cause of decline in blood pressure a few years before the dementia diagnosis in old age. Another possibility is that lower blood pressure could be due to the incipient neurodegenerative process that disturbs brain homeostatic mechanisms (Qiu et al., 2004). Whatever the case, the findings indicate a need for constant re-evaluation of HBP treatment in old adults. Thus, if we take into consideration the exponential increase in dementia incidence at 5-year intervals after age 70 years, re-evaluation of antihypertensive treatment every 5 years would be recommended in individuals with essential HBP who enter age categories with increased risk of developing a neurodegenerative process, in order to eliminate the risk of reduced brain perfusion.

Obesity

Obesity has also been associated with an increased risk of developing DAT. Earlier longitudinal studies with short follow-up and those with long follow-up, but few waves at long time intervals apart, have shown an association between weight loss and incident DAT (Barrett-Connor et al., 1996; White et al., 1996). On the contrary, in a longitudinal study of 70-year-olds, the H70 study in Gothenburg, Sweden, women showed 36% increased risk of developing DAT 10–18 years later for every point increment in body mass index (BMI) at baseline, and no significant change in BMI was detected during follow-up (Gustafson et al., 2003). These findings were confirmed in another study with long follow-up of 36 years in a diverse population gathered from a health delivery system in the USA, which showed that overweight individuals (BMI 25–29.9) had double the risk of developing DAT and obese individuals (BMI 30 and over) had threefold increased risk of DAT than those with normal BMI (BMI 18.5–24.9), independent of stroke, cardiovascular disease, and diabetes (Whitmer et al., 2007). Also, in the 21-year longitudinal study Cardiovascular Risk Factors Aging and Dementia from Finland, obesity at midlife conferred an increased risk for DAT later in life (OR 2.4, 95% confidence interval (CI) 1.2–5.1) (Kivipelto et al., 2005).

Moreover, cholesterol level, a blood parameter related to obesity, was found to be associated with the development of DAT in late life. Moderately high (200–239 mg/dL) to high (above 240 mg/dL) blood levels of cholesterol in midlife increased the risk of DAT in old age (Solomon et al., 2009). Interestingly, though, in a longitudinal study of 70-year-olds followed up at regular intervals to age 88 for the development of dementia, high levels of cholesterol at ages 70, 75, and 79 years were protective against incident dementia between 79 and 88 years (Mielke et al., 2005). This apparent mismatch between findings regarding the association between cholesterol levels (and obesity and HBP, as mentioned above) and dementia development is due to a complex interplay between lifestyle factors and age as determinants of DAT, and mortality as a confounder of this relationship (survival bias). In a study with a 32-year follow-up of women in Gothenburg, Sweden, the Prospective Population Study of Women, it was clearly demonstrated that only by excluding those individuals who died during follow-up did the cholesterol level in midlife become significantly associated with increased risk of developing dementia 32 years later (Mielke et al., 2010). On the contrary, if all participants at baseline were included in the analyses, no association could be demonstrated between cholesterol level at baseline and incident DAT over the 32 years of follow-up. However, a decline of more than 50 mmol/L in levels of blood cholesterol between subsequent waves of the study doubled the risk of dementia at follow-up (hazard ratio (HR) 2.4, 95% CI 1.2–4.6), and a similar trend was observed for DAT (Mielke et al., 2010). Thus, yet another aspect to be taken into consideration in epidemiologic studies of lifestyle factors with very long follow-up for the outcome of dementia is that repeated measurements may be a better predictor of incident dementia than one timepoint measurement at baseline.

Diabetes mellitus

DM type 2, a metabolic condition that may be acquired in adulthood, is another major risk factor for cognitive decline and dementia (DAT and VaD) (Biessels et al., 2006). As with other lifestyle factors, it seems that midlife DM incurs a higher risk of developing DAT later in life compared to DM that develops in old adults (Hassing et al., 2002; Xu et al., 2009a). Since DM is associated with increased mortality, survival bias may attenuate the association of DM with dementia incidence in the very old. However, the presence of DM induces cognitive decline even in old populations without dementia; for example, rapid decline in Mini Mental State Examination (MMSE) scores in diabetic compared to nondiabetic individuals (Hassing et al., 2004). Especially therapeutically uncontrolled DM may be deleterious for the brain and cognitive function, as reported from a European cohort study (the Kungsholmen project) (Xu et al., 2009b). In the Kungsholmen project, DM was identified in a cohort aged 75 years or above (mean age at baseline 82 years) by using the inpatient register system, the use of hypoglycemic drugs, and blood glucose repeated measurements during a 9-year follow-up. Participants were followed up for the outcome of dementia (DAT and VaD). Individuals with controlled DM (blood glucose <7.8 mmol/L) did not show an increased risk of developing dementia, but those with uncontrolled DM (borderline or undiagnosed) had a higher risk of developing DAT (Xu et al., 2009a).

However, glycemic control seems not to play an important role in other populations, as exemplified by a Taiwanese nationwide register study in which the use of antidiabetic medication did not reduce the risk of developing DAT on 11-year follow-up and insulin therapy even increased the risk of DAT (Huang et al., 2014). This latter finding is in line with findings from a European cohort, the Rotterdam study, in which insulin therapy was associated with higher risk for DAT (Ott et al., 1999). Although, again, these are apparently discordant findings, all studies that analyzed the role of hypoglycemic medication in the relationship between DM and DAT converge on showing that severe, uncontrolled DM with complications incurs a higher risk of developing DAT later in life. Insulin resistance and consequent deficient regulation of amyloid metabolism in the brain by insulin is the generally accepted hypothesis to explain the reported associations between DM and dementias (Watson and Craft, 2003). Other explanatory alternatives are inflammatory mechanisms and systemic vascular damage induced by deficient blood glucose regulation.

In summary, epidemiologic studies have shown that HBP and measures related to obesity and lipid dysregulation, i.e., BMI and cholesterol levels, follow different trajectories throughout life in relation to the development of dementia: from high-end values in midlife in those who survive long enough to develop dementia later in life to a decline to lower than average values in the preclinical stages of dementia. These associations, as well as DM in midlife and development of dementia in late life, are all prone to survival bias in epidemiologic studies. Therefore, different study designs may result in apparently contradictory results and their interpretation requires careful consideration.

In many cases, HBP is comorbid with obesity and DM, a condition known as metabolic syndrome. A cumulative pathologic effect of these three conditions (HBP, DM, and obesity) seems to contribute to accumulation of neurodegenerative pathology of Alzheimer's type in the brain (Frisardi et al., 2010). In the 21-year longitudinal study Cardiovascular Risk Factors Aging and Dementia from Finland, obesity, high total cholesterol level, and HBP in midlife increased additively the risk of DAT later in life, from OR of 1.9–2.4 for each individual factor, respectively, to OR 6.2 in individuals who presented all three factors (Kivipelto et al., 2005). Too few individuals in this study presented with comorbid DM and thus the role of this condition could not be analyzed reliably, which points out the difficulties of studying comorbid conditions that confer high mortality in longitudinal studies of aging populations. These factors also increase the risk of stroke. Thus individuals presenting with a metabolic syndrome have increased risk of developing DAT, vascular or mixed dementia (vascular and Alzheimer's) (Panza et al., 2010). The main hypotheses that have been put forward to explain the association between metabolic syndrome and cognitive impairment are inflammatory pathway activation and insulin resistance (Gustafson, 2010; Kim and Feldman, 2015).

Nutritional deficiencies in B-vitamins as risk factor for DAT

The important role of B-vitamins in brain function has been recognized since the mid-1800s in conjunction with severe symptoms of delirium in cases of pernicious anemia due to vitamin B_{12} deficit. Even less severe deficits of vitamin B_{12}, especially if coupled with low folate levels (vitamin B_9), may impair neurologic and mental functioning, by affecting brain methylation processes essential in DNA replication, neurogenesis, and neurotransmitter and fatty acid syntheses (Morris, 2012). Levels of vitamins B_6, B_9, and B_{12} are related to cognitive test performances in nondemented elderly (Reay et al., 2013). Although findings from epidemiologic studies supporting an association between low blood levels of vitamin B_{12} and incident DAT were not homogeneous (Smith, 2008), they spurred on a good deal of RCTs, most likely as a result of vitamin B_{12} supplementation being a low-cost preventive action. Unfortunately, most of these trials have been too small and poorly designed to provide useful evidence for the use of vitamin B_{12} supplements against cognitive decline (Malouf and Areosa Sastre, 2003; Li et al., 2014). Moreover, well-designed epidemiologic studies with long follow-up did not support a relation between serum B_{12} and DAT (Seshadri et al., 2002; Morris et al., 2006).

The interpretation of results from epidemiologic studies together with negative RCT findings is difficult because in many cases the vitamin B_{12} deficit is accompanied by other micronutrient deficits, e.g., low folate and high homocysteine (the latter being the product of a methylation cycle that mainly involves B-vitamins, the so-called "one-carbon metabolism"). Currently, no consensus has been reached as to what parameter to measure, e.g., dietary intake in relation to circulating plasma levels of B-vitamins or to their functional markers (e.g., homocysteine, 2-methylcitric acid, and beta-cystathione that all acumulate in the blood stream in case of poor vitamin B_{12} function and folate deficit, or methylmalonic acid that only increases if vitamin B_{12} is functionally insufficient). Results of epidemiologic studies have been more consistent regarding the association between low blood folate and increased risk for DAT (Morris, 2012). Also the accumulation of homocysteine, even without deficits in B_{12} and folate, seems to affect brain functioning by exerting neurotoxicity and promoting atherosclerosis (Prins et al., 2002; Hooshmand et al., 2010; Zylberstein et al., 2011). However, hyperhomocysteinemia is not specific for vitamin B_{12} or folate deficits and to replenish these vitamins when not necessary may be deleterious for cognitive functioning, with high serum levels of folate related to impairments in information processing, memory, and MMSE (Morris, 2012). More clinical series and epidemiologic research are necessary in order to address the relationship between B-vitamins and prevention of cognitive decline.

Smoking and risk of DAT

Both smoking and alcohol habits are lifestyle factors that are strongly associated with mortality and different morbidities, including dementia (Letenneur et al., 2004). Contrary to the earlier opinion built on evidence from cross-sectional studies or longitudinal studies with short follow-up, that smoking could have a protective effect against cognitive decline (Graves et al., 1991), modern studies have shown that smoking is deleterious for cognition. For example, a more recent EURODEM study that collected data from four representative cohorts of nondemented older adults aged 65 years and above showed that the decline in MMSE was larger in smokers than in nonsmokers (0.13 point/year compared to 0.03 points/year, $p < 0.001$) (Ott et al., 2004). Heavy smoking in midlife, defined as at least two packs a day, has been associated with double the risk of developing DAT later in life compared to nonsmokers in a longitudinal study with a mean follow-up of 23 years (HR for DAT 2.6, 95% CI 1.6–4.0, independent of sociodemographic factors, other lifestyle factors, and comorbidities, i.e., hypertension, hyperlipidemia, diabetes, heart disease, stroke) (Rusanen et al., 2011). A similar study that analyzed midlife smoking habits in relation to incident dementia and DAT on a 21-year follow-up (mean age at dementia assessment 71.6 years) found that smoking was associated with incident DAT only in APOE ε4 carriers, which suggests that the association between lifestyle factors in midlife may vary by genotype (Rusanen et al., 2010). However, the modulating role of APOE ε4 status in the relation between smoking and incident DAT is not clearly established.

An earlier prospective study in a biracial community in Chicago, USA, in individuals 65 years and older (mean age at baseline 73.8 years) with 4-year follow-up for the outcome of DAT found higher risk for DAT in smokers who were noncarriers of APOE ε4 allele (Aggarwal et al., 2006). In the same study, the APOE ε4 carriers who were current or former smokers were less likely to develop DAT than nonsmokers (current smokers nonsignificant OR 0.57, 95% CI 0.11–3.12; and former smokers significant OR 0.28, 95% CI 0.08–0.93) (Aggarwal et al., 2006). The contradicting results between the latter two studies could be partially explained by methodologic differences: (1) long period of follow-up versus shorter follow-up incurs different survival effects in the studies; (2) different mean age at dementia assessment may affect the prevalence of DAT; (3) different APOE ε4 allele frequencies in populations, to name just a few discrepancies. Although such differences may preclude comparisons of findings from epidemiologic studies of smoking and DAT to date, the majority of studies concur that smoking is deleterious for cognition and increases the risk of developing DAT in late life in a dose–response manner. The negative effect of smoking on cognitive function is thought to be due to both vascular and neurodegenerative pathways affecting the brain (Rusanen et al., 2011).

Alcohol consumption and risk of DAT

The majority of epidemiologic studies have shown a J- or U-shaped relation between alcohol and risk for dementia, i.e., lower risk in moderate drinkers and increased risk in heavy drinkers than in those with no alcohol consumption (Orgogozo et al., 1997; Ruitenberg et al., 2002; Anttila et al., 2004). Adults who are moderate alcohol drinkers seem to be protected against cognitive decline in late life (Ganguli et al., 2005b; Stampfer et al., 2005). Although the protective effect of moderate alcohol consumption has been difficult to clarify in relation to cognitive function, some suggestions were made related to specific properties of red wine, i.e., flavonoids and other antioxidants in red wine that may reduce oxidative injuries (Truelsen et al., 2002), and a protective effect related to a healthier lifestyle and reduced burden of cerebrovascular insults in moderate drinkers (Skoog, 1998). Drinking wine in moderate quantities, as opposite to other alcoholic beverages, seems to have a general beneficial effect, as evidenced in a longitudinal study of a representative population of middle-aged women with a 32-year follow-up for the outcome of dementia, that found not only an inverse relationship between moderate wine drinking and incident dementia (HR 0.3, 95% CI 0.1–0.8), but also that moderate consumption of wine tended to predict longevity (Mehlig et al., 2008). Given the negative findings in an earlier large, well-designed epidemiologic study (Graves et al., 1991), raising questions about the association between moderate alcohol intake and reduced risk for DAT, coupled with the fact that alcohol is toxic for the brain and induces Wernicke–Korsakoff disease in heavy drinkers, more studies are necessary to clarify if moderate alcohol intake could be a viable preventive action against DAT.

Stress and risk of DAT

The limbic system and especially the hippocampus have a high density of corticosteroid receptors and, thus, are susceptible to disturbances of the hypothalamic–pituitary–adrenal axis (Magri et al., 2006). The hippocampus is also affected early in the neurodegeneration process of DAT. Thus, higher levels of cortisol, the main hormone released in the hypothalamic–pituitary–adrenal axis in response to stress, may increase the susceptibility of the hippocampus to Alzheimer's changes and precipitate dementia onset. Epidemiologic studies showed that stress increases the risk of developing DAT, both in studies with prospective follow-up of dementia outcome and stress in midlife (Johansson et al., 2010), and retrospective work-related stress (job control and job demands) in a representative aged population 75 years and older in relation to short-term incident dementia (Wang et al., 2012b). It has also been shown that the quantity of psychosocial stressors at a given time (e.g., divorce, widowhood, work-related problems, and illness in relative) is associated with later development of DAT in a dose-dependent fashion (HR 1.20, 95% CI 1.07–1.35) (Johansson et al., 2013).

Head injury and risk of DAT

Traumatic brain injury (TBI) has been associated with the development of Alzheimer's type of pathology (Barrio et al., 2015) and, by some accounts, even with the development of DAT, especially in men, as exemplified in case studies (Lepreux et al., 2015; Tateno et al., 2015) and by earlier epidemiologic accounts (Mortimer et al., 1991; O'Meara et al., 1997). The latest EURODEM pooled analysis that used data from four large epidemiologic studies in individuals 65 years and older who were followed up for an average of 2 years (Odense Study in Denmark, PAQUID in France, Rotterdam Study in The Netherlands, and MRC-ALPHA study in the UK) could not verify this association (relative risk 1.66, 95% CI 0.94–2.95 in men) (Launer et al., 1999). However, data from the MRC-ALPHA study seemed to drive these results as the only dataset with a significantly increased risk of DAT in men with TBI in the EURODEM pooled analyses, thus supporting previous studies that found male gender as a modulating factor of the association between TBI and DAT. Although not assessing the development of DAT specifically, but rather young-onset dementia before age 65, a large population case-control study of 18-year-old conscripts from Sweden ($n = 811\ 622$) who were followed up on average for 33 years for the outcome of dementia showed increased risk of developing dementia before age 65 and the relationship seemed dose-dependent on the severity of head injuries (HR 1.7 for one or at least two mild TBI, and HR 2.6 for one severe TBI) (Nordstrom et al., 2014). Thus, most available epidemiologic data seem to seem to support the association between head injuries and dementia, especially DAT, among men.

Psychiatric illnesses and risk of DAT

There is a large overlap between psychiatric illnesses (i.e., autism spectrum disorders, schizophrenia, psychotic disorders, and affective disorders) and cognitive impairments. So far, evidence has gathered pinpointing depression and PTSD as risk factors for the development of DAT. However, while most reports support the role of PTSD in army veterans as increasing the risk of DAT and other dementias (Qureshi et al., 2010; Yaffe et al., 2010), the role of depression is not as clear-cut.

Major depression

Major depression has been associated with dementia in numerous cross-sectional and longitudinal epidemiologic studies (Palsson et al., 1999; Byers and Yaffe, 2011). Some studies support depression, especially history of major depression in midlife, as a risk factor for the development of DAT (Ownby et al., 2006). Other studies support depression in late life representing symptoms due to Alzheimer's pathology (Li et al., 2011; Diniz et al., 2013) rather than a separate disease entity that increases the risk of dementia. Lately, even depressive symptoms in MCI have been conceptualized as being part of the clinical presentation of the underlying neurodegenerative process (Gabryelewicz et al., 2007; Houde et al., 2008). More important for therapeutic implications, MCI patients with depression seem to respond poorly to antidepressants and have double the risk of developing dementia at follow-up compared to those without depression (Modrego and Ferrandez, 2004).

However, some recent neuroimaging studies have brought support for chronic depressive symptomatology in MCI representing a risk factor for faster decline to DAT (Lee et al., 2012; Xie et al., 2012; Sacuiu et al., 2016). There is still a need for epidemiologic studies in large, representative populations to clearly establish if major depression and depressive symptomatology not of the severity or frequency to meet criteria for major depression may reflect progressive DAT neuropathology or a risk factor for DAT. This is especially important since depression is a treatable disease and thus may represent a modifiable risk factor.

INTERVENTION EPIDEMIOLOGY

The discovery of a disease-modifying therapy for dementia has been difficult due to numerous phenotypic presentations of the syndrome, even for the same etiologic type of dementia. Only epidemiologic studies of DAT with early debut and of typical sporadic DAT that occurs later in life have provided consistent outcomes which have made it possible to devise protocols for therapeutic agents against DAT. Pharmacologic therapies that help prevent rapid cognitive decline in DAT have been available since the 1990s in the form of acetylcholine esterase inhibitors, followed by the NMDA receptor antagonist memantine a decade later, as discussed previously in this chapter. Unfortunately, promising agents against brain amyloid, like monoclonal antibodies targeting Aβ, i.e., bapineuzumab and solanezumab, have not been successful in phase III clinical trials (Vellas et al., 2013), as also emphasized in this chapter. While antibody therapies and other avenues addressing disease-modifying mechanisms are intensely explored, other researchers are beginning to turn their attention toward primary preventive therapies.

Preventive intervention trials

To reduce the impact of diseases at individual and population level, preventive measures should be initiated as early as possible in the development of the underlying pathology. Prevention could thus be directed toward preventing the disease developing (primary prevention), arresting the development of the disease in its initial stages (secondary prevention), or reducing the impact of the disease once it started (tertiary prevention). The role of lifestyle factors in the development of DAT is supported by findings at molecular and systemic (Gustafson et al., 2015) and at population level (Deckers et al., 2015). Due to a more profound understanding of the role of lifestyle as risk factor for DAT during the last decades, primary prevention trials aimed to reduce the burden of dementia at population level have started to gain momentum. The evidence gathered so far has made possible the initiation of large-scale primary intervention studies against DAT. In the EU, there are currently three ongoing large prevention studies, namely the Finnish Geriatric Intervention Study to Prevent Cognitive Impairment and Disability (FINGER), Prevention of Dementia by Intensive Vascular Care (PreDIVA), and Multidomain Alzheimer Prevention Study (MAPT), all part of the European Dementia Prevention Initiative (EDPI) (Richard et al., 2012). A large clinical randomized trial was also initiated in the USA to test the effect of lifestyle intervention on cognitive outcome, the Lifestyle Interventions and Independence for Elderly (LIFE) study (Sink et al., 2014).

FINGER was the first EDPI study to report findings from its 2-year follow-up. FINGER is a double-blind RCT in individuals aged 60–77 years (mean age 69.3 years, SD 4.7), randomly assigned in a 1:1 ratio to a 2-year multidomain intervention (diet, exercise, cognitive training, vascular risk monitoring) or a control group (general health advice). The primary outcome was change in cognition as measured by a composite Z-score based on a standardized battery of 14 neurocognitive tests that covered executive domain, processing speed, and memory (trial registered at ClinicalTrials.gov, identifier NCT01041989). The mean difference in improvement of total cognitive Z-score per year was significantly higher in the intervention group (data available in $n=554$ of 631 enrolled participants) than in controls (data available in $n=565$ of 629 participants) (0.022, 95% CI 0.002–0.042, $p=0.030$) (Ngandu et al., 2015). Interestingly enough, the intervention effect on secondary outcomes showed 83% improvement in executive functioning and 150% improvement in processing speed, but no significant change in the memory domain. Considering the negative results from previous single or limited domain interventions, these results strengthen the multidomain approach in prevention trials, both in intervention and at outcome level.

Another aspect to be considered is the difficulty level of the cognitive tests used as outcome measure, since *post hoc* analyses in FINGER showed an improvement after intervention in memory tests with a higher level of complexity (associative memory and logical memory tasks), but not in less complex memory tasks such as word list (learning and delayed recall). In support of these results come previous pathologic studies *in vivo* that have shown that only high-complexity tasks could detect cognitive impairments in individuals with presumably higher cognitive reserve who were carrying brain amyloid (Rentz et al., 2010). Therefore it is paramount to adjust the cognitive primary outcome in prevention trials to different levels of complexity of cognitive tests in order to avoid ceiling effects.

LIFE is a multicenter, randomized clinical trial that involves eight centers in the USA, with the primary outcome physical disability measured as the inability to walk 400 meters (trial registered at ClinicalTrials.gov, identifier NCT01072500) (Fielding et al., 2011; Sink et al., 2014). Cognitive function is one of the secondary outcomes and, thus, the decline in scores of a test battery that focuses on memory and executive functions will be assessed as exploratory outcomes at 2-year follow-up (Modified MMSE as a global test of cognition, tests of executive function such as the Eriksen Flanker Task for response inhibition, the Task-Switching test for attention and flexibility, the Digit Symbol Substitution Test as a measure of attention and perceptual speed, and memory tests such as the N-Back Test for working memory and the Hopkins Verbal Learning Test, a learning and memory test) (Fielding et al., 2011). The development of dementia is a tertiary outcome. The trial selected sedentary elderly people, ages 70–89 years, who could walk 400 meters in 15 minutes without breaks and support, but who scored less than 10 points in the Short Physical Performance Battery (SPPB, score range 0–12) that assesses mobility of lower extremities. A study group of 1600 individuals, of whom at least 45% have SPPB score ≤ 7 points (high-risk individuals), received either an intervention program based on supervised physical activity (center- and home-based activities) or age-specific successful aging health education information in supervised sessions (the comparison group) (Fielding et al., 2011). The cognitive outcomes have not yet been published. Similarly, results from other recently initiated large population preventive trials worldwide are to be expected in the following decade (Carrillo et al., 2013).

FUTURE DEVELOPMENTS

Refinement of diagnosis criteria

Dementia and DAT diagnostic criteria have been refined through the years with the publication of updated versions of the DSM by the American Psychiatric Association, as well as the ICD, issued by the World Health Organization. The most recent version of the DSM manual, DSM-5, proposes a radical change in the field of cognition in psychiatry by replacing the syndrome of dementia with the syndrome of neurocognitive disorder (APA, 2013). The newly defined syndrome may better help researchers to harmonize protocols for the study of MCI (comprising mild neurocognitive disorder) and dementia (comprising major neurocognitive disorder). Another important aspect that has been brought up with the syndrome of neurocognitive disorder is the inclusion of social cognition (theory of mind) as a new cognitive domain that has to be assessed in order to exhaustively cover cognitive changes that may occur in neurocognitive disorders (Sachdev et al., 2014). Moreover, the DSM-5 criteria for major neurocognitive disorder have shifted toward requiring impairment in at least one cognitive domain, not necessarily memory, if the other criteria are met, in order to increase early recognition of other etiologic types of dementia that may clinically be expressed by impairments in executive abilities, social cognition, or language (e.g., dementia due to cerebrovascular lesions or frontotemporal-lobe degeneration) (Sachdev et al., 2014). Thus, the DSM-5 criteria are also intended to better help clinicians worldwide to come nearer a standardized diagnosis of other etiologic types of dementia than DAT, e.g., VaD, frontotemporal-lobe, dementia, and Lewy body dementia.

Biomarkers

The medical and research community has been engaged in a race to find sensitive parameters to accurately monitor the results of intervention trials against DAT. Although Aβ has been established as one of the biomarkers of the disease in the CSF, a pathologic staging of DAT based on amyloid accumulation in the brain similar to NFT-defined Braak stages has not been possible. Instead, amyloid depletion in CSF was shown to be associated with the early stages of the disease, reaching a plateau later in moderate-to-severe stages. CSF-amyloid together with CSF-tau proteins, especially the phosphorylated tau protein, are currently considered the strongest candidates for reliable biomarkers of DAT (McKhann et al., 2011). However, it has been emphasized that in the incipient DAT stages "the recommendations of the preclinical DAT workgroup are intended purely for research purposes" (Jack et al., 2011). Research directed toward finding less invasive biomarkers, for example, in blood plasma, have been less successful so far, although the latest results from the Framingham Study were encouraging in this respect: in a large population study of 2189 nondemented individuals followed up on average for 7.6 years for the outcome of dementia and DAT, it was shown that 1 SD increase in plasma levels of the toxic variant of amyloid $A\beta_{1-42}$ was associated with a 21% lower risk of DAT, after correcting the regression survival models for age, gender and the appropriate confounders related to comorbidities (hypertension, diabetes, cardiovascular disease, total cholesterol levels, and waist-to-hip ratio), APOE ε4 status, education, and current smoking (Chouraki et al., 2015). However, the authors could not confirm these results in a meta-analysis including all available representative epidemiologic studies that provided valid plasma amyloid measurements (published and unpublished data) (Chouraki et al., 2015). Moreover, these studies could not investigate the correlation between plasma and CSF and brain amyloid, both of the latter being currently established

DAT biomarkers. Intensive work is being undertaken instead to find correlates of CSF biomarkers with structural and functional brain changes in initial stages of cognitive decline (MCI stage).

References

Acrescent - Assessment Report (2012). European Public Assessment Report. European Medicines Agency/ Committee for Medicinal Products for Human Use, London, UK.

Aevarsson O, Skoog I (1996). A population-based study on the incidence of dementia disorders between 85 and 88 years of age. J Am Geriatr Soc 44: 1455–1460.

Aggarwal NT, Wilson RS, Beck TL et al. (2005). Mild cognitive impairment in different functional domains and incident Alzheimer's disease. J Neurol Neurosurg Psychiatry 76: 1479–1484.

Aggarwal NT, Bienias JL, Bennett DA et al. (2006). The relation of cigarette smoking to incident Alzheimer's disease in a biracial urban community population. Neuroepidemiology 26: 140–146.

Alexopoulos P, Grimmer T, Perneczky R et al. (2006). Progression to dementia in clinical subtypes of mild cognitive impairment. Dement Geriatr Cogn Disord 22: 27–34.

Alzheimer's Association (2015). 2015 Alzheimer's disease facts and figures. Alzheimers Dement 11: 332–384.

Amieva H, Letenneur L, Dartigues JF et al. (2004). Annual rate and predictors of conversion to dementia in subjects presenting mild cognitive impairment criteria defined according to a population-based study. Dement Geriatr Cogn Disord 18: 87–93.

Andreasen N, Blennow K (2005). CSF biomarkers for mild cognitive impairment and early Alzheimer's disease. Clin Neurol Neurosurg 107: 165–173.

Andreasen N, Vanmechelen E, Vanderstichele H et al. (2003). Cerebrospinal fluid levels of total-tau, phospho-tau and A beta 42 predicts development of Alzheimer's disease in patients with mild cognitive impairment. Acta Neurol Scand Suppl 179: 47–51.

Anttila T, Helkala EL, Viitanen M et al. (2004). Alcohol drinking in middle age and subsequent risk of mild cognitive impairment and dementia in old age: a prospective population based study. BMJ 329: 539.

APA (2013). Diagnostic and Statistical Manual of Mental Disorders, American Psychiatric Association, Arlington, VA.

Apostolova LG, Thompson PM (2008). Mapping progressive brain structural changes in early Alzheimer's disease and mild cognitive impairment. Neuropsychologia 46: 1597–1612.

Artero S, Petersen R, Touchon J et al. (2006). Revised criteria for mild cognitive impairment: validation within a longitudinal population study. Dement Geriatr Cogn Disord 22: 465–470.

Bachman DL, Wolf PA, Linn RT et al. (1993). Incidence of dementia and probable Alzheimer's disease in a general population: the Framingham Study. Neurology 43: 515–519.

Ballard C, Gauthier S, Corbett A et al. (2011). Alzheimer's disease. Lancet 377: 1019–1031.

Banerjee S (2012). The macroeconomics of dementia – will the world economy get Alzheimer's disease? Arch Med Res 43: 705–709.

Barrett-Connor E, Edelstein SL, Corey-Bloom J et al. (1996). Weight loss precedes dementia in community-dwelling older adults. J Am Geriatr Soc 44: 1147–1152.

Barrio JR, Small GW, Wong KP et al. (2015). In vivo characterization of chronic traumatic encephalopathy using [F-18]FDDNP PET brain imaging. Proc Natl Acad Sci U S A 112: E2039–E2047.

Bartus RT, Dean 3rd RL, Beer B et al. (1982). The cholinergic hypothesis of geriatric memory dysfunction. Science 217: 408–414.

Bateman RJ, Aisen PS, De Strooper B et al. (2011). Autosomal-dominant Alzheimer's disease: a review and proposal for the prevention of Alzheimer's disease. Alzheimers Res Ther 3: 1.

Beach TG, Monsell SE, Phillips LE et al. (2012). Accuracy of the clinical diagnosis of Alzheimer disease at National Institute on Aging Alzheimer Disease Centers, 2005–2010. J Neuropathol Exp Neurol 71: 266–273.

Bennett DA, Leurgans S (2010). Is there a link between cancer and Alzheimer disease? Neurology 74: 100–101.

Berchtold NC, Cotman CW (1998). Evolution in the conceptualization of dementia and Alzheimer's disease: Greco-Roman period to the 1960s. Neurobiol Aging 19: 173–189.

Biessels GJ, Staekenborg S, Brunner E et al. (2006). Risk of dementia in diabetes mellitus: a systematic review. Lancet Neurol 5: 64–74.

Birks J (2006). Cholinesterase inhibitors for Alzheimer's disease. Cochrane Database Syst Rev. CD005593.

Boller F, Forbes MM (1998). History of dementia and dementia in history: an overview. J Neurol Sci 158: 125–133.

Borjesson-Hanson A, Edin E, Gislason T et al. (2004). The prevalence of dementia in 95 year olds. Neurology 63: 2436–2438.

Braak H, Braak E (1995). Staging of Alzheimer's disease-related neurofibrillary changes. Neurobiol Aging 16: 271–278. discussion 278–84.

Braak H, Del Tredici K (2015). The preclinical phase of the pathological process underlying sporadic Alzheimer's disease. Brain 138: 2814–2833.

Braak H, Thal DR, Ghebremedhin E et al. (2011). Stages of the pathologic process in Alzheimer disease: age categories from 1 to 100 years. J Neuropathol Exp Neurol 70: 960–969.

Braak H, Thal DR, Matschke J et al. (2013). Age-related appearance of dendritic inclusions in catecholaminergic brainstem neurons. Neurobiol Aging 34: 286–297.

Brookmeyer R, Gray S, Kawas C (1998). Projections of Alzheimer's disease in the United States and the public health impact of delaying disease onset. Am J Public Health 88: 1337–1342.

Buckley JS, Salpeter SR (2015). A risk-benefit assessment of dementia medications: systematic review of the evidence. Drugs Aging 32: 453–467.

Bunch TJ, Weiss JP, Crandall BG et al. (2010). Atrial fibrillation is independently associated with senile, vascular, and Alzheimer's dementia. Heart Rhythm 7: 433–437.

Burns LC, Minster RL, Demirci FY et al. (2011). Replication study of genome-wide associated SNPs with late-onset Alzheimer's disease. Am J Med Genet B 156B: 507–512.

Byers AL, Yaffe K (2011). Depression and risk of developing dementia. Nat Rev Neurol 7: 323–331.

Cacabelos R (2007). Donepezil in Alzheimer's disease: from conventional trials to pharmacogenetics. Neuropsychiatr Dis Treat 3: 303–333.

Carrillo MC, Brashear HR, Logovinsky V et al. (2013). Can we prevent Alzheimer's disease? Secondary "prevention" trials in Alzheimer's disease. Alzheimers Dement 9: 123–131. e1.

Chartier-Harlin MC, Crawford F, Houlden H et al. (1991). Early-onset Alzheimer's disease caused by mutations at codon 717 of the beta-amyloid precursor protein gene. Nature 353: 844–846.

Chaudhry SI, Wang Y, Gill TM et al. (2010). Geriatric conditions and subsequent mortality in older patients with heart failure. J Am Coll Cardiol 55: 309–316.

Chiang GC, Insel PS, Tosun D et al. (2010). Hippocampal atrophy rates and CSF biomarkers in elderly APOE2 normal subjects. Neurology 75: 1976–1981.

Chouraki V, Beiser A, Younkin L et al. (2015). Plasma amyloid-beta and risk of Alzheimer's disease in the Framingham Heart Study. Alzheimers Dement 11: 249–257. e1.

Chui HC, Zheng L, Reed BR et al. (2012). Vascular risk factors and Alzheimer's disease: are these risk factors for plaques and tangles or for concomitant vascular pathology that increases the likelihood of dementia? An evidence-based review. Alzheimers Res Ther 4: 1.

Cobb JL, Wolf PA, Au R et al. (1995). The effect of education on the incidence of dementia and Alzheimer's disease in the Framingham Study. Neurology 45: 1707–1712.

Cooper C, Li R, Lyketsos C et al. (2013). Treatment for mild cognitive impairment: systematic review. Br J Psychiatry 203: 255–264.

Corey-Bloom J (2002). The ABC of Alzheimer's disease: cognitive changes and their management in Alzheimer's disease and related dementias. Int Psychogeriatr 14 (Suppl 1): 51–75.

CSHA Working-Group (2000). The incidence of dementia in Canada. The Canadian Study of Health and Aging Working Group. Neurology 55: 66–73.

Cummings JL, Isaacson RS, Schmitt FA et al. (2015). A practical algorithm for managing Alzheimer's disease: what, when, and why? Ann Clin Transl Neurol 2: 307–323.

Decarli C (2003). Mild cognitive impairment: prevalence, prognosis, aetiology, and treatment. Lancet Neurol 2: 15–21.

Deckers K, Van Boxtel MP, Schiepers OJ et al. (2015). Target risk factors for dementia prevention: a systematic review and Delphi consensus study on the evidence from observational studies. Int J Geriatr Psychiatry 30: 234–246.

Ding D, Zhao Q, Guo Q et al. (2014). The Shanghai Aging Study: study design, baseline characteristics, and prevalence of dementia. Neuroepidemiology 43: 114–122.

Diniz BS, Butters MA, Albert SM et al. (2013). Late-life depression and risk of vascular dementia and Alzheimer's disease: systematic review and meta-analysis of community-based cohort studies. Br J Psychiatry 202: 329–335.

Dodge HH, Buracchio TJ, Fisher GG et al. (2012). Trends in the prevalence of dementia in Japan. Int J Alzheimers Dis 2012: 956354.

Driver JA, Beiser A, Au R et al. (2012). Inverse association between cancer and Alzheimer's disease: results from the Framingham Heart Study. BMJ 344: e1442.

Duthey B (2013). Update on 2004 Background Paper, BP 6.11 Alzheimer Disease [Online]. World Health Organization. Available: http://www.who.int/medicines/areas/priority_medicines/BP6_11Alzheimer.pdf [Accessed 2015 April, 21 2015].

Ehret MJ, Chamberlin KW (2015). Current practices in the treatment of Alzheimer disease: where is the evidence after the phase III trials? Clin Ther 37: 1604–1616.

Elias MF, Wolf PA, D'Agostino RB et al. (1993). Untreated blood pressure level is inversely related to cognitive functioning: the Framingham Study. Am J Epidemiol 138: 353–364.

Ercoli L, Siddarth P, Huang SC et al. (2006). Perceived loss of memory ability and cerebral metabolic decline in persons with the apolipoprotein E-IV genetic risk for Alzheimer disease. Arch Gen Psychiatry 63: 442–448.

Eriksdotter-Jonhagen M, Linderoth B, Lind G et al. (2012). Encapsulated cell biodelivery of nerve growth factor to the basal forebrain in patients with Alzheimer's disease. Dement Geriatr Cogn Disord 33: 18–28.

Erkinjuntti T, Ostbye T, Steenhuis R et al. (1997). The effect of different diagnostic criteria on the prevalence of dementia. N Engl J Med 337: 1667–1674.

Evans D, Ganguli M, Harris T et al. (1999). Women and Alzheimer disease. Alzheimer Dis Assoc Disord 13: 187–189.

Farlow MR, Andreasen N, Riviere ME et al. (2015). Long-term treatment with active Abeta immunotherapy with CAD106 in mild Alzheimer's disease. Alzheimers Res Ther 7: 23.

Fielding RA, Rejeski WJ, Blair S et al. (2011). The Lifestyle Interventions and Independence for Elders Study: design and methods. J Gerontol A Biol Sci Med Sci 66: 1226–1237.

Fitzpatrick AL, Kuller LH, Ives DG et al. (2004). Incidence and prevalence of dementia in the Cardiovascular Health Study. J Am Geriatr Soc 52: 195–204.

Fitzpatrick AL, Kuller LH, Lopez OL et al. (2005). Survival following dementia onset: Alzheimer's disease and vascular dementia. J Neurol Sci 229–230: 43–49.

Fouquet M, Besson FL, Gonneaud J et al. (2014). Imaging brain effects of APOE4 in cognitively normal individuals across the lifespan. Neuropsychol Rev 24: 290–299.

Fox NC, Black RS, Gilman S et al. (2005). Effects of Abeta immunization (AN1792) on MRI measures of cerebral volume in Alzheimer disease. Neurology 64: 1563–1572.

Fratiglioni L, Viitanen M, Von Strauss E et al. (1997). Very old women at highest risk of dementia and Alzheimer's disease: incidence data from the Kungsholmen Project, Stockholm. Neurology 48: 132–138.

Fratiglioni L, Launer LJ, Andersen K et al. (2000). Incidence of dementia and major subtypes in Europe: A collaborative study of population-based cohorts. Neurologic Diseases in the Elderly Research Group. Neurology 54: S10–S15.

Frisardi V, Solfrizzi V, Seripa D et al. (2010). Metabolic-cognitive syndrome: a cross-talk between metabolic syndrome and Alzheimer's disease. Ageing Res Rev 9: 399–417.

Gabryelewicz T, Styczynska M, Luczywek E et al. (2007). The rate of conversion of mild cognitive impairment to dementia: predictive role of depression. Int J Geriatr Psychiatry 22: 563–567.

Ganguli M (2015). Cancer and dementia: it's complicated. Alzheimer Dis Assoc Disord 29: 177–182.

Ganguli M, Dodge HH, Chen P et al. (2000). Ten-year incidence of dementia in a rural elderly US community population: the MoVIES Project. Neurology 54: 1109–1116.

Ganguli M, Dodge HH, Shen C et al. (2005a). Alzheimer disease and mortality: a 15-year epidemiological study. Arch Neurol 62: 779–784.

Ganguli M, Vander Bilt J, Saxton JA et al. (2005b). Alcohol consumption and cognitive function in late life: a longitudinal community study. Neurology 65: 1210–1217.

Garibotto V, Borroni B, Kalbe E et al. (2008). Education and occupation as proxies for reserve in aMCI converters and AD: FDG-PET evidence. Neurology 71: 1342–1349.

Gatz M, Svedberg P, Pedersen NL et al. (2001). Education and the risk of Alzheimer's disease: findings from the study of dementia in Swedish twins. J Gerontol B Psychol Sci Soc Sci 56: P292–P300.

Gazzaley A, Cooney JW, Rissman J et al. (2005). Top-down suppression deficit underlies working memory impairment in normal aging. Nat Neurosci 8: 1298–1300.

Genin E, Hannequin D, Wallon D et al. (2011). APOE and Alzheimer disease: a major gene with semi-dominant inheritance. Mol Psychiatry 16: 903–907.

Gilman S, Koller M, Black RS et al. (2005). Clinical effects of Abeta immunization (AN1792) in patients with AD in an interrupted trial. Neurology 64: 1553–1562.

Gold G, Giannakopoulos P, Montes-Paixao Junior C et al. (1997). Sensitivity and specificity of newly proposed clinical criteria for possible vascular dementia. Neurology 49: 690–694.

Graves AB, Van Duijn CM, Chandra V et al. (1991). Alcohol and tobacco consumption as risk factors for Alzheimer's disease: a collaborative re-analysis of case-control studies. EURODEM Risk Factors Research Group. Int J Epidemiol 20 (Suppl 2): S48–S57.

Grinberg LT, Rub U, Ferretti RE et al. (2009). The dorsal raphe nucleus shows phospho-tau neurofibrillary changes before the transentorhinal region in Alzheimer's disease. A precocious onset? Neuropathol Appl Neurobiol 35: 406–416.

Grossberg GT (2002). The ABC of Alzheimer's disease: behavioral symptoms and their treatment. Int Psychogeriatr 14 (Suppl 1): 27–49.

Guo Z, Viitanen M, Winblad B et al. (1999). Low blood pressure and incidence of dementia in a very old sample: dependent on initial cognition. J Am Geriatr Soc 47: 723–726.

Guo X, Waern M, Sjogren K et al. (2007). Midlife respiratory function and Incidence of Alzheimer's disease: a 29-year longitudinal study in women. Neurobiol Aging 28: 343–350.

Gustafson B (2010). Adipose tissue, inflammation and atherosclerosis. J Atheroscler Thromb 17: 332–341.

Gustafson D, Rothenberg E, Blennow K et al. (2003). An 18-year follow-up of overweight and risk of Alzheimer disease. Arch Intern Med 163: 1524–1528.

Gustafson D, Morris MC, Scarmeas N et al. (2015). New perspectives on Alzheimer's disease and nutrition. J Alzheimers Dis 46: 1111–1127.

Haass C, De Strooper B (1999). The presenilins in Alzheimer's disease – proteolysis holds the key. Science 286: 916–919.

Hardy JA, Higgins GA (1992). Alzheimer's disease: the amyloid cascade hypothesis. Science 256: 184–185.

Hassing LB, Johansson B, Nilsson SE et al. (2002). Diabetes mellitus is a risk factor for vascular dementia, but not for Alzheimer's disease: a population-based study of the oldest old. Int Psychogeriatr 14: 239–248.

Hassing LB, Grant MD, Hofer SM et al. (2004). Type 2 diabetes mellitus contributes to cognitive decline in old age: a longitudinal population-based study. J Int Neuropsychol Soc 10: 599–607.

Henderson VW (2010). Action of estrogens in the aging brain: dementia and cognitive aging. Biochim Biophys Acta 1800: 1077–1083.

Heneka MT, Carson MJ, El Khoury J et al. (2015). Neuroinflammation in Alzheimer's disease. Lancet Neurol 14: 388–405.

Herrup K (2015). The case for rejecting the amyloid cascade hypothesis. Nat Neurosci 18: 794–799.

Heyman A, Peterson B, Fillenbaum G et al. (1997). Predictors of time to institutionalization of patients with Alzheimer's disease: the CERAD experience, part XVII. Neurology 48: 1304–1309.

Holmes C, Boche D, Wilkinson D et al. (2008). Long-term effects of Abeta42 immunisation in Alzheimer's disease: follow-up of a randomised, placebo-controlled phase I trial. Lancet 372: 216–223.

Honda K, Casadesus G, Petersen RB et al. (2004). Oxidative stress and redox-active iron in Alzheimer's disease. Ann N Y Acad Sci 1012: 179–182.

Hooshmand B, Solomon A, Kareholt I et al. (2010). Homocysteine and holotranscobalamin and the risk of Alzheimer disease: a longitudinal study. Neurology 75: 1408–1414.

Houde M, Bergman H, Whitehead V et al. (2008). A predictive depression pattern in mild cognitive impairment. Int J Geriatr Psychiatry 23: 1028–1033.

Howard R, Mcshane R, Lindesay J et al. (2012). Donepezil and memantine for moderate-to-severe Alzheimer's disease. N Engl J Med 366: 893–903.

Huang CC, Chung CM, Leu HB et al. (2014). Diabetes mellitus and the risk of Alzheimer's disease: a nationwide population-based study. PLoS One 9: e87095.

Hurd MD, Martorell P, Langa K (2015). Future monetary costs of dementia in the United States under alternative dementia prevalence scenarios. J Popul Ageing 8: 101–112.

Irie F, Fitzpatrick AL, Lopez OL et al. (2008a). Enhanced risk for Alzheimer disease in persons with type 2 diabetes and APOE epsilon4: the Cardiovascular Health Study Cognition Study. Arch Neurol 65: 89–93.

Irie F, Masaki KH, Petrovitch H et al. (2008b). Apolipoprotein E epsilon4 allele genotype and the effect of depressive symptoms on the risk of dementia in men: the Honolulu-Asia Aging Study. Arch Gen Psychiatry 65: 906–912.

Jack Jr CR, Albert MS, Knopman DS et al. (2011). Introduction to the recommendations from the National Institute on Aging-Alzheimer's Association workgroups on diagnostic guidelines for Alzheimer's disease. Alzheimers Dement 7: 257–262.

Jack Jr CR, Knopman DS, Jagust WJ et al. (2013). Tracking pathophysiological processes in Alzheimer's disease: an updated hypothetical model of dynamic biomarkers. Lancet Neurol 12: 207–216.

Jagger C, Andersen K, Breteler MM et al. (2000). Prognosis with dementia in Europe: A collaborative study of population-based cohorts. Neurologic Diseases in the Elderly Research Group. Neurology 54: S16–S20.

Jellinger KA, Attems J (2010). Prevalence of dementia disorders in the oldest-old: an autopsy study. Acta Neuropathol 119: 421–433.

Johansson L, Guo X, Waern M et al. (2010). Midlife psychological stress and risk of dementia: a 35-year longitudinal population study. Brain 133: 2217–2224.

Johansson L, Guo X, Hallstrom T et al. (2013). Common psychosocial stressors in middle-aged women related to longstanding distress and increased risk of Alzheimer's disease: a 38-year longitudinal population study. BMJ open 3: e003142.

Jorm AF, Jolley D (1998). The incidence of dementia: a meta-analysis. Neurology 51: 728–733.

Jorm AF, Korten AE, Henderson AS (1987). The prevalence of dementia: a quantitative integration of the literature. Acta Psychiatr Scand 76: 465–479.

Jorm AF, Dear KB, Burgess NM (2005). Projections of future numbers of dementia cases in Australia with and without prevention. Aust N Z J Psychiatry 39: 959–963.

Kalaria RN, Maestre GE, Arizaga R et al. (2008). Alzheimer's disease and vascular dementia in developing countries: prevalence, management, and risk factors. Lancet Neurol 7: 812–826.

Karp A, Kareholt I, Qiu C et al. (2004). Relation of educa tion and occupation-based socioeconomic status to incident Alzheimer's disease. Am J Epidemiol 159: 175–183.

Karran E, Mercken M, De Strooper B (2011). The amyloid cascade hypothesis for Alzheimer's disease: an appraisal for the development of therapeutics. Nat Rev Drug Discov 10: 698–712.

Kawas C, Gray S, Brookmeyer R et al. (2000). Age-specific incidence rates of Alzheimer's disease: the Baltimore Longitudinal Study of Aging. Neurology 54: 2072–2077.

Kemppainen NM, Aalto S, Karrasch M et al. (2008). Cognitive reserve hypothesis: Pittsburgh Compound B and fluorodeoxyglucose positron emission tomography in relation to education in mild Alzheimer's disease. Ann Neurol 63: 112–118.

Kim B, Feldman EL (2015). Insulin resistance as a key link for the increased risk of cognitive impairment in the metabolic syndrome. Exp Mol Med 47: e149.

Kivipelto M, Ngandu T, Fratiglioni L et al. (2005). Obesity and vascular risk factors at midlife and the risk of dementia and Alzheimer disease. Arch Neurol 62: 1556–1560.

Kivipelto M, Ngandu T, Laatikainen T et al. (2006). Risk score for the prediction of dementia risk in 20 years among middle aged people: a longitudinal, population-based study. Lancet Neurol 5: 735–741.

Kokmen E, Beard CM, O'Brien PC et al. (1993). Is the incidence of dementing illness changing? A 25-year time trend study in Rochester, Minnesota (1960–1984). Neurology 43: 1887–1892.

Kua EH, Ho E, Tan HH et al. (2014). The natural history of dementia. Psychogeriatrics 14: 196–201.

Kukull WA, Higdon R, Bowen JD et al. (2002). Dementia and Alzheimer disease incidence: a prospective cohort study. Arch Neurol 59: 1737–1746.

Lambert JC, Amouyel P (2011). Genetics of Alzheimer's disease: new evidences for an old hypothesis? Curr Opin Genet Dev 21: 295–301.

Lambert MA, Bickel H, Prince M et al. (2014). Estimating the burden of early onset dementia; systematic review of disease prevalence. Eur J Neurol 21: 563–569.

Lambracht-Washington D, Rosenberg RN (2013). Anti-amyloid beta to tau-based immunization: developments in immunotherapy for Alzheimer disease. Immuno Targets Ther 2013: 105–114.

Lane RM, Kivipelto M, Greig NH (2004). Acetylcholinesterase and its inhibition in Alzheimer disease. Clin Neuropharmacol 27: 141–149.

Langa KM, Larson EB, Karlawish JH et al. (2008). Trends in the prevalence and mortality of cognitive impairment in the United States: is there evidence of a compression of cognitive morbidity? Alzheimers Dement 4: 134–144.

Launer LJ, Andersen K, Dewey ME et al. (1999). Rates and risk factors for dementia and Alzheimer's disease: results from EURODEM pooled analyses. EURODEM Incidence Research Group and Work Groups. European Studies of Dementia. Neurology 52: 78–84.

Lee GJ, Lu PH, Hua X et al. (2012). Depressive symptoms in mild cognitive impairment predict greater atrophy in Alzheimer's disease-related regions. Biol Psychiatry 71: 814–821.

Lepreux S, Auriacombe S, Vital C et al. (2015). Dementia pugilistica: a severe tribute to a career. Clin Neuropathol 34: 193–198.

Letenneur L, Larrieu S, Barberger-Gateau P (2004). Alcohol and tobacco consumption as risk factors of dementia: a review of epidemiological studies. Biomed Pharmacother 58: 95–99.

Li G, Wang LY, Shofer JB et al. (2011). Temporal relationship between depression and dementia: findings from a large community-based 15-year follow-up study. Arch Gen Psychiatry 68: 970–977.

Li MM, Yu JT, Wang HF et al. (2014). Efficacy of vitamins B supplementation on mild cognitive impairment and Alzheimer's disease: a systematic review and meta-analysis. Curr Alzheimer Res 11: 844–852.

Liao WC, Lin CL, Chang SN et al. (2015). The association between chronic obstructive pulmonary disease and dementia: a population-based retrospective cohort study. Eur J Neurol 22: 334–340.

Llewellyn DJ, Matthews FE (2009). Increasing levels of semantic verbal fluency in elderly English adults. Neuropsychol Dev Cogn B Aging Neuropsychol Cogn 16: 433–445.

Llibre Rodriguez J, Valhuerdi A, Sanchez II et al. (2008). The prevalence, correlates and impact of dementia in Cuba. A 10/66 group population-based survey. Neuroepidemiology 31: 243–251.

Lobo A, Saz P, Marcos G et al. (2007). Prevalence of dementia in a southern European population in two different time periods: the ZARADEMP Project. Acta Psychiatr Scand 116: 299–307.

Lucca U, Tettamanti M, Logroscino G et al. (2015). Prevalence of dementia in the oldest old: the Monzino 80-plus population based study. Alzheimers Dement 11: 258–270. e3.

Magri F, Cravello L, Barili L et al. (2006). Stress and dementia: the role of the hypothalamic–pituitary-adrenal axis. Aging Clin Exp Res 18: 167–170.

Mahley RW, Huang Y (2012). Apolipoprotein e sets the stage: response to injury triggers neuropathology. Neuron 76: 871–885.

Malouf R, Areosa Sastre A (2003). Vitamin B_{12} for cognition. Cochrane Database Syst Rev: CD004326.

Mangialasche F, Solomon A, Winblad B et al. (2010). Alzheimer's disease: clinical trials and drug development. Lancet Neurol 9: 702–716.

Matthews FE, Arthur A, Barnes LE et al. (2013). A two-decade comparison of prevalence of dementia in individuals aged 65 years and older from three geographical areas of England: results of the Cognitive Function and Ageing Study I and II. Lancet 382: 1405–1412.

McGuinness B, Todd S, Passmore P et al. (2009). Blood pressure lowering in patients without prior cerebrovascular disease for prevention of cognitive impairment and dementia. Cochrane Database Syst Rev. CD004034.

McKhann G, Drachman D, Folstein M et al. (1984). Clinical diagnosis of Alzheimer's disease: report of the NINCDS-ADRDA Work Group under the auspices of Department of Health and Human Services Task Force on Alzheimer's Disease. Neurology 34: 939–944.

McKhann GM, Knopman DS, Chertkow H et al. (2011). The diagnosis of dementia due to Alzheimer's disease: recommendations from the National Institute on Aging-Alzheimer's Association workgroups on diagnostic guidelines for Alzheimer's disease. Alzheimers Dement 7: 263–269.

McShane R, Areosa Sastre A, Minakaran N (2006). Memantine for dementia. Cochrane Database Syst Rev CD003154.

Mehlig K, Skoog I, Guo X et al. (2008). Alcoholic beverages and incidence of dementia: 34-year follow-up of the prospective population study of women in Goteborg. Am J Epidemiol 167: 684–691.

Mielke MM, Zandi PP, Sjogren M et al. (2005). High total cholesterol levels in late life associated with a reduced risk of dementia. Neurology 64: 1689–1695.

Mielke MM, Zandi PP, Shao H et al. (2010). The 32-year relationship between cholesterol and dementia from midlife to late life. Neurology 75: 1888–1895.

Mirra SS, Heyman A, McKeel D et al. (1991). The Consortium to Establish a Registry for Alzheimer's Disease (CERAD). Part II. Standardization of the neuropathologic assessment of Alzheimer's disease. Neurology 41: 479–486.

Misiak B, Cialkowska-Kuzminska M, Frydecka D et al. (2013). European studies on the prevalence of dementia in the elderly: time for a step towards a methodological consensus. Int J Geriatr Psychiatry 28: 1211–1221.

Mitchell TW, Mufson EJ, Schneider JA et al. (2002). Parahippocampal tau pathology in healthy aging, mild cognitive impairment, and early Alzheimer's disease. Ann Neurol 51: 182–189.

Modrego PJ, Ferrandez J (2004). Depression in patients with mild cognitive impairment increases the risk of developing dementia of Alzheimer type: a prospective cohort study. Arch Neurol 61: 1290–1293.

Morris JC (1993). The Clinical Dementia Rating (CDR): current version and scoring rules. Neurology 43: 2412–2414.

Morris MS (2012). The role of B vitamins in preventing and treating cognitive impairment and decline. Adv Nutr 3: 801–812.

Morris MC, Evans DA, Schneider JA et al. (2006). Dietary folate and vitamins B-12 and B-6 not associated with incident Alzheimer's disease. J Alzheimers Dis 9: 435–443.

Mortimer JA, Van Duijn CM, Chandra V et al. (1991). Head trauma as a risk factor for Alzheimer's disease: a collaborative re-analysis of case-control studies. EURODEM Risk Factors Research Group. Int J Epidemiol 20 (Suppl 2): S28–S35.

Mosconi L, Tsui WH, Herholz K et al. (2008). Multicenter standardized 18 F-FDG PET diagnosis of mild cognitive impairment, Alzheimer's disease, and other dementias. J Nucl Med 49: 390–398.

Nelson PT, Alafuzoff I, Bigio EH et al. (2012). Correlation of Alzheimer disease neuropathologic changes with cognitive status: a review of the literature. J Neuropathol Exp Neurol 71: 362–381.

Neuropathology Group of the Medical Research Council Cognitive Function and Ageing Study (MRC-CFAS) (2001). Pathological correlates of late-onset dementia in a multicentre, community-based population in England and Wales. Lancet 357: 169–175.

Newell KL, Hyman BT, Growdon JH et al. (1999). Application of the National Institute on Aging (NIA)-Reagan Institute criteria for the neuropathological diagnosis of Alzheimer disease. J Neuropathol Exp Neurol 58: 1147–1155.

Ngandu T, Lehtisalo J, Solomon A et al. (2015). A 2 year multidomain intervention of diet, exercise, cognitive training, and vascular risk monitoring versus control to prevent cognitive decline in at-risk elderly people (FINGER): a randomised controlled trial. Lancet 385: 2255–2263.

Nierenberg J, Pomara N, Hoptman MJ et al. (2005). Abnormal white matter integrity in healthy apolipoprotein E epsilon4 carriers. Neuroreport 16: 1369–1372.

Nobili F, Frisoni GB, Portet F et al. (2008). Brain SPECT in subtypes of mild cognitive impairment. Findings from the DESCRIPA multicenter study. J Neurol 255: 1344–1353.

Nordstrom P, Michaelsson K, Gustafson Y et al. (2014). Traumatic brain injury and young onset dementia: a nationwide cohort study. Ann Neurol 75: 374–381.

Norton S, Matthews FE, Brayne C (2013). A commentary on studies presenting projections of the future prevalence of dementia. BMC Public Health 13: 1.

Norton S, Matthews FE, Barnes DE et al. (2014). Potential for primary prevention of Alzheimer's disease: an analysis of population-based data. Lancet Neurol 13: 788–794.

Olazaran J, Reisberg B, Clare L et al. (2010). Nonpharmacological therapies in Alzheimer's disease: a systematic review of efficacy. Dement Geriatr Cogn Disord 30: 161–178.

O'Meara ES, Kukull WA, Sheppard L et al. (1997). Head injury and risk of Alzheimer's disease by apolipoprotein E genotype. Am J Epidemiol 146: 373–384.

Orgogozo JM, Dartigues JF, Lafont S et al. (1997). Wine consumption and dementia in the elderly: a prospective community study in the Bordeaux area. Rev Neurol 153: 185–192.

Ott A, Stolk RP, Van Harskamp F et al. (1999). Diabetes mellitus and the risk of dementia: the Rotterdam Study. Neurology 53: 1937–1942.

Ott A, Andersen K, Dewey ME et al. (2004). Effect of smoking on global cognitive function in nondemented elderly. Neurology 62: 920–924.

Ownby RL, Crocco E, Acevedo A et al. (2006). Depression and risk for Alzheimer disease: systematic review, meta-analysis, and metaregression analysis. Arch Gen Psychiatry 63: 530–538.

Palmer K, Backman L, Winblad B et al. (2008). Mild cognitive impairment in the general population: occurrence and progression to Alzheimer disease. Am J Geriatr Psychiatry 16: 603–611.

Palsson S, Aevarsson O, Skoog I (1999). Depression, cerebral atrophy, cognitive performance and incidence of dementia. Population study of 85-year-olds. Br J Psychiatry 174: 249–253.

Panza F, Frisardi V, Capurso C et al. (2010). Metabolic syndrome and cognitive impairment: current epidemiology and possible underlying mechanisms. J Alzheimers Dis 21: 691–724.

Perls T (2004). Dementia-free centenarians. Exp Gerontol 39: 1587–1593.

Petersen RC (2011). Clinical practice. Mild cognitive impairment. N Engl J Med 364: 2227–2234.

Petersen RC, Smith GE, Waring SC et al. (1999). Mild cognitive impairment: clinical characterization and outcome. Arch Neurol 56: 303–308.

Pimouguet C, Le-Goff M, Rizzuto D et al. (2015). Effect of early referral to specialist in dementia on institutionalization and functional decline: findings from a population-based study. J Alzheimers Dis 49: 819–828.

Plassman BL, Langa KM, Fisher GG et al. (2007). Prevalence of dementia in the United States: the aging, demographics, and memory study. Neuroepidemiology 29: 125–132.

Polvikoski T, Sulkava R, Myllykangas L et al. (2001). Prevalence of Alzheimer's disease in very elderly people: a prospective neuropathological study. Neurology 56: 1690–1696.

Polvikoski T, Sulkava R, Rastas S et al. (2006). Incidence of dementia in very elderly individuals: a clinical, neuropathological and molecular genetic study. Neuroepidemiology 26: 76–82.

Poon LW, Woodard JL, Stephen Miller L et al. (2012). Understanding dementia prevalence among centenarians. J Gerontol A Biol Sci Med Sci 67: 358–365.

Potkin SG (2002). The ABC of Alzheimer's disease: ADL and improving day-to-day functioning of patients. Int Psychogeriatr 14 (Suppl 1): 7–26.

Prince M, Acosta D, Chiu H et al. (2003). Dementia diagnosis in developing countries: a cross-cultural validation study. Lancet 361: 909–917.

Prince M, Ferri CP, Acosta D et al. (2007). The protocols for the 10/66 dementia research group population-based research programme. BMC Public Health 7: 165.

Prince M, Acosta D, Ferri CP et al. (2012). Dementia incidence and mortality in middle-income countries, and associations with indicators of cognitive reserve: a 10/66 Dementia Research Group population-based cohort study. Lancet 380: 50–58.

Prince M, Bryce R, Albanese E et al. (2013). The global prevalence of dementia: a systematic review and metaanalysis. Alzheimers Dement 9: 63–75. e2.

Prins ND, Den Heijer T, Hofman A et al. (2002). Homocysteine and cognitive function in the elderly: the Rotterdam Scan Study. Neurology 59: 1375–1380.

Putter H, Fiocco M, Geskus RB (2007). Tutorial in biostatistics: competing risks and multi-state models. Stat Med 26: 2389–2430.

Puzzo D, Gulisano W, Arancio O et al. (2015). The keystone of Alzheimer pathogenesis might be sought in Abeta physiology. Neuroscience 307: 26–36.

Qiu C, Von Strauss E, Fastbom J et al. (2003). Low blood pressure and risk of dementia in the Kungsholmen project: a 6-year follow-up study. Arch Neurol 60: 223–228.

Qiu C, Von Strauss E, Winblad B et al. (2004). Decline in blood pressure over time and risk of dementia: a longitudinal study from the Kungsholmen project. Stroke 35: 1810–1815.

Qiu C, Winblad B, Fratiglioni L (2005). The age-dependent relation of blood pressure to cognitive function and dementia. Lancet Neurol 4: 487–499.

Qureshi SU, Kimbrell T, Pyne JM et al. (2010). Greater prevalence and incidence of dementia in older veterans with posttraumatic stress disorder. J Am Geriatr Soc 58: 1627–1633.

Reay JL, Smith MA, Riby LM (2013). B vitamins and cognitive performance in older adults: review. ISRN Nutr 2013: 650983.

Reiman EM, Chen K, Liu X et al. (2009). Fibrillar amyloid-beta burden in cognitively normal people at 3 levels of genetic risk for Alzheimer's disease. Proc Natl Acad Sci U S A 106: 6820–6825.

Rentz DM, Locascio JJ, Becker JA et al. (2010). Cognition, reserve, and amyloid deposition in normal aging. Ann Neurol 67: 353–364.

Richard E, Andrieu S, Solomon A et al. (2012). Methodological challenges in designing dementia prevention trials – the European Dementia Prevention Initiative (EDPI). J Neurol Sci 322: 64–70.

Ritchie K, Kildea D (1995). Is senile dementia "age-related" or "ageing-related"? Evidence from meta-analysis of dementia prevalence in the oldest old. Lancet 346: 931–934.

Ritchie K, Artero S, Touchon J (2001). Classification criteria for mild cognitive impairment: a population-based validation study. Neurology 56: 37–42.

Rocca WA, Petersen RC, Knopman DS et al. (2011). Trends in the incidence and prevalence of Alzheimer's disease, dementia, and cognitive impairment in the United States. Alzheimers Dement 7: 80–93.

Rockwood K, Lindsay J, McDowell I (1996). High blood pressure and dementia. Lancet 348: 65. author reply 65–6.

Roe CM, Fitzpatrick AL, Xiong C et al. (2010). Cancer linked to Alzheimer disease but not vascular dementia. Neurology 74: 106–112.

Rolinski M, Fox C, Maidment I et al. (2012). Cholinesterase inhibitors for dementia with Lewy bodies, Parkinson's disease dementia and cognitive impairment in Parkinson's disease. Cochrane Database Syst Rev 3: CD006504.

Ruitenberg A, Van Swieten JC, Witteman JC et al. (2002). Alcohol consumption and risk of dementia: the Rotterdam Study. Lancet 359: 281–286.

Rusanen M, Rovio S, Ngandu T et al. (2010). Midlife smoking, apolipoprotein E and risk of dementia and Alzheimer's disease: a population-based cardiovascular risk factors, aging and dementia study. Dement Geriatr Cogn Disord 30: 277–284.

Rusanen M, Kivipelto M, Quesenberry Jr CP et al. (2011). Heavy smoking in midlife and long-term risk of Alzheimer disease and vascular dementia. Arch Intern Med 171: 333–339.

Sachdev P (2000). Is it time to retire the term "dementia"? J Neuropsychiatry Clin Neurosci 12: 276–279.

Sachdev PS, Blacker D, Blazer DG et al. (2014). Classifying neurocognitive disorders: the DSM-5 approach. Nat Rev Neurol 10: 634–642.

Sacuiu S, Sjogren M, Johansson B et al. (2005). Prodromal cognitive signs of dementia in 85-year-olds using four sources of information. Neurology 65: 1894–1900.

Sacuiu S, Gustafson D, Johansson B et al. (2009). The pattern of cognitive symptoms predicts time to dementia onset. Alzheimers Dement 5: 199–206.

Sacuiu S, Gustafson D, Sjogren M et al. (2010). Secular changes in cognitive predictors of dementia and mortality in 70-year-olds. Neurology 75: 779–785.

Sacuiu S, Insel PS, Mueller S et al. (2016). Chronic depressive symptomatology in mild cognitive impairment is associated with frontal atrophy rate which hastens conversion to Alzheimer dementia. Am J Geriatr Psychiatry 24 (2): 126–135.

Sadowski M, Pankiewicz J, Scholtzova H et al. (2004). Links between the pathology of Alzheimer's disease and vascular dementia. Neurochem Res 29: 1257–1266.

Salloway S, Sperling R, Fox NC et al. (2014). Two phase 3 trials of bapineuzumab in mild-to-moderate Alzheimer's disease. N Engl J Med 370: 322–333.

Sarter M, Bruno JP (2004). Developmental origins of the age-related decline in cortical cholinergic function and associated cognitive abilities. Neurobiol Aging 25: 1127–1139.

Schaie KW, Willis SL (1993). Age difference patterns of psychometric intelligence in adulthood: generalizability within and across ability domains. Psychol Aging 8: 44–55.

Schaie KW, Willis SL, Pennak S (2005). An historical framework for cohort differences in intelligence. Res Hum Dev 2: 43–67.

Schneider JA, Wilson RS, Cochran EJ et al. (2003). Relation of cerebral infarctions to dementia and cognitive function in older persons. Neurology 60: 1082–1088.

Schuff N, Zhu XP (2007). Imaging of mild cognitive impairment and early dementia. Br J Radiol 80 (Spec No 2): S109–S114.

Selkoe DJ (2000). Toward a comprehensive theory for Alzheimer's disease. Hypothesis: Alzheimer's disease is caused by the cerebral accumulation and cytotoxicity of amyloid beta-protein. Ann N Y Acad Sci 924: 17–25.

Seshadri S, Beiser A, Selhub J et al. (2002). Plasma homocysteine as a risk factor for dementia and Alzheimer's disease. N Engl J Med 346: 476–483.

Severson MA, Smith GE, Tangalos EG et al. (1994). Patterns and predictors of institutionalization in community-based dementia patients. J Am Geriatr Soc 42: 181–185.

Sherrington R, Rogaev EI, Liang Y et al. (1995). Cloning of a gene bearing missense mutations in early-onset familial Alzheimer's disease. Nature 375: 754–760.

Sink KM, Espeland MA, Rushing J et al. (2014). The LIFE Cognition Study: design and baseline characteristics. Clin Interv Aging 9: 1425–1436.

Skoog I (1998). Status of risk factors for vascular dementia. Neuroepidemiology 17: 2–9.

Skoog I, Nilsson L, Palmertz B et al. (1993). A population-based study of dementia in 85-year-olds. N Engl J Med 328: 153–158.

Skoog I, Lernfelt B, Landahl S et al. (1996). 15-year longitudinal study of blood pressure and dementia. Lancet 347: 1141–1145.

Slooter AJ, Cruts M, Kalmijn S et al. (1998a). Risk estimates of dementia by apolipoprotein E genotypes from a population-based incidence study: the Rotterdam Study. Arch Neurol 55: 964–968.

Slooter AJ, Van Duijn CM, Bots ML et al. (1998b). Apolipoprotein E genotype, atherosclerosis, and cognitive decline: the Rotterdam Study. J Neural Transm Suppl 53: 17–29.

Small SA, Duff K (2008). Linking Abeta and tau in late-onset Alzheimer's disease: a dual pathway hypothesis. Neuron 60: 534–542.

Small BJ, Rosnick CB, Fratiglioni L et al. (2004). Apolipoprotein E and cognitive performance: a meta-analysis. Psychol Aging 19: 592–600.

Small GW, Kepe V, Ercoli LM et al. (2006). PET of brain amyloid and tau in mild cognitive impairment. N Engl J Med 355: 2652–2663.

Smith AD (2008). The worldwide challenge of the dementias: a role for B vitamins and homocysteine? Food Nutr Bull 29: S143–S172.

Solomon A, Kivipelto M, Wolozin B et al. (2009). Midlife serum cholesterol and increased risk of Alzheimer's and vascular dementia three decades later. Dement Geriatr Cogn Disord 28: 75–80.

Stampfer MJ, Kang JH, Chen J et al. (2005). Effects of moderate alcohol consumption on cognitive function in women. N Engl J Med 352: 245–253.

Stern Y (1998). Increased risk of mortality in AD patients with higher education? Neurology 51: 1238.

Stern Y, Gurland B, Tatemichi TK et al. (1994). Influence of education and occupation on the incidence of Alzheimer's disease. JAMA 271: 1004–1010.

Stern Y, Tang MX, Denaro J et al. (1995). Increased risk of mortality in Alzheimer's disease patients with more advanced educational and occupational attainment. Ann Neurol 37: 590–595.

Stern Y, Zarahn E, Hilton HJ et al. (2003). Exploring the neural basis of cognitive reserve. J Clin Exp Neuropsychol 25: 691–701.

Stewart R, Xue QL, Masaki K et al. (2009). Change in blood pressure and incident dementia: a 32-year prospective study. Hypertension 54: 233–240.

Storey E, Slavin MJ, Kinsella GJ (2002). Patterns of cognitive impairment in Alzheimer's disease: assessment and differential diagnosis. Front Biosci 7: e155–e184.

Tanzi RE, Bertram L (2005). Twenty years of the Alzheimer's disease amyloid hypothesis: a genetic perspective. Cell 120: 545–555.

Tateno A, Sakayori T, Takizawa Y et al. (2015). A case of Alzheimer's disease following mild traumatic brain injury. Gen Hosp Psychiatry 37: 97 e7–97 e9.

Terry RD, Katzman R (2001). Life span and synapses: will there be a primary senile dementia? Neurobiol Aging 22: 347–348. discussion 353–4.

Tierney MC, Fisher RH, Lewis AJ et al. (1988). The NINCDS-ADRDA Work Group criteria for the clinical diagnosis of probable Alzheimer's disease: a clinicopathologic study of 57 cases. Neurology 38: 359–364.

Tola-Arribas MA, Yugueros MI, Garea MJ et al. (2013). Prevalence of dementia and subtypes in Valladolid, northwestern Spain: the DEMINVALL study. PLoS One 8: e77688.

Truelsen T, Thudium D, Gronbaek M (2002). Amount and type of alcohol and risk of dementia: the Copenhagen City Heart Study. Neurology 59: 1313–1319.

Velazquez-Brizuela IE, Ortiz GG, Ventura-Castro L et al. (2014). Prevalence of dementia, emotional state and physical performance among older adults in the metropolitan area of Guadalajara, Jalisco, Mexico. Curr Gerontol Geriatr Res 2014: 387528.

Vellas B, Carrillo MC, Sampaio C et al. (2013). Designing drug trials for Alzheimer's disease: what we have learned from the release of the phase III antibody trials: a report from the EU/US/CTAD Task Force. Alzheimers Dement 9: 438–444.

Verghese J, Lipton RB, Hall CB et al. (2003). Low blood pressure and the risk of dementia in very old individuals. Neurology 61: 1667–1672.

Vickland V, Mcdonnell G, Werner J et al. (2010). A computer model of dementia prevalence in Australia: foreseeing outcomes of delaying dementia onset, slowing disease progression, and eradicating dementia types. Dement Geriatr Cogn Disord 29: 123–130.

Villemagne VL, Burnham S, Bourgeat P et al. (2013). Amyloid beta deposition, neurodegeneration, and cognitive decline in sporadic Alzheimer's disease: a prospective cohort study. Lancet Neurol 12: 357–367.

Visser PJ, Kester A, Jolles J et al. (2006). Ten-year risk of dementia in subjects with mild cognitive impairment. Neurology 67: 1201–1207.

Visser PJ, Verhey F, Knol DL et al. (2009). Prevalence and prognostic value of CSF markers of Alzheimer's disease pathology in patients with subjective cognitive impairment or mild cognitive impairment in the DESCRIPA study: a prospective cohort study. Lancet Neurol 8: 619–627.

Von Strauss E, Viitanen M, De Ronchi D et al. (1999). Aging and the occurrence of dementia: findings from a population-based cohort with a large sample of nonagenarians. Arch Neurol 56: 587–592.

Wancata J, Borjesson-Hanson A, Ostling S et al. (2007). Diagnostic criteria influence dementia prevalence. Am J Geriatr Psychiatry 15: 1034–1045.

Wang HX, Gustafson DR, Kivipelto M et al. (2012a). Education halves the risk of dementia due to apolipoprotein epsilon4 allele: a collaborative study from the Swedish brain power initiative. Neurobiol Aging 33: 1007 e1–1007 e7.

Wang HX, Wahlberg M, Karp A et al. (2012b). Psychosocial stress at work is associated with increased dementia risk in late life. Alzheimers Dement 8: 114–120.

Watson GS, Craft S (2003). The role of insulin resistance in the pathogenesis of Alzheimer's disease: implications for treatment. CNS Drugs 17: 27–45.

White H, Pieper C, Schmader K et al. (1996). Weight change in Alzheimer's disease. J Am Geriatr Soc 44: 265–272.

Whitmer RA, Gunderson EP, Quesenberry Jr CP et al. (2007). Body mass index in midlife and risk of Alzheimer disease and vascular dementia. Curr Alzheimer Res 4: 103–109.

WHO (2012). Dementia cases set to triple by 2050 but still largely ignored. News Release [Online]. World Health Organization, Geneva Available: http://www.who.int/mediacentre/news/releases/2012/dementia_20120411/en/ [Accessed 4 May 2015 2015].

Wiberg P, Waern M, Billstedt E et al. (2013). Secular trends in the prevalence of dementia and depression in Swedish septuagenarians 1976–2006. Psychol Med 43: 2627–2634.

Wilson RS, Schneider JA, Barnes LL et al. (2002). The apolipoprotein E epsilon 4 allele and decline in different cognitive systems during a 6-year period. Arch Neurol 59: 1154–1160.

Winblad B, Palmer K, Kivipelto M et al. (2004). Mild cognitive impairment – beyond controversies, towards a consensus: report of the International Working Group on Mild Cognitive Impairment. J Intern Med 256: 240–246.

Winblad B, Andreasen N, Minthon L et al. (2012). Safety, tolerability, and antibody response of active Abeta immunotherapy with CAD106 in patients with Alzheimer's disease: randomised, double-blind, placebo-controlled, first-in-human study. Lancet Neurol 11: 597–604.

Wisniewski T, Konietzko U (2008). Amyloid-beta immunisation for Alzheimer's disease. Lancet Neurol 7: 805–811.

Withall A, Draper B, Seeher K et al. (2014). The prevalence and causes of younger onset dementia in Eastern Sydney, Australia. Int Psychogeriatr 26: 1955–1965.

World Health Organization (2010). The ICD-10 classification of mental and behavioural disorders: clinical descriptions and diagnostic guidelines, World Health Organization, Geneva.

Wu YT, Lee HY, Norton S et al. (2013). Prevalence studies of dementia in mainland china, Hong Kong and taiwan: a systematic review and meta-analysis. PLoS One 8e66252.

Wu YT, Fratiglioni L, Matthews FE et al. (2016). Dementia in western Europe: epidemiological evidence and implications for policy making. Lancet Neurol 15: 116–124.

Xie J, Brayne C, Matthews FE (2008). Survival times in people with dementia: analysis from population based cohort study with 14 year follow-up. BMJ 336: 258–262.

Xie C, Li W, Chen G et al. (2012). The co-existence of geriatric depression and amnestic mild cognitive impairment detrimentally affect gray matter volumes: voxel-based morphometry study. Behav Brain Res 235: 244–250.

Xu W, Qiu C, Gatz M et al. (2009a). Mid- and late-life diabetes in relation to the risk of dementia: a population-based twin study. Diabetes 58: 71–77.

Xu WL, Von Strauss E, Qiu CX et al. (2009b). Uncontrolled diabetes increases the risk of Alzheimer's disease: a population-based cohort study. Diabetologia 52: 1031–1039.

Yaffe K, Vittinghoff E, Lindquist K et al. (2010). Posttraumatic stress disorder and risk of dementia among US veterans. Arch Gen Psychiatry 67: 608–613.

Yan R, Vassar R (2014). Targeting the beta secretase BACE1 for Alzheimer's disease therapy. Lancet Neurol 13: 319–329.

Yang Z, Zhou X, Zhang Q (2013). Effectiveness and safety of memantine treatment for Alzheimer's disease. J Alzheimers Dis 36: 445–458.

Zaccai J, Ince P, Brayne C (2006). Population-based neuropathological studies of dementia: design, methods and areas of investigation–a systematic review. BMC Neurol 6: 2.

Zhu L, Fratiglioni L, Guo Z et al. (2000). Incidence of dementia in relation to stroke and the apolipoprotein E epsilon4 allele in the very old. Findings from a population-based longitudinal study. Stroke 31: 53–60.

Ziegler-Graham K, Brookmeyer R, Johnson E et al. (2008). Worldwide variation in the doubling time of Alzheimer's disease incidence rates. Alzheimers Dement 4: 316–323.

Zylberstein DE, Lissner L, Bjorkelund C et al. (2011). Midlife homocysteine and late-life dementia in women. A prospective population study. Neurobiol Aging 32: 380–386.

Handbook of Clinical Neurology, Vol. 138 (3rd series)
Neuroepidemiology
C. Rosano, M.A. Ikram, and M. Ganguli, Editors
http://dx.doi.org/10.1016/B978-0-12-802973-2.00009-4

Chapter 9

Epidemiology of alpha-synucleinopathies: from Parkinson disease to dementia with Lewy bodies

R. SAVICA[1]*, B.F. BOEVE[1], AND G. LOGROSCINO[1,2]
[1]*Department of Neurology, Mayo Clinic, Rochester, MN, USA*
[2]*Neurodegenerative Diseases Unit, Department of Basica Medicine, Neuroscience and Sense Organs, University of Bari, Italy*

Abstract

The epidemiology of the diagnosis of Parkinson's disease and dementia with Lewy bodies is still based on clinical criteria and the definition of the different diseases is still a challenge for clinician and researcher. The epidemiologic estimates of prevalence and incidence are highly affected by differences in diagnostic criteria, geographic location, and methodologic limitations.

Studies of prevalence and incidence show increases with advancing age and a higher rate of Parkinson's disease and dementia with Lewy bodies in men compared to women.

INTRODUCTION: THE CHANGING FACE OF THE EPIDEMIOLOGY OF NEURODEGENERATIVE DISORDERS

Neurodegenerative diseases are probably the most important public health challenge in the near future both for clinical medicine and public health. This is mainly due to the huge demographic changes that have characterized the second half of the 20th century. The results of the Global Burden of Disease study (Global Burden of Disease Study Collaborators, 2015) clearly indicate the presence of two related trends: (1) decline in global age and mortality; and (2) older age at death.

This complex demographic phenomenon has produced a change in the observable general pattern of diseases and deaths, with increasing prevalence of neurologic diseases related to aging, like dementias and some movement disorders. This is paralleled by the increased mortality due to the same group of diseases.

The increased prevalence of neurodegenerative diseases is also due to the decline of possible competing causes of morbidity and mortality observed in previous decades, mainly cardiovascular and cerebrovascular diseases (Cooper et al., 2000).

Neurodegenerative disorders have three main characteristics in common (Giancarlo Logroscino, 2014): (1) incidence and prevalence increase with age; (2) only a minority of cases have a dominant genetic causation, (generally monogenetic forms with early age of onset under age 50). Most neurodegenerative disease cases are sporadic, with complex models of inheritance, multiple genes, and environmental risk factors as determinants; and (3) the clinical phenotype of neurodegenerative diseases is highly variable, strongly conditioned by the age of onset of clinical diseases and by the presence of other concomitant pathologies. The typical phenotypes fulfilling established diagnostic criteria are probably only a fraction of clinical cases, while overlapping presentations and transitional phenotypes across the whole spectrum of neurodegeneration represent probably a significant part of the load of cases in the community. Changes in clinical features that create different phenotypes of the disease have been substantially in Parkinson's disease (PD). The recognition of the importance of nonmotor symptoms in the natural history of disease, both in the premotor stage and in the motor phase, has produced a substantial change in defining patterns of disease that are more complex than previously thought. Hyposmia, rapid-eye-movement (REM) behavioral disorders (RBD) and sleepiness,

*Correspondence to: Rodolfo Savica, MD, Department of Neurology, Mayo Clinic, 200 First St. SW, Rochester MN 55905, USA. E-mail: Savica.Rodolfo@mayo.edu

depression and behavioral disorders, cognitive impairment and dementia, constipation, and dysautonomic disorders are part of the set of symptoms and signs at which we need to look in a patient with suspected PD, beyond the set of cardinal signs established in the UK Brain Bank criteria (Anonymous, 2006). This added complexity to the identification of cases at the beginning of disease is also important in the definition of the prognosis. The cluster of symptoms may define different courses of PD, as recently established in a clinically based study in Canada. The presence of cognitive impairment, especially when multiple-domain, is one of the three elements characterizing the symptom cluster of the malignant subtype, along with orthostatic hypotension and RBD (Fereshtehnejad et al., 2015). The most distinctive feature is the presence of cognitive impairment at baseline, even when compared to clusters of symptoms that present an intermediate prognosis.

The spectrum of parkinsonism with cognitive impairment needs further characterization because of the diagnostic difficulties in identifying dementia with Lewy bodies (DLB) in the early phase of disease.

ALPHA-SYNUCLEINOPATHIES

Alpha-synuclein is a protein that is normally present in synapses. It is encoded by chromosome 4 (McKeith et al., 2005) and seems to be localized specifically in presynaptic terminals (Lee et al., 2005). Beta-synuclein and gamma-synuclein have also been described, but these last two molecules do not seem to be involved in neurodegenerative disease (Eller and Williams, 2011). The function of alpha-synuclein is not totally understood; however, it seems to play a role in vesicle production. When aggregated it is insoluble, depositing in the brain and becoming the main component of the fibrils constituting Lewy bodies (Luis et al., 1999). Lewy body deposition is the pathologic hallmark of a number of diseases: PD, DLB, and multiple-system atrophy are the most common. These diseases are named alpha-synucleinopathies to highlight the major role of this protein deposit in the pathophysiology of the disease.

The factors underlying the dysregulation of α-synuclein and the subsequent development of Lewy bodies are still unknown. It is possible that there is an upregulation of normal alpha-synuclein production due to changes in mRNA expression. It is also possible that the presence of Lewy bodies in the brain is a normal response of the neurons to toxic fibrils that have been formed by dysfunctional proteosomes (McKeith et al., 2004).

CLASSIFICATION OF MOTOR AND COGNITIVE PHENOTYPES OF PARKINSONISM

Can we define parkinsonism first?

The classification of different forms of parkinsonism is strongly conditioned by the presence of both minor and major cognitive impairment and by the strictness with which the diagnostic criteria are applied (Bower et al., 2000).

Parkinsonism has been defined as the presence of two of four cardinal signs: resting tremor, bradykinesia, rigidity, and impaired postural reflexes (Bower et al., 1999).

The presence of cognitive impairment may be one of the additional features of parkinsonism. The relationship between criteria and incidence is clearly illustrated using data from the Rochester Epidemiological Project in Olmsted County in Minnesota. The addition of cognitive impairment, making the criteria broader, changes completely the incidence curve of parkinsonism, as follows (Bower et al., 2000):

1. strict criteria using laterality and response to levodopa: the incidence curve deflects after age 75, especially for men
2. diagnostic criteria including two out of four cardinal signs: the incidence curve is flat after age 70
3. parkinsonism criteria plus dementia: the incidence curve rises even after age 80 for both men and women.

The change in diagnostic criteria is relevant for DLB and eventually for PD dementia. After a certain age the presence of dementia or cognitive impairment is probably a common feature of subjects who develop parkinsonism. This is concordant with other data showing that changing the criteria of motor parkinsonism changes completely the incidence curve.

If the criteria of parkinsonism are limited to two or three out of four cardinal signs, as was shown in population-based data from Sicily (Italy), the incidence curve of parkinsonism tends to rise with increasing age, in a steeper way compared to data from the Rochester Epidemiology Program. With the introduction of response to levodopa or asymmetry as a criterion, the incidence curve becomes flat (de Rijk et al., 1992). Similar data were produced in the Rotterdam study. The incidence of PD is quite stable after age 75, while the incidence of parkinsonism continues to rise (de Lau et al., 2004). Epidemiologic data show concordantly that the phenotype of parkinsonism shifts with aging, from a primary motor disease with a different cluster of nonmotor symptoms to a motor-plus-cognitive phenotype. The motor phenotype in older ages is more likely an axial parkinsonism characterized by postural instability, facial impassivity, gait difficulty, and no or scarce tremor (McKeith, 2004). Axial parkinsonism is positively associated with age (Levy et al., 2000). DLB and the DLB prodromal stages are probably the more common clinical form of parkinsonism after 80 years of age.

PREVALENCE AND INCIDENCE OF PARKINSONS DISEASE

The prevalence of a disease refers to the proportion of events in a given population at a designated point in time. The incidence of the disease is defined as the rate at which new cases of a given disease develop during a defined period of time in a specified population (Last and International Epidemiological Association, 2001).

The epidemiology of PD has been extensively studied over decades using different methodologic approaches and in many different geographic areas worldwide.

A recent meta-analysis compared a number of studies to estimate the prevalence of PD (Pringsheim et al., 2014). The prevalence of PD varies from 41 per 100 000 in individuals 40–49 years, increasing to 1087 per 100 000 in individuals 70–79 years, and reaching 1903 per 100 000 over age 80. In addition, when comparing geographic regions, there was a lower PD prevalence in Asia (646 per 100 000) compared with populations of the same age in Europe, North America, and Australia.

Despite the large methodologic and geographic differences, the different studies pointed out a number of differences and common characteristics. First of all, the use of different clinical criteria plays a major role in the possible difference that may occur between studies. De Rijk et al. (1997) reported that a change in diagnostic criteria might account for a 36% change in the identification of cases. In addition, many studies do not use direct examination but clinical questionnaires that may have vary in their ability to detect PD cases. Second, large variability can be seen with different expertise of the medical personnel in diagnosing cases of PD; indeed, it is plausible that tertiary centers can more accurately differentiate between the different forms of parkinsonism and PD. Third, geographic characteristics have to be taken into account when worldwide prevalence is analyzed. In fact, there seem to be a lower prevalence of PD in Asia as compared with North America, Europe, and Australia (Pringsheim et al., 2014). In order to take into account differences in exposure to environmental risk factors and differences in survival of the geographic regions it would be better to report the prevalence of PD separately by region.

On the other hand, a number of characteristics seem to be common across studies. Prevalence of PD increases with age, keeping a steady increase with aging, with a higher increase later in life as compared with the younger population. Women seem to have a consistently lower prevalence then men, especially in the 50–59-year group.

The same common characteristics and differences may be evoked for incidence studies. In fact, the incidence of PD increases with age, is higher in men, and there are substantial geographic differences that resemble the prevalence differences (Twelves et al., 2003).

The incidence of PD has been reported to be 16–19/100 000/year; in our recent population-based study we confirmed these findings, taking advantage of the Rochester Epidemiological Project in Olmsted County, MN, we reported an incidence of 21/100 000/year that increased with age and was higher in men than in women (Savica et al., 2013a).

PREVALENCE AND INCIDENCE OF PD DEMENTIA AND DLB

DLB is currently considered the second most common type of degenerative dementia in older people, accounting for 10–15% of cases at autopsy (McKeith et al., 2004).

In population studies of people older than 65, the prevalence of DLB ranged from 0.1% (de Silva et al., 2003) to 2.0% of the population (Stevens et al., 2002). The prevalence of DLB in dementia varies between 2.8% (Yamada et al., 2001) and 30.5% (Stevens et al., 2002) in different studies using the same clinical criteria.

When the study population is older than 70 or 75 years of age, the prevalence of DLB ranges from 0% (Yamada et al., 2002) to 5.0% (Rahkonen et al., 2003) of the general population and 0% (Yamada et al., 2002) to 21.9% (Rahkonen et al., 2003) among demented individuals. Notably, the higher prevalence has been reported in studies conducted in Europe, including prevalence among dementia patients of 30.5% in north London, England (Stevens et al., 2002) and 21.9% in Finland (Rahkonen et al., 2003).

Only two studies have reported the incidence of DLB. The Cache County Study (Cache County, UT) estimated an incidence rate of 0.1% of DLB cases a year in the general population of people older than 65 years of age, corresponding to a 3.2% rate of cases of DLB among cases of dementia (Miech et al., 2002).

In Olmsted County, MN, the incidence of DLB was 3.5 per 100 000 person-years, increased with age, and was higher in men than in women, especially after 60 years of age (Savica et al., 2013b).

The incidence of PD dementia was 2.5 per 100 000 person-years overall, increased with age, and was similar in men and women (2.3 vs. 2.7). However, it seemed to be markedly increased in men in the oldest age group. When the incidence of DLB and PD dementia was combined it was 5.9 overall, and was higher in men than in women. Interestingly, men had consistently higher incidence rates than women across the age spectrum.

A study on the projected prevalence of dementia for 2030 and 2050 reported that the prevalence of DLB in 2050 will increase by 131% (Bach et al., 2011) in the USA. Despite the limitations of combining heterogeneous studies, the projections signal increased clinical identification and knowledge about the disease.

Considering the difficulty in diagnosis and case identification, there are no studies in the literature, that we are aware of, that have explored the role of risk or protective factors in the development of DLB. However, it is possible that the findings of PD may apply to some extent to DLB, considering the clinical resemblance between these two diseases and the sharing of a common hallmark, the deposition of alpha-synuclein.

The frequency of Lewy body pathology may be considerably more common than the prevalence of the clinical disorder. The Honolulu Study of Aging, which is a population-based study, showed that about 12.2% of brains examined of normal men older than 60 years of age had Lewy body pathology (Abbott et al., 2007). These findings have been confirmed by other authors (Frigerio et al., 2011). It is possible that incidental findings of DLB may be present during life.

SPECTRUM OF CLINICAL PHENOTYPES OF DLB

The clinical criteria for DLB proposed in the 2005 consensus conference (McKeith, 2006) have the principal features of cognitive disorder and parkinsonism. The possible overlap between DLB, PD, mild cognitive impairment, and dementia of Alzheimer's disease (AD) can cause difficulties in the identification and treatment of DLB. Given that DLB is a dementing illness, progressive memory loss may be considered the most common initial presentation of DLB. Cognitive fluctuations may be recurrent; however, they may not always be present (McKeith, 2006).

Parkinsonism is the other main feature of the DLB phenotype. The presence of parkinsonism may help to differentiate DLB from AD (Ferman et al., 2004). However, the parkinsonism associated with DLB seems to have different clinical characteristics from the PD cases; indeed, resting tremor does not seem to be consistently associated with DLB. However, an early observation pointed out that resting tremor was the most common motor symptom of DLB (Crystal et al., 1990). A number of other symptoms may help to identify DLB: myoclonus, psychiatric symptoms (depression, anxiety, paranoia, etc.), dysautonomia, sleep and wakefulness disturbances, and hallucinations. A detailed analysis of these features is beyond the scope of this article; however, complex visual hallucinations may be present in the early stages of the disease (Manford and Andermann, 1998), and they seem to be associated with the presence of Lewy bodies in the temporal lobes (Harding et al., 2002).

ENVIRONMENTAL FACTORS IN DLB AND METHODOLOGIC CHALLENGES

Environmental risk factors have been considered to play a role in all the neurodegenerative disorders. Numerous studies have been associated with PD but limited information is available for DLB. No studies, so far, have explored the risk or protective factors for DLB. The reasons can be explained by the relatively low prevalence, and the challenges in diagnosis make traditional population-based epidemiologic investigations very challenging. Phenotypic heterogeneity adds to the complexity. Case-control studies that use prevalent cases from tertiary centers (memory or movement disorder clinics) (Boot et al., 2013) are likely to be subject to various biases that make conclusions uncertain. It will be some time before studies on risk factors for DLB from unbiased studies will be available.

On the other hand, there is evidence available on incidental Lewy body disease. Interestingly, the risk factors with a significant association with PD have been reported to have the same association in incidental DLB. Caffeine intake, occupation, years of education, head injury, and increased number of children seem to be risk factors not only for PD but also for incidental DLB (Frigerio et al., 2009). Unfortunately, the data were limited and warrant caution in the interpretation, but it is very likely that patients with LBD may have the same environmental risk factors forf PD, although a different individual/environmental predisposition can lead to a different phenotype of degeneration.

CHALLENGES IN EPIDEMIOLOGY OF PD, PD DEMENTIA, AND DLB

1. Changing load of cases. The number of cases of all three diseases will increase in the near future. We have to consider that the oldest old are the population segment with the fastest rate of expansion and by 2050 will become five times larger. The major consequences of these changes will be the substantial increase in the number of cases with dementia or cognitive involvement, and the decline in subjects with primarily motor diseases (PD). Most of the cases will be in the segment of 85+ years. According to these changes, because DLB is the dementia of the oldest old, the prevalence and incidence of DLB will strongly increase.
2. The number of underdiagnosed cases of DLB will increase. Both in clinical series and epidemiologic studies the sensitivity of the PD diagnosis is high, whereas it is low when DLB is diagnosed. This probably indicates a significant underdiagnosis of DLB (Litvan et al., 1998) and perhaps an

overdiagnosis of PD. Many cases of cognitive disorders with mild parkinsonism, especially in the oldest age group, are labeled as AD. This is due to the lack of familiarity with this diagnosis of many physicians and less awareness of this specific diagnosis in the wider public. The diagnosis is possible only after prolonged observation and usually after many specialist visits (Galvin et al., 2010). One consequence is the limited possibility of detecting true incident cases in epidemiologic studies. The common symptomatology of DLB and AD is however an additional important factor. This is due to the presence of AD-like pathology in 80% of cases of DLB (presence of beta-amyloid plaques in the same cortical areas). In typical AD there is Lewy body pathology in almost 50% of cases. Thus, the neuropathology as diagnostic gold standard is not very different between primary physicians and movement disorder specialists.

3. Complexity of case diagnosis and management of PD, PD dementia, and DLB because of comorbidity and frailty. Compared to AD, DLB has more comorbid stroke, migraine, and depression (Fereshtehnejad et al., 2014). Frailty is a complex condition characterized by reduced response to acute and chronic conditions of stress, making the individual more vulnerable. Physical frailty is based on timed walk, grip strength, body mass composition, fatigue, and low energy expenditure (Erekson et al., 2015). Many of these five characteristics are altered in PD, PD dementia, and DLB. Frailty is associated with older age and is conceptually different from comorbidity. The Lewy body pathology and neuronal loss in the substantia nigra are both associated with a more rapid progression of frailty together with AD pathology and vascular pathology. The combination of multiple pathologies is responsible both for the total amount of frailty at baseline and the rapidity of progression (Buchman et al., 2013). There is still a paucity of studies addressing the role of physical frailty in the onset of cognitive impairment and dementia in parkinsonism, particularly in the oldest age group, where the prevalence and incidence of DLB are higher. Different clinical forms of PD dementia and DLB may be associated with different forms of physical frailty.

SUMMARY

The diagnosis of PD and DLB is still based on clinical criteria. The diagnostic challenge of the clinical and phenotypic identification between the different types of synucleinopathies is a major cause of difficulty in providing consistent data on prevalence and incidence across studies.

The heterogeneity of the studied populations further increases the instability of the epidemiologic results.

On the other hand, the majority of studies report that both the prevalence and incidence of PD and DLB increase with advancing age, with a higher rate in men as compared to women.

We anticipate a further increase in PD and DLB in the older population due to the aging of the general population and the continuous effort in improving the diagnosis and advancement in medical training to manage these disorders by a larger group of medical professionals.

References

Abbott RD, Ross GW, Petrovitch H et al. (2007). Bowel movement frequency in late-life and incidental Lewy bodies. Mov Disord 22: 1581–1586.

Anonymous (2006). Parkinson's Disease: National Clinical Guideline for Diagnosis and Management in Primary and Secondary Care. London. https://www.nice.org.uk/guidance/cg35/evidence/full-guideline-194930029.

Bach JP, Ziegler U, Deuschl G et al. (2011). Projected numbers of people with movement disorders in the years 2030 and 2050. Mov Disord 26: 2286–2290.

Boot BP, Orr CF, Ahlskog JE, Ferman TJ et al. (2013). Risk factors for dementia with Lewy bodies: a case-control study. Neurology 81 (9): 833–840. http://dx.doi.org/10.1212/WNL.0b013e3182a2cbd1. Epub 2013, Jul 26.

Bower JH, Maraganore DM, Mcdonnell SK et al. (1999). Incidence and distribution of parkinsonism in Olmsted County, Minnesota, 1976–1990. Neurology 52: 1214–1220.

Bower JH, Maraganore DM, Mcdonnell SK et al. (2000). Influence of strict, intermediate, and broad diagnostic criteria on the age- and sex-specific incidence of Parkinson's disease. Mov Disord 15: 819–825.

Buchman AS, Yu L, Wilson RS et al. (2013). Association of brain pathology with the progression of frailty in older adults. Neurology 80: 2055–2061.

Cooper R, Cutler J, Desvigne-Nickens P et al. (2000). Trends and disparities in coronary heart disease, stroke, and other cardiovascular diseases in the United States: findings of the national conference on cardiovascular disease prevention. Circulation 102: 3137–3147.

Crystal HA, Dickson DW, Lizardi JE et al. (1990). Antemortem diagnosis of diffuse Lewy body disease. Neurology 40: 1523–1528.

De Lau LM, Giesbergen PC, De Rijk MC et al. (2004). Incidence of parkinsonism and Parkinson disease in a general population: the Rotterdam Study. Neurology 63: 1240–1244.

De Rijk EP, Van Strien FJ, Roubos EW (1992). Demonstration of coexisting catecholamine (dopamine), amino acid (GABA), and peptide (NPY) involved in inhibition of melanotrope cell activity in *Xenopus laevis*: a quantitative ultrastructural, freeze-substitution immunocytochemical study. J Neurosci 12: 864–871.

De Rijk MC, Rocca WA, Anderson DW et al. (1997). A population perspective on diagnostic criteria for Parkinson's disease. Neurology 48: 1277–1281.

De Silva HA, Gunatilake SB, Smith AD (2003). Prevalence of dementia in a semi-urban population in Sri Lanka: report from a regional survey. Int J Geriatr Psychiatry 18: 711–715.

Eller M, Williams DR (2011). alpha-Synuclein in Parkinson disease and other neurodegenerative disorders. Clin Chem Lab Med 49: 403–408.

Erekson EA, Fried TR, Martin DK et al. (2015). Frailty, cognitive impairment, and functional disability in older women with female pelvic floor dysfunction. Int Urogynecol J 26: 823–830.

Fereshtehnejad SM, Damangir S, Cermakova P et al. (2014). Comorbidity profile in dementia with Lewy bodies versus Alzheimer's disease: a linkage study between the Swedish Dementia Registry and the Swedish National Patient Registry. Alzheimers Res Ther 6: 65.

Fereshtehnejad SM, Romenets SR, Anang J et al. (2015). Clinical subtypes of Parkinson's disease and disease progression: Data-driven comparison of new phenotypes with other cluster solutions [abstract]. Mov Disord 30 (Suppl 1): 685.

Ferman TJ, Smith GE, Boeve BF et al. (2004). DLB fluctuations: specific features that reliably differentiate DLB from AD and normal aging. Neurology 62: 181–187.

Frigerio R, Fujishiro H, Maraganore DM et al. (2009). Comparison of risk factor profiles in incidental Lewy body disease and Parkinson disease. Arch Neurol 66: 1114–1119.

Frigerio R, Fujishiro H, Ahn TB et al. (2011). Incidental Lewy body disease: do some cases represent a preclinical stage of dementia with Lewy bodies? Neurobiol Aging 32: 857–863.

Galvin JE, Duda JE, Kaufer DI et al. (2010). Lewy body dementia: the caregiver experience of clinical care. Parkinsonism Relat Disord 16: 388–392.

Giancarlo Logroscino RT (2014). Imaging IN Neurodegenerative Disorders, Oxford Press, Oxford.

Global Burden of Disease Study Collaborators (2015). Global, regional, and national incidence, prevalence, and years lived with disability for 301 acute and chronic diseases and injuries in 188 countries, 1990–2013: a systematic analysis for the Global Burden of Disease Study 2013. Lancet 386: 743–800.

Harding AJ, Broe GA, Halliday GM (2002). Visual hallucinations in Lewy body disease relate to Lewy bodies in the temporal lobe. Brain 125: 391–403.

Last JM, International Epidemiological Association (2001). A dictionary of epidemiology, Oxford University Press, New York.

Lee HJ, Patel S, Lee SJ (2005). Intravesicular localization and exocytosis of alpha-synuclein and its aggregates. J Neurosci 25: 6016–6024.

Levy G, Tang MX, Cote LJ et al. (2000). Motor impairment in PD: relationship to incident dementia and age. Neurology 55: 539–544.

Litvan I, Macintyre A, Goetz CG et al. (1998). Accuracy of the clinical diagnoses of Lewy body disease, Parkinson disease, and dementia with Lewy bodies: a clinicopathologic study. Arch Neurol 55: 969–978.

Luis CA, Barker WW, Gajaraj K et al. (1999). Sensitivity and specificity of three clinical criteria for dementia with Lewy bodies in an autopsy-verified sample. Int J Geriatr Psychiatry 14: 526–533.

Manford M, Andermann F (1998). Complex visual hallucinations. Clinical and neurobiological insights. Brain 121 (Pt 10): 1819–1840.

Mckeith I (2004). Dementia in Parkinson's disease: common and treatable. Lancet Neurol 3: 456.

Mckeith IG (2006). Consensus guidelines for the clinical and pathologic diagnosis of dementia with Lewy bodies (DLB): report of the Consortium on DLB International Workshop. J Alzheimers Dis 9: 417–423.

Mckeith I, Mintzer J, Aarsland D et al. (2004). Dementia with Lewy bodies. Lancet Neurol 3: 19–28.

Mckeith IG, Dickson DW, Lowe J et al. (2005). Diagnosis and management of dementia with Lewy bodies: third report of the DLB Consortium. Neurology 65: 1863–1872.

Miech RA, Breitner JC, Zandi PP et al. (2002). Incidence of AD may decline in the early 90s for men, later for women: The Cache County study. Neurology 58: 209–218.

Pringsheim T, Jette N, Frolkis A et al. (2014). The prevalence of Parkinson's disease: a systematic review and meta-analysis. Mov Disord 29: 1583–1590.

Rahkonen T, Eloniemi-Sulkava U, Rissanen S et al. (2003). Dementia with Lewy bodies according to the consensus criteria in a general population aged 75 years or older. J Neurol Neurosurg Psychiatry 74: 720–724.

Savica R, Grossardt BR, Bower JH et al. (2013a). Incidence and pathology of synucleinopathies and tauopathies related to parkinsonism. JAMA Neurol 70: 859–866.

Savica R, Grossardt BR, Bower JH et al. (2013b). Incidence of dementia with Lewy bodies and Parkinson disease dementia. JAMA Neurol 70: 1396–1402.

Stevens T, Livingston G, Kitchen G et al. (2002). Islington study of dementia subtypes in the community. Br J Psychiatry 180: 270–276.

Twelves D, Perkins KS, Counsell C (2003). Systematic review of incidence studies of Parkinson's disease. Mov Disord 18: 19–31.

Yamada T, Hattori H, Miura A et al. (2001). Prevalence of Alzheimer's disease, vascular dementia and dementia with Lewy bodies in a Japanese population. Psychiatry Clin Neurosci 55: 21–25.

Yamada T, Kadekaru H, Matsumoto S et al. (2002). Prevalence of dementia in the older Japanese-Brazilian population. Psychiatry Clin Neurosci 56: 71–75.

Handbook of Clinical Neurology, Vol. 138 (3rd series)
Neuroepidemiology
C. Rosano, M.A. Ikram, and M. Ganguli, Editors
http://dx.doi.org/10.1016/B978-0-12-802973-2.00010-0

Chapter 10

Epidemiology of epilepsy

S. ABRAMOVICI[1]* AND A. BAGIĆ
University of Pittsburgh Comprehensive Epilepsy Center (UPCEC), University of Pittsburgh Medical School, Pittsburgh, PA, USA

Abstract

Modern epidemiology of epilepsy maximizes the benefits of advanced diagnostic methods and sophisticated techniques for case ascertainment in order to increase the diagnostic accuracy and representativeness of the cases and cohorts studied, resulting in better comparability of similarly performed studies. Overall, these advanced epidemiologic methods are expected to yield a better understanding of diverse risk factors, high-risk populations, seizure triggers, multiple and poorly understood causes of epilepsy, including the increasing and complex role of genetics, and establish the natural course of treated and untreated epilepsy and syndromes – all of which form the foundation of an attempt to prevent epileptogenesis as the primary prophylaxis of epilepsy. Although data collection continues to improve, epidemiologists still need to overcome definition and coding variability, insufficient documentation, as well as the interplay of socioeconomic factors and stigma. As most of the 65–70 million people with epilepsy live outside of resource-rich countries, extensive underdiagnosis, misdiagnosis, and undertreatment are likely. Epidemiology will continue to provide the necessary information to the medical community, public, and regulators as the foundation for improved health policies, targeted education, and advanced measures of prevention and prognostication of the most common severe brain disorder.

DEFINITIONS

Achieving homogeneity regarding nomenclature, taxonomy, and definitions is an essential prerequisite for accurate documentation of epileptic phenomena. Using heterogeneous taxonomy may lead to a mismatch in the identification of key diagnostic features and pathophysiologic entities and thus lead to results that cannot be compared regarding the key epidemiologic parameters, such as incidence, prevalence, and risk factors. It is important therefore to clearly distinguish an epileptic seizure from other seizure-like events and mimics, such as syncope, tics, breath-holding spells, transient ischemic attacks, and mental health-related events. Once the clinical entity has been identified as a true epileptic seizure (i.e., symptom), it is important to determine if that seizure was a provoked seizure, such as neonatal, febrile, or acute symptomatic seizure, whether it was a single unprovoked seizure, or whether it was a recurrent unprovoked seizure, mandating the diagnosis of epilepsy (i.e., disease).

Epileptic seizure

An epileptic seizure is now generally accepted as being defined as "the transient occurrence of signs and/or symptoms due to abnormal excessive or synchronous neuronal activity in the brain" (Fisher et al., 2005). When including seizures when attempting to diagnose epilepsy (i.e., disease), the suspected event must be distinguished from other paroxysmal neurologic entities. Other events that are often mistaken for seizures include

[1]Current address: Sergiu Abramovici, MD, Tel-Aviv Sourasky Medical Center, 6 Weizmann St., Tel-Aviv, 6423906, Israel. Cell-phone number: +972-52-7360535, Email: sergiua@tlvmc.gov.il

*Correspondence to: S. Abramovici, University of Pittsburgh Comprehensive Epilepsy Center (UPCEC), University of Pittsburgh Medical School, Suite 811, Kaufmann Medical Building, 3471 Fifth Ave, Pittsburgh PA 15213, USA. Tel: +1-412-877-2582, E-mail: sergiu.a@gmail.com

transient ischemic attacks, migraine-related transient neurologic deficits or visual phenomena, syncope, acute psychosis-related hallucinations, parasomnias, especially rapid-eye-movement sleep behavioral disorder, hyperkinetic movement disorders, like tremor, chorea, and athetosis, metabolic and toxic derangements that cause movement and awareness-related changes, and nonepileptic behavioral spells (Smith, 2012). The latter is common in conversion disorders and was diagnosed in 25–26% of epilepsy monitoring unit (EMU) admissions (Salinsky et al., 2011). Misidentification of the specific episode may lead to false diagnosis and misdiagnosis of epilepsy, thus skewing the data towards over- or underestimation of incidence and prevalence.

Acute symptomatic seizures

An acute symptomatic seizure, or provoked seizure, is a seizure that was provoked by a direct insult to the brain. This insult may be focal, such as stroke, encephalitis, a tumor, or any focal process exerting direct irritation over the cortex. Other insults may be generalized and exert their effect indirectly by a systemically diffuse toxic, metabolic, inflammatory, or paraneoplastic process. In a review of several studies conducted between 1990 and 2005, on diverse populations from different parts of the world, the incidence of acute symptomatic seizures ranged between 20 and 39 cases per 100 000 persons per year (Loiseau et al., 1990; Hauser et al., 1993; Forsgren et al., 1996; Jallon et al., 1997; MacDonald et al., 2000; Olafsson et al., 2005).

Unprovoked seizure

An unprovoked seizure is a seizure without any identified focal or systemic immediate cause, although it frequently has such an event as an antecedent, serving as a risk factor. A classic example is the emergence of seizures after a significant latent period of several months or even years following a stroke or encephalitis. In an analysis of several studies conducted between 1990 and 2005, on diverse populations from different parts of the world, the incidence of isolated unprovoked seizures ranged between 42.3 and 61 cases per 100 000 persons per year (Loiseau et al., 1990; Hauser et al., 1993; Forsgren et al., 1996; Jallon et al., 1997; MacDonald et al., 2000; Olafsson et al., 2005). In general, these studies revealed that the incidence of acute symptomatic seizures, although variable, tends to be lower than that of epilepsy. Acute symptomatic seizures are predominant in men, in patients younger than 1 year, and patients older than 65 years. The variability in results regarding the incidence of acute symptomatic seizures may be explained by different methodologies in case validation and definitions used (Hauser and Beghi, 2008).

Epilepsy

Epilepsy is defined as "a disorder of the brain characterized by an enduring predisposition to generate epileptic seizures, and by the neurobiologic, cognitive, psychologic, and social consequences of this condition (Fisher et al., 2014)." A diagnosis of epilepsy should be made in any of the following conditions: (1) at least two unprovoked (or reflex) seizures occurring > 24 hours apart; (2) one unprovoked (or reflex) seizure and a probability of further seizures similar to the general recurrence risk after two unprovoked seizures, occurring over the next 10 years (at least 60%); or (3) diagnosis of an epilepsy syndrome (Fisher et al., 2014). The definition of epilepsy, as adopted by the International League Against Epilepsy (ILAE) in 2014, requires the occurrence of at least one epileptic seizure. This detail is extremely important in the current and future study of epilepsy epidemiology, since previous definitions of epilepsy required two seizures at least 24 hours apart. As many cohorts studied after this declaration are expected to include more cases diagnosed with epilepsy, the future incidence and prevalence of epilepsy are expected to rise. Nevertheless, there are some circumstances under which recurrent seizures do not necessarily mandate the diagnosis of epilepsy and are seldom considered as such. A cluster of multiple seizures occurring in a period of 24 hours, a single episode of status, or febrile seizures should not be considered epilepsy indefinitely (Commission on Epidemiology and Prognosis, 1993). While epilepsy can begin in the neonatal period, neonatal seizures are usually considered to be reactive to an acute cerebral insult and usually do not persist beyond a few weeks after onset.

Drug-resistant epilepsy

Drug-resistant epilepsy, also named intractable epilepsy or refractory epilepsy, is defined as failure of adequate trials of two tolerated, appropriately chosen and used antiepileptic drug schedules (whether as monotherapy or in combination) to achieve sustained seizure freedom (Kwan et al., 2009). This definition requires expert epilepsy care, since there are numerous cases of patients using multiple agents in combination but often inadequate dosing, which have as yet, or unrightfully, been declared as drug-resistant epilepsy cases. This, of course, has a direct implication on quality of care and timely referral to presurgical evaluation of adequate cases.

Active epilepsy and epilepsy in remission

Active epilepsy and epilepsy in remission are opposite but elusive and poorly defined terms. These terms, however, become important for understanding that different types of epilepsy behave in different ways and a patient's epilepsy may be active or in remission, depending on an elaborate matrix of factors. It is complicated to define when to consider a patient's epilepsy as in remission, but it is clear that these patients may not be included in some epidemiologic studies.

Epilepsy resolved

Epilepsy resolved is a term adopted recently by the ILAE Task Force to define patients with age-dependent epilepsy syndromes that have passed the applicable age and who no longer have seizures, as well as those who remained seizure-free for over 10 years, of which at least 5 years were without any seizure medication (Fisher et al., 2014).

Therefore, since epilepsy represents a heterogeneous group of diseases with variable duration, typical age of onset (and resolution times in age-limited entities), severity, and manifestations, there is great variability in the registration and inclusion of patients in various epidemiologic studies. Variability and unpredictability of epilepsy dynamics are important factors responsible for the skewing of the incidence and prevalence of epilepsy and must be taken into account when studying epilepsy epidemiology.

EPIDEMIOLOGIC METHODOLOGIES AND CHALLENGES

A recent study estimated that there are roughly 65 million people living with epilepsy worldwide (Ngugi et al., 2010); however, since then, the number has surely increased. In developed countries, the annual incidence of epilepsy was estimated to be around 50 per 100 000 population and the prevalence was estimated around 700 per 100 000 population (Hirtz et al., 2007). In the developing world, with middle and low income, the numbers are estimated to be higher (Hauser, 1995; Ngugi et al., 2010). Proper study and analysis of epilepsy and epidemiologic characteristics pose an enormous challenge, partially due to heterogeneous definitions, different methods of comparison, and different resources. There are two mechanisms used for the epidemiologic study of epilepsy: public health surveillance and epidemiologic research.

Point prevalence

Point prevalence is calculated as the total number of patients diagnosed with active epilepsy on the prevalence day, divided by the total population under study on that prevalence day. It is of note that active prevalence differs from lifetime prevalence, in that the latter includes patients with a past history of epilepsy, epilepsy in remission, or even resolved epilepsy (Banerjee et al., 2009).

Incidence

Incidence of epilepsy is defined as the number of new cases of epilepsy over a specific time period, as part of the investigated susceptible population. This is usually expressed as number of new cases per 100 000 persons per year.

Public health surveillance

Public health surveillance is defined as the ongoing systematic collection, analysis, and interpretation of health data, necessary for the design, implementation, and evaluation of public health prevention programs (German et al., 2001). This ongoing process provides researchers with a dynamic view of incidence, prevalence, morbidity, and mortality related to epilepsy, utilizing sources such as national mortality registries and hospital discharge data. This study permits constant monitoring of healthcare program effectiveness and medical policy success. Public health surveillance involves the systematic identification of epilepsy cases at multiple points of contact (hospitals, clinics, physician offices, electroencephalography (EEG) laboratories, schools, etc.) using a broad definition, or inclusion criteria matrix, with high sensitivity, followed by a more specialized and refined criteria matrix for high specificity, thus assuring broad and correct identification of cases (Banerjee et al., 2009).

Epidemiologic research

Epidemiologic research is typically a more elaborate and detailed study, conducted over a relatively short period of time, usually for deliberate testing of a novel hypothesis regarding a specific healthcare paradigm. These studies describe the incidence rate, prevalence, and mortality rate of epilepsy, as well as defining high-risk groups, associated risk factors, comorbidities, causes, severity, and outcome (Thurman et al., 2011).

A successful epidemiologic study or surveillance program should be economic and avoid collecting unnecessary data. It should be accepted and easily tolerated by subjects. It should be accurate, with sensitivity, specificity, and positive predictive value for measured information as high as possible. It should reliably represent the general population of patients – since some epileptic patients will never be admitted to hospital, basing collected data solely on discharge registries will poorly represent that population (Thurman et al., 2011).

There are several commonly used data sources for seizures and epilepsy-related surveillance and epidemiologic studies:

Direct population surveys

These studies are more advantageous in small populations, where data can be gathered via door-to-door interview of patients and their proxies. In recent years these studies have been replaced in part by telephone- and web-based questionnaires; however, as modern life is ever more demanding, people do not find the time to participate and samples may be underrepresentative quantitatively as well as qualitatively. Another disadvantage of this data source is occasional cognitive difficulty the patients may have. This factor may cause poor recall of important information like age of onset and date of last seizure. There is also limited detection of nonconvulsive seizures, which leads to skewed reports.

Existing coded data

This information may be based on *International Classification of Diseases*, ninth edition (ICD-9), ICD-10, and Systematized Nomenclature of Medicine – Clinical Terms (SNOMED-CT)-coded discharge diagnosis. The data comes from national or regional registries, hospital discharge registries, and coronary registries. Studies based on coded data are prone to error due to occasional lack of training and individual coding interpretation and coder knowledge. There is often limited and poor identification of nonepilepsy behavioral spells, due to lack of readily available EMUs (Salinsky et al., 2011), leading to erroneous diagnosis of epilepsy that may skew registry data as well.

Existing uncoded data

These data may be gathered from additional nonmedical figures such as teachers, religious leaders, and traditional healers in some populations. Pharmacy prescription records regarding antiseizure medication represent another data source. However, since there is growing use of antiseizure medication in the treatment of depression, migraine, and some pain disorders, a growing number of patients without epilepsy are being prescribed and may be erroneously identified. EEG laboratories represent another commonly used source of information.

STIGMA

Special emphasis should be placed on the role of stigma in the skewing of epidemiologic studies and treatment gaps. What we understand nowadays as stigma leads to the regrettable, unnecessary, and preventable suffering of patients with epilepsy due to social factors which we all experience. Over 50 years ago, William G. Lennox (1884–1960) conceptualized his infamous "epilepsy pyramid" (Lennox, 1960). The foundation and main structure of the pyramid are built of "social-emotional conditions" created by society and family that may be healthy and lead to "public understanding and acceptance," along with family support instilling "security, courage and hope." Alternatively, the structure may be broken, leading to "misconceptions and ostracism," with well-known negative consequences: shame, withdrawal, isolation, limited access to resources, and even various forms of discrimination prevalent even in modern society (Link and Phelan, 2006; Hatzenbuehler et al., 2013). Ironically, what modern society focuses most on – "good medical and surgical treatment" – represents only a smaller part of the Lennox pyramid (Lennox, 1960). Stigma as a sociopsychologic construct was originally conceptualized, best defined, and most profoundly studied by the sociologist Erving Goffman (1922–1982), who defined it as "the process by which the reaction of others spoils normal identity" (Goffman, 1986). The concept has been further developed (Link and Phelan, 2001), in particular with a more comprehensive understanding of its public health implications (Link and Phelan, 2006) and recognition as "a fundamental cause of population health inequalities" (Hatzenbuehler et al., 2013). Just two short decades ago, it was eloquently stated that "the history of epilepsy can be summarized as 4000 years of ignorance, superstition, and stigma followed by 100 years of knowledge, superstition, and stigma" (Kale, 1997). Considering the profound effects of stigma on public and overall visibility of affected persons, modern epidemiologic studies have to incorporate these realizations in order to achieve more accurate results.

DESCRIPTIVE EPIDEMIOLOGY OF EPILEPSY

The epidemiologic study of epilepsy mandates the constant gathering of information regarding patient demographics, incidence, prevalence, seizure types, etiology, and risk factors. Cross-checking information from multiple sources may improve study accuracy, by increasing case ascertainment. Epidemiologists increasingly deploy advanced automated data-mining algorithms and database analytics for better case ascertainment and enhanced data accuracy, although case ascertainment and validation still represent major challenges. Nevertheless, epilepsy was estimated to affect 65 million people worldwide, with roughly 80% of these people living in the developing world (Ngugi et al., 2010; Thurman et al., 2011). According to the fact sheet on epilepsy (no. 999) published by the World Health Organization in May 2015, epilepsy accounts for 0.75% of the global burden of disease (http://www.who.int/mediacentre/factsheets/fs999/en/), as measured by combining years of life lost due to premature mortality and years lived in less than full health (Fig. 10.1). In 2012 epilepsy alone was responsible for 20.6 million disability-adjusted life-years (DALYs) lost, and was comparable to DALY of breast cancer in woman and lung cancer in men (Murray et al., 2013).

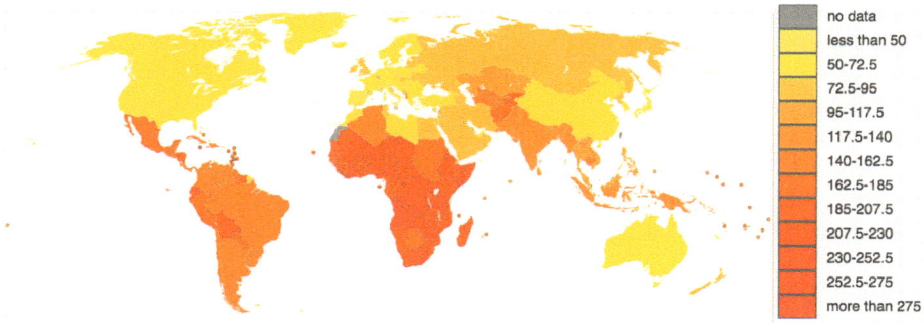

Fig. 10.1. Disability-adjusted life-year rates from epilepsy by country (per 100 000 persons) in 2004. (Reproduced from https://commons.wikimedia.org/wiki/File:Epilepsy_world_map_-_DALY_-_WHO2002.svg).

Incidence of all unprovoked seizures

The incidence of all unprovoked seizures (age-adjusted) ranged from 41 cases per 100 000 persons in New York, USA (Benn et al., 2008) to 69 cases per 100 000 person-years in Minnesota, USA (Hauser et al., 1993). The crude incidence in Stockholm, Sweden was 33.9 per 100 000 person-years (Adelöw et al., 2009). In general, the incidence of unprovoked seizures was reported to be higher in male when compared to female patients, both in a US study as well as in a study conducted in Iceland (Hauser et al., 1993; Olafsson et al., 2005).

Incidence of acute symptomatic seizures

The incidence of acute symptomatic seizures was estimated between 29 and 39 cases per 100 000 person-years, with increased incidence in men as well as in the extremity age groups of the very young and the elderly (Hauser and Beghi, 2008).

Door-to-door and record-based epilepsy prevalence studies

As reviewed by Banerjee et al. (2009), door-to-door and record-based epilepsy prevalence studies reveal dynamic and diverse worldwide age-adjusted epilepsy prevalence, with numbers ranging from 2.2 per 1000 population in India to 32.8 per 1000 population in Mexico. It is important to take into account that the studies were performed years apart and therefore cannot accurately serve as substrate for comparative studies between different regions.

North American studies

North American studies revealed an age-adjusted prevalence of epilepsy estimated to be 5.0 per 1000 persons in a self-reported epilepsy study in New York (Kelvin et al., 2007), 6.8 per 1000 persons in 1980 in Rochester, MN (Hauser et al., 1991, 1996), and 7.1 per 1000 persons in Mississippi (Haerer et al., 1986).

Central and South American studies

In Central and South America prevalence studies reveal results ranging from 3.7 per 1000 persons in Argentina (Melcon et al., 2007), through 11.1 per 1000 persons in rural Bolivia (Nicoletti et al., 1999), to 17.1 per 1000 population in Ecuador (Cruz et al., 1985). The latter study was performed on a population with endemic goiter and therefore may not represent accurately the general population of Ecuador. A rate as high as 38.2 per 1000 in Mexico has been reported (San-Juan et al., 2015). A record-based age-adjusted prevalence of active epilepsy in Chile was 17.6 per 1000 persons (Lavados et al., 1992).

European studies

European studies reveal an age-adjusted prevalence of 5.4 per 1000 persons in a medium-sized city in France (Picot et al., 2008), 2.71–3.3 per 1000 persons in Sicilian municipalities (Reggio et al., 1996; Rocca et al., 2001), and a crude prevalence as high as 14.2 per 1000 persons in Albania (Kruja et al., 2012).

Asian studies

Asian studies documented a prevalence of 2.2 per 1000 persons in India (Koul et al., 1988). The crude prevalence rate of epilepsy was 9.98 per 1000 persons in Pakistan and 7.0 per 1000 persons in Turkey. In both countries the prevalence was roughly twice as high in rural compared to urban areas (Aziz et al., 1997). Lifetime prevalence of 7.0 per 1000 persons and active prevalence of 4.6 per 1000 persons has been reported in China (Wang et al., 2003).

African studies

African studies documented the prevalence of epilepsy in Nigeria ranging from 41.0 per 1000 persons in a pilot study (Osuntokun, 1982) to 5.3 per 1000 in 2011 (Osuntokun et al., 2011) in the most recent and updated study. Other studies conducted across the continent revealed age-adjusted prevalence of 3.6 per 1000 in Tunisia, 12.5–14.3 per 1000 in Zambia (Birbeck and Kalichi, 2004), an age-adjusted prevalence ratio of 7.4 per 1000 person in south Tanzania (Dent et al., 2005) vs. 8.7 per 1000 persons in north Tanzania (Winkler et al., 2009). In Benin, the age-adjusted prevalence was 7.33 per 1000 persons (Yemadje et al., 2012).

Incidence of epilepsy

Incidence of epilepsy is usually reported as the incidence of epilepsy cases according to the old definition requiring recurrent (more than one) unprovoked seizures. In North America, age-adjusted incidence of epilepsy was estimated to be 44 cases per 100 000 person-years (Hauser et al., 1993). In South America, a study conducted in rural Chile reported 111 cases per 100 000 person-years (Lavados et al., 1992), representing the highest epilepsy incidence reported. In Europe, age-adjusted incidence of epilepsy in the London area based on general practice records combined with hospital records revealed an incidence of 46 per 100 000 person-years (MacDonald et al., 2000). In Tanzania, Africa, Rwiza et al. (1992) reported a crude incidence of 73 cases per 100 000 person-years and an age-adjusted incidence of 51 per 100 000 person-years. In Ethiopia, Africa, Tekle-Haimanot et al. (1997) reported an age-adjusted epilepsy incidence of 43 per 100 000 and a crude incidence of 64 cases per 100 000 person-years. In India, a study conducted in 1998 reported an epilepsy incidence of 49.3 cases per 100 000 person-years (Mani et al., 1998).

Sudden unexpected death in epilepsy patients (SUDEP)

SUDEP is defined as the unexpected death of a patient with epilepsy, who otherwise seemed healthy, where no cause of death can be found. SUDEP represents one of epilepsy's most devastating outcomes. It represents a growing focus of interest and awareness when it comes to the prevention of epilepsy complications, as it represents a measure for tertiary prevention. This clinical entity is a good example of the importance of epidemiologic studies in identifying and preventing epilepsy complications. The incidence of SUDEP was first estimated to be 0.09–9 per 1000 patient-years, depending on the population studied (Devinsky, 2011).

A more recent study, based on epidemiologic studies conducted in the USA, estimates the incidence of SUDEP to be around 1.16 cases per 1000 patient-years. Identification of SUDEP represents a diagnostic challenge and is occasionally overlooked. It was found that most of the reported cases occurred in the third and fourth decades of life, but this interpretation may be skewed by underdiagnosis of SUDEP in infants, attributing their death to sudden infant death syndrome, and in the elderly population, attributing their death to other age-dependent comorbidities like cardiovascular disease. Nonetheless, SUDEP was found to have a significant mortality burden as calculated in years of potential life loss, second only to stroke when compared to other neurologic diseases (Thurman et al., 2014). It is important not only to monitor SUDEP but also to raise awareness of this devastating epilepsy complication from an ethical and patient education point of view, since patients and their caregivers do want to know about this, and prefer to learn about it from their treating physician rather than incidentally while browsing the internet (Ramachandrannair et al., 2013; Kroner et al., 2014). The increasing awareness and active epidemiologic study of SUDEP yielded important insights regarding its risk factors, among which the most important and modifiable one is represented by uncontrolled generalized tonic-clonic seizures (Hesdorffer et al., 2011). Having these in mind, efforts are being made to explore possible preventive interventions and formulation of preventive recommendations, such as continuous respiratory monitoring in EMUs as well as outside of them, since a possible implicated mechanism may be airway- and breathing-related (Tao et al., 2010), overnight supervision via direct or indirect methods, or the selective use of selective serotonin reuptake inhibitors (Massey et al., 2014).

ETIOLOGIC CONSIDERATIONS

Although in most cases (60–75%), the etiology of epilepsy remains unknown, and according to modern classifications may probably be attributed to complex-inheritance, single-gene mutation on susceptibility alleles (Thomas and Berkovic, 2014), the epidemiologic study and surveillance of known etiologies is particularly important, since some of these etiologies may be preventable and thus offer a unique opportunity for intervention, providing primary prevention of epilepsy. In a review of studies from the USA, Europe, and Africa, the etiology of epilepsy, whether with a progressive course (such as encephalopathy), a remote cause (such as stroke or other self-limited process) or undetermined, was only identified in 14–39% of cases (Banerjee et al., 2009). Traumatic brain injury (TBI), cerebrovascular disease, and brain infection represent the majority of epileptogenic brain insults (Herman, 2002). Other causes include neurodegenerative disease, central nervous system demyelinating disease, primary brain neoplasms, or metastasis (Sperling and Ko, 2006). Other less frequent etiologies are prenatal and perinatal injuries, and preterm birth (Crump et al., 2011). Additional comorbidities have been extensively reported in relation to epilepsy diagnosis and include autistic spectrum disorder, developmental delay, intellectual or learning disability, psychiatric comorbidities, especially the mood disorders anxiety and depression, and other less specified persistent or evolving neurologic or cognitive deficits that may coexist with epilepsy from birth, or appear in early childhood (like Lennox–Gastaut syndrome and other static encephalopathy conditions), or even later in life. These comorbid conditions and epilepsy may appear simultaneously or in tandem, with up to years of latency between the onset of epilepsy and onset of associated neurocognitive deficit (Rudzinski and Meador, 2013). The combined pathogenesis may also have a multiphasic course, as opposed to the afore-mentioned group of disorders that include stroke and TBI, where there is a clear causal relation between the cause and epilepsy. In this last group of comorbidities, these do not represent the cause, but probably coexist and are caused by a common precedent pathologic process or brain insult, whether remote static or progressive, that may even be timing-dependent to a crucial ontogenetic period or phase (Rakhade and Jensen, 2009). As the common aggressor that causes epilepsy and these neurocognitive deficits may represent a complex matrix of contributing forces that span indeterminable time periods, it is as yet poorly understood in the majority of cases (Pitkänen et al., 2007). Intensive research will probably eventually explain some of these mechanisms, providing treatment and prevention opportunities in the future. For the purpose of this chapter, we will focus on primary and secondary prevention of epilepsy by preventing the most prevalent known causes, namely cerebrovascular disease, TBI, and infections affecting the brain parenchyma or immediate overlying meningeal structures.

Cerebrovascular disease

Cerebrovascular disease represents a good example of a partially preventable etiology of epilepsy (Kotila and Waltimo, 2010). By reducing the risk factors for stroke (e.g., quitting smoking, controlling blood pressure, lipid profile and diabetes, as well as avoiding a sedentary lifestyle), the risk for possible future secondary epilepsy can be reduced. In a retrospective follow-up study of 200 patients who suffered from stroke, epilepsy developed in 17%. Seizures appeared in 14% of patients after ischemic stroke, and in 15% of patients after intracerebral hemorrhage. A significantly larger proportion (35%) developed seizures after subarachnoid hemorrhage. Most of these patients suffered from an arterial spasm that caused an ischemic stroke. Among those with seizures, 15% had their first seizure within 2 weeks and 55% developed epilepsy within 6 months. Of note, 17 of the 28 patients who were started on anticonvulsive therapy still had seizures (Kotila and Waltimo, 2010).

Traumatic brain injury

TBI represents another major and preventable cause of epilepsy. It was found that the relative risk for seizures after TBI ranged between 1.5, after a mild injury, and 17.2 after a severe injury. The risk factors implicated in increased risk for seizures after TBI were loss of consciousness, amnesia, skull fracture, brain contusion, subdural hematoma, and age above 65 (Annegers and Coan, 2000). In a prospective longitudinal follow-up on a large cohort of Vietnam War veterans with TBI, posttraumatic epilepsy was found in 45–53% of patients, roughly 35 years after the injury (Raymont et al., 2010). Military veterans represent a special population at high risk for a multitude of neurologic and psychiatric conditions, among which posttraumatic epilepsy deserves additional consideration, as with recent advances in protective technologies, an increasing number of veterans survive their head trauma (Goldberg, 2010) and subsequently more live to be at risk of developing epilepsy. Importantly, veterans who suffer from posttraumatic stress disorder also have a high risk of suffering from nonepilepsy behavioral spells. While the prevalence of these spells among EMU admissions was similar in veterans (25%) and civilians (26%), the delay from spells onset to diagnosis was significantly longer in veterans (60.5 months) when compared to civilians (12.5 months) and the cumulative antiseizure medication treatment was four times higher in veterans than in civilians (Salinsky et al., 2011). Thus, veterans represent a vulnerable population that is at increased risk of being unnecessarily treated with antiseizure medication for prolonged periods, exposing them to unnecessary toxic effects and waste of limited resources. Active efforts at reducing TBI are being made worldwide, with mass-media campaigns to encourage seatbelt and helmet use, with some documented success (Coronado et al., 2011). Although intuitively these measures should reduce the prevalence of posttraumatic epilepsy, this relation has yet to be appropriately proven in high-evidence-level studies.

Brain infections

Brain infections represent the third major category of modifiable risk factors for epilepsy. Neurocysticercosis remains an important cause of infection-related epilepsy worldwide and was found to be the most common cause of epilepsy in Zambia in a study from 2004 (Birbeck and Kalichi, 2004). This condition occurs when humans become an intermediate host for *Taenia solium*, a geohelminth, by ingesting contaminated food, especially undercooked contaminated meat. While this condition was considered endemic in the developing world due to poor sanitation conditions and education, it is increasingly seen in the developed world due to population shifting and immigration (Del Brutto, 2012). Neurocysticercosis can cause seizures in the infection phase, or remotely due to scar tissue formation in the brain. Additionally, bacterial meningitis, especially pneumococcal meningitis, represents another important risk factor for remote postinfectious epilepsy. It was found that pneumococcal vaccination in the USA decreased pneumococcal meningitis-related hospitalization by 66.9% in children up to 2 years of age, and by 51.5% in children 2–4 years of age (Tsai et al., 2008). A multitude of other bacterial, viral, fungal, and parasitic pathogens are also implicated in the pathogenesis of immediate-progressive or delayed-onset epilepsy, and represent ongoing prevention challenges using diverse management strategies (Pitkänen and Lukasiuk, 2011). Therefore, brain infection represents an important and, to some degree, relatively easily preventable epilepsy etiology, whether by proper vaccination, hygiene, avoiding raw meat or food in contact with raw meat, and avoiding other vectors like mosquito bite prevention.

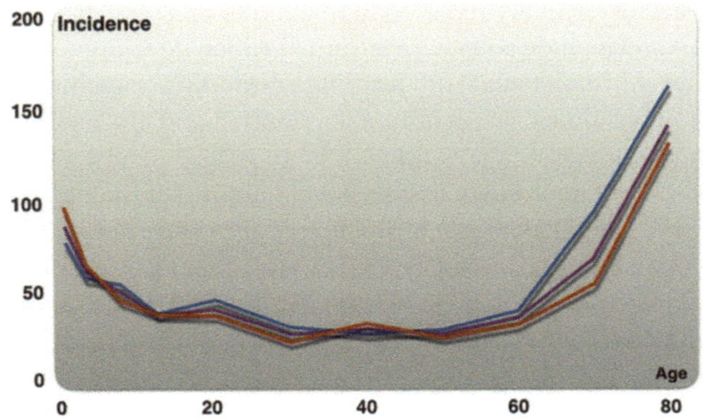

Fig. 10.2. Age- and gender-specific incidence per 100 000 of epilepsy in Rochester, MN, 1935–1984. Purple, total; blue, male; red, female. (Adapted with permission from Hauser et al., 1993.)

DEMOGRAPHIC AND GEOSOCIAL CONSIDERATIONS

The population of people with epilepsy is far from being homogeneous. Significant differences continue to perpetuate between the two genders, different age groups, with increased vulnerability among children and the elderly, different social or occupational classes, such as military veterans, which are significantly more exposed to TBI, and differences arising from social, economic, and geographic considerations, that manifest through heterogeneous access to education, sanitation, and treatment opportunities. These differences manifest as variability in access to treatment, personal, occupational or economic-driven exposure to various etiologic risk factors, level of education, and intricate cultural and other geosocial factors.

Gender differences

The incidence of acute symptomatic seizures is higher in men (Hauser and Beghi, 2008). This may be related to increased preponderance of risk-taking behavior, increased incidence of TBI, with and without association to alcohol or illicit drug consumption (Savic, 2014). Additionally, the cumulative incidence of unprovoked seizures in Iceland and the USA shows significant male preponderance (Fig. 10.2; Hauser et al., 1993; Olafsson et al., 2005). Risk-taking behavior may also increase the prevalence of late-onset epilepsy related to remote causes. The prevalence of epilepsy was measured to be slightly lower in women compared to men, and this was partially explained by the higher prevalence of lesion-associated epilepsy in men (Christensen et al., 2005). This result is congruent with the results of an animal model, in which male rats were more vulnerable to develop epilepsy in a combined lesion and hyperthermia or lesion and early life stress (Desgent et al., 2012).

Although, in general, epilepsy prevalence was measured to be somewhat lower in women, a slightly higher prevalence of some types of idiopathic generalized epilepsy (although some of these syndromes are now classified as generalized genetic epilepsy; Berg et al., 2010) was found in women, as typical childhood absence epilepsy was found in a ratio of 19:11 favoring female gender (Asadi-Pooya et al., 2012). A recent review of sex differences in human epilepsy concluded that, although the differences are not prominent, they do exist in some forms of epilepsy. In childhood absence epilepsy and juvenile myoclonic epilepsy, there seems to be a higher prevalence in females; however, the mechanism is still unclear and speculations revolve around differences in gender-dependent seizure network and propagation network formation. Nevertheless, the information regarding gender differences in epileptogenesis is still limited and warrants additional research for better understanding of the pathogenesis (Savic, 2014).

Age considerations

Most studies agree that epilepsy has an age incidence distribution with two peaks. The first peak occurs in the early years of life and rapidly declines, as the incidence of perinatal complications, infections, and birth defect-related epilepsy etiology influences decline. In the sixth decade of life, epilepsy incidence starts to rise again independently of other demographic factors, probably due to increased incidence of pathologies causing secondary, structural-related epilepsy, led by cerebrovascular diseases (Stephen and Brodie, 2000).

CHILDREN

Children may represent an epilepsy diagnostic challenge, age-related exclusions, like neonatal seizures, occurring in the first 28 days, as well as acute symptomatic seizures and febrile seizures must be taken into account (Commission on Epidemiology and Prognosis, 1993). Nevertheless, several types of epilepsy can start in the neonatal period and should count as epilepsy if seizures persist after that period, like KCNQ2 related epilepsy (Singh et al., 2003). Febrile seizures are also not to be confused with epilepsy, as they are considered benign unless they are atypical: focal features, prolonged duration (>15 minutes), or with associated postictal neurologic deficits – all warranting additional investigations. There are specific age-related types of epilepsy, including infantile spasms and West syndrome, Lennox–Gastaut syndrome (evolving from or overlapping with infantile spasms), and Landau–Kleffner syndrome. There are types of epilepsy secondary to inborn errors of metabolism, resulting in severe myoclonic epilepsy and other neurologic and systemic deficits, such as sialidosis, Lafora disease, Unverricht–Lundborg, and ceroid lipofuscinosis. Other myoclonic types of epilepsy may be caused by mitochondrial mutations like myoclonic epilepsy with ragged red fibers (MERRF). These forms of epilepsy also become symptomatic in early stages of life and are usually catastrophic (Kullmann, 2002b). Additionally, epilepsies due to genetic abnormalities tend to manifest in early childhood. Examples of types of epilepsy with gene mutations that cause ion channel dysfunctions include Dravet syndrome, a severe myoclonic epilepsy of childhood that is caused by mutations in the SCN1A gene and KCNQ mutation-related epilepsy (Kullmann, 2002a). Malformations of cortical development may be X-linked, or may be caused by mutations in LIS1 (in subcortical band heterotopia), XLIS, Filamin A, mTOR (in hemimegalencephaly), TSC1, TSC2 (tuberous sclerosis), and other mutations (Kullmann, 2002b). Other examples of age-related syndromes with typical onset in childhood are childhood absence epilepsy, and later in life, juvenile absence epilepsy, juvenile myoclonic epilepsy, benign rolandic epilepsy (with centrotemporal spikes) – 13–24% of all childhood epilepsy (Cavazzuti, 1980), and benign occipital epilepsy – 6–13% of childhood epilepsy (Panayiotopoulos, 2000).

ELDERLY POPULATION

Epilepsy in the elderly population represents challenges of its own. The prevalence of epilepsy in this age group (over 60 years old) is expected to grow with the aging of the population and the increased rate of survival after brain insults. Although the estimated annual incidence of epilepsy in people aged over 60 is approximated to be around 134 per 100 000, this may be an underestimation, as epilepsy in the elderly is occasionally misdiagnosed (de la Court et al., 1996). One of the possible reasons may be because elderly patients do not always show neurologic signs. Most new seizures in elderly patients are with focal onset, with or without secondary generalization, probably because they are secondary to a focal brain insult, such as cerebrovascular disease – representing the single most common pathologic factor, as well as TBI, and encephalitis.

The annual incidence of status epilepticus in elderly patients reached 86 per 100 000 – roughly twice that of the general population. Roughly 60% of status epilepticus in this age group was remote, symptomatic to a stroke. The mortality rate was directly proportional to patient age, and etiology-dependent, approaching 50% in patients over 80 years of age, and approaching 100% in patients with anoxia (Towne, 2007). Seizures are also more likely to recur in association with neurodegenerative diseases, tumors, and due to medication, toxic and metabolic causes. Other causes for seizures in the elderly are hypertensive encephalopathy and anoxic brain injury (Stephen and Brodie, 2000). Patients with nonketotic hyperglycemia may develop focal motor seizures (epilepsia partialis continua), especially if there is an underlying focal structural abnormality (Singh and Strobos, 1980).

A recently published study that used machine learning on a prospective cohort analysis, stratified by age, aimed to predict age of onset according to predefined risk factors and to analyze their significance. The study identified 14 risk factors with absolute predictive values that were maximally predictive in patients who were 65–70 years old. These factors were: current chronic medical conditions, stroke, events precipitated by stress, family history of epilepsy, generalized tonic-clonic seizure, abnormal magnetic resonance imaging (MRI), developmental delay, prior psychiatric referral, learning disorder, regular use of recreational drugs, nonepileptic events, birth trauma, and precipitation of seizures by flashing lights. In contrast, for example, vascular malformations had a maximal predictive value at 35 years of age (Josephson et al., 2016). This study demonstrates the use of machine learning tools in creating models that may enable clinicians to predict epilepsy onset in older age, according to present risk factors in a younger patient, which, through targeted counseling, may reduce the future incidence of epilepsy in the elderly population.

Racial, ethnic, and socioeconomic considerations

There are few studies that provide insight into racial and ethnic difference in epilepsy and unprovoked seizure incidence. These studies were unable to show a statistically significant difference between non-Hispanic white, African-Americans, Hispanics and Asians in Huston, Texas (Annegers et al., 1999). There was also no significant difference in the incidence of a first unprovoked seizure between Hispanic, white, and African Americans (Benn et al., 2008). However, several studies demonstrated a significant correlation between low socioeconomic status and increased incidence of epilepsy, reaching almost double incidence when compared to high-socioeconomic-status population (Heaney et al., 2002). Presumably, people with high socioeconomic status are less likely to be exposed to certain residential habitat, occupational, and lifestyle-related risk factors that predispose to acquiring epilepsy (Banerjee et al., 2009), and their epilepsy is probably less severe due to better accessibility to the medical system, including better access to diagnostic and treatment possibilities.

Developing world and treatment gap

The treatment gap can be measured directly in a prevalence study and reflects how many of detected cases do not receive treatment, or indirectly by calculating the proportion of patients who can be treated with monotherapy during a year, using prevalence data and drug supply data. Most treatment gap studies show a significant gap in the developing world, with up to 98% not receiving treatment in rural Pakistan (Aziz et al., 1994) and Ethiopia (Tekle-Haimanot et al., 1990), and 74.5% in India (Shorvon and Farmer, 1988), compared with much lower rates in the developed world, with only 10% in Great Britain (Goodridge and Shorvon, 1983), 7% in the USA (Haerer et al., 1986), and almost 0% in Sweden (Sidenvall et al., 1996). Most of the difference in treatment gap is mostly attributed to poor education, economic possibilities, social factors (stigma), and deficiencies in health services (missed diagnosis, unavailable drugs, poor organization) (Kale, 2002). The treatment gap in the developing world, as studied in China in 2003, recorded that 41% of patients have never received appropriate treatment and 63% of patients with active epilepsy did not receive antiepileptic treatment in the week prior to the survey (Wang et al., 2003). According to the fact sheet on epilepsy (No. 999) published by the World Health Organization in May 2015 (http://www.who.int/mediacentre/factsheets/fs999/en/), nearly 80% of people with epilepsy live in low- and middle-income countries, and about 75% of them do not get the treatment they need. Efforts are being made, both on a local and global scale, to increase funds, education, and availability of treatment, in order to minimize this treatment gap. However, in regions where potable water represents a huge challenge, epilepsy treatment may be somewhat less prioritized.

FUTURE PERSPECTIVES

The continuing epidemiologic studies and surveillance of epilepsy, along with all the implied studies of this vast spectrum of clinical phenomena, its dynamic pathogenic mesh, and comorbidities, will continue to serve clinicians, caregivers, patients, and their communities worldwide. Epidemiologic study and surveillance allow better understanding of the mechanisms involved in epileptogenesis, creating new intervention opportunities and permitting the creation of new healthcare policies and individualized recommendations that will allow for better primary, secondary, and tertiary epilepsy prevention modalities. These studies will improve the ability to predict short-term outcome and long-term prognosis and foresee possible complications. The epidemiologic study of epilepsy still faces the challenge of unification and standardization of definitions among different medical organizations worldwide, in the vision of a unified definition and classification system that will permit more efficient studies and unification of cohorts, larger samples, and better data quality. With advances in genetic research and more accurate and increasingly available full exome sequencing, and as the number of known mutations involved in epilepsy pathogenesis increases exponentially and their mechanisms of action are better understood, new possible therapeutic opportunities will arise. Genetic counseling will become more accurate, readily available, and personalized and may facilitate the reduction of fatal or severely debilitating encephalopathic-epileptic syndromes and diseases. In recent years we have witnessed the exponential growth of worldwide availability and use of the internet, smartphones, and the recent introduction of even more personal wearable computing devices like smart watches, and wristbands with biometric features. The increasing availability of commercial EEG devices with variable accuracy, as well as the emerging field named "the internet of things" (Tan and Wang, 2010), that basically represents the capability of household devices, like the fridge, television, shower, and bed to connect to the internet and transmit data, and the increasing connectivity between all these devices, mark an era in which much more versatile, abundant, and diverse information will be collected from patients' biometric devices while awake, asleep, running outside, or having their lunch. Emerging "big-data" management (O'Leary, 2013) and analytic technologies (such as Watson by IBM), (Wagle, 2013), deployment of machine learning algorithms (Josephson et al., 2016), and using large-scale computing platforms will help depersonalize and translate this colossal volume of data into predictive conclusions that will allow better understanding of epileptogenesis at individual and social levels. This will also facilitate the creation of more personalized medicine and better planning of public health policies. Although new patient privacy challenges will surely emerge, this ubiquitous and diverse data collection and analysis will expand our perspective exponentially and most likely lead to better, new-generation surveillance of epilepsy and a better quality of epidemiologic studies. Ultimately, it is hoped, prerequisites for more widely available and much improved care will emerge.

References

Adelöw C, Åndell E, Åmark P et al. (2009). Newly diagnosed single unprovoked seizures and epilepsy in Stockholm, Sweden: First report from the Stockholm Incidence Registry of Epilepsy (SIRE). Epilepsia 50: 1094–1101.

Annegers JF, Coan SP (2000). The risks of epilepsy after traumatic brain injury. Seizure 9: 453–457.

Annegers JF, Dubinsky S, Coan SP et al. (1999). The incidence of epilepsy and unprovoked seizures in multiethnic, urban health maintenance organizations. Epilepsia 40: 502–506.

Asadi-Pooya AA, Emami M, Nikseresht A (2012). Early-onset versus typical childhood absence epilepsy; clinical and electrographic characteristics. Seizure 21: 273–275.

Aziz H, Ali SM, Frances P et al. (1994). Epilepsy in Pakistan: a population-based epidemiologic study. Epilepsia 35: 950–958.

Aziz H, Güvener A, Akhtar SW et al. (1997). Comparative epidemiology of epilepsy in Pakistan and Turkey: population-based studies using identical protocols. Epilepsia 38: 716–722.

Banerjee PN, Filippi D, Hauser WA (2009). The descriptive epidemiology of epilepsy – a review. Epilepsy Res 85 (1): 31–45.

Benn EKT, Hauser WA, Shih T et al. (2008). Estimating the incidence of first unprovoked seizure and newly diagnosed epilepsy in the low-income urban community of Northern Manhattan, New York City. Epilepsia 49: 1431–1439.

Berg AT, Berkovic SF, Brodie MJ et al. (2010). Revised terminology and concepts for organization of seizures and epilepsies: Report of the ILAE Commission on Classification and Terminology, 2005–2009. Epilepsia 51: 676–685.

Birbeck GL, Kalichi EMN (2004). Epilepsy prevalence in rural Zambia: a door-to-door survey. Trop Med Int Heal 9: 92–95.

Cavazzuti GB (1980). Epidemiology of different types of epilepsy in school age children of Modena, Italy. Epilepsia 21: 57–62.

Christensen J, Kjeldsen MJ, Andersen H et al. (2005). Gender differences in epilepsy. Epilepsia 46: 956–960.

Commission on Epidemiology and Prognosis ILAE (1993). Guidelines for Epidemiologic Studies on Epilepsy. Epilepsia 34: 592–596.

Coronado VG, Xu L, Basavaraju SV et al. (2011). Surveillance for traumatic brain injury-related deaths – United States, 1997–2007. MMWR Surveill Summ 60: 1–32.

Crump C, Sundquist K, Winkleby MA et al. (2011). Preterm birth and risk of epilepsy in Swedish adults. Neurology 77: 1376–1382.

Cruz ME, Schoenberg BS, Ruales J et al. (1985). Pilot study to detect neurologic disease in Ecuador among a population with a high prevalence of endemic goiter. Neuroepidemiology 4: 108–116.

de la Court A, Breteler MM, Meinardi H et al. (1996). Prevalence of epilepsy in the elderly: the Rotterdam Study. Epilepsia 37: 141–147.

Del Brutto OH (2012). Neurocysticercosis. Continuum (Minneap Minn) 18: 1392–1416.

Dent W, Helbok R, Matuja WBP et al. (2005). Prevalence of active epilepsy in a rural area in south Tanzania: a door-to-door survey. Epilepsia 46: 1963–1969.

Desgent S, Duss S, Sanon NT et al. (2012). Early-life stress is associated with gender-based vulnerability to epileptogenesis in rat pups. PLoS One 7: e42622.

Devinsky O (2011). Sudden, unexpected death in epilepsy. N Engl J Med 365: 1801–1811.

Fisher RS, Van Emde Boas W, Blume W et al. (2005). Epileptic seizures and epilepsy: Definitions proposed by the International League Against Epilepsy (ILAE) and the International Bureau for Epilepsy (IBE). Epilepsia 46: 470–472.

Fisher RS, Acevedo C, Arzimanoglou A et al. (2014). ILAE Official Report: a practical clinical definition of epilepsy. Epilepsia 55: 475–482.

Forsgren L, Bucht G, Eriksson S et al. (1996). Incidence and clinical characterization of unprovoked seizures in adults: a prospective population-based study. Epilepsia 37: 224–229.

German RR, Lee LM, Horan JM et al. (2001). Updated guidelines for evaluating public health surveillance systems: recommendations from the Guidelines Working Group. MMWR Recomm Rep 50: 1–35. quiz CE1–E7.

Goffman G (June 15, 1986). Stigma: Notes on the Management of Spoiled Identity, Reissue edn. Touchstone, New York.

Goldberg MS (2010). Death and injury rates of U.S. military personnel in Iraq. Mil Med 175: 220–226.

Goodridge DM, Shorvon SD (1983). Epileptic seizures in a population of 6000. II: Treatment and prognosis. Br Med J (Clin Res Ed) 287: 645–647.

Haerer AF, Anderson DW, Schoenberg BS (1986). Prevalence and clinical features of epilepsy in a biracial United States population. Epilepsia 27: 66–75.

Hatzenbuehler ML, Phelan JC, Link BG (2013). Stigma as a fundamental cause of population health inequalities. Am J Public Health 103: 813–821.

Hauser WA (1995). Recent developments in the epidemiology of epilepsy. Acta Neurol Scand Suppl 162: 17–21.

Hauser WA, Beghi E (2008). First seizure definitions and worldwide incidence and mortality. Epilepsia 49: 8–12.

Hauser WA, Annegers JF, Kurland LT (1991). Prevalence of epilepsy in Rochester, Minnesota: 1940–1980. Epilepsia 32: 429–445.

Hauser WA, Annegers JF, Kurland LT (1993). Incidence of epilepsy and unprovoked seizures in Rochester, Minnesota: 1935–1984. Epilepsia 34: 453–468.

Hauser WA, Annegers JF, Rocca WA (1996). Descriptive epidemiology of epilepsy: contributions of population-based

studies from Rochester. Minnesota Mayo Clin Proc 71: 576–586.
Heaney DC, MacDonald BK, Everitt A et al. (2002). Socioeconomic variation in incidence of epilepsy: prospective community based study in south east England. BMJ 325: 1013–1016.
Herman ST (2002). Epilepsy after brain insult: targeting epileptogenesis. Neurology 59: S21–S26.
Hesdorffer DC, Tomson T, Benn E et al. (2011). Combined analysis of risk factors for SUDEP. Epilepsia 52: 1150–1159.
Hirtz D, Thurman DJ, Gwinn-Hardy K et al. (2007). How common are the "common" neurologic disorders? Neurology 68: 326–337.
Jallon P, Goumaz M, Haenggeli C et al. (1997). Incidence of first epileptic seizures in the canton of Geneva, Switzerland. Epilepsia 38: 547–552.
Josephson CB, Engbers JDT, Sajobi TT et al. (2016). Towards a clinically informed, data-driven definition of elderly onset epilepsy. Epilepsia 57: 298–305.
Kale R (1997). Bringing epilepsy out of the shadows. BMJ 315: 2–3.
Kale R (2002). The treatment gap. Epilepsia 43: 31–33.
Kelvin EA, Hesdorffer DC, Bagiella E et al. (2007). Prevalence of self-reported epilepsy in a multiracial and multiethnic community in New York City. Epilepsy Res 77: 141–150.
Kotila M, Waltimo O (2010). Epilepsy after stroke. Epilepsia 33: 495–498.
Koul R, Razdan S, Motta A (1988). Prevalence and pattern of epilepsy (Lath/Mirgi/Laran) in Rural Kashmir, India. Epilepsia 29: 116–122.
Kroner BL, Wright C, Friedman D et al. (2014). Characteristics of epilepsy patients and caregivers who either have or have not heard of SUDEP. Epilepsia 55: 1486–1494.
Kruja J, Beghi E, Zerbi D et al. (2012). High prevalence of major neurological disorders in two Albanian communities: results of a door-to-door survey. Neuroepidemiology 38: 138–147.
Kullmann DM (2002a). The neuronal channelopathies. Brain 125: 1177–1195.
Kullmann DM (2002b). Genetics of epilepsy. J Neurol Neurosurg Psychiatry 73 (Suppl II): ii32–ii35.
Kwan P, Arzimanoglou A, Berg AT et al. (2009). Definition of drug resistant epilepsy: consensus proposal by the ad hoc Task Force of the ILAE Commission on Therapeutic Strategies. Epilepsia 51: 1069–1077.
Lavados J, Germain L, Morales A et al. (1992). A descriptive study of epilepsy in the district of El Salvador, Chile, 1984–1988. Acta Neurol Scand 85: 249–256.
Lennox WG (1960). Epilepsy and related disorders, Little Brown, Boston.
Link BG, Phelan JC (2001). On stigma and its public health implications. Paper presented at US NIH conference on stigma and global health: developing a research agenda. Bethesda, MD. http://www.stigmaconference.nih.gov (accessed March 4, 2011).
Link BG, Phelan JC (2006). Stigma and its public health implications. Lancet 367: 528–529.
Loiseau J, Loiseau P, Guyot M et al. (1990). Survey of seizure disorders in the French Southwest. I. Incidence of epileptic syndromes. Epilepsia 31: 391–396.
MacDonald BK, Cockerell OC, Sander JW et al. (2000). The incidence and lifetime prevalence of neurological disorders in a prospective community-based study in the UK. Brain 123: 665–676.
Mani KS, Rangan G, Srinivas HV et al. (1998). The Yelandur study: a community-based approach to epilepsy in rural south India—epidemiological aspects. Seizure 7: 281–288.
Massey CA, Sowers LP, Dlouhy BJ et al. (2014). Mechanisms of sudden unexpected death in epilepsy: the pathway to prevention. Nat Rev Neurol 10: 271–282.
Melcon MO, Kochen S, Vergara RH (2007). Prevalence and clinical features of epilepsy in Argentina: a community-based study. Neuroepidemiology 28: 8–15.
Murray C, Vos T, Lozano R et al. (2013). Disability-adjusted life years (DALYs) for 291 diseases and injuries in 21 regions, 1990–2010: a systematic analysis. Lancet 380 (9859): 2197–2223.
Ngugi AK, Bottomley C, Kleinschmidt I et al. (2010). Estimation of the burden of active and life-time epilepsy: a meta-analytic approach. Epilepsia 51: 883–890.
Nicoletti A, Reggio A, Bartoloni A et al. (1999). Prevalence of epilepsy in rural Bolivia: a door-to-door survey. Neurology 53: 2064–2069.
O'Leary DE (2013). "Big data", the "internet of things" and the "internet of signs". Intell Syst Account Finance Manag 20: 53–65.
Olafsson E, Ludvigsson P, Gudmundsson G et al. (2005). Incidence of unprovoked seizures and epilepsy in Iceland and assessment of the epilepsy syndrome classification: a prospective study. Lancet Neurol 4: 627–634.
Osuntokun BO (1982). Research protocol for measuring the prevalence of neurologic disorders in developing countries: results of a pilot study in Nigeria. Neuroepdemiology 1: 143–153.
Osuntokun BO, Adeuja AO, Nottidge VA et al. (2011). Prevalence of the epilepsies in Nigerian Africans: a community-based study. Epilepsia 28: 272–279.
Panayiotopoulos CP (2000). Benign childhood epileptic syndromes with occipital spikes: new classification proposed by the International League Against Epilepsy. J Child Neurol 15: 548–552.
Picot MC, Baldy-Moulinier M, Daurès JP et al. (2008). The prevalence of epilepsy and pharmacoresistant epilepsy in adults: a population-based study in a Western European country. Epilepsia 49: 1230–1238.
Pitkänen A, Lukasiuk K (2011). Mechanisms of epileptogenesis and potential treatment targets. Lancet Neurol 10: 173–186.
Pitkänen A, Kharatishvili I, Karhunen H et al. (2007). Epileptogenesis in experimental models. Epilepsia 48 (Suppl 2): 13–20.
Rakhade SN, Jensen FE (2009). Epileptogenesis in the immature brain: emerging mechanisms. Nat Rev Neurol 5: 380–391.

Ramachandrannair R, Jack SM, Meaney BF et al. (2013). SUDEP: what do parents want to know? Epilepsy Behav 29: 560–564.

Raymont V, Salazar AM, Lipsky R et al. (2010). Correlates of posttraumatic epilepsy 35 years following combat brain injury. Neurology 75: 224–229.

Reggio A, Failla G, Patti F et al. (1996). Prevalence of epilepsy. A door-to-door survey in the Sicilian community of Riposto. Ital J Neurol Sci 17: 147–151.

Rocca WA, Savettieri G, Anderson DW et al. (2001). Door-to-door prevalence survey of epilepsy in three Sicilian municipalities. Neuroepidemiology 20: 237–241.

Rudzinski LA, Meador KJ (2013). Epilepsy and neuropsychological comorbidities. Contin Lifelong Learn Neurol 19: 682–696.

Rwiza HT, Kilonzo GP, Haule J et al. (1992). Prevalence and incidence of epilepsy in Ulanga, a rural Tanzanian district: a community-based study. Epilepsia 33: 1051–1056.

Salinsky M, Spencer D, Boudreau E et al. (2011). Psychogenic nonepileptic seizures in US veterans. Neurology 77: 945–950.

San-Juan D, Alvarado-León S, Barraza-Díaz J et al. (2015). Prevalence of epilepsy, beliefs and attitudes in a rural community in Mexico: a door-to-door survey. Epilepsy Behav 46: 140–143.

Savic I (2014). Sex differences in human epilepsy. Exp Neurol 259: 38–43.

Shorvon SD, Farmer PJ (1988). Epilepsy in developing countries: a review of epidemiological, sociocultural, and treatment aspects. Epilepsia 29 (Suppl 1): S36–S54.

Sidenvall R, Forsgren L, Heijbel J (1996). Prevalence and characteristics of epilepsy in children in northern Sweden. Seizure 5: 139–146.

Singh BM, Strobos RJ (1980). Epilepsia partialis continua associated with nonketotic hyperglycemia: clinical and biochemical profile of 21 patients. Ann Neurol 8: 155–160.

Singh NA, Westenskow P, Charlier C et al. (2003). KCNQ2 and KCNQ3 potassium channel genes in benign familial neonatal convulsions: expansion of the functional and mutation spectrum. Brain 126: 2726–2737.

Smith PEM (2012). Epilepsy: mimics, borderland and chameleons. Pract Neurol 12: 299–307.

Sperling MR, Ko J (2006). Seizures and brain tumors. Semin Oncol 33: 333–341.

Stephen LJ, Brodie MJ (2000). Epilepsy in elderly people. Lancet 355: 1441–1446.

Tan L, Wang N (2010). Future internet: The Internet of Things. In: 2010 3rd International Conference on Advanced Computer Theory and Engineering (ICACTE), pp. 376–380.

Tao JX, Qian S, Baldwin M et al. (2010). SUDEP, suspected positional airway obstruction, and hypoventilation in postictal coma. Epilepsia 51: 2344–2347.

Tekle-Haimanot R, Forsgren L, Abebe M et al. (1990). Clinical and electroencephalographic characteristics of epilepsy in rural Ethiopia: a community-based study. Epilepsy Res 7: 230–239.

Tekle-Haimanot R, Forsgren L, Ekstedt J (1997). Incidence of epilepsy in rural central Ethiopia. Epilepsia 38: 541–546.

Thomas RH, Berkovic SF (2014). The hidden genetics of epilepsy – a clinically important new paradigm. Nat Rev Neurol 10: 283–292.

Thurman DJ, Beghi E, Begley CE et al. (2011). Standards for epidemiologic studies and surveillance of epilepsy. Epilepsia 52: 2–26.

Thurman DJ, Hesdorffer DC, French JA (2014). Sudden unexpected death in epilepsy: assessing the public health burden. Epilepsia 55: 1479–1485.

Towne AR (2007). Epidemiology and outcomes of status epilepticus in the elderly. Int Rev Neurobiol 81: 111–127.

Tsai CJ, Griffin MR, Nuorti JP et al. (2008). Changing epidemiology of pneumococcal meningitis after the introduction of pneumococcal conjugate vaccine in the United States. Clin Infect Dis 46: 1664–1672.

Wagle K (2013). IBM Watson: Revolutionizing healthcare? Young Sci J 6: 17.

Wang WZ, Wu JZ, Wang DS et al. (2003). The prevalence and treatment gap in epilepsy in China: an ILAE/IBE/WHO study. Neurology 60: 1544–1545.

Winkler AS, Kerschbaumsteiner K, Stelzhammer B et al. (2009). Prevalence, incidence, and clinical characteristics of epilepsy – a community-based door-to-door study in northern Tanzania. Epilepsia 50: 2310–2313.

Yemadje L-P, Houinato D, Boumédiène F et al. (2012). Prevalence of epilepsy in the 15 years and older in Benin: a door-to-door nationwide survey. Epilepsy Res 99: 318–326.

Handbook of Clinical Neurology, Vol. 138 (3rd series)
Neuroepidemiology
C. Rosano, M.A. Ikram, and M. Ganguli, Editors
http://dx.doi.org/10.1016/B978-0-12-802973-2.00011-2

Chapter 11

The epidemiology of multiple sclerosis: insights to a causal cascade†

D.S. GOODIN*
Multiple Sclerosis Center at the University of California, San Francisco, San Francisco, CA, USA

Abstract

MS-pathogenesis involves both genetic-susceptibility and environmental determinants. Three (or more) sequential environmental-factors are implicated. The first acts near birth, the second acts during childhood/adolescence, and the third acts subsequently. Two candidate factors (vitamin D deficiency and Epstein-Barr viral infection) seem particularly well-suited to the first two environmental-events but other factors (e.g., obesity and smoking behavior) seem also to be involved in the causal scheme.

MS-pathogenesis can be modeled by incorporating both the environmental and genetic-factors into a causal scheme, which can then help to explain some of the changes in MS-epidemiology (e.g., increasing disease-prevalence, changing sex-ratio, and regional-variations in monozygotic-twin-concordance-rates), which have been taking place recently. This model suggests that genetic-susceptibility is overwhelmingly the most important determinant of MS and that, at least, 92.5% of individuals (and likely much more) are, essentially, incapable of developing MS, regardless of their specific environmental-exposures. Nevertheless, the genetics is complex and the contribution of any specific gene to MS-susceptibility seems to be quite modest. Thus, even for the DRB1*1501 allele (the strongest known MS-susceptibility marker), most carriers are not in the genetically-susceptible group. Moreover, 45–50% of individuals with MS lack this allele entirely and some of the haplotypes that carry this allele don't also confer any disease–risk.

Finally, because the prevalence of genetic-susceptibility seems to be so similar throughout North America and Europe, and despite the crucial importance of a person's genetic make-up to disease pathogenesis, it is the environmental-factors, which largely responsible for the observed regional variations in disease-characteristics. Thus, despite MS being more common in women, men are more likely to be genetically-susceptible. This apparent paradox seems to relate to the fact that women are much more responsive than men to the recent changes in environmental-exposure (whatever these have been). These gender-differences may help to explain changes in the sex-ratio and the increasing disease-prevalence, which have both been observed recently. The potential importance of these conclusions regarding the role of environment in MS-pathogenesis is that they open the door to the possibility of pursuing strategies for primary primary disease prevention in the future.

†This chapter has been partially adapted from Goodin DS (2014). The epidemiology of multiple sclerosis: insights to disease pathogenesis. In Goodin DS (ed.) *Multiple Sclerosis and Related Disorders. Handbook of Clinical Neurology*, vol. 122. Amsterdam: Elsevier.

*Correspondence to: Douglas S. Goodin, MD, MS Center at the University of California, San Francisco, 350 Parnassus Ave., Suite #908, San Francisco CA 94117, USA. Tel: +1-415-514-2464, Fax: +1-415-514-2470, E-mail: douglas.goodin@ucsf.edu

OVERVIEW

Multiple sclerosis (MS) is a chronic recurrent inflammatory disease of the central nervous system (CNS) and it is one of the most disabling neurologic diseases of young adults. Unpredictable inflammatory episodes, lasting days to months, result in injury to the myelin sheaths, to the oligodendrocytes themselves and, to a somewhat lesser extent, to the nerve cells and axons (Weinshenker et al., 1989; Runmarker and Andersen, 1993; Liguori et al., 2000; Confavreux et al., 2003; Pittock et al., 2004; Confavreux and Vukusic, 2006; Compston et al, 2006; Kremenchutsky et al., 2006; Tremlett et al., 2006). Immune mechanisms are definitely involved in MS pathogenesis, either primarily or secondarily, and, in fact, most authorities favor primary autoimmunity as the pathogenic basis for this disease (Compston et al., 2006). Evidence of both acute and chronic inflammation is typically found in the cerebrospinal fluid (CSF) of individuals, especially during acute clinical attacks (Compston et al., 2006). This evidence includes the findings of an elevation in total CSF protein concentration, an increased number of mononuclear cells in the CSF, and an increase in the gamma-globulin (IgG) fraction in the CSF, as reflected by the presence of unique oligoclonal CSF IgG bands on gel electrophoresis and/or an increase in the so-called CSF IgG index (Compston et al., 2006; Housley et al., 2015).

In the northern parts of North America and Europe, where the disease is most common, the prevalence is between 0.1 and 0.2% of the population and the incidence is approximately 5–6 per 100 000 population per year (Wynn et al., 1990). There is, however, considerable variation in the prevalence of MS in different countries around the world (Fig. 11.1). Women are affected two to three times more often than men. Typically, the disease becomes clinically apparent between the ages of 20 and 40 years (mean age at onset: ~28 years), although it can begin as early as the first or second year of life (Ruggieri et al., 1999; Gadoth, 2003; Lee and Chitnis, 2016) or as late as the seventh decade of life (Confavreux et al., 2003; Confavreux and Vukusic, 2006). In addition, it is now recognized that the onset of the disease generally precedes the first clinical manifestation, often by many years, if not decades (Okuda et al., 2009, 2014). Indeed, some individuals with pathologically verified MS may be discovered incidentally at autopsy. In fact, based on several large autopsy series performed before the era of magnetic resonance imaging (MRI), the prevalence of such "asymptomatic" MS was reported to be approximately 0.1% (Vost et al., 1964; Georgi, 1966; Gilbert and Sadler, 1983; Engell, 1989). If so, this suggests that, over their entire lifetime, as many as half of pathologically proven MS cases will never experience sufficient clinical symptoms to bring them to medical attention. The experience in the post-MRI era seems to confirm these earlier autopsy observations. Thus, with the ever-increasing availability of MRI, the appearance of radiographic disease in persons who have never had neurologic symptoms is now a common finding (e.g., Okuda et al., 2009, 2014). Nevertheless, after 5 years, approximately a third

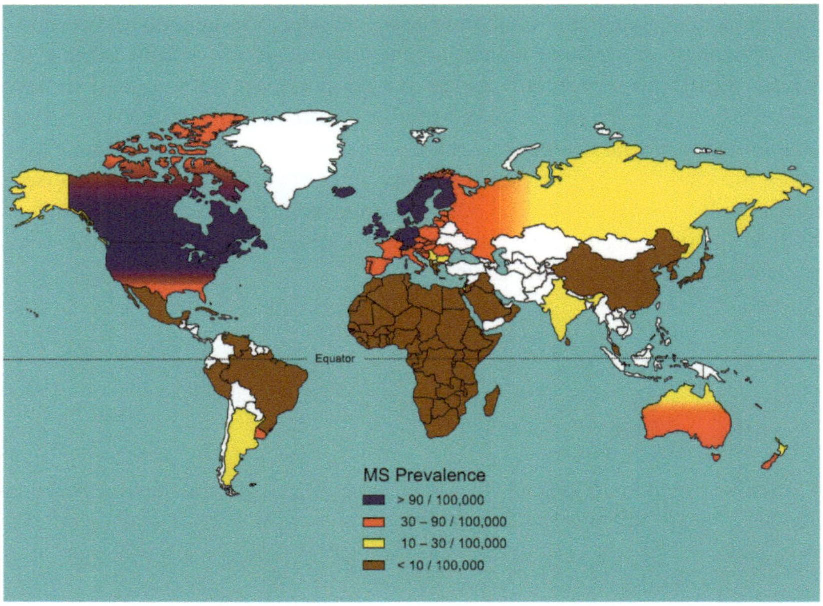

Fig. 11.1. Global distribution of multiple sclerosis (MS) prevalence. Dotted black line indicates the Equator. Data taken from reviews of worldwide epidemiology of MS (Rosati, 2001; Compston et al., 2006).

of these so-called radiographically isolated syndrome patients will have experienced clinical evidence of MS disease activity (Okuda et al., 2009, 2014).

THE PATHOLOGY OF MS

Pathologically, MS is characterized by multifocal patches of demyelination and gliosis (plaques) within the CNS white matter. Indeed, within these plaques there is evidence of white-matter tissue injury of differing ages both to the myelin sheaths (surrounding the axons) and to the oligodendrocytes themselves (Bar-Or et al., 1999; Conlon et al., 1999; Compston et al., 2006; Hauser and Oksenberg, 2006). In addition, there is evidence of axonal injury – especially within active MS lesions; and gray-matter demyelination – especially later in the disease course (Ferguson et al., 1997; Trapp et al., 1998; Peterson et al., 2001; Bø et al., 2003; Lucchinetti et al., 2011). Within acute lesions, presumably guided by cellular adhesion molecules and proinflammatory cytokines, autoreactive immune cells cross the blood–brain barrier (BBB) and enter into the CNS – these invading cells include cluster of differentiation (CD)4+, thymic-derived lymphocytes (T cells), CD20+ bone marrow-derived lymphocytes (B cells), CD8+ cytotoxic lymphocytes, and CD68+ macrophages (Bar-Or et al., 1999; Conlon et al., 1999; Compston et al., 2006; Hauser and Oksenberg, 2006). These activated cells are thought to contribute to the CNS tissue damage that occurs in acute MS lesions.

On histopathologic examination, lesions that are characterized on MRI as having only T2 hyperintensity (i.e., T2-only lesions) are much more likely to demonstrate myelin preservation compared to lesions that, on MRI, are characterized by T2 hyperintensity in addition to persistent T1 hypointensity and a reduced magnetization transfer ratio (MTR: i.e.,T2/T1/MTR lesions) – in fact, only 20–45% of T2-only lesions are associated with demyelination on pathologic examination compared to 80–83% of the T2/T1/MTR lesions (Fisher et al., 2002; Moll et al., 2009). Nevertheless, regardless of the state of myelin preservation, most of the T2-only MRI lesions still contain activated microglia and evidence of BBB breakdown and, thus, would show up as gadolinium-enhanced (Gd+) lesions on routine imaging.

Even in the so-called normal-appearing white matter, 30% of the regions examined histopathologically were still found to contain activated microglia (Fisher et al., 2002; Moll et al., 2009). It is unknown whether activated microglia within myelinated T2-only lesions are causing new damage. Thus, it is possible that these microglial cells are simply responding to a cytokine release associated with either the breakdown of the BBB, or Wallerian degeneration, or both. Indeed, it may well be that the microglia may actually be involved in removing debris from lesions and possibly even promoting repair, whereas the macrophages may be cells responsible for the actual tissue damage (Yamasaki et al., 2014). Regardless, it is clear that MS lesions demonstrate a considerable histopathologic diversity – ranging from chronic gliotic demyelinated scars, to highly inflamed demyelinating lesions, to less inflamed regions in which the myelin seems to be completely preserved.

The endothelial cells in the CNS are not fenestrated and have extraordinarily tight junctions between them. For many years, these tight junctions were thought to be primarily responsible for creating and maintaining the BBB. Nevertheless, it is now understood that this barrier is actually produced by a complex interaction between the vascular system and the CNS, which, collectively, is referred to as the neurovascular unit (Holman et al., 2011; Engelhardt et al., 2014; Muoio et al., 2014). This unit includes the endothelial cells, the extracellular matrix, the basement membrane, and also the cells surrounding the endothelial cells, notably the pericytes and the astrocytes. Together, this unit works both to provide mutual trophic support and to make the CNS entry by hydrophilic molecules (either by active transport or by diffusion) and entry by transcytosis extremely selective. A focal breakdown of the BBB can be caused by any one of a variety of CNS insults, including inflammation, toxic exposure, ischemia, trauma, and neoplasia. In MS, the breakdown of the BBB is thought to represent both a critical step in the development of a new MS lesion and the pathogenic basis of an acute MS attack. Nevertheless, whether this BBB breakdown is the initial event in lesion formation is not entirely clear (e.g., Filippi et al., 1998; Goodkin et al., 1998). Thus, using the MTR, focal changes in the relative concentrations of free and bound water can be detected in regions of otherwise normal-appearing CNS white matter that, months later, are destined to become Gd+lesions seen on MRI. Presumably, these MTR changes reflect biochemical alterations, which are the initial events in lesion formation. It is nonetheless possible that these early events represent a selective breakdown in the BBB, which is not detectable by conventional MRI and, in this view, the more general breakdown of the BBB, which is reflected by Gd+lesions, would be a secondary phenomenon.

THE CLINICAL COURSE OF MS

The symptoms of MS depend, to a large extent, upon the location of plaques within the CNS and commonly include visual impairment, weakness, muscle spasms, sensory disturbances (including pain and paresthesias), incoordination, gait abnormalities, bladder dysfunction, and double vision (Compston et al., 2006). Three different disease courses of MS have been defined (Lublin et al., 2014), although it is unclear whether these different courses reflect a variable clinical expression of a single disease or whether these represent different diseases with different pathophysiologic underpinnings. The first, and by far the most common course, is relapsing/remitting MS (RRMS). About 85–90% of MS cases begin with this course. RRMS is characterized by self-limited "attacks" of neurologic dysfunction, (variously referred to as "attacks," "relapses," "bouts," "exacerbations," or "flares"). These attacks develop acutely, evolve over hours to days, last days to months, and are followed by a recovery of neurologic function (sometimes incomplete). Between these episodes, the patient is neurologically stable. On MRI, these patients often have new lesions on T2-weighted images (new T2 lesions) or they may have Gd+lesions, which reflect a focal breakdown of the BBB due to the new inflammatory activity, or they may have both (Compston et al., 2006).

The second defined clinical course is secondary progressive MS (SPMS), which always begins as typical RRMS but, at some point in time, it changes course such that the acute inflammatory episodes become less frequent and the patient begins to experience an insidious and progressive decline in neurologic function, independent of acute attacks. On MRI, the evidence of acute inflammation (new T2 lesions and Gd+lesions) becomes less conspicuous and the occurrence of brain atrophy becomes more prominent. SPMS ultimately develops in the majority of RRMS patients and causes most of the disability experienced by patients. For example, longitudinal population-based studies have estimated that 50% of RRMS patients will require the use of a cane to assist with their ambulation after 15–30 years of disease (Weinshenker et al., 1989; Runmarker and Andersen, 1993; Liguori et al., 2000; Confavreux et al., 2003; Pittock et al., 2004; Confavreux and Vukusic, 2006; Kremenchutsky et al., 2006; Tremlett et al., 2006; Scalfari et al., 2010). Nevertheless, predicting the clinical course for any individual patient is difficult, and, as noted earlier, many patients with pathologic MS are discovered only incidentally at autopsy (Vost et al., 1964; Georgi, 1966; Gilbert and Sadler, 1983; Engell, 1989). Men, patients with high initial attack rates, patients with early involvement of motor or cerebellar pathways, patients with moderate disability after 5 years of illness, and patients with a large MRI disease burden are more likely to become disabled compared to patients without these risk factors (Weinshenker et al., 1989; Runmarker and Andersen, 1993; Confavreux et al., 2003; Scalfari et al., 2010). However, even among patients who have experienced little disability in the first 10 years of their illness, significant disability can still develop subsequently (Hawkins and McDonnell, 1999) and, for this reason, the diagnosis of so-called "benign MS" can only be made in retrospect. Moreover, any definition of "benign" MS must include a consideration of the patient's cognitive function, which can be affected by MS, independently from a person's physical abilities (Correale et al., 2012).

The third clinical type, primary progressive MS (PPMS), accounts for only about 10% of cases. In PPMS, the onset of illness is insidious and, from the beginning, the patient experiences a steady decline in neurologic function without acute clinical attacks (although these may develop subsequently). Similarly to SPMS, these patients have less evidence of active inflammation on MRI compared to patients with RRMS. Also, PPMS patients have a more equal sex ratio, a later age of onset, and a worse prognosis for ultimate disability compared to patients with RRMS. In fact, within the first decade of the illness, over 50% of PPMS patients will require a cane to assist with ambulation (Weinshenker et al., 1989; Runmarker and Andersen, 1993; Liguori et al., 2000; Confavreux et al., 2003; Pittock et al., 2004; Confavreux and Vukusic, 2006; Kremenchutsky et al., 2006; Tremlett et al., 2006; Scalfari et al., 2010).

It is possible that the poor long-term prognosis for untreated MS (which is often assumed) may be overestimated. For example, in patients with attacks of idiopathic optic neuritis (ON), a condition closely associated with MS and having similar genetic determinants, the conversion rate to clinically definite MS (CDMS) may be as low as 40% after 40 years (Rodriguez et al., 1994). Even in the long-term follow-up of the patient cohort who participated in the ON treatment trial (Optic Neuritis Study Group, 2008), the conversion rate to CDMS was reported to be only 50%. Because this cohort is known to include individuals who already had MS at baseline (Goodin, 1999), this 50% number undoubtedly overestimates the true risk of MS following an episode of isolated ON. Thus, for some patients with clinically isolated syndromes such as ON, it seems that benign forms of demyelinating disease may be much more common than is currently believed by many authors. Nevertheless, in patients presenting with a clinically isolated syndrome, certain laboratory or imaging features such as characteristic MRI abnormalities, evidence of inflammation in the CSF (i.e., either the presence of oligoclonal IgG bands or an elevation in the IgG index), or abnormalities on evoked potential testing significantly increase the likelihood of that individual developing MS in the future (Sharief and Thompson, 1991; O'Riordan et al., 1998; Gronseth and Ashman, 2000). For example, over 60% of patients with a monosymptomatic clinically isolated syndrome will have MRI abnormalities consistent with MS and, of these, more than 80% will develop CDMS within the next 20 years

(Fisniku et al., 2008). By contrast, in the absence of such MRI abnormalities, the 20-year risk of developing CDMS is only about 20% (Fisniku et al., 2008). The spinal MRI may have special prognostic significance. Indeed, one study reported that the odds ratio (OR) for conversion from radiographically isolated syndrome to CDMS was more than 75 when a spinal cord lesion was present (Okuda et al., 2011).

CONSIDERATIONS REGARDING THE CAUSES OF MS AND ITS EXACERBATION

In considering the causation of MS, two possibly separate questions need to be addressed. The first is to consider what is known about those factors that may trigger an attack of the disease in an individual who already has MS. The second, and perhaps more important, question is to ask what is known about the causal factors that lead to the disease in the first place.

Factors associated with the exacerbation of MS

The timing of acute MS attacks seems to be largely (although not entirely) a chance event. If factors exist that do trigger MS attacks, these are, almost certainly, environmental events (or environment–gene interactions). They are not likely to be primarily genetic events because the genetic make-up of an individual is fixed. Indeed, certain environmental factors do seem to be consistently associated with an increased or decreased likelihood of a person experiencing an MS attack. For example, three high-quality studies have looked at the effect of pregnancy on the likelihood of MS exacerbations (Confavreux et al., 1998; Salemi et al., 2004; Vukusic et al., 2004). Each of these studies reported a significant reduction in the likelihood of a woman experiencing an attack during pregnancy (especially during the last trimester) compared to her prepregnancy risk (Confavreux et al., 1998; Salemi et al., 2004; Vukusic et al., 2004). Also, during the period 3–6 months postpartum, the risk of a women with MS experiencing an attack is increased compared to her prepregnancy state (Confavreux et al., 1998; Salemi et al., 2004; Vukusic et al., 2004). The basis for these observations is not known, but one possibility is that fluctuations in hormonal levels (possibly in estrogen), which occur during and after pregnancy, influence the likelihood of MS attacks. Nevertheless, there are many other (nonhormonal) physiologic changes that occur during pregnancy (Houtchens, 2007) and there is no basis on which to discount these factors as being responsible for the observed pregnancy-related changes in attack risk.

Another factor that seems to influence the attack risk consistently is the occurrence of nonspecific infectious syndromes (e.g., rhinorrhea, fever, cough, malaise, nausea, abdominal pain, diarrhea, and so forth). Four high-quality studies have looked at the association between these clinical syndromes (often attributed to nonspecific upper respiratory or gastrointestinal viral infections) and each has reported that there is an increased risk of an MS attack around the symptomatic period compared to other times during the year (Sibley et al., 1985; Andersen et al., 1993; Panitch, 1994; Edwards et al., 1998).

Some authors have suggested that vaccinations (e.g., to influenza or hepatitis B) can be related to the timing of MS attacks, although the available data are unconvincing (Merelli and Casoni, 2000). Other factors have also been suggested as both causal factors and as influencing the likelihood of an MS attack (e.g., trauma and psychologic stress), and these are considered below.

Factors associated with the development of MS

Before considering the different candidate factors that might be in the causal pathway leading to MS, it is important to review the implications of several of the general epidemiologic observations that have been made with regard to MS pathogenesis.

General considerations

Chronic diseases such as MS typically have complex etiologic bases (Rothman and Greenland, 1998). Both individual genetic background and the environmental events that they experience during their lives are critical to whether they will ultimately develop the disease. For example, as noted earlier, an individual from northern North America or northern Europe has a lifetime risk of developing MS of approximately 0.1–0.2% (Compston et al., 2006). The risk for individuals with an affected family member increases roughly in proportion to the genetic similarity between themselves and the proband (French Research Group on Multiple Sclerosis, 1992; Mumford et al., 1994; Ebers et al., 1995, 2004; Robertson et al., 1996; Sadovnick et al., 1996; Compston and Coles, 2002; Willer et al., 2003; Nielsen et al., 2005; Compston et al., 2006; Islam et al., 2006; Ristori et al., 2006). Siblings of an MS proband (~50% similarity) have a 20–30-fold increased risk compared to the general population. By contrast, monozygotic twins (~100% genetic similarity) have more than 200 times the general population risk.

Despite this strong genetic predisposition, however, it is clear that genetics is not the only factor. If it were, the proband-wise concordance rate for monozygotic twins – an estimate of the lifetime risk of MS for an individual with a monozygotic twin who has MS (Witte et al., 1999) – would be much closer to 100% than to the

20–30% reported in these northern populations (Mumford et al., 1994; Ebers et al., 1995; Willer et al., 2003). In southern populations, where the proband-wise concordance rate for monozygotic twins is approximately half that in the north (French Research Group on Multiple Sclerosis, 1992; Islam et al., 2006; Ristori et al., 2006), this conclusion is even more evident. Consequently, it is clear that, in addition to any genetic contributions, there must be environmental and/or epigenetic factors that contribute in an important ways to MS pathogenesis.

When considering the environmental events that might relate to MS pathogenesis, although anticipated on theoretic grounds (Goodin, 2016), it is noteworthy that the observed microenvironmental contributions to MS risk seem to be minimal. Thus, studies in conjugal couples, brothers and sisters of different birth order, adopted individuals, and in siblings and half-siblings raised together or apart have generally indicated that MS risk is unaffected by these microenvironmental influences (Sadovnick et al., 1996, 2005; Ebers et al., 2000, 2004; Bager et al., 2006; Dyment et al., 2006). If so, then the relevant environmental events in MS pathogenesis must be acting at a population level. Moreover, if, in addition to a genetic predisposition, one or more population-level environmental events is necessary for MS to develop, then it is only natural to enquire as to how many such events there are and whether these events need to occur at any particular time or in any particular order. Several published epidemiologic findings bear on these issues.

Environmental factors near birth

The first finding is the presence of the so-called "maternal effect" in MS (Ebers et al., 2004). Epidemiologic support for such "maternal effect" is provided by three independent observations. The first is that half-siblings (i.e., siblings who share one, but not both, biologic parents), who are concordant for MS, are twice as likely to share the mother as they are to share the father (Sadovnick et al., 1996; Ebers et al., 2004). Such a circumstance suggests that MS susceptibility is transmitted from mother to child through some mechanism other than the passage of nuclear genes. An environmental exposure, occurring either in the intrauterine period or soon thereafter, is one possibility. Once the child is born and becomes independent of the mother, however, such a maternal effect would be not be expected from an environmental event.

This maternal effect, however, need not be environmental. It could equally be the result of mitochondrial inheritance, from genetic imprinting favoring expression of certain maternal genes, or from other epigenetic factors (Bartolomei and Tilghman, 1997). With respect to these other possibilities, however, there has been an interesting discussion in the literature about the possible existence of a so-called "Carter effect" in MS (Kantarci et al., 2006; Herrera et al., 2007). This hypothetic effect occurs because men (thought to be less susceptible to MS than women) are presumed to have more "potent" susceptibility genes when they actually develop the disease. In such a circumstance, one would anticipate paternal transmission of MS to be more common when the father's side is "genetically loaded" compared to maternal transmission when the mother's side is similarly "loaded." One report found weak evidence ($p=0.032$) for such a "Carter effect" (Kantarci et al., 2006), whereas another (larger) study did not (Herrera et al., 2007). Importantly, however, neither study provided evidence for the excessive maternal transmission expected if mitochondrial genes, genetic imprinting, or epigenetic factors were the basis of the "maternal effect" in MS (Ebers et al., 2004). By contrast, any potentially responsible environmental factor would not be expected to produce a maternal effect in these studies because the intrauterine and early postnatal environments are the same irrespective of which parent transmits the MS risk.

The second observation is that the MS concordance rate for fraternal twins seems to be greater than that for full siblings. For example, in a large cohort from Canada (Willer et al., 2003), the concordance rate for MS in full-siblings was 2.9%, with a standard error (SE) of 0.6%, compared to a concordance rate in dizygotic twins of 5.4%. Although few studies directly compare these rates, other large studies (Robertson et al., 1996; Islam et al., 2006), including a review (Compston and Coles, 2002), generally support the same conclusion, although, in a population-based study from Sweden, dizygotic twins and siblings seemed to have a similar risk (Hansen et al., 2005a, b). Such a disparity in recurrence risk between siblings and dizygotic twins (if it exists) cannot be attributed to mitochondrial inheritance, genetic imprinting, or epigenetic factors because, on average, these factors should be similar for both siblings and fraternal twins sharing the same biologic parents. Rather, this discrepancy must be due to environmental events occurring during the shared intrauterine or in the early post-natal period.

The third observation relates to the possible month-of-birth effect for MS, which has now been reported in studies from Canada, northern Europe, and Australia (Templer et al., 1992; Willer et al., 2005; Sadovnick et al., 2007; Staples et al., 2010; Torkildsen et al., 2014). Thus, combining patients from the northern hemisphere (Canada, Denmark, and Sweden), significantly more MS patients were reportedly born in May and fewer were born in November, compared to other months of the year (Willer et al., 2005). Another study found more

RRMS patients born in May than November (Sadovnick et al., 2007). Finally, in 67 Canadian patients, born in the southern hemisphere, this month-of-birth effect seemed reversed (Willer et al., 2005). Recently, this reversal has been fully documented in Australia, where MS risk peaks for babies born in November/December and has its nadir for children born in May/June (Staples et al., 2010).

There has been, however, some controversy regarding the possibility that this month-of-birth effect may be due to an analytic bias (e.g., Fiddes et al., 2013, 2014; Torkildsen et al., 2014). Nevertheless, if these month-of-birth observations cannot be explained away as simple artifacts (Torkildsen et al., 2014), such an effect would provide unequivocal evidence for an early environmental event involved in MS pathogenesis that is time-locked to birth. As the interval between the birth and any environmental event increases, the coupling between birth and the event will inevitably become less precise and, as a result, the observed birth signal will become increasingly less distinct. The fact that this signal remains so clear (Willer et al., 2005; Staples et al., 2010) would then indicate that the responsible environmental event underlying this effect must occur very near to the birth itself. In addition, the timing of this environmental event is periodic and it is coupled to the solar cycle (Willer et al., 2005; Staples et al., 2010). Perhaps importantly, mothers of May babies in the northern hemisphere (or November babies in the southern hemisphere) spend much of their pregnancy during the winter months (with less sun exposure) compared to mothers who are pregnant over the summer months. This *circa annum* periodicity to MS susceptibility might be due to variations in vitamin D levels from differences in maternal sun exposure while the child is *in utero* (Chaudhuri, 2005; Willer et al., 2005). Seasonal infections might also produce such periodicity although, because intrauterine infections of the child are uncommon, any such association would probably need to be secondary.

Environmental factors during adolescence

A second environmental factor is suggested by observations in people who migrate from one geographic region to another with differing MS risks (Dean and Kurtzke, 1971; Alter et al., 1978; Elian et al., 1990; Kahana et al., 1994; Cabre et al., 2005; Compston et al., 2006). For example, when individuals move (prior to their adolescent years) from an area of high MS prevalence to an area of low prevalence (or vice versa), their MS risk becomes similar to that of the region to which they moved. By contrast, when they make the same move after adolescence, their MS risk remains similar to that of the region from which they moved. Moreover, the children of immigrants from low-MS-prevalence areas who are born in a high-MS-prevalence area have an MS risk similar to their birth country rather than their country of ethnic origin (Elian et al., 1990). These observations (if correct) indicate that there is some environmental event, which is involved in MS pathogenesis, that occurs some time between birth and adolescence.

Environmental factors during adult life

Third, the initial clinical symptoms in MS are generally delayed considerably (often by decades) following the period when the maternal factor and the migratory factor take place. It is possible that these early environmental events, by themselves, are sufficient to cause MS although, in that case, the long delay between these events and MS onset seems somewhat difficult to rationalize. Consequently, it seems likely that a subsequent environmental event (or events) is responsible for the timing of symptom onset.

ENVIRONMENTAL FACTORS

Many potential environmental triggers, including trauma, stress, vaccinations, obesity, tobacco, typhoid, smallpox, Epstein–Barr virus (EBV), human herpesvirus (HHV)-6, chickenpox, *Chlamydia*, other infections, vitamin deficiencies, low sunlight, cosmic rays, occupational hazards, living with domesticated animals, dietary habits, and toxic exposures, have been postulated to be linked to MS pathogenesis (Compston et al., 2006). Of these, EBV infection, vitamin D deficiency, tobacco, and obesity have attracted the greatest current interest for their potential role in MS pathogenesis. Nonetheless, several of these other factors continue to have strong proponents and no single factor has been proven conclusively to be related. Neither has any factor been completely excluded, although many of the proposed associations lack credible scientific evidence, biologic plausibility, or both.

Epstein–Barr virus

EBV is a double-stranded linear DNA virus of the herpes family. It is a very common infection of humans, with over 90% of the population becoming infected (Ascherio et al., 2001; Goldacre et al., 2004; Sundström et al., 2004; Ponsonby et al., 2005; Thacker et al., 2006; Ascherio and Munger, 2007, 2016; Farrell, 2007; Nielsen et al., 2007; Serafini et al., 2007). Following birth, as the antibody protection provided by the mother subsides, infants become susceptible to EBV infection. In many parts of the world the initial EBV infection occurs during early childhood and is either asymptomatic or produces nonspecific symptoms indistinguishable from many other childhood illnesses. However, if the initial infection is delayed until adolescence or young adulthood (which often happens in highly developed regions such as North America and Europe), the syndrome of infectious mononucleosis (glandular fever) develops in 35–50% of cases. The viral infection seems to specifically target the epithelial cells of the oropharynx and the B cells.

Once a cell is infected, the viral genome becomes circularized and persists within the cell as an episome. Latent rather than lytic infection ultimately predominates (probably due to immune responses by the host). Latently infected B lymphocytes proliferate and are directed to specific sites (e.g., the bone marrow) where the virus persists indefinitely, periodically becoming reactivated, resulting in further cell lysis and producing fresh viral particles.

During either the late incubation period or early in the acute illness, antibodies to antigens associated with the process of viral replication, such as the viral capsid antigen (VCA) and the diffuse and restricted early antigens (EA), are found in the serum (Henle et al., 1987). Antibodies to VCA are initially of the IgM class. However, this response lasts only 1–2 months, after which time the anti-VCA response shifts to the IgG class. These antibodies persist for the lifetime of the individual. Antibodies to EA are also of the IgG class but usually drop to undetectable levels after 3–6 months. Thus, the EA antibodies are generally taken as a sign of active infection, although, in approximately 20–30% of patients, these antibody titers may persist for years. Moreover, these antibodies can be found in patients with chronic active infections or with secondary complications such as nasopharyngeal carcinoma or Burkitt's lymphoma. The EBV nuclear antigens (EBNA 1–5) are expressed in latently infected B cells, and antibodies to these antigens typically appear 3–6 weeks following the initial infection. These too persist for the lifetime of the individual. In addition to any possible role in MS pathogenesis, EBV has been implicated in the pathogenesis both of certain malignancies (e.g., EBV-positive Hodgkin lymphoma, nasopharyngeal carcinoma, and Burkitt's lymphoma) and of several autoimmune diseases, such as rheumatoid arthritis, Sjögren's syndrome, and systemic lupus erythematosus (Vaughan, 1995).

Despite the fact that EBV infection has been consistently linked to MS, especially when it causes symptomatic mononucleosis infection (Ascherio et al., 2001; Goldacre et al., 2004; Sundström et al., 2004; Ponsonby et al., 2005; Thacker et al., 2006; Ascherio and Munger, 2007, 2016; Farrell, 2007; Nielsen et al., 2007; Serafini et al., 2007), it is notable that fewer than one in 900 individuals with an EBV infection and only a small fraction of patients with mononucleosis will ever develop MS. Nevertheless, the evidence that EBV plays some role in MS pathogenesis seems quite compelling (Ascherio et al., 2001; Goldacre et al., 2004; Levin et al., 2005; Ponsonby et al., 2005; Thacker et al., 2006; Ascherio and Munger, 2007, 2016; Farrell, 2007; Nielsen et al., 2007; Serafini et al., 2007). Thus, even though EBV infection occurs in over 90% of the non-MS population (Sumaya et al., 1980, 1985; Bray et al., 1983; Larsen et al., 1985; Shirodaria et al., 1987; Munch et al., 1998; Myhr et al., 1998; Wagner et al., 2000; Ascherio et al., 2001; Haahr et al., 2004; Sundström et al., 2004; Ponsonby et al., 2005), the evidence for prior EBV infection in adult-onset MS is essentially 100% and is significantly more likely in cases than in controls (Table 11.1). Even in those rare MS patients who test negatively for prior exposure to EBV, this finding could easily be a false-negative result because, in every such case, the antibody response was not measured to the entire set of EBV antigens (Table 11.1). Also, the prior nature of the EBV infection is supported both by the presence of IgG (not IgM) antibodies to EBV antigens and by the unequivocal evidence (when it has been assessed) of infection years prior to the onset of clinical symptoms (Ascherio et al., 2001; Goldacre et al., 2004; Sundström et al., 2004; Ponsonby et al., 2005; Thacker et al., 2006; Ascherio and Munger, 2007, 2016; Farrell, 2007; Nielsen et al., 2007). It should be noted that, here, the term "prior" is being used to mean "prior to the symptomatic onset of MS." As discussed earlier, disease onset frequently (perhaps generally) occurs several years before symptom onset. Nevertheless, such a strong association between EBV and MS is very difficult to ignore.

Moreover, this high prevalence of EBV antibodies in adult-onset MS does not seem to be the consequence of either false-negative tests within the general population or false-positive tests in MS patients. Certainly, as shown in Table 11.2, the near-100% prevalence cannot be due to a general hyperimmune state in MS patients because their antibody responses to other common pathogens (e.g., mumps, measles, chickenpox, cytomegalovirus, herpes simplex) are not similarly increased (Bray et al., 1983; Ascherio et al., 2001; Haahr et al., 2004; Sundström et al., 2004; Ponsonby et al., 2005). Interestingly, one report found evidence of EBV infection in a substantial proportion of those B lymphocytes infiltrating the CNS in 21 of 22 MS cases examined at postmortem (Serafini et al., 2007). This isolated report, however, has not been replicated. Finally, the increased risk of MS, either with delayed exposure to EBV (Ascherio et al., 2001; Haahr et al., 2004; Ponsonby et al., 2005; Thacker et al., 2006) or following symptomatic mononucleosis (Goldacre et al., 2004; Sundström et al., 2004; Thacker et al., 2006; Nielsen et al., 2007), strongly

Table 11.1

Prevalence of antibodies to Epstein–Barr virus (EBV) in the sera of patients and controls

Study	EBV+ multiple sclerosis cases (%)	EBV+ controls (%)	*p*-value
Sumaya et al., 1980‡	155/157 (98.7%)	76/81 (93.8%)	0.05
Bray et al., 1983‡	309/313 (98.7%)	363/406 (89.4%)	0.0001
Larsen et al., 1985‡	93/93 (100%)	78/93 (83.9%)	0.0001
Sumaya et al., 1985*	104/104 (100%)	23/26 (88.5%)	0.007
Shirodaria et al., 1987‡‡	26/26 (100%)	24/26 (92.3%)	–
Munch et al., 1998†	137/138 (99.3%)	124/138 (89.9%)	0.0004
Myhr et al., 1998*	144/144 (100%)	162/170 (95.3%)	0.008
Wagner et al., 2000†	107/107 (100%)	153/163 (93.9%)	0.01
Ascherio et al., 2001††	143/144 (99.3%)	269/287 (93.7%)	0.008
Sundström et al., 2004	234/234 (100%)	693/702 (98.7%)	NS
Haahr et al., 2004†	153/153 (100%)	50/53 (94.3%)	0.05
Ponsonby et al., 2005‡‡	136/136 (100%)	252/261 (96.6%)	0.05
Total	1741/1749 (99.5%)	2267/2406 (94.2%)	$p < 10–23$

*Study measured antibodies against the Epstein–Barr nuclear antigens (EBNA), the viral capsid antigen (VCA), and the early antigens (EA).
†Study measured antibodies only against EBNA and EA.
‡Study measured antibodies only against VCA.
††Study measured antibodies only against EBNA and VCA. One person was antibody-negative to each antigen, but it is unclear from the text whether this was the same person. The review by Haahr et al. (2004) suggests it was not.
‡‡Study measured antibodies only against EBNA and VCA.
NS, not significant.

suggests that the association between MS and this particular pathogen is genuine. Taken at face value, the near-100% association with a "prior" EBV infection would seem to indicate that EBV is a necessary (but not a sufficient) condition for adult MS to develop and, therefore (if this is correct), that EBV must be a part of the causal pathway leading to MS.

Even in childhood-onset MS, high-quality evidence indicates that MS is associated with a prior EBV exposure (Alotaibi et al., 2004; Pohl et al., 2006; Krone et al., 2008; Makhani et al., 2016). However, despite such an unequivocal association, the actual prevalence of prior EBV infection reported in children with MS has varied considerably. Thus, in an early study from Canada, 83% of MS cases had evidence of prior EBV infection (Alotaibi et al., 2004). A more recent Canadian study also made a very similar observation (Makhani et al., 2016). Thus, in this second study, 23.5% of 247 patients with acquired demyelinating syndrome (ADS) were diagnosed as MS and, of these MS cases, 84.5% had evidence of a "prior" EBV infection. This rate was markedly different from the rate of prior EBV exposure (44.4%) found in patients with monophasic forms of ADS (Makhani et al., 2016). Both of these observations suggest that, unlike adult-onset MS, a sizable proportion of patients with childhood MS do not have a previous exposure to EBV. By contrast, a large German study, which included even younger children than those studied in Canada, found that virtually all (98.6%) of childhood MS cases have evidence of a prior EBV infection (Pohl et al., 2006). In fact, these authors argue that, based on the distribution of the antibody concentrations, their two EBV-negative MS patients were actually the expected number of "false negatives" (Krone et al., 2008).

It is unclear how to account for these differences between reports – either those between geographic regions in children or those between adults and children in general. Nevertheless, if the observations out of Canada are correct (as seems likely), it must be the case that some children can develop MS independently from any EBV exposure. Perhaps some children develop different demyelinating diseases, with a different pathogeneses, from typical adult MS or perhaps children can develop the disease through different pathways, which are unavailable to adults. Alternatively, as noted above, the disease onset frequently precedes, often by many years, the symptom onset (Okuda et al., 2009, 2014) so that, perhaps, the disease onset actually begins very much earlier than is currently suspected and that the EBV association is due to the disease making EBV infection both more likely to occur in general and more likely to occur at an earlier age. This latter explanation, however, would not easily account either for the migration data (Dean and Kurtzke, 1971; Alter et al., 1978; Elian et al., 1990; Kahana et al., 1994; Cabre et al., 2005) or for the apparent association of MS with mononucleosis – the manifestations of a late EBV infection (Goldacre et al., 2004; Thacker et al., 2006; Farrell,

Table 11.2

Prevalence of antibodies to different viruses in the sera of patients with multiple sclerosis (MS) and controls†

Virus	Virus+ MS cases (%)	Virus+ controls (%)	*p*-value	OR
Adults with MS				
EBV‡	1741/1749 (99.5%)	2267/2406 (94.2%)	$p<10^{-23}$	13.3
HSV*	507/666 (76.1%)	997/1287 (77.4%)	NS	0.93
HSV1*	109/138 (79.0%)	159/205 (77.6%)	NS	1.09
HSV2*	40/138 (29.0%)	29/205 (14.1%)	$p<0.01$	2.5
VZV*	516/637 (81.0%)	1004/1179 (85.2%)	NS	0.74
CMV*	312/693 (45.0%)	447/854 (52.3%)	NS	0.74
Measles*	661/749 (88.3%)	1309/1609 (81.4%)	NS	1.72
Mumps*	224/386 (58.0%)	305/524 (58.2%)	NS	0.99
Rubella*	226/235 (96.2%)	299/307 (97.4%)	NS	0.67
Chlamydia pneumoniae‡‡	85/129 (65.9%)	160/258 (62.0%)	NS	1.18
Children with MS				
EBV**	145/147 (98.6%)	106/147 (72.1%)	$p<0.001$	27.3
HSV-1/2	68/133 (51.0%)	(?)/152 (46.7%)	NS	1.19
HSV-2**	65/149 (43.5%)	(?)/152 (36.2%)	$p<0.0001$	1.36
VZV**	88/99 (88.4%)	147/152 (96.7%)	$p<0.0001$	0.26
Measles**	70/82 (85.6%)	122/152 (90.1%)	$p<0.0001$	0.65
Mumps**	74/90 (82.0%)	(?)/152 (73.0%)	NS	1.68
Rubella**	101/122 (82.5%)	123/152 (80.9%)	NS	1.11
HHV-6**	126/132 (95.6%)	144/152 (94.7%)	NS	1.22
Adenovirus**	8/127 (6.3%)	20/152 (13.2%)	NS	0.44
Influenza A**	24/127 (18.9%)	15/152 (9.9%)	$p<0.05$	2.12
Parainfluenza 2**	27/123 (22.0%)	30/152 (19.7%)	$p<0.05$	1.15

*Data from Ascherio and Munger (2007).
†OR, odds ratio; EBV, Epstein–Barr virus; HSV, herpes simplex virus; VZV, varicella-zoster virus; CMV, cytomegalovirus; HHV, human herpesvirus; NS, not significant.
‡Data from Table 11.1.
**Data from Krone et al. (2008). This is data from children with MS. A (?) appears when the numbers provided don't make sense completely. Data on EBV are partly derived from Pohl et al. (2006).
‡‡Data from Munger et al. (2004a).

2007). Consequently, some variant of the first set of explanations seems more likely.

Presumably, if EBV is in the (or a) causal pathway to MS, it cannot be the factor responsible for the "maternal effect" in MS because EBV infection typically occurs neither *in utero* nor during the early postpartum period. If EBV infection is permissive in the manner suggested then, like the second environmental factor in MS pathogenesis, it probably acts during adolescence (when both late infection and mononucleosis occur) or thereafter and, consequently, it would be a much better candidate for the second (or a later) environmental event. Nevertheless, although the timing of the EBV infection and the second environmental event seems to occur at similar times, the migration data are difficult to explain on the basis of a late infection by a single strain of EBV. Thus, individuals from low-prevalence regions for MS (presumably) would have already acquired EBV early so that moving after early childhood to a high-prevalence region should not matter. Similarly, why children from a high-prevalence region (who presumably have not yet acquired EBV) should adopt a low prevalence when they move is equally unclear. Nevertheless, it is known that there are at least two different strains of EBV (Zimber et al., 1986; Young et al., 1987) with different geographic distributions and, perhaps, a better understanding of the complexity of EBV biology might help to rationalize the migration data.

Regardless of these complexities, however, on the basis of abundant and very consistent high-quality evidence, it seems clear that EBV infection plays some role in MS pathogenesis.

Other infectious agents

Over the years, many other common infectious agents have been suggested as possible causes of MS. Certainly, the evidence is consistent that MS attacks occur more frequently than expected around the time that patients are experiencing nonspecific infectious syndromes (Sibley et al., 1985; Andersen et al., 1993; Panitch, 1994; Edwards et al., 1998). Nevertheless, as causal agents involved in MS pathogenesis, other than for EBV, the evidence is considerably less compelling (Munger et al., 2004a; Pohl et al., 2006; Ascherio and Munger, 2007, 2016; Krone et al., 2008). For example, considering the OR for the seroprevalence of antibodies against many common pathogens in both MS and non-MS populations reveals that, with the notable exception of EBV (in both adults and children with MS), there is not much evidence of an association between specific agents and the likelihood of MS (Table 11.2). Similarly, even for recently proposed agents such as HHV-6 and *Chlamydia pneumoniae*, the findings are inconclusive (Moses and Sriram, 2001; Kaufman et al., 2002) and the seroprevalence data are unimpressive (Table 11.2).

The fact that the seroprevalence data do not particularly suggest a role for infectious agents other than EBV, however, does not exclude the possibility that these other agents might be involved in MS pathogenesis. For example, in the pediatric population (Krone et al., 2008), it is of interest that, although IgG antibodies against *C. pneumoniae* were no different between patients and controls (Table 11.2), there was a highly significant increase in the presence of IgM antibodies (presumably indicating a more recent infection) in patients. However, for unclear reasons, this excess of IgM antibodies was not replicated in an adult series of MS cases (Munger et al., 2004a). Also, the reported excess occurrence of *C. pneumoniae* DNA in the CSF of MS patients (Bagos et al., 2006) and reports of an excess of *C. pneumoniae*-specific immune complexes in the serum of MS patients, especially early in the disease the course (Parratt et al., 2008), if consistently confirmed, will require explanation.

Similarly, in the case of HHV-6 infection, where the prevalence of infection in the general population is high, the seroprevalence data could be misleading. Thus, if the critical factor was either the timing of the HHV-6 infection or a penetration of the CNS by the organism, which occurs randomly in infected individuals, then the seroprevalence data may well be nonsuggestive despite the organism playing a critical causal role. Consequently, it is hard to exclude any of these infections as being potential contributors to MS pathogenesis. Moreover, as noted above, there are at least three environmental factors involved in the cause of MS and, at the moment, there are no leading candidates for the third environmental event.

Vitamin D deficiency

The production of active vitamin D by mammals *in vivo* requires the two-step conversion of 7-dehydro-cholesterol into vitamin D_3 (Holick, 1998; Hayes et al., 2003; Nagpal et al., 2005; Lips, 2006). The first step – the synthesis of previtamin D_3 – is catalyzed by the exposure of 7-dehydro-cholesterol in the skin to ultraviolet B (UVB) radiation (wavelength = 280–320 nm). The second step – a rearrangement of the internal double-bond structure of the previtamin D_3 molecule – forms vitamin D_3. Vitamin D_3 is then hydroxylated, first to 25(OH) D_3 by 25-hydroxylase (primarily in the liver) and second by 1-α hydroxylase in the tissues (in the kidneys and many others) to form $1,25(OH)_2D_3$ (active vitamin D). The dietary intake of vitamin D_3 can circumvent the UVB-dependent part of this pathway and can allow an individual to have normal vitamin D_3 serum levels even in the absence of adequate UVB radiation. Vitamin D, however, is found in only a few natural dietary sources. Sufficient quantities are only present in a few animals (Gillie, 2006) such as oily fishes and reindeer, which derive it or its precursors from their diet (fish from phytoplanctonic algae in the sea and reindeer from lichen on the tundra). Interestingly, two human populations with a notably low MS risk (Sinclair, 1977; Koch-Henderson, 1995; Grønlie et al., 2000; Gillie, 2006) are the Inuit or Eskimos (who consume large quantities of oily fish) and the Sami or Lapps (who eat reindeer meat regularly). In both instances, the principal source of vitamin D for the population comes from their respective diets (Gillie, 2006). Other human populations, by contrast, require sufficient exposure of the skin to UVB radiation in order to maintain adequate vitamin D_3 serum levels throughout the year.

Furthermore, it should be noted that the grouping of vitamin D together with the other vitamins is probably inappropriate. Thus, typically, vitamins are organic molecules, which are involved in specific chemical reactions, and which humans (during the course of their evolution) have lost the ability to synthesize. Presumably, humans were able to do this because these compounds (in general) are widely available in our foods and, consequently, most of our diets provide a sufficient supply, even without specific supplementation. By contrast, for truly essential compounds such as cholesterol (critical for the function of every cell), humans have preserved the ability to synthesize them, even when they are abundant in our diet. Clearly, vitamin D also falls into this second category. Humans have the ability to synthesize it, it is not prevalent in our diet, and it does not participate in any specific biochemical reaction. Rather, vitamin D acts (together with its receptor and the retinoid X receptor) as a transcription factor that controls the expression of thousands of nuclear genes throughout the body and it has become increasingly apparent that its deficiency is associated with a number of medical conditions (Tavera-Mendoza and White, 2007). Moreover, it seems very likely that, for diverse human populations living in temperate (and more extreme) regions of the earth, the occurrence of lighter skin tones represents a convergent evolutionary adaptation to the need for adequate vitamin D in these areas (Jablonski and Chaplin, 2000, 2002).

In this context, it is interesting that, as latitude increases (both north and south of the equator), the angle at which the arriving sunlight strikes the earth's atmosphere becomes more oblique. Such a circumstance causes the light to travel for a longer path through the atmosphere before reaching the earth's surface, thereby reducing the amount of UVB radiation available to human populations for vitamin D synthesis. In central and north-east Africa (where *Homo sapiens* evolved) there is plenty of UVB radiation for vitamin D_3 synthesis throughout the year (Jablonski and Chaplin, 2000, 2002; Kimlin et al., 2007). By contrast, once human populations left Africa and began to inhabit the temperate (and even more extreme) regions of the earth, UVB exposure became inadequate for vitamin D_3 synthesis during some (or many) months of the year. For example, it has been estimated that the level of UVB radiation at the US–Canadian border during most months of the year (Fig. 11.2) is insufficient to produce an adequate amount of vitamin D_3 (Adams, 1989; Jablonski and Chaplin, 2000, 2002; Kimlin et al., 2007). Moreover, as noted by several authors (Adams, 1989; Jablonski and Chaplin, 2000, 2002; Rosati, 2001; Compston et al., 2006) and consistent with a putative role for vitamin D deficiency in MS pathogenesis, worldwide distribution maps of reduced UVB availability (Fig. 11.2) are strikingly similar to comparable maps of MS prevalence (Fig. 11.1).

Vitamin D acts by binding to its receptor (VDR), which is located on the surface of cells throughout the body. This receptor–ligand complex is then internalized to form a heterodimer together with the retinoid X receptor, a complex which is then translocated to the cell nucleus, where it binds to a short sequence of DNA bases (the vitamin D response element or VDRE) located in the promoter region of several (many) nuclear genes (Nagpal et al., 2005; Tavera-Mendoza and White, 2007). Interestingly, the VDRE has recently been identified in the promoter region adjacent to the human leukocyte antigen (HLA) DRB1*1501 allele (Ramagopalan et al., 2009). This observation is notable because, as discussed below, this genomic region (and this allele in particular) has been consistently linked to MS pathogenesis for decades (Compston et al., 2006).

The roles of vitamin D in calcium homeostasis and the maintenance of bone health have been widely known for years (Holick, 1998; Lips, 2006). Its role in other cellular processes, including a variety of immune functions, such as cell proliferation, differentiation, and immunomodulation, as well as its antineoplastic actions, has been less widely recognized (Cantorna, 2000; Deluca and

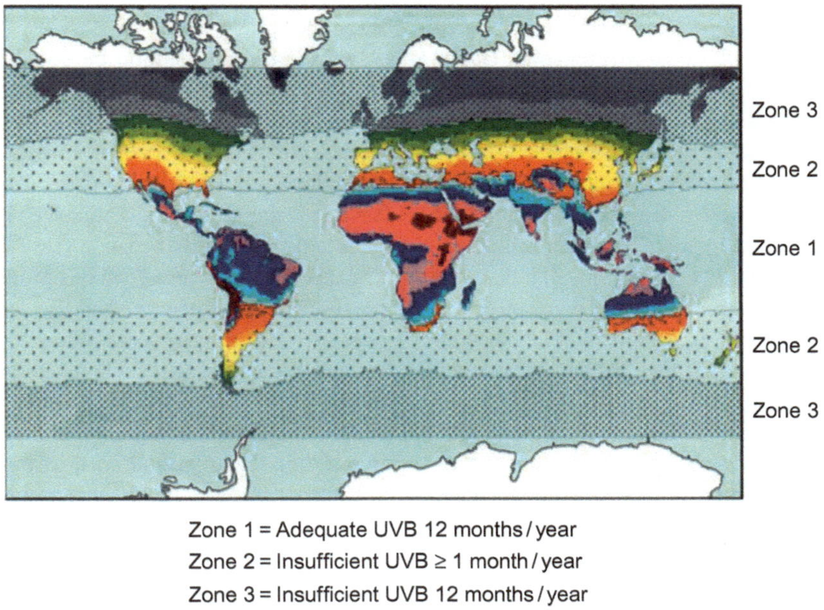

Fig. 11.2. Global distribution of ultraviolet B (UVB) radiation expressed as the number of months estimated to be insufficient for adequate vitamin D synthesis. In zone 2 (where there is a variable insufficiency), an increasing number of insufficient months is indicated by the coloration changing from reds to oranges to yellows to greens. The darkest green indicates insufficient UVB for 11/12 months. The amount of UVB radiation also varies in zones 1 and 3, but is either always adequate (zone 1) or always inadequate (zone 3). (Reproduced from Jablonski and Chaplin, 2000, with permission.)

Cantorna, 2001; Griffin et al., 2001; Hayes et al., 2003; Nagpal et al., 2005; Tavera-Mendoza and White, 2007). Nevertheless, the VDR is expressed on cells throughout the body, including activated T and B cells and on macrophages (Cantorna, 2000; Deluca and Cantorna, 2001; Nagpal et al., 2005). Vitamin D has also been implicated in the maturation of dendritic cell and in the modulation of antigen-specific immune responses *in vivo* (Cantorna, 2000; Deluca and Cantorna, 2001; Griffin et al., 2001). Human decidual cells synthesize active vitamin D during gestation (especially in early pregnancy), suggesting that vitamin D may play a role in the regulation of both acquired and innate immune responses at the fetal–maternal interface (Evans et al., 2006). Finally, vitamin D deficiency seems to play a role in the pathogenesis of several autoimmune diseases, such as insulin-dependent diabetes mellitus, rheumatoid arthritis, experimental autoimmune encephalomyelitis, and inflammatory bowel disease (Cantorna, 2000; Deluca and Cantorna, 2001; Cantorna and Mahon, 2004). Because autoimmune diseases, in general, are more common in women, it is possible that there exist gender-specific differences in the physiologic responses to vitamin D and, indeed, there is some evidence for this proposition (Suarez et al., 1998; Spach and Hayes, 2005). Notably, in one study (Spach and Hayes, 2005) vitamin D supplementation seemed to confer protection against experimental autoimmune encephalomyelitis only to surgically naïve female mice but not to male mice or to ovariectomized females.

With this background, there have been several notable studies, which have explored, more directly, the possible relationship of vitamin D to MS. For example, one group of investigators (Van der Mei et al., 2003) retrospectively interviewed 136 Tasmanian MS cases (through a recruitment campaign that included informational meetings, physician referrals, and posted fliers) as well as 272 sex- and birth year-matched controls recruited from the same source population (using the roll of registered electors). Participants filled in calendars (using a validated questionnaire) and were interviewed about the amount of time they spent in the sun during weekends and holidays during summer for each year of their life. They were also asked about measures taken to protect against the sun and the use of vitamin D supplementation. Higher sun exposure during childhood and early adolescence (6–15 years), as documented by calendar, was associated with a decreased risk of MS ($p < 0.01$). No attempt was made in this study to measure sun exposure in the 0–5-year age range.

Another study (Munger et al., 2004b) reported results from two prospectively established cohorts of nurses, the first established in 1976 (consisting of 121 700 nurses aged 30–55 years) and the second established in 1989 (consisting of 116 671 nurses aged 25–42 years). Both cohorts filled out validated semiquantitative food frequency questionnaires on several occasions. Total vitamin D intake (dietary plus supplemental) was estimated and persons were ranked according to their quintile of vitamin D consumption. Estimates of MS cases per person-year were calculated for each quintile of these cohorts, both collectively and separately. In this study, total vitamin D intake at baseline was found to be inversely associated with risk of MS. Thus, the age-adjusted pooled relative risk (RR) of MS comparing the highest with the lowest quintile of vitamin D consumption was 0.67 ($p = 0.03$). When dietary and supplemental vitamin D were examined separately, the effect seemed to be largely related to the amount of supplemental vitamin D intake.

In another report (Munger et al., 2006), the investigators retrospectively studied 315 MS cases who had previously served in the US armed services. Cases were identified from all active duty personnel (using medical records from the Physical Evaluation Boards' database) and participants were required to have at least one stored serum sample from entry into the military available for analysis. Controls (two per case) were randomly selected from the same population for each case and were matched to the cases on the bases of age, sex, race/ethnicity, dates of sample collection, and branch of military service. Using regression analysis, among whites (but not blacks), there was a significant 41% decrease in MS risk for every 50 nmol/L increase in 25-(OH)-vitamin D ($p = 0.04$), Comparing the top quintile of 25-(OH)-vitamin D level to the bottom quintile yielded an RR of 0.38 ($p = 0.006$).

As a result of these considerations, vitamin D deficiency would seem to be a good candidate for the "maternal" factor in MS pathogenesis. Not only is vitamin D (like this maternal factor) coupled to the solar cycle in temperate regions (Willer et al., 2003), it is known to be involved in immune system maturation (Cantorna, 2000; Deluca and Cantorna, 2001; Griffin et al., 2001; Hayes et al., 2003; Nagpal et al., 2005), its deficiency has been associated with other autoimmune disorders (Cantorna, 2000; Deluca and Cantorna, 2001; Cantorna and Mahon, 2004), its worldwide distribution mirrors that of reduced UVB radiation (Figs 11.1 and 11.2), extreme northern populations with high dietary intake of vitamin D (e.g., the Inuit and Sami people) have a very low prevalence of MS (Sinclair, 1977; Koch-Henderson, 1995; Grønlie et al., 2000; Gillie, 2006), and there are known interactions between the physiologic effects of vitamin D and gender in some mammals (Suarez et al., 1998; Spach and Hayes, 2005). This last aspect of vitamin D physiology might provide insight to the gender specificity of MS and, perhaps, also, to the basis of the increase in MS incidence among women (Hernán et al., 1999; Koch-Henriksen, 1999; Celius and Vandvik, 2001; Barnett et al., 2003; Sarasoja et al., 2004; Orton et al., 2006).

However, regardless of its possible connection to the maternal factor in MS, if vitamin D deficiency acts at all, it could also act during childhood, during adolescence, later in life, or even at multiple different times. In fact, as noted above, the available direct data supporting a role of vitamin D in MS pathogenesis actually suggest that there may be an effect during childhood or adolescence (Van der Mei et al., 2003; Munger et al., 2004b, 2006). This uncertainty about when vitamin D deficiency might play a critical role has important implications for the design of any clinical trial meant to test the so-called "vitamin D hypothesis." Thus, if, as has been suggested previously (Ascherio and Munger, 2007), a clinical trial is designed to enrich the study cohort for individuals at high risk of developing MS by including only first-degree relatives of MS probands, then the study will be a failure if the critical time for the environmental exposure occurs *in utero*, during the early postnatal period, or even during childhood. Thus, in such a circumstance, by the time that MS probands are identified, most of their brothers and sisters will have already passed their window of therapeutic opportunity. Obviously, similar concerns also apply to any therapeutic trial targeting any other potentially early-acting environmental events (e.g., EBV exposure).

Physical trauma

The idea that physical trauma or psychologic stress might play a role in either the causation or exacerbation of MS dates back to the late 19th century and to the earliest descriptions of MS. Since that time there has been an ongoing debate about these possible relationships (Goodin et al., 1999). Nevertheless, actual experimental evidence from well-controlled clinical studies has generally been missing. For example, it was not until 1952 that the first controlled study of the effects of trauma on MS was reported (see Goodin et al., 1999). The authors of this study interviewed 250 MS patients and 250 controls and reported that 36 (14.4%) MS patients had a history of trauma within the 3 months prior to the clinical onset of their MS (which could have taken place years before the interview). By contrast, only 13 control subjects (5.2%) reported a history of trauma in the 3 months prior to the time of the interview ($p < 0.01$). However, because the MS patients were interviewed about remote events, whereas controls were interviewed about recent events, there is likely a difference in recall between the groups and, thus, this study is probably biased. Moreover, the study definition of trauma included events (e.g., peripheral injuries and dental procedures) which are not considered to be biologically plausible antecedents of MS even by the most ardent advocates for an association between MS and trauma (see Goodin et al., 1999). Most importantly, however, every one of the other six controlled trials (of varying levels of quality) have not found any evidence to support a relationship between MS and trauma (Goodin et al., 1999). The most definitive of these studies was a population-based study (Goldacre et al., 2006), which compared the incidence of MS in a cohort of 110 993 patients admitted to hospital following a head injury to the same incidence in a reference cohort of 534 600 admitted to hospital for other reasons. These authors found no significant difference in MS risk between cohorts at any time interval after head injury or with any length of hospital stay. This study effectively excludes any clinically important relationship between the clinical onset of MS and head trauma. In another, more recent study, Spitzer et al. (2012) administered the Childhood Trauma Questionnaire to 234 adult MS cases and 885 control adults from the general population. After adjusting for sociodemographic factors and current depression, these authors found that adult MS patients reported significantly more emotional abuse, sexual abuse, and emotional neglect during childhood compared to adult controls. Notably, however, the MS patients did not report more physical abuse, suggesting that physical trauma during childhood is not associated with MS.

Some authors have suggested that, perhaps, the relationship is most evident following cervical rather than head trauma although, even here, the data (such as they exist) are quite unconvincing. For example, the first study to suggest this relationship was a case series of 16 MS patients who also had cervical spondylosis (Brain and Wilkinson, 1957). In this study, the authors provided pathologic evidence from just two autopsied patients in whom the worst cord demyelination was not at the level of maximum disk disease. Remarkably, despite this apparently negative finding, and without marshalling either other evidence or reasoned argument, these authors concluded that MS lesions within the spinal cord were "associated with the presence of the spondylotic bars" (Brain and Wilkinson, 1957). In a subsequent uncontrolled case study of the distribution of MS lesions within the cervical spinal cord in 18 patients with MS (Oppenheimer, 1978), the author reported on three autopsied patients with severe degenerative changes in their cervical spine and noted that his findings were in conflict with the spondylotic bar "hypothesis" proposed earlier (Oppenheimer, 1978). In fact this author specifically concluded that this hypothesis "loses its force" because the lesions of MS "do not appear to be related to points of compression by spondylotic bars" (Oppenheimer, 1978).

Another uncontrolled case series of 39 MS patients (Chaudhuri and Behan, 2001) examined the relationship between whiplash injury and MS attacks and concluded that such injuries were related to the occurrence of MS symptoms. Nevertheless, there are several serious concerns regarding this study and these conclusions. Thus, the cervical trauma was quite mild, as evidenced by the fact that no patient suffered cervical vertebral fracture, dislocation, or spinal cord compression. Moreover, the nature and severity of the injury were judged retrospectively over a period ranging from 1 to 10 years, a circumstance leading to a high probability of recall bias. In addition, it is unclear on what basis the authors concluded that certain clinical syndromes (e.g., ON, oscillopsia, and internuclear ophthalmoplegia) were precipitated by minor cervical trauma. Most tellingly, however, the population of patients was compiled from a series of patients who were referred for medicolegal consultation specifically because they were known to have a history of both MS and antecedent cervical trauma. Such a "study" cannot provide any useful information about a possible causal relationship between cervical trauma and MS. Indeed, based on all of these considerations, it seems clear that none of these studies provide any substantive evidence that either cervical whiplash injury or trauma from cervical spondylosis can cause (or exacerbate) MS.

Psychologic stress

A possible association between psychologic stress and MS is more difficult to assess. Part of this difficulty is related to the lack of any consistent, agreed-upon measure or definition of stress in the literature. Part, also, is related to the fact that the relationship between psychologic stress and MS (if any) is likely to be different for different types and different severities of life stress. Indeed, the number of potential variations is quite large. Thus, life stress may be acute and severe but self-limited, it may be chronic and mild but long-lasting, or it can be any combination of these attributes. It can range in severity from only a minor disturbance of a person's life to a life-threatening, psychologically traumatic event. Finally, it seems almost certain that the nature and severity of stress produced by similar life events (e.g., marriage, loss of job, emotional abuse, a financial reversal, or the death of a spouse) will vary considerably both between individuals and for individuals at different times of their life. All of these factors combine to make this a particularly difficult area of scientific inquiry.

In a review of this topic, the American Academy of Neurology concluded that a relationship between MS and psychologic stress was, at best, only possible (Goodin et al., 1999). Since then, several further investigations have been undertaken. In 1999, a study of the relationship between stress and MS exacerbation in 61 patients was published (Kroencke and Denney, 1999). Patients retrospectively filled out questionnaires for the Hassles scale, the Uplifts scale, and Ways of Coping scale regarding the previous 6 months. Patients were classified as being in remission, in exacerbation, or in the chronic phase of the illness at the time they filled out the questionnaires. These authors reported that the exacerbation subgroup experienced more hassles and used more passive-avoidant or aggressive coping compared to the chronic subgroup ($p = 0.02$ and $p = 0.05$ respectively) but not compared to the remission subgroup. There are several problems with this study. The retrospective assessment of hassles and uplifts introduces the possibility of recall bias, which can substantially contaminate studies of this type (Teschke et al., 2000; Kip et al., 2001; McIntosh et al., 2002; Ebers, 2004). Also, the absence of a difference between the exacerbation and remission subgroups is hard to rationalize if the apparent effect of stress on MS exacerbations is genuine. Lastly, the failure to statistically adjust for the numerous between-group comparisons made in this study, together with the marginal significance of the reported observations, makes it quite likely that some, if not all, of these observations are actually type I errors.

A prospective study explored the relationship between life stress and new lesion formation on brain MRI in 36 patients (Mohr et al., 2003). Life stress was measured using the Holmes-Rahe Social Readjustment Rating Scale (SRRS), the Hassles Scale, and the Profile of Mood States. This study reported that the number of new MRI lesions was increased after a lag of 8 weeks ($p < 0.001$) following an increase in "conflicts and disruption in routine" subscale of the SRRS (e.g., family or job conflict, changes in routine). No such increase in MRI activity was seen following "major negative" life events (e.g., death of a family member), "positive stress" (e.g., outstanding personal achievement), or daily hassles. Nor did the authors find a similar increase in MRI activity after lags of 0, 4, or 12 weeks. Although the use of MRI is an important methodologic improvement compared to earlier studies, it is noteworthy that: (1) the 8-week lag was not an *a priori* prediction; (2) no adjustment was made to the statistical significance of this *ad hoc* observation despite the large number of between-group comparisons undertaken; (3) there was no evidence of a dose–response (i.e., major stressful life events did not impact on MRI activity); and (4) there was no clinical accompaniment to the reported MRI changes. All of these facts, taken together, raise substantial concern about the reliability of the reported findings.

In a 2003 study of the relationship between stress and MS (Buljevac et al., 2003), 110 eligible patients were identified, 37 of whom refused to participate because of the intense follow-up required. Another 13 dropped out during the course of the study. Thus, only 60 patients (55%) completed the trial. Weekly diaries were collected which catalogued stressful events ranging from "stress related to a holiday" to "death of close family member." The authors excluded 48/505 (9.5%) events, which were thought to be "caused by multiple sclerosis itself." These authors found that patients with stress in the preceding 4 weeks were more likely to experience a first or second exacerbation compared to patients who were stress-free ($p = 0.01–0.02$). Although no analysis of the relationship of stress severity to exacerbation is provided in the text, there was no difference between the effect of a single stressful event compared to the effect of multiple such events in the prior 4 weeks. This study has several sources of potential bias. First, despite the fact that patient diaries were completed on Sunday of each week, there is still a substantial probability of recall bias. For example, if an MS exacerbation began prior to the Sunday survey, it seems likely that patients, aware of the research hypothesis being tested, would report more stress in the prior week than if they had been symptom-free during the same interval. Indeed, a substantial impact of such recall bias has already been well documented in a prospective study of similar design (Kip et al., 2001). Moreover, the large number of dropouts and excluded events may have biased the findings. As a result of these considerations, this study represents only weak evidence in favor of an association. Moreover, the 4-week (or shorter) interval between the stress and its apparent clinical effect is inconsistent with the 8-week lag between the stress and its measurable MRI effect reported by others (Mohr et al., 2003).

Also in 2003, the results of another study examining the relationship between stress and MS exacerbations were reported (Ackerman et al., 2003). This study included 50 patients (all women) who were followed for a period of up to 1 year. Subjects completed weekly questionnaires regarding life events, which may or may not have occurred during the previous week. These events were rated on a four-point scale as either severe (levels 1 or 2) or nonsevere (levels 3 or 4). As in the earlier study (Buljevac et al., 2003), events "potentially related to MS disease activity (e.g., losing a job following an MS attack) were excluded from analysis." The number of excluded events is not clear from the text but, in a preliminary report of the findings in the first 23 subjects (Ackerman et al., 2002), 63% of the severe events and 26% of the total events were excluded on this basis (Ackerman, 2004; Goodin, 2004). In this study the apparent risk of MS exacerbation in the 6 weeks following level 1–3 life events was greater than following level 4 events ($p < 0.05$). The authors found no difference in risk between severe and nonsevere events. Interestingly, no comparison seems to have been made between patients with and without any events. Also, these authors found that the density of life events (i.e., number of events/week) was "positively correlated with the proportion of weeks ill with MS-exacerbations" ($p < 0.05$). This trial also has a substantial risk of recall bias for the same reasons discussed in connection with the earlier study (Buljevac et al., 2003). Moreover, bias may have been introduced by the exclusion of such a large proportion of life events and the failure to compare the findings in patients with and without any life events in the preceding 6 weeks (Ackerman, 2004; Goodin, 2004). As a result this study represents only weak evidence of an association. Most notably, however, the results from this trial seem inconsistent with earlier reports. Thus, the finding that both severe and nonsevere events are associated with MS exacerbations is at odds with one of the reports cited earlier (Mohr et al., 2003), and the finding that event density is associated with MS attacks is at odds with another (Buljevac et al., 2003).

In 2004, the results of a nationwide Danish cohort study were reported (Li et al., 2004). This study retrospectively examined the subsequent risk of MS in parents who had lost a child between 1980 and 1996 compared to parents who had not sustained such a loss. These authors found that the hazard ratio for the development of MS was significantly greater ($p < 0.05$) in bereaved parents. However, this effect was significant only in the subgroup of patients followed for more than 8 years but not in those followed for less time. The major difficulty with this study, therefore, is its apparent marked inconsistency (compared to earlier reports) with respect to the time lag between the occurrence of stress and the onset of disease (Goodin et al., 1999). Moreover, even though not directly analogous to a possible effect of stress on MS exacerbations, a lag of more than 8 years between the stressor and its clinical impact seems quite unexpected based on the reports discussed earlier (Ackerman et al., 2002, 2003; Buljevac et al., 2003; Mohr et al., 2003).

In 2008, Golan and colleagues reported what appeared to be markedly increased occurrence of MS exacerbations requiring steroids in 156 RRMS patients residing in northern Israel during the 33-day Israel–Hezbollah war in 2006. Although there is no question that this war was extremely stressful for the civilian populations of both northern Israel and southern Lebanon and although chance is a very unlikely explanation for the reported findings, the nature of the relationship is still very difficult to decide. Without objective MRI evidence, it is hard to be certain that the observed "neurologic events" actually represented new areas of focal CNS inflammation, even though these attacks were "confirmed" by a neurologist and steroid treatment recommended. It is well described (Compston et al., 2006) that external heat, internal fever, and exercise are capable of causing the re-emergence of previously recovered symptoms without any with new inflammation. These episodes are termed "pseudo-attacks" and, obviously, the comparable mechanisms might conceivably produce symptoms under conditions of extreme stress.

Moreover, this study did not find any time lag between the stress and the exacerbation (Golan et al., 2008). Because the stress in this study was far more severe than that reported in the earlier study (Mohr et al., 2003), it is possible that this difference altered the reported association. If so, however, then it is noteworthy that, during the first Persian Gulf War of 1991, Israeli MS patients experienced fewer MS exacerbations (compared to baseline) both during and after the stress of the Scud missile attacks (Nisipeanu and Korczyn, 1993). Unfortunately, such confusing (and conflicting) results are typical of the literature in this area of research.

Finally, as discussed earlier, in a recent study regarding traumatic events during childhood (Spitzer et al., 2012), the authors found that, even after making adjustments for sociodemographic factors and current depression, adult MS patients still reported significantly more emotional abuse, sexual abuse, and emotional neglect during their childhood compared to adult controls from the same population.

In summary, therefore, although there is some evidence for an association between antecedent psychologic stress and the onset or exacerbation of MS, even the recent data on this subject are inconclusive. At present, there are important conflicts between reports, both with respect to the effect of different degrees of stress and about the timing of these relationships, which prevent any strong conclusions from being drawn. Therefore, a relationship between stress and either MS onset or MS exacerbation remains, at best, only a possibility.

Tobacco

The first case-control study suggesting a possible association between the occurrence of MS and tobacco smoking was published in the mid-1960s (Antonovsky et al., 1965). Subsequent to this initial report, there have now been several similar case-control studies of this relationship (Villard-Macintosh and Vessey, 1993; Thorogood and Hannaford, 1998; Hernán et al., 2001, 2005; Riise et al., 2003; Sundström et al., 2008; Da Silva et al., 2009; Rodríguez Regal et al., 2009) and, even though some individual studies have not achieved statistical significance, each demonstrates an excess of MS cases in smokers compared to nonsmokers. Consequently, in aggregate, this association seems reasonably well established. The basis for this relationship, however, is more difficult to understand. Smoking has been linked to other autoimmune diseases such as systemic lupus erythematosus, Crohn's disease, rheumatoid arthritis, and thyroid disease (Harel-Meir et al., 2007) and, perhaps, the association is due to a direct or indirect physiologic effect of nicotine on the CNS (e.g., immunomodulatory changes or altered vascular permeability). However, it might also be due to an interaction with certain MS-susceptibility genes. In fact, an interesting report in rheumatoid arthritis has provided evidence of an interaction between smoking behavior and the presence of one or more of the HLA DRB1 susceptibility alleles, which are associated with MS (Padyukov et al., 2004). Thus, the risk of rheumatoid arthritis was substantially greater in smokers who carried one or more of these susceptibility alleles compared not only to nonsmokers who also carried these alleles but also to smokers who did not carry them (Padyukov et al., 2004).

In addition, smoking might be associated with disease progression in MS patients, although these reports have been somewhat inconsistent and, therefore, the evidence provided by them is inconclusive. Interestingly, one report claimed that passive exposure to tobacco smoke during childhood was associated with a greater risk of subsequently developing MS (Mikaeloff et al., 2007). Nevertheless, a very detailed subsequent report failed to confirm this finding (Montgomery et al., 2008).

Obesity

In 2009, the cohorts of women who took part in the Nurses' Health Study ($n = 121\ 700$) and the Nurses' Health Study II ($n = 116\ 671$) were analyzed and, after 40 years, 593 MS cases were identified (Munger et al., 2009). In this study the body mass index (BMI) of the participants was estimated at ages 5, 10, 18, and 20 years. These authors reported that patients at age 18 with a BMI of ≥ 30 had a greater than twofold increase in the RR of MS compared to individuals with a BMI of $18.5-23$. A similar risk was found for individuals who were obese at age 20 but not for obesity at ages 5, 10, and during adulthood, suggesting that adolescence was the principal period of risk. A study from Sweden also found a similar relationship (Hedström et al., 2012). Thus, in this study, individuals (men and women) with a BMI of >27 at age 20 years were at twice the risk of MS compared to their nonobese controls. A study out of Denmark (Munger et al., 2013) reported that, among school-aged children (7–13 years), girls in the 95th percentile for BMI had an RR of MS ($RR = 1.61$), which was significantly greater than for nonobese individuals. By contrast, the risk in obese boys seemed to be attenuated compared to girls (Munger et al., 2013). A similar result was found in a study utilizing the large Kaiser database in California

(Gianfrancesco et al., 2014), where girls who were obese at ages 10 and 20 had a greater MS risk than their non-obese counterparts. Finally, in a study from Norway and Italy, the authors found that in Norway (but not Italy) obesity in the period of childhood through young adulthood (in both sexes) increased the likelihood of MS subsequently (Wesnes et al., 2015). Thus, in sum, there seems to be quite consistent evidence that obesity during childhood and young adulthood is a risk factor for MS. Whether this effect is confined to (or most conspicuous in) females is unclear, but possible. Also, the biologic basis for this apparent association is unknown, although it has been suggested that it might be due to the induction of a chronic inflammatory state (with elevated levels of proinflammatory cytokines and with lower circulating serum levels of vitamin D) in obese individuals (Munger et al., 2009; Wesnes et al., 2015).

Dietary factors

More than 60 years ago, Swank (1950) put forth the hypothesis that variations in the frequency of MS seen worldwide could be accounted for by variations in the amount (and possibly the nature) of fat consumption. For example, in North America and Europe, which have high levels (>100 g/day) of per capita fat consumption, there is a considerably greater prevalence of MS compared to Asia, Latin America, and Africa, where this a low level (<60 g/day) of per capita fat consumption (Swank, 1950; Swank et al., 1952). Also, Swank thought that some of the variations in MS incidence both during and after World War II could be explained by variations in fat consumption during that period. And lastly, he noted the greater incidence of MS in northern compared to southern Switzerland and speculated that this was related to differences in the fat content of the diets between these two regions.

In a second paper (Swank and Grimsgaard, 1988), the group studied the occurrence of MS in different regions of Norway. These authors reported a fourfold increase in MS prevalence in the inland (farming) regions compared to coastal regions and related this to a lower fat consumption in the coastal areas where residents depended upon fishing for their livelihoods. The differences in consumption of butter and animal fat between regions, however, were most conspicuous when the authors compared the contemporary fat consumption in the inland areas to fat consumption in the coastal areas determined from an earlier study of diet. When they made the same comparison using simultaneously acquired dietary information, the difference in fat consumption between regions was not as apparent and, thus, their conclusions regarding dietary fats are of uncertain validity. Moreover, these authors, themselves, noted that their observations could not exclude the possibility that fat consumption was correlated with other factors having the same geographic distribution. Also, it is possible that a difference in the incidence of MS might relate to dietary factors other than fats (e.g., the consumption of vitamin D from fish in the coastal regions).

Nevertheless, Swank continued to follow a cohort of 156 patients over the period from 1949 to 1984 (Swank and Dugan, 1990; Swank, 1997). This cohort was derived from a larger group of 264 MS patients seen at the Montreal Neurological Institute between December 1948 and April 1954, 108 of whom declined to participate in the prospective study (41% initial dropout rate). All patients were encouraged to participate in the dietary recommendations. Only 8 patients were lost to follow-up and, of these, 4 were subsequently relocated. Initially, only the total fat intake was restricted, resulting in a drop in daily fat consumption from 125 to 20–30 g/day. Beginning in 1951, however, the recommended diet was modified. Butter fats and hydrogenated oils were eliminated. The consumption of cod liver oil (5 g/day) and vegetable oil (10–40 g/day) was added. In addition, the consumption of animal fats was limited to 15 g/day. During the follow-up interval, commercially processed foods and pen-fattened beef were also restricted because of the increased fat content of these products. Protein intake was limited to 60–90 g/day. This protein was mostly in the form of fish, seafood, and the white meat from chicken and turkey (with the skins removed), although skim milk, lean meats, an occasional egg, vegetables, cereals, and nuts were also allowed. Carbohydrates were consumed as necessary for energy. Not everyone in the cohort elected to follow the diet strictly, and patients were, therefore, divided into three groups based on their total fat consumption (≤20 g/day, 20–30 g/day, and > 30 g/day).

These authors report significant differences in outcome based upon the degree of patient adherence to the diet ($p < 0.001$). Thus, the "good dieters" (≤20 g/day) had a 31% mortality (mostly due to MS) and, after 34 years of follow-up, had an increase in disability of approximately one point on the Expanded Disability Status Scale (EDSS). By contrast, the mortality in the "poor dieters" (the other two groups combined) was 80% (also mostly due to MS) and their disability increased by five to six points on the EDSS scale at the completion of the study (Swank and Dugan, 1990; Swank, 1997). Although these results suggest that total fat consumption might influence the course of MS, there are important limitations to the study design that prevent any strong conclusions from being drawn from its results. Thus, it is even unclear which aspect of the diet might have been responsible for the apparently beneficial effect. For example, the consumption of cod liver oil (very high in vitamin D) was also recommended in addition to the restriction of total fat consumption, so that there is no way to determine which, if either, of these two elements to the diet was important. The same ambiguity is present for each individual aspect of the recommended diet and, importantly, we are not provided with enough information in the published papers to sort out these potential relationships. Moreover, because the study was not randomized, there are concerns that the two groups may have differed in some important respects other than their daily fat consumption. For example, there is evidence that disability levels (possibly from self-selection) were maldistributed between the "good" and "poor" dieters at entry into the study (Swank, 1997), and it is possible that this maldistribution has biased the trial results in favor of the "good dieters." Thus, 32% (23/72) of the "good dieters" who entered the study had minimal disability (neurologic grade of 1) on examination. By contrast, significantly fewer (8% or 6/72) of the "poor dieters" has such a low level of disability at study entry ($z=3.59$; $p<0.001$).

Two other small clinical studies (Alter et al., 1974; Fitzgerald et al., 1987) also suggest that dietary factors may play a role in the causation or exacerbation of the illness. Thus, the short-term trial of Fitzgerald et al. (1987), studying a diet low in fat and high in polyunsaturated oils in 83 MS patients, found that those patients who complied with their recommended diet had fewer clinical attacks and less neurologic disability compared to patients who did not comply with the diet. A study by Alter et al. (1974) examined the covariance of MS prevalence and diet in 22 countries with good information on both parameters. For inclusion, the countries had to have both good medical facilities and fairly standardized case-finding methods. Dietary information was derived from information supplied by the United Nations. These authors found that there were significant linear correlations ($p<0.01$) between the consumption of either animal fat (expressed as the total calories) or fats and oils and the prevalence of MS in the different countries. In a case-control study from Croatia (Sepcic et al., 1993), the ORs for the consumption of full-fat milk (OR = 21.7) and of potatoes with lard and fresh or smoked meat (OR = 20.7) were significantly greater in MS patients compared to controls. In another case-control study in Canada, the authors reported that there was a significantly increased risk of MS in persons who consumed more animal fats and a significantly lower risk in persons consuming more vegetable protein and dietary fiber (Ghadirian et al., 1998).

In summary, although far from conclusive, there is some indication in the literature that dietary factors may possibly be involved (at least partially) in either the causation or the exacerbation of MS. Such a conclusion, however, does not distinguish the dietary hypothesis from the potential role that vitamin D might play in MS pathogenesis. Nevertheless, because of the established (non-MS) health benefits for the so-called "Mediterranean" diet (Estruch et al., 2013) and because of the similarity of this diet to that proposed by Swank, it is probably appropriate to recommend the Mediterranean diet to patients who express an interest in dietary approaches to the management of their disease.

Other factors

As noted earlier, there have been a wide variety of other factors suggested as having some role in either the cause or exacerbation of MS. In general, however, these reports have generally been inconsistent and inconclusive and often the mechanisms by which these factors are proposed to operate in MS pathogenesis lack biologic plausibility.

CHANGING ENVIRONMENTAL EXPOSURES

MS epidemiology has changed in important ways over the past several decades. Thus, the incidence (prevalence) of MS is increasing, especially in women (Hernán et al., 1999; Koch-Henriksen, 1999; Freedman et al., 2000; Celius and Vandvik, 2001; Barnett et al., 2003; Ranzato et al., 2003; Sundström et al., 2003; Sarasoja et al., 2004; Orton et al., 2006). As a consequence of this, the gender ratio has been altered (Orton et al., 2006) and a switch in the latitude gradient for MS incidence has been reported (Hernán et al., 1999). Because MS genetics seems unlikely to have shifted in so short an interval, these observations presumably relate to a change in the environmental determinants of MS. Although many widespread environmental changes are known to be taking place (e.g., increasing atmospheric concentrations of CO_2, CH_4, and other pollutants; increasing global temperatures; a depletion of stratospheric ozone; a greater dietary consumption of *trans* fats and processed foods), one recent change (potentially relevant to the possible role of vitamin D deficiency) is that people are increasingly encouraged to avoid prolonged sun exposure and to use sunblock to prevent skin cancer (Emmons and Colditz, 1999). Nevertheless, sunblock with sun-protective factor (SPF) 15 blocks approximately 94% of the incoming UVB radiation and higher SPF levels block even more (Emmons and Colditz, 1999). As a result, any widespread use of sunblock and/or sun avoidance will exacerbate any population deficiency of vitamin D synthesis and, presumably, will increase the likelihood of diseases related to vitamin D deficiency. By contrast, the pattern of EBV infection seems to have changed little over this interval (Müller et al., 2005).

In summary, the current epidemiologic evidence seems to support the existence of three (or more) environmental events which contribute to MS pathogenesis. The first event occurs near birth, the second occurs during childhood, and the third occurs long after the first two have already taken place. At present, the two best candidate factors identified are vitamin D deficiency and EBV infection. Indeed, these two factors seem particularly well suited to the first two environmental events in MS pathogenesis. However, because there are at least three distinct events involved in MS pathogenesis, there is no need to choose between vitamin D and EBV. They could, easily, both be involved. Nevertheless, even if these two environmental events are implicated as part of a pathway to adult MS, this does not guarantee that they are on the same or the only pathway. Indeed, assuming that each factor is part of some causal pathway, there are several possible ways these two environmental events might interact to produce MS (Goodin, 2009). No pathway can be excluded entirely, although, if prior EBV infection is a necessary condition for MS, then this implies that these two factors act sequentially and are part of a causal cascade of environmental events, which leads to adult MS (Goodin, 2009).

GENETIC FACTORS

As noted earlier, the risk of developing MS for individuals who have an affected family member increases in rough proportion to the amount of shared genetic information between themselves and the proband (Mumford et al., 1994; Ebers et al., 1995, 2004; Robertson et al., 1996; Sadovnick et al., 1996; Compston and Coles, 2002; Willer et al., 2003; Nielsen et al., 2005; Compston et al., 2006; Islam et al., 2006). Thus, for example, siblings of an MS proband (50% genetic similarity) have a 20–30-fold increased risk compared to the general population, whereas monozygotic twins (100% genetic similarity) have a risk more than 200 times that in the general population (e.g., Mumford et al., 1994; Willer et al., 2003; Hansen et al., 2005a,b; Islam et al., 2006; Ristori et al., 2006; Kuusisto et al., 2008). These observations, by themselves, unequivocally implicate genetic factors in the pathogenesis of MS.

General considerations

The field of the genetic epidemiology of MS has been under particularly active investigation for the past two decades and several large genomewide association studies (GWAS) have now been completed and have identified almost 200 MS-susceptibility genes, in addition to the long-established linkage to the HLA class II region on chromosome 6, which is thought to be due to an association with the DRB1*1501 allele (Baranzini et al., 2009; De Jager et al., 2009; International Multiple Sclerosis Genetics Consortium and the Wellcome Trust Case Control Consortium, 2011; International Multiple Sclerosis Genetics Consortium, 2014). However, such GWAS investigations focus their attention upon the analysis of approximately 500 000 single-SNPs (single-nucleotide polymorphisms: located in scattered regions throughout the genome) and these analyses identify those single-SNPs, which are significantly associated with the disease. Although, rarely, an SNP identified in this manner may actually represent the genetic alteration responsible for the disease association, in most cases these SNPs exists on haplotypes within a particular genomic region and merely "tag" an allele of some nearby gene, which is the actual basis of the observed disease relationship. In fact, most of these single-SNPs are located in either intronic or intergenic regions of the genome (e.g., Fig. 11.3). This type of association has been referred to as a "synthetic" association

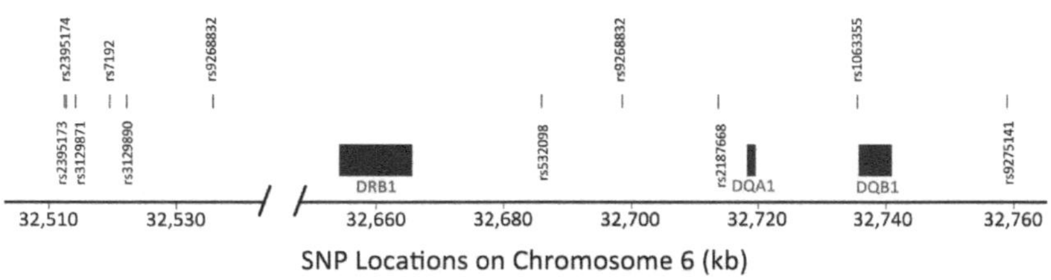

Fig. 11.3. Location of the 11 single-nucleotide polymorphisms (SNPs) in the SNP haplotype surrounding the DRB1 gene on chromosome 6, which has the most significant disease association (see text). The two single-SNPs in this particular SNP set, which have the greatest disease association, are: rs2395173 ($\rho = 0.63$) and rs3129871 ($\rho = 0.59$). Both are located on the extreme left-hand edge of the SNP haplotype, far from the DRB1 locus. The blue bars represent the location of the start and stop points for the DRB1, DQA1, and DQB1 genes. Table 11.3 lists some of the SNP haplotypes (composed of these 11 SNPs) found in this genomic region. The locations of other human leukocyte antigen (HLA) genes discussed in the text (not depicted) are as follows:

HLA A: 30 018–30 021 kb
HLA C: 31 345–31 348 kb
HLA B: 31 430–31 433 kb

(Adapted from Khankhanian et al., 2015.)

(Dikson et al., 2010). However, because a single-SNP can tag more than one haplotype, these synthetic associations require a very large number of patients to uncover, their SNPs typically have a greater allelic frequency than the underlying susceptibility allele, these SNP associations are generally weak and underestimate the strength of the true genetic association (Dikson et al., 2010; Goodin and Khankhanian, 2014), and typically, the GWAS analysis using single-SNPs leaves a large residual "gap" of unexplained heritability (Gourraud et al., 2012; Hofker et al., 2014). As a consequence, these SNP associations only identify genomic regions of interest and do not really help to identify the actual basis for the association (e.g., Fig. 11.3). By contrast, using these SNPs to create SNP haplotypes in the different genetic regions can markedly reduce the heritability gap compared with single-SNP methods and make it clear that these associations identify genetic regions rather than particular genes (Khankhanian et al., 2015).

In fact, these haplotype methods can be used to fine-map the genetic associations within regions of interest previously identified by a single-SNP GWAS (Goodin and Khankhanian, 2014; Khankhanian et al., 2015). Thus, multi-SNP haplotypes detected disease associations in 32 of the 110 regions that were at least 1000-fold more significant than those detected by single-SNPs (Khankhanian et al., 2015). By contrast, single-SNPs were never similarly more significant than the multi-SNP haplotypes in detecting disease associations. Moreover, the nature of the disease associations identified was altered, even within a previously defined genomic region, and the amount of the heritability of MS explained by the observed associations was markedly improved using multi-SNP haplotype methods compared to the use of only single-SNPs (Khankhanian et al., 2015).

In addition, the use of these SNP haplotype methods can help to clarify some of the underlying genetic relationships of MS. For example, Khankhanian and coworkers (2015) reported that, in the class II region of chromosome 6, the most disease-associated haplotype was an 11-SNP haplotype consisting of SNPs (rs2395173_A; rs2395174_C; rs3129871_A; rs7192_A; rs3129890_G; rs9268832_A; rs532098_A; rs17533090_A; rs2187668_A; rs1063355_A; and rs9275141_C), which includes 246 kb of DNA surrounding the DRB1 gene (Fig. 11.3). The final letter in each SNP name designates the minor allele nucleotide (in the control population) at that SNP location. This minor allele is coded as equal to "1" in the haplotype, whereas the major allele is coded as "0" (Table 11.3). Several of the 20 major haplotypes in this region are listed in Table 11.3. These 20 haplotypes account for over 95% of all haplotypes identified and were found at similar frequencies (Table 11.3) in both the EPIC and WTCCC datasets (Baranzini et al., 2009; International Multiple Sclerosis Genetics Consortium, 2014; Khankhanian et al., 2015). The EPIC dataset consisted of 882 controls and 975 MS cases; the WTCCC dataset consisted of 18 872 controls and 11 376 MS cases. Each SNP haplotype has been named arbitrarily according to its order of identification in the EPIC dataset (Goodin and Khankhanian, 2014), using the SNP set presented above and in Figure 11.3. There was a notable consistency between the EPIC and WTCCC control populations with regard to the haplotype identity and frequency, both of which consisted, largely, of persons having a northern European ancestry (Table 11.3). Moreover, each of the SNP haplotypes in this region was very specific with regard to its HLA haplotype association (Fig. 11.4) and each of these SNP haplotypes (as well as the others not listed) had the same HLA association in both datasets (Khankhanian et al., 2015).

Table 11.3

Single-nucleotide polymorphism (SNP) haplotypes in the class II region of chromosome 6*

		Haplotype frequency	
Name	SNP haplotype	EPIC	WTCCC
a1	10110100010	0.11	0.12
a2	00000000100	0.02	0.02
a3	00000010001	0.21	0.19
a4	00000000001	0.13	0.11
a5	10100010001	0.08	0.09
a6	01011100100	0.09	0.10
a8	10110100011	0.00	0.00
a9	01000001010	0.11	0.11
a11	00000010010	0.03	0.02
a27	10100100011	0.00	0.00
a36	10100100010	0.00	0.00

*The "name" is arbitrary and indicates the order of haplotype identification in the EPIC dataset (Goodin and Khankhanian, 2014). The SNP haplotype represents the haplotypes identified using the set of 11 SNPs, shown in Figure 11.1. The number "0" indicates the presence of the major allele (in the control population) at the particular SNP location. By contrast, the number "1" indicates the presence of the minor allele (in the control population) at this particular location. Only 11 selected SNP haplotypes are listed. Haplotype frequencies found in two independent datasets (EPIC and WTCCC) are shown (Goodin and Khankhanian, 2014; Khankhanian et al., 2015).

In addition, as expected from previous observations in Caucasian populations (e.g., Ahmad et al., 2003; Wennerström et al., 2013; Zúñiga et al., 2013), several of the haplotypes in this region were extremely extended. Nevertheless, the frequency of such extended haplotypes in this dataset seems somewhat surprising (Fig. 11.5). Thus, for example, considering only extended

Analysis of DRB1/DQB1 Haplotypes (by SNP-haplotype)

	a1	a2	a3	a4	a5	a6	a8	a9	a11	a27	a36
101/501								0.91			
301/201		1.00				1.00					
401/301		0.00	0.25	0.00	0.01						
401/302			0.26	0.00	0.01						
404/302			0.22	0.00	0.00						
701/202			0.17		0.61						
701/303					0.35						
1101/301			0.00	0.71							
1104/301				0.15							
1301/603	0.00								0.98		0.04
1501/502	0.00						0.94			0.31	
1501/602	0.99						0.06			0.58	0.91
	a1	a2	a3	a4	a5	a6	a8	a9	a11	a27	a36
% Identified	1.00	1.00	1.00	1.00	1.00	1.00	1.00	1.00	0.98	0.90	1.00
Total #	11093	1182	10398	5804	5021	5905	121	5928	1297	48	57

A

Analysis of DRB1/DQB1 Haplotypes (by HLA-haplotype)

	a1	a2	a3	a4	a5	a6	a8	a9	a11	a27	a36	% Idnetified	Total #
101/501								0.99				1.00	5484
301/201		0.17				0.83						0.99	7135
401/301		0.00	0.96	0.00	0.01							0.99	2680
401/302			0.96	0.00	0.01							0.99	2861
404/302			0.96	0.00	0.01							0.99	2400
701/202			0.36		0.62							0.99	4993
701/303					0.98							0.99	1769
1101/301			0.00	0.96								0.98	4305
1104/301				0.98								0.98	911
1301/603	0.00								0.37		0.00	0.99	3441
1501/502	0.02						0.83			0.11		0.99	137
1501/602	0.99						0.00			0.00	0.00	1.00	11172
	a1	a2	a3	a4	a5	a6	a8	a9	a11	a27	a36		

B

Fig. 11.4. The human leukocyte antigen (HLA) haplotype–single-nucleotide polymorphism (SNP)haplotype associations – both by SNP haplotype (**A**) and also by HLA-haplotype (**B**) – for the selected SNP haplotypes presented in Table 11.3. Haplotypes not presented also had very specific haplotype–haplotype associations (Khankhanian et al., 2015).

haplotypes that are composed of a specific SNP haplotype, together with a specific combination of alleles at each of three HLA class I loci (A, B, and C) and two HLA class II loci (DRB1 and DQB1) – i.e., haplotypes which span a genomic region of more than 2.5 mb of DNA – there are more than 4 billion possible unique combinations of these SNP haplotypes together with these five HLA alleles. Despite this large number of possibilities, however, there were only 10 076 unique combinations represented within the set of 59 882 haplotypes of the WTCCC population. Moreover, of this total of 59 882 haplotypes, 22% were accounted for by only 10 unique combinations, 30% were accounted for by 25 combinations, and 71% were accounted for by 810 combinations (Table 11.4). Furthermore, for each of the top 25 haplotypes, the association of the class I component with the class II component of the haplotype was markedly greater than chance ($p << 10^{-300}$). By contrast, 6014 (60%) of the combinations were unique (i.e., had only a single representative) in the WTCCC dataset. Similarly, 7410 (74%) of the combinations had two or fewer representatives but accounted for only 8806 (15%) of the total number of combinations in the WTCCC dataset.

Of the 14 extended haplotypes in Caucasians described by Ahmad and coworkers (2003), six were also in the top 25 in the WTCCC dataset (Table 11.4). Another two haplotypes in these top 25 were extremely similar to those described previously (Table 11.4); both of the haplotypes actually reported were very rare in the WTCCC – one was not present and the other had only two representations. Similarly, of the 17 extended haplotypes for the A, B, and DRB1 loci in Caucasians described by Wennerström and coworkers (2013), nine were in the top 25 in the WTCCC dataset (Table 11.4). And finally, in the small Caucasian population from

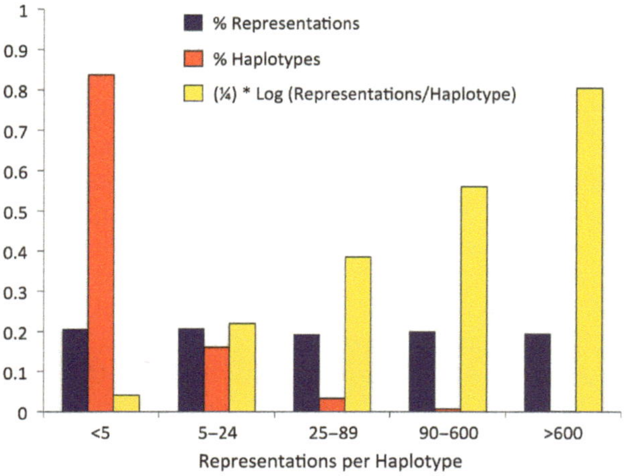

Fig. 11.5. Plots of the number of different haplotypes and the number of representations for each haplotype in the WTCCC dataset. Groups have been divided into quintiles based on the total number of haplotypes represented in the WTCCC data (blue). Also plotted are the number of haplotypes per group (red) and one-quarter of the logarithm (base 10) of the ratio of representations to the number of haplotypes represented (yellow).

the report by Zúñiga and coworkers (2013), three of the six haplotypes were in the top 25 (Table 11.4). Of the remaining 17 described haplotypes, all but the three from Zúñiga and coworkers (2013) had at least some representation and 11 of them had 10 or more representations in the WTCCC dataset.

The reason for these differences between the Caucasians studied is not known, although it seems likely to be a consequence of differences (possibly subtle) between populations. For example, a recent study from Nigeria also reported a high prevalence of extended HLA haplotypes (Testi et al., 2015). Thus, in this study, 14 extended 5-locus West African haplotypes (A-C-B-DRB1-DQB1) accounted for 16% of the haplotypes present in their population. Also, in an Amerindian population, a high prevalence of 5-locus extended HLA haplotypes was, again, reported (Zúñiga et al., 2013). Thus, in their mixed population (which included Amerindians, Caucasians, Africans, and persons of undetermined ancestry), 16 Amerindian haplotypes accounted for 13% of the haplotypes present. Presumably, if the different groups had been segregated in this report (Zúñiga et al., 2013), this percentage would be much higher in the Amerindian population. Interestingly, however, there was essentially no overlap between any of these different populations. Common haplotypes in West Africa and Amerindians were completely distinct from each other (Zúñiga et al., 2013; Testi et al., 2015). Moreover, in the WTCCC dataset, only a single individual carried a single copy of the most common West African haplotype. The remainder of the WTCCC was

Table 11.4

Most common (top) 25 extended haplotypes in the WTCCC

Haplotype (A_C_B_DRB1_DQB1_SNP)†	Frequency	%††
01:01_07:01_08:01_03:01_02:01_a6*§	3782	0.063
03:01_07:02_07:02_15:01_06:02_a1*§	2961	0.049
02:01_07:02_07:02_15:01_06:02_a1¶§	1465	0.024
03:01_04:01_35:01_01:01_05:01_a9§	1086	0.018
02:01_05:01_44:02_04:01_03:01_a3*§	906	0.015
29:02_16:01_44:03_07:01_02:02_a5**¶	730	0.012
24:02_07:02_07:02_15:01_06:02_a1	728	0.012
02:01_03:04_15:01_04:01_03:02_a3**§	556	0.009
01:01_06:02_57:01_07:01_03:03_a5*	554	0.009
11:01_04:01_35:01_01:01_05:01_a9*	539	0.009
25:01_12:03_18:01_15:01_06:02_a1	440	0.007
02:01_03:04_40:01_13:02_06:04_a19§	416	0.007
01:01_07:02_07:02_15:01_06:02_a1	405	0.007
02:01_07:01_08:01_03:01_02:01_a6	397	0.007
02:01_06:02_13:02_07:01_02:02_a3§	361	0.006
01:01_07:01_08:01_15:01_06:02_a1	320	0.005
23:01_04:01_44:03_07:01_02:02_a5	312	0.005
02:01_06:02_57:01_07:01_03:03_a5	293	0.005
02:01_05:01_44:02_15:01_06:02_a1	289	0.005
30:01_06:02_13:02_07:01_02:02_a3*	285	0.005
31:01_03:04_40:01_04:04_03:02_a3	253	0.004
11:01_07:02_07:02_15:01_06:02_a1§	229	0.004
30:02_05:01_18:01_03:01_02:01_a2¶	212	0.004
02:01_01:02_27:05_01:01_05:01_a9	211	0.004
03:01_08:02_14:02_13:02_06:09_a25	201	0.003
Total	17931	0.299

†Extended haplotypes including the human leukocyte antigen (HLA) loci (A, C, B, DRB1, and DQB1), together with a particular SNP haplotype (see Table 11.3).

††Percentage (%) of the 59 882 haplotypes in the WTCCC dataset represented by each extended haplotype.

*Extended haplotypes (in Caucasians) described previously by Ahmad et al. (2003) without inclusion of the SNP haplotypes.

**Very similar haplotypes to those described by Ahmad et al. (2003), except that A_29:01 is replaced by A_29:02; and C_03:03 is replaced by C_03:04.

¶Extended haplotypes (in Caucasians) described by Zúñiga et al. (2013), again without the SNP haplotypes.

§Extended haplotypes (for only the A, B, and DRB1 loci) from Finnish (and other) populations, described by Wennerström et al. (2013).

completely distinct from the common haplotypes in both the Amerindian and West African populations.

Regardless of the explanation for these differences, however, the fact remains that the large majority of haplotypes in the WTCCC dataset are represented multiple (≥ 3) times and that a high prevalence of extended HLA haplotypes also characterizes other populations. These observations suggest that, while different haplotype combinations frequently do arise from one generation to the next, most of these combinations are under very strong negative selection pressure. Such a circumstance further suggests that natural selection is operating upon specific combinations of alleles that "work" together (whatever this means) rather than on the specific alleles within each combination (e.g., Ahmad et al., 2003). If so, this also raises the possibility that any observed disease association in this genomic region similarly relates to the combination of alleles rather than to a particular allele – a consideration which has obvious relevance to the discussions that follow.

DRB1*1501 – DQB1*0602 HLA HAPLOTYPE

This two-locus HLA haplotype is strongly associated with the (a1) SNP haplotype (Table 11.3; Fig. 11.4). It also occurs in specific association with several other SNP haplotypes, including (a27), (a34), and (a36). For example, among the 52 individuals who carried the (a36) SNP haplotype, 48 also carried the DRB1*1501 –DQB1*0602 HLA haplotype but did not carry an (a1) SNP haplotype. Of these persons, 38 (79%) were controls. Consequently, the OR for those rare individuals who carry a single-copy of the (a36) DRB1*1501 – DQB1*0602 haplotype but not the (a1) SNP haplotype (compared to those who do not carry either the (a36) SNP haplotype or the DRB1*1501 – DQB1*0602 HLA haplotype) was significantly different from the disease association found for carriers of a single copy of the (a1) DRB1*1501/DQB1*0602 haplotype (Fig. 11.6).

Similarly, 14/19 (74%) of the carriers of the (a27) SNP haplotype, who also carry the DRB1*1501/DQB1*0602 haplotype but not the (a1) SNP haplotype, were controls. Thus, those non-(a1) individuals who carry a single-copy of the (a27) DRB1*1501 – DQB1*0602 haplotype (compared to those who do not carry either the (a27) SNP-haplotype or the DRB1*1501 – DQB1*0602 HLA haplotype) also had a significantly different disease association from that found for single-copy carriers of the (a1) DRB1*1501 – DQB1*0602 haplotype (Fig. 11.4). Indeed, considering all non-(a1) DRB1*1501 – DQB1*0602 carriers, the OR for disease association was both not significantly different from unity and was significantly different lower than the OR in (a1) DRB1*1501 – DQB1*0602 carriers (Fig. 11.6).

By contrast, those rare individuals who carry the (a1) SNP haplotype but not the DRB1*1501/DQB1*0602 HLA haplotype, have a disease risk (compared to noncarriers of either haplotype) which does not differ significantly from the risk in single-copy carriers of the (a1) DRB1*1501 – DQB1*0602 haplotype (Fig. 11.6). Also, the disease risk in single-copy carriers of the (a1) DRB1*1501 – DQB1*0602 haplotype is unaffected by having a second copy of the DRB1*1501 allele (Fig. 11.6).

As anticipated from earlier publications (e.g., Ahmad et al., 2003; Wennerström et al., 2013; Zúñiga et al., 2013), the (a1) SNP haplotype resides on one of several very long extended haplotypes, which stretch over a genomic distance of more than 2.5 mb of DNA (Table 11.4). Many of these haplotypes share the motif of C*0702 – B*0702, in addition to having the DRB1*1501 – DQB1*0602 class II haplotype (Table 11.4). Nevertheless, dividing the (a1) carriers into those who have this motif and those who do not, there was no difference between groups with regard to their association with disease risk (Khankhanian et al., 2015).

Taken together, these observations suggest that the apparent association of MS with the DRB1*1501 – DQB1*0602 HLA haplotype is not due to either of these HLA alleles but, rather, is due to something else about extended (a1) SNP haplotype. If so, then the apparent association of these two HLA alleles and MS, which has been observed for decades (Baranzini et al., 2009; De Jager et al., 2009; International Multiple Sclerosis Genetics Consortium and the Wellcome Trust Case Control Consortium, 2011; International Multiple Sclerosis Genetics Consortium, 2014), is presumably an artifact that is caused by the extremely tight coupling between the (a1) SNP haplotype and this particular extended HLA haplotype (Figs 11.4 and 11.6).

DRB1*0301 – DQB1*0201 HLA HAPLOTYPE

This particular HLA haplotype is very tightly coupled to both the (a2) and the (a6) SNP haplotypes (Fig. 11.4). These two SNP haplotypes, however, clearly have different disease associations from each other. In the case of (a6), heterozygotes carry no risk and the pattern seems to be recessive (Fig. 11.6). By contrast, in the case of (a2), the heterozygotes do carry a significant risk and the pattern suggests either a dominant or a dose-dependent effect (Fig. 11.6).

Similar to the (a1) SNP haplotype, both the (a2) and the (a6) SNP haplotypes reside on very long extended haplotypes (Khankhanian et al., 2015). Thus, in the case of (a6), this extended haplotype includes the A*0101, C*0701, and B*0801 alleles, whereas, in the case of (a2), the extended haplotype includes the A*3002, C*0501, and B*1801 alleles (only a single case was homozygous). If only these extended haplotypes are considered, the patterns of disease association (noted above) are exaggerated (Fig. 11.6). This, again, suggests that the disease association is not due to either the DRB1*0301 – DQB1*0201 HLA haplotype or to one of these two HLA alleles individually. Rather, it seems that the disease risk must be conferred by something else about the (a2) and (a6) SNP haplotypes.

In addition, there are several general points about the genetics of MS that deserve mention, as they are material to the interactions that take place between genetic

Relationships of HLA Haplotypes with the (a1)-SNP-haplotype

SNP Haplotypes	HLA Haplotype {DRB1*1501 / DQB1*0602}	OR (1 v. 0)	OR (2 v. 0)
All		3.00 (CI = 2.85 - 3.16)	6.38 (CI = 5.59 - 7.28)
a1		3.04 (CI = 2.88 - 3.20)	6.52 (CI = 5.70 - 7.45)
a27, non-a1		0.91 (CI = 0.33 - 2.52)	na
a36, non-a1		0.67 (CI = 0.33 - 1.34)	na
non-a1		1.42 (CI = 0.98 - 2.06)	na
	*2 Copies: {DRB1*1501 / DQB1*0602}*		
a1		2.77 (CI = 1.22 - 6.29)	6.52 (CI = 5.70 - 7.45)
	*No Copies: {DRB1*1501 / DQB1*0602}*		
a1		1.97 (CI = 1.15 - 3.36)	na
	*1 Copy: {DRB1*1501 / * }*		
All		2.99 (CI = 2.84 - 3.15)	
a1		3.03 (CI = 2.87 - 3.19)	6.54 (CI = 5.72 - 7.47)
non-a1		1.52 (CI = 1.18 - 1.95)	
	*2 Copies: {DRB1*1501 / * }*		
a1		2.76 (CI = 1.59 - 4.82)	6.52 (CI = 5.76 - 7.52)

Relationships of HLA Haplotypes with the (a2) and (a6) SNP-haplotypes

SNP Haplotypes	HLA Haplotype *DRB1*0301 / DQB1*0201*	OR (1 v. 0)	OR (2 v. 0)
All		0.98 (CI = 0.92 -1.04)	1.67 (CI = 1.41 -1.97)
a2		1.31 (CI = 1.16 -1.48)	1.91 (CI = 0.93 -3.91)
a6		0.94 (CI = 0.89 -1.01)	1.65 (CI = 1.35 -2.02)
a2 + a1		4.61 (CI = 3.44 -6.16)	na
a6 + a1		3.89 (CI = 3.45 -4.89)	na
a2, non-a1		2.10 (CI = 1.84 -2.41)	3.64 (CI = 1.78 -7.47)
a6, non-a1		1.08 (CI = 1.00 -1.16)	2.60 (CI = 2.13 -3.19)
non-a2, non-a6		0.75 (CI = 0.40 -1.42)	na
	HLA-hyplotype including *A*3002; C*0501; B*1801*		
a2		1.73 (CI = 1.33 -2.33)	na
non-a2		0.67 (CI = 0.21 -2.13)	na
	HLA-hyplotype including *A*0101; C*0701; B*0801*		
a6		0.98 (CI = 0.91 -1.06)	2.11 (CI = 1.55 -2.89)
non-a6		0.71 (CI = 0.36 -1.41)	na

Fig. 11.6. Relationships between and the a1, a2, and a6 single-nucleotide polymorphism (SNP) haplotypes and multiple sclerosis (MS) in conjunction with different combinations of human leukocyte antigen (HLA) haplotypes. The odds ratios (OR) are given for comparing cases and controls carrying either one or two copies of the risk SNP haplotype to cases and controls carrying zero copies. In both circumstances, the disease association varied markedly, depending upon which SNP haplotype carried the HLA haplotype. Such an observation indicates that the observed disease associations were not due to these specific HLA alleles but, rather, to something else about these SNP haplotypes (see text for a discussion). CI, confidence interval.

susceptibility and environmental exposure. First, it seems that most MS cases occur in genetically susceptible individuals who experience the necessary environmental exposures (Goodin, 2016). MS very rarely develops (if at all) in the remainder of the population (i.e., in nonsusceptible individuals). Second, the prevalence of such susceptibility is, at most, 8.5% in the general population, and likely to be much less (Goodin, 2016). By contrast, the environmental factors must occur at a largely population-wide exposure level (Sadovnick et al., 1996, 2005; Ebers et al., 2000, 2004; Bager et al., 2006; Dyment et al., 2006; Goodin, 2016).

Consequently, a person's genetic make-up is, by far, the most important component of that risk. Thus, it seems that more than 92.5% of the individuals (and likely a lot more) in these different populations are genetically incapable of getting MS, regardless of what environmental events they experience during their lives (Goodin, 2016). Moreover, because the prevalence of genetic susceptibility seems to be so similar in the different regions of Europe and North America (Goodin, 2009), it must be a difference in the environmental events which underlies the observed variations in MS epidemiology between these geographic regions. In addition, the mechanisms underlying genetic susceptibility to MS are likely to be quite complex. Thus, despite the fact that the (a1) SNP haplotype (which usually carries the DRB1*1501 allele) has the largest and most consistent association with MS susceptibility of any genetic marker (Baranzini et al., 2009; De Jager et al., 2009; International Multiple Sclerosis Genetics Consortium and the Wellcome Trust Case Control Consortium, 2011; International Multiple Sclerosis Genetics Consortium, 2014), the probability of being in the genetically susceptible subset is still quite small, even when this haplotype is present (Goodin, 2016). Also, 45–50% of individuals with MS lack this haplotype entirely (International Multiple Sclerosis Genetics Consortium and the Wellcome Trust Case Control Consortium, 2011; International Multiple Sclerosis Genetics Consortium, 2014). And, finally, the likelihood of developing MS in "genetically" susceptible individuals is approximately the same regardless of whether the susceptible genotype includes or does not include this haplotype (Goodin, 2016). As a consequence, based on all of these considerations, it seems that this particular haplotype may not be as uniquely important for MS-susceptibility as is currently believed by many.

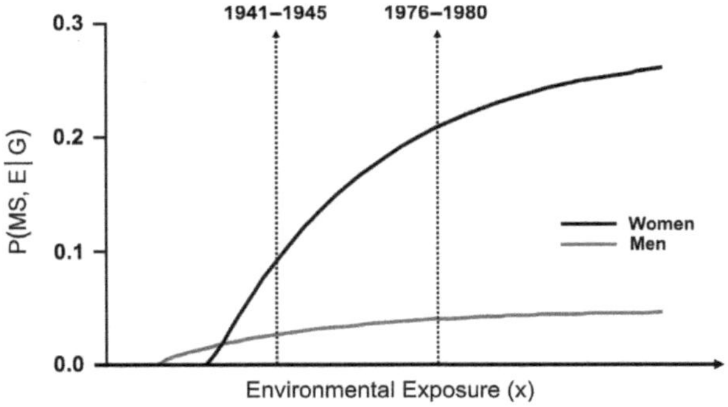

Fig. 11.7. Response curves for developing multiple sclerosis (MS), derived from a simple mathematic model (Goodin, 2009, 2016), in men and women to an increasing likelihood of a "sufficient" environmental exposure (E). Proportionate (but not necessarily equal or constant) hazards have been assumed for the two genders. The conditional probability of developing MS in a genetically susceptible individual (G) is shown on the *y*-axis. The transformed exposure level (x) is shown on the *x*-axis (Goodin, 2009, 2016). One environmental unit is defined (arbitrarily) as the change in the environmental exposure level (whatever this represents and however many factors this involves), which has taken place in Canada between the time periods of 1941–1945 and 1976–1980. Response curves are derived solely based on published epidemiologic observations (Willer et al., 2003; Orton et al., 2006). The response curve for men begins at a smaller value of x than women, but their response is almost at its plateau by 1941–1945. By contrast, women are nowhere near their (much higher) plateau in 1941–1945 and have a much steeper rise in incidence as a response to environmental changes during the interval of observation compared to men. (Adapted from Goodin, 2009, 2016.)

Genetic and environmental interactions

There are important interactions between the genetic make-up of individuals and their environment, which are indicated by the observation that the proband-wise monozygotic twin concordance rate for women (0.34) is significantly larger ($p<0.001$) than the same rate (0.065) for men (Willer et al., 2003). The consequence of this observation (if correct) is that men are 60% more likely to be genetically susceptible to getting MS than are women (Goodin, 2016). Also, a direct consequence of this difference, together with the observation that the greater current prevalence of MS is greater in women than men, is that women must be more likely to experience the necessary environmental events, more likely to get MS in response to these events, or both (Goodin, 2016). These relationships can be derived mathematically from the available epidemiologic data (Goodin, 2016) and are illustrated in Figure 11.7. Considering some of the potential environmental factors involved in MS pathogenesis, it might be that women use more sunblock or sun avoidance than men and, therefore, they experience greater vitamin D deficiency. Or it might be women have better hygiene as children than men and, thus, acquire EBV infections later in life. Alternatively, it could be that there are gender-specific differences in vitamin D metabolism, such that men and women experience a deficiency at different levels of absolute exposure (Suarez et al., 1998; Spach and Hayes, 2005). Alternatively, it could be that women have a greater likelihood of actually developing MS once the necessary environmental and genetic events have come together or it could be that some combination of these factors contributes to the observed gender-specific differences.

Importantly, however, not all of the current epidemiologic observations are consistent with this formulation derived from the Canadian data. Thus, a US study also looked at the monozygotic twin concordance rates in 95 men and 323 women (Islam et al., 2006). These authors identified 55 twin pairs concordant for MS (11 men and 45 women), none of whom were "doubly ascertained" (Islam et al., 2006). They estimated that their sampling methods had identified "approximately 27% of the North American twin cases prevalent at any time during the [study] period". In contrast to the Canadian data, however, these authors reported that the pair-wise concordance rate in female twin pairs of

13.9% (45/323) was not significantly different from the rate of 11.6% (11/95) in male twin pairs (Islam et al., 2006). Thus, unlike the Canadian data, these results would imply that women are substantially more likely to be genetically susceptible to MS than men (Goodin, 2016). Although such a conclusion might superficially seem to fit well with certain preconceived notions about MS susceptibility, the actual findings of this particular study are suspect for two reasons. First, the finding that the twin concordance rate in men and women is approximately equal indicates that the penetrance (i.e., the probability of actually developing MS for a susceptible individual) is similar for men and women (Goodin, 2016). Such a conclusion is hard to rationalize with the repeated observation that the prevalence of MS is increasing around the world and that this increase is predominantly or exclusively found in women (Hernán et al., 1999; Koch-Henriksen, 1999; Celius and Vandvik, 2001; Barnett et al., 2003; Sarasoja et al., 2004; Orton et al., 2006). Second, using their sampling methods, the authors found an extremely low rate of double ascertainment (i.e., zero). Low rates of double ascertainment are recognized as an indicator of a biased method of sampling (Witte et al., 1999) and, in this study, the probability of not doubly ascertaining a single twin pair is extremely unlikely by chance alone ($p < 10^{-6}$). Consequently, the sample used in this study is unquestionably biased and, therefore, the data out of Canada (where the authors report 54% double ascertainment) are likely to provide more reliable estimates (Willer et al., 2003).

Also, the data out of Canada are notable for the marked and continuous change in the gender ratio that has taken place over the 35-year interval of the study (Orton et al., 2006). This observation, together with the difference in twin concordance rate between men and women, has important implications with respect to the interaction between the gender of the individual and the environmental factors involved in MS pathogenesis (Goodin, 2016). From the gender ratio change alone, it is evident that there must have been a greater than 32% increase in the prevalence of MS in Canada over the 35-year period (Goodin, 2016). Although better diagnostic imaging and laboratory methods could contribute to such an increase, the fact that both the minimum increase in MS prevalence and the change in gender ratio are direct reflections of a disproportionate increase in MS prevalence among women (Goodin, 2016) suggests that better diagnostic techniques are an unlikely explanation. Moreover, the change in gender ratio also indicates that men begin to get MS at lower levels of environmental exposure than women (i.e., men have a lower threshold of exposure for getting the disease compared to women). Importantly, the parameter estimates for each of these effects (i.e., the increased prevalence of MS in the population and the lower threshold in men) are quite stable using any of the observed data points during the 35-year period to estimate them (Goodin, 2009).

From the observed difference in twin concordance rate between men and women, together with the two conclusions regarding disease prevalence and threshold, and assuming a proportionate hazard in men and women, the response curves for men and women to changes in their environmental exposure can be derived (Goodin, 2016). From these response curves it is apparent that, although men are more likely to be genetically susceptible than women and that they begin to get the disease at lower levels of environmental exposure, women are far more responsive the environmental changes (whatever these are), which have been taking place and that they approach a substantially higher response plateau for their probability of getting the disease (Goodin, 2016).

In addition, assuming environmental exposure levels have been increasing for longer than just the period of this study, the fact that men have a lower threshold than women suggests the possibility that, at some time in the past, MS may have been more prevalent in men than in women. Viewed in this context, it is quite interesting that a 1922 survey of MS cases in Europe and the USA (Wechsler, 1922) found that approximately 58% of cases (in each region) were men.

SUMMARY

Based on known epidemiology, it seems clear that the causal pathway leading to adult MS involves both multiple genetic susceptibility factors and, at least, three necessary (and sequential) environmental events or factors. If this is indeed the case, then, potentially, the disruption or modification of any one of these environmental factors might prevent the disease from ever occurring, regardless of an individual's genetic make-up. Of the promising candidate factors considered, the easiest to modify would be vitamin D deficiency, because supplemental vitamin D_3 is both widely available and very inexpensive. Nevertheless, any test of such a therapeutic strategy (or another similar approach directed at other factors) will require large numbers of persons to be followed for long time periods (>30 years). Randomized, placebo-controlled designs in this context are not feasible, so that testing such a notion will require the use of nonrandomized, open-label studies and applying statistical methods for bias mitigation (e.g., Rosenbaum and Rubin, 1983; Trojano et al., 2007; Goodin et al., 2011). The only requirement would be to monitor prospectively the vitamin D intake of all participating persons and, ideally, monitor their blood levels as well. Also, data could be simultaneously collected about a variety of other candidate factors, such as EBV infection, other viral exposures, birth location, ethnicity, migration history, smoking, the occurrence of traumas, and so forth. Because such a study would require persons (both adults and their children) to be followed prospectively, realistically, it is only feasible in places that have universal healthcare (and, thus, have large population-based centralized medical records) or in communities where the population is nonmobile and where complete ascertainment and longitudinal follow-up can be accomplished. Nevertheless, such a design would be cost-effective (only information easily available needs to be captured), it could be accomplished without much difficulty (including everyone who wants to participate), and it poses no ethical dilemmas (each person, together with their physician, is free to choose what they feel is best for themselves). It also seems like extremely important information to begin to collect. Certainly, such a longitudinal observational study carries minimal risk, is inexpensive, and, potentially, could provide inestimable benefits to future patients.

References

Ackerman K (2004). Relationship between multiple sclerosis exacerbations and stress. Psychosom Med 66: 288–289.

Ackerman KD, Heyman R, Rabin BS et al. (2002). Stressful life events precede exacerbations of multiple sclerosis. Psychosom Med 64: 916–920.

Ackerman KD, Stover A, Heyman R et al. (2003). Relationship of cardiovascular reactivity, stressful life events, and multiple sclerosis disease activity. Brain Behav Immun 17: 141–151.

Adams CWM (1989). A Colour Atlas of Multiple Sclerosis and Other Myelin Disorders, Wolfe Medical Publications, Ipswich, Suffolk, UK, p.101.

Ahmad T, Neville M, Marshall SE et al. (2003). Haplotype-specific linkage disequilibrium patterns. Hum Mol Genet 12: 647–656.

Alotaibi S, Kennedy J, Tellier R et al. (2004). Epstein–Barr virus in pediatric multiple sclerosis. JAMA 291: 1875–1879.

Alter M, Yamoor M, Harshe M (1974). Multiple sclerosis and nutrition. Arch Neurol 31: 267–272.

Alter M, Kahana E, Loewenson R (1978). Migration and risk of multiple sclerosis. Neurology 28: 1089–1093.

Andersen O, Lygner PE, Bergstrom T et al. (1993). Viral infections trigger multiple sclerosis relapses: a prospective seroepidemiological study. J Neurol 240: 417–422.

Antonovsky A, Leibowitz U, Smith HA et al. (1965). Epidemiological study of multiple sclerosis in Israel. I. An overall review of methods and findings. Arch Neurol 13: 183–193.

Ascherio A, Munger KL (2007). Environmental risk factors for multiple sclerosis. Part I: The role of infection. Ann Neurol 61: 288–299.

Ascherio A, Munger KL (2016). Epidemiology of multiple sclerosis: from risk factors to prevention – an update. Semin Neurol 36: 103–114.

Ascherio A, Munger KL, Lennette ET et al. (2001). Epstein–Barr virus antibodies and the risk of multiple sclerosis: a prospective study. Am Med Assoc J 286: 3083–3088.

Bager P, Nielsen NM, Bihrmann K et al. (2006). Sibship characteristics and risk of multiple sclerosis: a nationwide cohort study in Denmark. Am J Epidemiol 163: 1112–1117.

Bagos PG, Ioannidid, Nikolopoulos A (2006). *Chlamydia pneumoniae* infection and the risk of multiple sclerosis: a meta-analysis. Mult Scler 12: 397–411.

Baranzini SE, Wang J, Gibson RA et al. (2009). Genome-wide association analysis of susceptibility and clinical phenotype in multiple sclerosis. Hum Mol Genet 18: 767–778.

Barnett MH, William DB, Day S et al. (2003). Progressive increase in incidence and prevalence of multiple sclerosis in Newcastle, Australia: a 35-year study. J Neurol Sci 213: 1–6.

Bar-Or A, Oliviera EML, Anderson DE et al. (1999). Molecular pathogenesis of multiple sclerosis. J Neuroimmunol 100: 252–259.

Bartolomei MS, Tilghman SM (1997). Genomic imprinting in mammals. Annu Rev Genet 31: 493–525.

Bø L, Vedeler CA, Nyland HI et al. (2003). Subpial demyelination in the cerebral cortex of multiple sclerosis patients. J Neuropathol Exp Neurol 62: 723–732.

Brain R, Wilkinson M (1957). The association of cervical spondylosis and disseminated sclerosis. Brain 80: 456–478.

Bray PF, Bloomer LC, Salmon VC et al. (1983). Epstein–Barr virus infection and antibody synthesis in patients with multiple sclerosis. Arch Neurol 40: 406–408.

Buljevac D, Hop WCJ, Reedeker W et al. (2003). Self reported stressful life events and exacerbations in multiple sclerosis: prospective study. Br Med J 327: 646–651.

Cabre P, Signate A, Olindo S et al. (2005). Role of return migration in the emergence of multiple sclerosis in the French West Indies. Brain 128: 2899–2910.

Cantorna MT (2000). Vitamin D and autoimmunity: is vitamin D status an environmental factor affecting autoimmune disease prevalence? Proc Soc Exp Biol Med 223: 230–233.

Cantorna MT, Mahon BD (2004). Mounting evidence for vitamin D as an environmental factor affecting autoimmune disease prevalence. Exp Biol Med 229: 1136–1142.

Celius EG, Vandvik B (2001). Multiple sclerosis in Oslo, Norway: prevalence on 1 January 1995 and incidence over a 25-year period. Eur J Neurol 8: 463–469.

Chaudhuri A (2005). Why we should offer routine vitamin D supplementation in pregnancy and childhood to prevent multiple sclerosis. Med Hypotheses 64: 608–618.

Chaudhuri MD, Behan PO (2001). Acute cervical hyperextension-hyperflexion injury may precipitate and/or exacerbate symptomatic multiple sclerosis. Eur J Neurol 8: 659–664.

Compston A, Coles A (2002). Multiple sclerosis. Lancet 359: 1221–1231.

Compston A, Confavreux C, Lassmann H et al. (2006). McAlpine's Multiple Sclerosis, 4th edn. Churchill Livingstone, London.

Confavreux C, Vukusic S (2006). Natural history of multiple sclerosis: a unifying concept. Brain 129: 606–616.

Confavreux C, Hutchinson M, Hours MM et al. (1998). Rate of pregnancy-related relapse in multiple sclerosis. N Engl J Med 339: 285–291.

Confavreux C, Vukusic S, Adeleine P (2003). Early clinical predictors and progression of irreversible disability in multiple sclerosis: an amnesic process. Brain 126: 770–782.

Conlon P, Oksenberg JR, Zhang J et al. (1999). The immunobiology of multiple sclerosis: an autoimmune disease of the central nervous system. Neurobiol Dis 6: 149–166.

Correale J, Peirano I, Romano L (2012). Benign multiple sclerosis: a new definition of this entity is needed. Mult Scler 18: 210–218.

Da Silva KRP, Papais RP, Fernandez y Fernandez O et al. (2009). Potenciais fatores de risco para o desenvolvimento de esclerose multipla no Rio de Janeiro: um estudo caso controle. Arq Neuropsiquiatr 67: 229–234.

De Jager PL, Jia X, Wang J et al. (2009). Meta-analysis of genome scans and replication identify CD6, IRF8 and TNFRSF1A as new multiple sclerosis susceptibility loci. Nature Genet 41: 776–782.

Dean G, Kurtzke JF (1971). On the risk of multiple sclerosis according to age at immigration to South Africa. Br Med J 3: 725–729.

Deluca HF, Cantorna MT (2001). Vitamin D: Its role and uses in immunology. FASEB J 15: 2579–2585.

Dikson SP, Wang K, Krantz I et al. (2010). Rare variants create synthetic genome-wide associations. PLoS Biol 8 (1): e1000294.

Dyment DA, Yee IML, Ebers GC et al. (2006). Multiple sclerosis in stepsiblings: recurrence risk and ascertainment. J Neurol Neurosurg Psychiatry 77: 258–259.

Ebers G (2004). Multiple sclerosis, stress and the rules of evidence. Practical Neurol 4: 66–69.

Ebers GC, Sadovnick AD, Risch NJ et al. (1995). A genetic basis for familial aggregation in multiple sclerosis. Nature 377: 150–151.

Ebers GC, Yee IML, Sandovnick AD et al. (2000). Conjugal multiple sclerosis: population-based prevalence and recurrence risks in offspring. Ann Neurol 48: 927–931.

Ebers GC, Sadovnick AD, Dyment DA et al. (2004). Parent-of-origin effect in multiple sclerosis: observations in half-siblings. Lancet 363: 1773–1774.

Edwards S, Zvartau M, Clarke H et al. (1998). Clinical relapses and disease activity on magnetic resonance imaging associated with viral upper respiratory tract infections in multiple sclerosis. J Neurol Neurosurg Psychiatry 64: 736–741.

Elian M, Nightingale S, Dean G (1990). Multiple sclerosis among United Kingdom-born children of immigrants from the Indian subcontinent, Africa, and the West Indies. J Neurol Neurosurg Psychiatry 53: 906–911.

Emmons KM, Colditz GA (1999). Preventing excess sun exposure: it is time for a national policy. J Natl Cancer Inst 91: 1269–1270.

Engelhardt S, Patkar S, Ogunshola OO (2014). Cell-specific blood–brain barrier regulation in health and disease: a focus on hypoxia. Br J Pharmacol 171: 1210–1230.

Engell T (1989). A clinical patho-anatomical study of clinically silent multiple sclerosis. Acta Neurol Scand 79: 428–430.

Estruch R, Ros E, Salas-Salvadó J et al. (2013). Primary prevention of cardiovascular disease with a Mediterranean diet. N Engl J Med 368: 1279–1290.

Evans KN, Nguyen L, Chan J et al. (2006). Effects of 25-hydroxyvitamin D3 and 1,25-dihydroxyvitamin D3 on cytokine production by human decidual cells. Biol Reprod 75: 816–822.

Farrell PJ (2007). Role for HLA in susceptibility to infectious mononucleosis. J Clin Invest 117: 2756–2758.

Ferguson B, Matyszak MK, Esiri MM et al. (1997). Axonal damage in acute multiple sclerosis lesions. Brain 120: 393–399.

Fiddes B, Wason J, Kemppinen A et al. (2013). Confounding underlies the apparent month of birth effect in multiple sclerosis. Ann Neurol 73: 714–720.

Fiddes B, Wason J, Sawcer S (2014). Confounding in association studies: month of birth and multiple sclerosis. J Neurol 261: 1851–1856.

Filippi M, Rocca MA, Martino G et al. (1998). Magnetization transfer changes in the normal appearing white matter precede the appearance of enhancing lesions in patients with multiple sclerosis. Ann Neurol 43: 809–814.

Fisher E, Rudick RA, Simon JH et al. (2002). Eight-year follow-up study of brain atrophy in patients with MS. Neurology 59: 1412–1420.

Fisniku LK, Brex PA, Altmann DR et al. (2008). Disability and T2 MRI lesions: a 20-year follow-up of patients with relapse onset of multiple sclerosis. Brain 131: 808–817.

Fitzgerald G, Harbige LS, Forti A et al. (1987). The effect of nutritional counseling on diet and plasma EFA status in multiple sclerosis patients over 3 years. Hum Nutr Appl Nutr 41A: 297–310.

Freedman DM, Mustafa M, Alavanja MCR (2000). Mortality from multiple sclerosis and exposure to residential and occupational solar radiation: a case-control study based on death certificates. Occup Environ Med 57: 418–421.

French Research Group on Multiple Sclerosis (1992). Multiple sclerosis in 54 twinships: concordance rate is independent of zygosity. Ann Neurol 32: 724–727.

Gadoth N (2003). Review article: Multiple sclerosis in children. Brain Dev 25: 229–232.

Georgi VW (1966). Multiple Sklerose: Pathologisch-Anamtomische Befunde multiple Sklerose bei klinisch nicht diagniostizierte Krankbeiten. Schweiz Med Wochenschr 20: 605–607.

Ghadirian P, Jain M, Ducic S et al. (1998). Nutritional factors in the aetiology of multiple sclerosis: a case-control study in Montreal, Canada. Int J Epidemiol 27: 845–852.

Gianfrancesco MA, Acuna B, Shen L et al. (2014). Obesity during childhood and adolescence increases susceptibility to multiple sclerosis after accounting for established genetic and environmental risk factors. Obes Res Clin Pract 8 (5): e435–e447.

Gilbert J, Sadler M (1983). Unsuspected multiple sclerosis. Arch Neurol 40: 533–536.

Gillie O (2006). A new government policy is needed for sunlight and vitamin D. Br J Dermatol 154: 1052–1061.

Goodkin DE, Rooney WD, Sloan R et al. (1998). A serial study of new MS lesions and the white matter from which they arise. Neurology 51: 1689–1697.

Golan D, Somer E, Dishon S et al. (2008). The impact of exposure to war stress on exacerbations of multiple sclerosis. Ann Neurol 64: 143–149.

Goldacre MJ, Wotton CJ, Seagroatt V et al. (2004). Multiple sclerosis after infectious mononucleosis: record linkage study. J Epidemiol Community Health 58: 1032–1035.

Goldacre MJ, Abisgold JD, Yeates DGR et al. (2006). Risk of multiple sclerosis after head injury: record linkage study. J Neurol Neurosurg Psychiatry 277: 351–353.

Goodin DS (1999). Perils and pitfalls in the interpretation of clinical trials: a reflection on the recent experience in multiple sclerosis. Neuroepidemiology 18: 53–63.

Goodin DS (2004). Relationship between multiple sclerosis exacerbations and stress. Psychosom Med 66: 287–288.

Goodin DS (2009). The causal cascade to multiple sclerosis: a model for MS pathogenesis. PLoS One 4: e4565.

Goodin DS (2016). The nature of genetic susceptibility to multiple sclerosis: constraining the possibilities. BMC Neurol 16 (1): 56.

Goodin DS, Khankhanian P (2014). Single nucleotide polymorphism (SNP)-strings: an alternative method for assessing genetic associations. PLoS One 9 (4): e90034.

Goodin DS, Ebers GC, Johnson KP et al. (1999). The relationship of MS to physical trauma and psychological stress: report of the Therapeutics and Technology Assessment Subcommittee of the American Academy of Neurology. Neurology 52: 1737–1745.

Goodin DS, Jones J, Li D et al. (2011). Establishing long-term efficacy in chronic disease: the use of recursive partitioning and propensity score adjustment to estimate long-term outcome in multiple sclerosis. PLoS One 6 (11): e22444.

Gourraud PA, Harbo HF, Hauser SL et al. (2012). The genetics of multiple sclerosis: an up-to-date review. Immunol Rev 248: 87–103.

Griffin MD, Lutz W, Phan VA et al. (2001). Dendritic cell modulation by 1-alpha,25 dihydroxyvitamin D3 and its analogs: a vitamin D receptor-dependent pathway that promotes a persistent state of immaturity in vitro and in vivo. Proc Natl Acad Sci 98: 6800–6805.

Grønlie SA, Myrvoll E, Hansen G et al. (2000). Multiple sclerosis in North Norway, and first appearance in an indigenous population. J Neurol 247: 129–133.

Gronseth GS, Ashman EJ (2000). Practice parameter: the usefulness of evoked potentials in identifying clinically silent lesions in patients with suspected multiple sclerosis (an evidence-based review): report of the Quality Standards Subcommittee of the American Academy of Neurology. Neurology 54: 1720–1725.

Haahr S, Plesner AM, Vestergaard BF et al. (2004). A role of late Epstein–Barr virus infection in multiple sclerosis. Acta Neurol Scand 109: 270–275.

Hansen T, Skytthe A, Stenager E et al. (2005a). Risk for multiple sclerosis in dizygotic and monozygotic twins. Mult Scler 11: 500–503.

Hansen T, Skytthe A, Stenager E et al. (2005b). Concordance for multiple sclerosis in Danish twins: an update of a nationwide study. Mult Scler 11: 504–510.

Harel-Meir M, Sherer Y, Shoenfeld Y (2007). Tobacco smoking and autoimmune rheumatic diseases. Nat Clin Pract Rheumatol 3: 707–715.

Hauser SL, Oksenberg JR (2006). The neurobiology of multiple sclerosis: genes, inflammation, and neurodegeneration. Neuron 52: 61–76.

Hawkins SA, McDonnell GV (1999). Benign multiple sclerosis? Clinical course, long term follow up, and assessment of prognostic factors. J Neurol Neurosurg Psychiatry 67: 148–152.

Hayes CE, Nashold FE, Spach KM et al. (2003). The immunological functions of the vitamin D endocrine system. Cell Molec Biol 49: 277–300.

Hedström AK, Olsson O, Alfredsson L (2012). High body mass index before age 20 is associated with increased risk for multiple sclerosis in both men and women. Mult Scler 18: 1334–1336.

Henle W, Henle G, Andersson J et al. (1987). Antibody responses to Epstein–Barr virus-determined nuclear antigen (EBNA)-1 and EBNA-2 in acute and chronic Epstein–Barr virus infection. Proc Natl Acad Sci U S A 84: 570–574.

Hernán MA, Olek MJ, Ascherio A (1999). Geographic variation of MS incidence in two prospective studies of US women. Neurology 53: 1711–1718.

Hernán MA, Olek MJ, Ascherio A (2001). Cigarette smoking and incidence of multiple sclerosis. Am J Epidemiol 154: 69–74.

Hernán MA, Jick SS, Logroscino G et al. (2005). Cigarette smoking and the progression of multiple sclerosis. Brain 128: 1461–1465.

Herrera BM, Ramagopalan SV, Orton S et al. (2007). Parental transmission of MS in a population-based Canadian cohort. Neurology 69: 1208–1212.

Hofker MH, Fu J, Wijmenga C (2014). The genome revolution and its role in understanding complex diseases. Biochim Biophys Acta 1842: 1889–1895.

Holick MF (1998). Vitamin D requirements for humans of all ages: new increased requirements for women and men 50 years and older. Osteoporos Int 8 (Suppl 2): S24–S29.

Holman DW, Klein RS, Ransohoff RM (2011). The blood–brain barrier, chemokines and multiple sclerosis. Biochim Biophys Acta 1812: 220–230.

Housley WJ, Pitt D, Hafler DA (2015). Biomarkers in multiple sclerosis. Clin Immunol 161: 51–58.

Houtchens MK (2007). Pregnancy and multiple sclerosis. Semin Neurol 27: 434–441.

International Multiple Sclerosis Genetics Consortium (IMSGC) (2014). Analysis of immune-related loci identifies 48 new susceptibility variants for multiple sclerosis. Nat Genet 45: 1353–1360.

International Multiple Sclerosis Genetics Consortium and the Wellcome Trust Case Control Consortium (2011). Genetic risk and a primary role for cell-mediated immune mechanisms in multiple sclerosis. Nature 476: 214–219.

Islam T, Gauderman WJ, Cozen W et al. (2006). Differential twin concordance for multiple sclerosis by latitude of birthplace. Ann Neurol 60: 56–64.

Jablonski NG, Chaplin G (2000). The evolution of human skin coloration. J Hum Evol 39: 57–106.

Jablonski NG, Chaplin G (2002). Skin deep. Sci Am 287: 74–81.

Kahana E, Zilber N, Abramson JH et al. (1994). Multiple sclerosis: genetic versus environmental aetiology: epidemiology in Israel updated. J Neurol 241: 341–346.

Kantarci OH, Barcellos LF, Atkinson EJ et al. (2006). Men transmit MS more often to their children vs women: the Carter effect. Neurology 67: 305–310.

Kaufman M, Gaydos CA, Sriram S et al. (2002). Is *Chlamydia pneumoniae* found in spinal fluid samples from multiple sclerosis patients? Conflicting results. Mult Scler 8: 289–294.

Khankhanian P, Gourraud PA, Lizee A et al. (2015). Haplotype-based approach to known MS-associated regions increases the amount of explained risk. J Med Genet 52 (9): 587–594.

Kimlin MG, Olds WJ, Moore MR (2007). Location and vitamin D synthesis: is the hypothesis validated by geophysical data? J Photochem Photobiol B Biol 86: 234–239.

Kip KE, Cohen F, Cole SR et al. (2001). Recall bias in a prospective cohort study of acute time-varying exposures: example from the herpetic eye disease study. J Clin Epidemiol 54: 482–487.

Koch-Henderson N (1995). Multiple sclerosis in Scandinavia and Finland. Acta Neurol Scand 161: 55–59.

Koch-Henriksen N (1999). The Danish Multiple Sclerosis Registry: a 50-year follow-up. Mult Scler 5: 293–296.

Kremenchutsky M, Rice GPA, Baskerville DM et al. (2006). The natural history of multiple sclerosis: a geographically based study 9: observations on the progressive phase of the disease. Brain 129: 584–594.

Kroencke DC, Denney DR (1999). Stress and coping in multiple sclerosis: exacerbation, remission and chronic subgroups. Mult Scler 5: 89–93.

Krone B, Pohl D, Rostasy K et al. (2008). Common infectious agents in multiple sclerosis: a case–control study in children. Mult Scler 14: 136–139.

Kuusisto H, Kaprio J, Kinnunen E et al. (2008). Concordance and heritability of multiple sclerosis in Finland: study on a nationwide series of twins. Eur J Neurol 15: 1106–1110.

Larsen PD, Bloomer LC, Bray PF (1985). Epstein–Barr nuclear antigen and viral capsid antigen antibody titers in multiple sclerosis. Neurology 35: 435–438.

Lee JY, Chitnis T (2016). Pediatric multiple sclerosis. Semin Neurol 36: 148–153.

Levin LI, Munger KL, Rubertone MV et al. (2005). Temporal relationship between elevation of Epstein–Barr virus antibody titers and initial onset of neurological symptoms in multiple sclerosis. Am Med Assoc J 293: 2496–2500.

Li J, Johansen C, Brønnum-Hansen H et al. (2004). The risk of multiple sclerosis in bereaved parents: a nationwide cohort study in Denmark. Neurology 62: 726–729.21.

Liguori M, Marrosu MG, Pugliatti M et al. (2000). Age at onset in multiple sclerosis. Neurol Sci 21: S825–S829.

Lips P (2006). Vitamin D physiology. Prog Biophys Molec Biol 92: 4–8.

Lublin FD, Reingold SC, Cohen JA et al. (2014). Defining the clinical course of multiple sclerosis: the 2013 revisions. Neurology 83: 278–286.

Lucchinetti CF, Popescu BF, Bunyan RF et al. (2011). Inflammatory cortical demyelination in early multiple sclerosis. N Engl J Med 365: 2188–2197.

Makhani N, Banwell B, Tellier R et al. (2016). Viral exposures and MS outcome in a prospective cohort of children with acquired demyelination. Mult Scler 22: 385–388.

McIntosh AM, Holmes S, Gleeson S et al. (2002). Maternal recall bias, obstetric history and schizophrenia. Br J Psychiatry 181: 520–525.

Merelli E, Casoni F (2000). Prognostic factors in multiple sclerosis: role of intercurrent infections and vaccinations against influenza and hepatitis B. Neurol Sci 21: S853–S856.

Mikaeloff Y, Caridade G, Tardieu M et al. (2007). Parental smoking at home and the risk of childhood-onset multiple sclerosis in children. Brain 130: 2589–2595.

Mohr DC, Goodkin DE, Bacchetti P et al. (2003). Psychological stress and the subsequent appearance of new brain lesions in MS. Neurology 55: 55–61.

Moll NM, Cossoy MB, Fisher E et al. (2009). Imaging correlates of leukocyte accumulation and CXCR4/CXCL12 in multiple sclerosis. Arch Neurol 66: 44–53.

Montgomery SM, Bahmanyar S, Hillert J et al. (2008). Maternal smoking during pregnancy and multiple sclerosis amongst offspring. Eur J Neurol 15: 1398–1402.

Moses H, Sriram S (2001). An infectious basis for multiple sclerosis: perspectives on the role of *Chlamydia pneumoniae* and other agents. Bio Drugs 15: 199–206.

Müller AMS, Ihorst G, Mertelsmann R et al. (2005). Epidemiology of non-Hodgkin's lymphoma (NHL): trends, geographic distribution, and etiology. Ann Hematol 84: 1–12.

Mumford CJ, Wood NW, Kellar-Wood H et al. (1994). The British Isles survey of multiple sclerosis in twins. Neurology 44: 11–15.

Munch M, Riisom K, Christensen T et al. (1998). The significance of Epstein–Barr virus seropositivity in multiple sclerosis patients? Acta Neurol Scand 97: 171–174.

Munger KL, DeLorenze GN, Levin LI et al. (2004a). A prospective study of *Chlamydia pneumoniae* infection and risk of MS in two US cohorts. Neurology 62: 1799–1803.

Munger KL, Zhang SM, O'Reilly E et al. (2004b). Vitamin D intake and incidence of multiple sclerosis. Neurology 62: 60–65.

Munger KL, Levin LI, Hollis BW et al. (2006). Serum 25-hydroxyvitamin D levels and risk of multiple sclerosis. Am Med Assoc J 296: 2832–2838.

Munger KL, Chitnis T, Ascherio A (2009). Body size and risk of MS in two cohorts of US women. Neurology 73: 1543–1550.

Munger KL, Bentzen J, Laursen B et al. (2013). Childhood body mass index and multiple sclerosis risk: a long-term cohort study. Mult Scler 19: 1323–1329.

Muoio V, Persson PB, Sendeski MM (2014). The neurovascular unit – concept review. Acta Physiol 210: 790–798.

Myhr KM, Riise T, Barrett-Connor E et al. (1998). Altered antibody pattern to Epstein–Barr virus but not to other herpesviruses in multiple sclerosis: a population based case-control study from western Norway. J Neurol Neurosurg Psychiatry 164: 539–542.

Nagpal S, Na S, Rathnachalam R (2005). Noncalcemic actions of vitamin D receptor ligands. Endocr Rev 26: 662–687.

Nielsen NM, Westergaard T, Rostgaard K et al. (2005). Familial risk of multiple sclerosis: a nationwide cohort study. Am J Epidemiol 162: 774–778.

Nielsen TR, Rostgaard K, Nielsen NM et al. (2007). Multiple sclerosis after infectious mononucleosis. Arch Neurol 64: 72–75.

Nisipeanu P, Korczyn AD (1993). Psychological stress as risk factor for exacerbations in multiple sclerosis. Neurology 43: 1311–1312.

O'Riordan JI, Thompson DP, Kingsley DP et al. (1998). The prognostic value of brain MRI in clinically isolated syndromes of the CNS. a 10-year follow-up. Brain 121: 495–503.

Okuda DT, Mowry EM, Beheshtian A et al. (2009). Incidental MRI anomalies suggestive of multiple sclerosis: the radiologically isolated syndrome. Neurology 72: 800–805.

Okuda DT, Mowry EM, Cree BA et al. (2011). Asymptomatic spinal cord lesions predict disease progression in radiologically isolated syndrome. Neurology 76: 686–692.

Okuda DT, Siva A, Kantarci O et al. (2014). Radiologically isolated syndrome: 5-year risk for an initial clinical event. PLoS One 9 (3): e90509.

Oppenheimer DR (1978). The cervical cord in multiple sclerosis. Neuropathol Appl Neurobiol 4: 151–162.

Optic Neuritis Study Group (2008). Multiple sclerosis risk after optic neuritis: final optic neuritis treatment trial follow-up. Arch Neurol 65: 727–732.

Orton SM, Herrera BM, Yee IM et al. (2006). Sex ratio of multiple sclerosis in Canada: a longitudinal study. Lancet Neurol 5: 932–936.

Padyukov L, Silva C, Stolt P et al. (2004). A gene–environment interaction between smoking and shared epitope genes in HLA–DR provides a high risk of seropositive rheumatoid arthritis. Arthritis Rheum 50: 3085–3092.

Panitch HS (1994). Influence of infection on exacerbations of multiple sclerosis. Ann Neurol 36: S25–S28.

Parratt J, Tavendale R, O'Riordan J et al. (2008). *Chlamydia pneumoniae*-specific serum immune complexes in patients with multiple sclerosis. Mult Scler 14: 292–299.

Peterson JW, Bø L, Mørk S et al. (2001). Transected neurites, apoptotic neurons and reduced inflammation in cortical MS lesions. Ann Neurol 50: 389–400.

Pittock SJ, Mayr WT, McClelland RL et al. (2004). Disability profile of MS did not change over 10 years in a population-based prevalence cohort. Neurology 62: 601–606.

Pohl D, Krone B, Rostasy K et al. (2006). High seroprevalence of Epstein–Barr virus in children with multiple sclerosis. Neurology 67: 2063–2065.

Ponsonby AL, van der Mei, Dwyer T et al. (2005). Exposure to infant siblings during early life and risk of multiple sclerosis. Am Med Assoc J 293: 463–469.

Ramagopalan SV, Maugeri NJ, Handunnetthi L et al. (2009). Expression of the multiple sclerosis-associated MHC class II allele HLA-DRB1*1501 is regulated by vitamin D. PLoS Genet 5 (2): e1000369.

Ranzato F, Perini P, Tzintzeva E et al. (2003). Increasing frequency of multiple sclerosis in Padova, Italy: a 30 year epidemiological survey. Mult Scler 9: 387–392.

Riise T, Nortvedt MW, Ascherio A (2003). Smoking is a risk factor for multiple sclerosis. Neurology 61: 1122–1124.

Ristori G, Cannoni S, Stazi MA et al. (2006). Multiple sclerosis in twins from continental Italy and Sardinia: a nationwide study. Ann Neurol 59: 27–34.

Robertson NP, Fraser M, Deans J et al. (1996). Age-adjusted recurrence risks for relatives of patients with multiple sclerosis. Brain 119: 449–455.

Rodríguez Regal A, del Campo Amigo M, Paz-Esquete J et al. (2009). Estudio de casos y controles sobre la influencia del hábito tabáquico en la esclerosis multiple. Neurologia 24: 177–180.

Rodriguez M, Siva A, Cross S et al. (1994). Optic neuritis: a population based study in Olmsted county, Minnesota. Neurology 44: A374.

Rosati G (2001). The prevalence of multiple sclerosis in the world: an update. Neurol Sci 22: 117–139.

Rosenbaum PR, Rubin DB (1983). The central role of propensity score in observational studies for causal effects. Biometrika 70: 41–55.

Rothman KJ, Greenland S (1998). Modern Epidemiology. 2nd edn, Lippincott Williams & Wilkins, Philadelphia, PA.

Ruggieri M, Polizzi A, Pavone L et al. (1999). Multiple sclerosis in children under 6 years of age. Neurology 53: 478–484.

Runmarker B, Andersen O (1993). Prognostic factors in a multiple sclerosis incidence cohort with twenty-five years of follow-up. Brain 116: 117–134.

Sadovnick AD, Ebers GC, Dyment DA et al. (1996). Evidence for genetic basis of multiple sclerosis. Lancet 347: 1728–1730.

Sadovnick AD, Yee IML, Ebers GC et al. (2005). Multiple sclerosis and birth order: a longitudinal cohort study. Lancet Neurol 4: 611–617.

Sadovnick AD, Duquette P, Herrera B et al. (2007). A timing-of-birth effect on multiple sclerosis clinical phenotype. Neurology 69: 60–62.

Salemi G, Callari G, Gammino M et al. (2004). The relapse rate of multiple sclerosis changes during pregnancy: a cohort study. Acta Neurol Scand 110: 23–26.

Sarasoja T, Wikström J, Paltamaa J et al. (2004). Occurrence of multiple sclerosis in central Finland: a regional and temporal comparison during 30 years. Acta Neurol Scand 110: 331–336.

Scalfari A, Neuhaus A, Degenhardt A et al. (2010). The natural history of multiple sclerosis: a geographically based study 10: relapses and long-term disability. Brain 133: 1914–1929.

Sepcic J, Mesaros E, Materljan E et al. (1993). Nutritional factors and multiple sclerosis in Gorski Kotar, Croatia. Neruoepidemiology 12: 234–240.

Serafini B, Roiscarelli B, Franciotta D et al. (2007). Dysregulated Epstein–Barr virus infection in multiple sclerosis. J Exp Med 204: 2899–2912.

Sharief MK, Thompson EJ (1991). The predictive value of intrathecal immunoglobulin synthesis and magnetic resonance imaging in acute isolated syndromes for subsequent development of multiple sclerosis. Ann Neurol 29: 147–151.

Shirodaria PV, Haire M, Fleming E et al. (1987). Viral antibody titers: comparison in patients with multiple sclerosis and rheumatoid arthritis. Arch Neurol 144: 1237–1241.

Sibley WA, Bamford CR, Clark K (1985). Clinical viral infections and multiple sclerosis. Lancet 1 (8441): 1313–1315.

Sinclair H (1977). Polyunsaturated fatty acids in multiple sclerosis. Br Med J 2: 1217.

Spach KM, Hayes CE (2005). Vitamin D3 confers protection from autoimmune encephalomyelitis only in female mice. J Immunol 175: 4119–4126.

Spitzer C, Bouchain M, Winkler LY et al. (2012). Childhood trauma in multiple sclerosis: a case-control study. Psychosom Med 74: 312–318.

Staples J, Ponsonby AL, Lim L (2010). Low maternal exposure to ultraviolet radiation in pregnancy, month of birth and risk of multiple sclerosis in offspring: a longitudinal analysis. Br Med J 340: c1640.

Suarez F, Rossignol C, Garabédian M (1998). Interactive effect of estradiol and vitamin D receptor gene polymorphisms as a possible determinant of growth in male and female infants. J Clin Endocrinol Metab 83: 3563–3568.

Sumaya CV, Myers LW, Ellison GW (1980). Epstein–Barr virus antibodies in multiple sclerosis. Arch Neurol 37: 94–96.

Sumaya CV, Myers LW, Ellison GW et al. (1985). Increased prevalence and titer of Epstein–Barr virus antibodies in patients with multiple sclerosis. Ann Neurol 17: 371–377.

Sundström P, Nyström L, Forsgren L (2003). Incidence (1988–97) and prevalence (1997) of multiple sclerosis in Västerbotten County in northern Sweden. J Neurol Neurosurg Psychiatry 74: 29–32.

Sundström P, Juto G, Wadell G et al. (2004). An altered immune response to Epstein–Barr virus in multiple sclerosis: a prospective study. Neurology 62: 2277–2282.

Sundström P, Nyström L, Hallmans G (2008). Smoke exposure increases the risk for multiple sclerosis. Eur J Neurol 15: 579–583.

Swank RL (1950). Multiple sclerosis: a correlation of its incidence with dietary fat. Am J Med Sci 220: 421–430.

Swank RL (1997). A prospective discussion of past international nutrition catastrophes – indications for the future. Nutrition 13: 344–348.

Swank RL, Dugan BB (1990). Viewpoint: effect of low saturated fat diet in early and late cases of multiple sclerosis. Lancet 336: 37–39.

Swank RL, Grimsgaard A (1988). Multiple sclerosis: the lipid relationship. Am J Clin Nutr 48: 1387–1393.

Swank RL, Lerstad O, Strom A et al. (1952). Multiple sclerosis in rural Norway: its geographic and occupational incidence in relation to nutrition. N Engl J Med 264: 721–728.

Tavera-Mendoza LE, White JH (2007). Cell defenses and the sunshine vitamin. Sci Am 97: 62–65. 68–70, 72.

Templer DI, Trent NH, Spencer DA et al. (1992). Season of birth in multiple sclerosis. Acta Neurol Scand 85: 107–109.

Teschke K, Smith JC, Olshan AF (2000). Evidence of recall bias in volunteered vs. prompted responses about occupational exposures. Am J Ind Med 38: 385–388.

Testi M, Battarra M, Lucarelli G et al. (2015). HLA-A-B-C-DRB1-DQB1 phased haplotypes in 124 Nigerian families indicate extreme HLA diversity and low linkage disequilibrium in Central-West Africa. Tissue Antigens 86: 285–292.

Thacker EL, Mirzaei F, Ascherio A (2006). Infectious mononucleosis and risk for multiple sclerosis: a meta-analysis. Ann Neurol 59: 499–503.

Thorogood M, Hannaford PC (1998). The influence of oral contraceptives on the risk of multiple sclerosis. Br J Obstet Gynecol 105: 1296–1299.

Torkildsen Ø, Aarseth J, Benjaminsen E et al. (2014). Month of birth and risk of multiple sclerosis: confounding and adjustments. Ann Clin Transl Neurol 1: 141–144.

Trapp BD, Peterson JP, Ransohoff RM et al. (1998). Axonal transection in the lesions of multiple sclerosis. N Engl J Med 338: 278–285.

Tremlett H, Paty D, Devonshire V (2006). Disability progression in multiple sclerosis is slower than previously reported. Neurology 66: 172–177.

Trojano M, Pellegrini F, Fuiani A et al. (2007). New natural history of interferon-beta-treated relapsing multiple sclerosis. Ann Neurol 61: 300–306.

Van der Mei IAF, Ponsonby A-L, Dwyer T et al. (2003). Past exposure to sun, skin phenotype, and risk of multiple sclerosis: case-control study. Br Med J 327: 316–322.

Vaughan JH (1995). The Epstein–Barr virus in autoimmunity. Springer Semin Immunopathol 17: 203–230.

Villard-Macintosh L, Vessey MP (1993). Oral contraceptives and reproductive factors in multiple sclerosis incidence. Contraception 47: 161–168.

Vost A, Wolochow D, Howell D (1964). Incidence of infarcts of the brain in heart diseases. J Path Bact 88: 463–470.

Vukusic S, Hutchinson M, Hours M et al. (2004). Pregnancy and multiple sclerosis (the PRIMS study): clinical predictors of post-partum relapse. Brain 127: 1353–1360.

Wagner HJ, Hennig H, Jabs WJ et al. (2000). Altered prevalence and reactivity of anti-Epstein–Barr virus antibodies in patients with multiple sclerosis. Viral Immunol 13: 497–502.

Wechsler IS (1922). Statistics of multiple sclerosis. Arch Neurol Psychiatry 8: 59–75.

Weinshenker BG, Bass B, Rice GPA et al. (1989). The natural history of multiple sclerosis: a geographically based study: II. Predictive value of the early clinical course. Brain 112: 133–146.

Wennerström A, Vlachopoulou E, Lahetla E et al. (2013). Diversity of extended HLA-DRB1 haplotypes in the Finnish population. PLoS One 8 (11): e79690.

Wesnes K, Riise T, Casetta I et al. (2015). Body size and the risk of multiple sclerosis in Norway and Italy: the EnvIMS study. Mult Scler 21: 388–395.

Willer CJ, Dyment DA, Risch NJ et al. (2003). Twin concordance and sibling recurrence rates in multiple sclerosis. Proc Natl Acad Sci U S A 100: 12877–12882.

Willer CJ, Dyment DA, Sadovnick AD et al. (2005). Timing of birth and risk of multiple sclerosis: population based study. Br Med J 330: 120–124.

Witte JS, Carlin JB, Hopper JL (1999). Likelihood-based approach to estimating twin concordance for dichotomous traits. Genetic Epidemiol 16: 290–304.

Wynn DR, Rodriguez M, O'Fallon M et al. (1990). A reappraisal of the epidemiology of multiple sclerosis in Olmsted County, Minnesota. Neurology 40: 780–786.

Yamasaki R, Lu H, Butovsky O et al. (2014). Differential roles of microglia and monocytes in the inflamed central nervous system. J Exp Med 211: 1533–1549.

Young LS, Yao QY, Rooney CM et al. (1987). New type B isolates of Epstein–Barr virus from Burkitt's lymphoma and from normal individuals in endemic areas. J Gen Virol 68: 2853–2862.

Zimber U, Adlinger HK, Lenoir GM et al. (1986). Geographical prevalence of two types of Epstein–Barr virus. Virology 154: 56–66.

Zúñiga J, Yu N, Barqera R et al. (2013). HLA class I and class II conserved extended haplotypes and their fragments or blocks in Mexicans: implications for the study of genetic diversity in admixed populations. PLoS One 8 (9): e74442.

Handbook of Clinical Neurology, Vol. 138 (3rd series)
Neuroepidemiology
C. Rosano, M.A. Ikram, and M. Ganguli, Editors
http://dx.doi.org/10.1016/B978-0-12-802973-2.00012-4

Chapter 12

Neuroepidemiology of traumatic brain injury

A.J. GARDNER[1]* AND R. ZAFONTE[2]

[1]*Hunter New England Local Health District Sports Concussion Program; Priority Research Centre for Stroke and Brain Injury, School of Medicine and Public Health, University of Newcastle, Callaghan, NSW, Australia*

[2]*Department of Physical Medicine and Rehabilitation, Harvard Medical School, Spaulding Rehabilitation Hospital; MassGeneral Hospital for Children Sport Concussion Program and Red Sox Foundation and Massachusetts General Hospital Home Base Program, Brigham and Women's Hospital, Boston, MA, USA*

Abstract

Traumatic brain injury (TBI) is a significant public-health concern. TBI is defined as an acute brain injury resulting from mechanical energy to the head from external physical forces. Some of the leading causes of TBI include falls, assaults, motor vehicle or traffic accidents, and sport-related concussion. Two of the most common identified risk factors are sex (males are nearly three times more likely to suffer a TBI than females); and a bimodal age pattern (persons 65 years and older, and children under 14 years old).

It is estimated that approximately 1.5–2 million Americans suffer from TBI annually. TBIs account for around 1.4 million emergency room visits, 275 000 hospital admissions, and 52 000 deaths in the USA each year. TBI contributes to approximately 30% of all deaths in the USA annually. In Australia, it is estimated that approximately 338 700 individuals (1.9% of the population) suffer from a disability related to TBI. Of these, 160 200 were severely or profoundly affected by acquired brain injury, requiring daily support. In the UK, TBI accounted for 3.4% of all emergency department attendances annually. An overall rate of 453 per 100 000 was found for all TBI severities, of which 40 per 100 000 (10.9%) were moderate to severe. TBI often results in residual symptoms that affect an individual's cognition, movement, sensation, and/or emotional functioning. Recovery and rehabilitation from TBI may require considerable resources and may take years. Some individuals never fully recover, and some require lifetime ongoing care and support. TBI has an enormous social and financial cost, with estimates of the annual financial burden associated with TBI ranging between 9 and 10 billion US dollars.

TRAUMATIC BRAIN INJURY DEFINITION, AND CLASSIFICATION

Head injury (HI), traumatic brain injury (TBI), and concussion are three terms that have frequently been used interchangeably in the literature (Table 12.1), which has resulted in some confusion for non-TBI experts and newcomers to the TBI field. HI is a nonspecific and antiquated term, which can encompasses both an external injury to the face or scalp (like a laceration, a contusion, an abrasion, or a fracture) and may or may not be associated with TBI, whereas a TBI is defined as an alteration in brain function as a result of mechanical energy transmitted to the head from external physical forces. The US Centers for Disease Control define a TBI as being caused by a bump, blow, or jolt to the head or penetrating HI that disrupts the normal function of the brain (Centers for Disease Control and Prevention, n.d.).

TBI is classified along a severity continuum, and may be classified as mild, moderate, or severe (Fig. 12.1).

TBI severity has typically been classified according to an individual's level of consciousness and/or duration of amnesia. TBI severity is commonly determined using the Glasgow Coma Scale (GCS), which has been

*Correspondence to: Andrew J. Gardner, Centre for Translational Neuroscience and Mental Health, Level 5, McCauley Building, Calvary Mater Hospital, Waratah NSW 2298, Australia. E-mail: andrew.gardner@neurogard.com.au

Table 12.1

Glossary of terms

Blast injury	Occurs as a result of direct or indirect exposure to an explosion. A blast injury is a common cause of TBI for soldiers in the battlefield
Chronic traumatic encephalopathy (CTE)	Considered to be a neurodegenerative disease associated with exposure to TBI or repetitive TBI
Diffuse axonal injury (DAI)	Widespread and extensive lesions to the brain's white-matter tracts
Loss of consciousness (LOC)	A lack of awareness of self and the environment, it typically includes a complete or near-complete lack of responsiveness to external stimuli
Glasgow Coma Scale (GCS)	An objective measure of an individual's conscious state, assessing eye, motor, and verbal responses
Glasgow Outcome Scale (GOS)	A measure allowing for standardized descriptions of the objective degree of recovery from brain injury
Head injury (HI)	A nonspecific term referring to clinically evident external injury to the face or scalp that may or may not be associated with a TBI
Hemorrhage	Extensive bleeding that is often difficult to stop
Mild traumatic brain injury (mTBI)	A TBI at the mildest end of the severity spectrum. Diagnostic criteria typically include: GCS 14–15/15, duration of PTA less than 60 minutes, and LOC less than 30 minutes. Symptoms are usually transient, and rapid recovery is common in the majority of cases
Neuroinflammation	A swelling and loss of function of nervous system tissue as a result of an insult
Neurometabolic cascade	A range of metabolic events that occur after TBI to protect and repair damaged neurologic systems and structures
Postconcussion syndrome (PCS)	Residual postconcussion symptoms that fail to resolve and persist for more than 3 months postinjury
Posttraumatic amnesia (PTA)	The period of time around the TBI (before and after the insult) for which memory of events is absent and new memories are unable to be created
Repetitive neurotrauma	Multiple TBIs sustained by an individual over time
Sport-related concussion (SRC)	An acute brain injury resulting from mechanical energy to the head from external physical forces during participation in sport
Traumatic brain injury (TBI)	An acute brain injury resulting from mechanical energy transmitted to the head from external physical forces

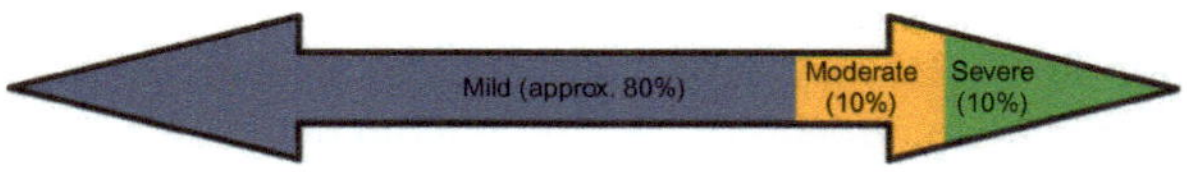

Fig. 12.1. Traumatic brain injury severity continuum.

demonstrated to be a useful classification system (Evans, 2006). The GCS may be administered at the scene of the injury and assists guiding the transport of a patient to the appropriate healthcare facility (Sasser et al., 2012), but is most commonly administered in the emergency department (ED). The GCS examines an individual's postinjury verbal and motor responses and eye movement (Teasdale and Jennett, 1974). Each of the three GCS scores provides a rudimentary measure of an individual's functional status (Table 12.2). The GCS score classify scores of 13–15 as a mild TBI (mTBI: approximately 80% of all TBIs fall into this category), scores of 9–12 as a moderate TBI (accounting for approximately 10% of all TBIs), and scores of 8 or less as a severe TBI (which also account for approximately 10% of all TBIs) (Kraus and Nourjah, 1988). In some instances it is not possible to obtain an accurate GSC score, for example, if an individual has used alcohol or other substances, or has been intubated. GCS can be impacted by numerous factors; as such, the best motor response GCS after resuscitation is often employed as the most reliable metric.

Other measures of TBI severity have included recording the duration of loss of consciousness (LOC) and posttraumatic amnesia (PTA); however, both measures may be of little value in some instances, where an individual has sustained a TBI but does not experienced LOC or PTA (LOC and/or PTA are not required to be present in order for a TBI to be diagnosed; i.e., LOC and PTA are not mandatory definitional criteria). When PTA is present, its duration is a good indicator of the extent of cognitive and functional deficits after TBI (Khan et al., 2003) and is considered to be a good predictor of TBI severity (Alexander, 1995; Bowen et al., 1999; Nakase-Richardson et al., 2011). PTA is essentially defined as the period of time that an individual is unable to reliably and consistently follow commands. The time to follow commands may take into account early complications, but it can be affected by early sedation and the patient requires monitoring over an extended period of time. This is one of the disadvantages in using PTA and/or time to follow commands, as it does not offer

Table 12.2

The Glasgow Coma Scale (GCS)

Best eye response	
No eye opening	1
Eye opening in response to pain	2
Eye opening to speech	3
Eye opening spontaneously	4
Best verbal response	
No verbal response	1
Incomprehensible sounds	2
Inappropriate words	3
Confused	4
Oriented	5
Best motor response	
No motor response	1
Extension to pain	2
Abnormal flexion to pain	3
Flexion/withdrawal to pain	4
Localizes to pain	5
Obeys commands	6
Total GCS score	/15
Teasdale and Jennett (1974) GCS score classifications	**GCS score**
Mild TBI	13–15
Moderate TBI	9–12
Severe TBI	≤ 8

TBI, traumatic brain injury.

an immediately available early prediction of outcome. However, both methods are considered to be good predictors of functional outcome (i.e., independent living and return to work) (Sherer et al., 2002).

The phenomena of anterograde and retrograde amnesia have been described in the laboratory and clinic for over 130 years (Ribot, 1881). Anterograde amnesia is characterized by a reduced ability to form new memories after a brain injury, which may lead to decreased attention and inaccurate perception. Anterograde memory is frequently the last function to return after the recovery from LOC. Following the recovery of consciousness patients may be unable to recall little or anything that occurred for days, weeks, or even months after their injury. In contrast, retrograde amnesia is characterized by the partial or total loss of the ability to recall events that occurred during the period prior to the brain injury. While retrograde amnesia is thought to be quite prevalent, severe retrograde amnesia is uncommon following TBI. The injured individual may be unable to recall events immediately preceding the injury, or sometimes during the entire day to several days prior to the injury; however, recall of information that was well learned (e.g., autobiographic information, well-known historic facts) prior to the injury is typically unimpaired. In contrast, anterograde amnesia occurs more frequently post-TBI (McMillan, 2015). Problems with acquisition/encoding, storage, and/or retrieval of new information are common TBI sequelae.

While many individuals with TBI remain capable of retaining information they encoded, even after a delay, their capacity is often diminished compared to their own premorbid level of function (Russell and Smith, 1961; McMillan, 2015). In one of the earliest studies of the association between memory impairment and TBI, Russell and Nathan (1946) raise a number of points pertaining to the relationship between anterograde and retrograde amnesia. Firstly, from a sample of 972 HI cases, where data related to memory loss were available, retrograde amnesia was typically brief (duration less than 30 minutes in 90% of cases). Secondly, the longer the duration of anterograde amnesia, the longer the retrograde amnesia. Lau and colleagues (2011) found that neither anterograde nor retrograde amnesia was predictive of protracted recovery (21 days or more to return to play in athletes suffering from concussion). This may, in part, be attributable to anterograde and retrograde amnesia being part of the normal acute response to a concussion, which resolves relatively quickly with few lasting effects. In the case of more severe TBI, the notion that duration of PTA is associated with predicting outcome (i.e., prognosis) has been proposed and supported for a number of decades (Ellenberg et al., 1996).

The Galveston Orientation and Amnesia Test (GOAT) was the first measure developed to assess the duration of PTA (Zasler et al., 2007), through the assessment of orientation and memory for events preceding and subsequent to a TBI (Levin et al., 1979). The GOAT and its modified version (MOAT) and children's versions (COAT) were developed to evaluate cognition serially during the subacute stage of recovery from closed HI. The GOAT comprises 10 questions that assess orientation, biographic recall, and memory (Table 12.3); a score greater than 75 (out of 100) for 3 consecutive days is considered the threshold for emergence from PTA (Levin et al., 1979).

The orientation log was designed for bedside use with rehabilitation inpatients (Jackson et al., 1998). It is a quantitative tool used to quickly measure orientation status (place, time, and situational domains). It is designed for serial administration to document change over time (Novack, 2000). Each item is scored as follows:

- 3 points = spontaneous free recall (i.e., first response)
- 2 points = correct upon logical cueing (i.e., "that was yesterday, so today must be…")
- 1 point = correct upon multiple choice or phonemic cueing
- 0 points = inccorrect response despite cueing, inappropriate response, or no response.

The Westmead PTA scale (WPTAS) and its abbreviated form (the A-WPTAS) are also valid measures for

Table 12.3

Galveston Orientation and Amnesia Test (GOAT)

Question	Error score	Notes
What is your name?	/2	Must give both first name and surname
When were you born?	/4	Must give day, month, and year
Where do you live?	/4	Town is sufficient
Where are you now?		
(a) City	/5	Must give actual town
(b) Building	/5	Usually in hospital or rehab center. Actual name necessary
When were you admitted to this hospital?	/5	Date
How did you get here?	/5	Mode of transport
What is the first event you can remember after the injury?	/5	Any plausible event is sufficient (record answer)
Can you give some detail?	/5	Must give relevant detail
Can you describe the last event you can recall before the accident?	/5	Any plausible event is sufficient (record answer)
What time is it now?	/5	1 for each half-hour error, etc.
What day of the week is it?	/3	1 for each day error, etc.
What day of the month is it (i.e., the date)?	/5	1 for each day error, etc.
What is the month?	/15	5 for each month error, etc.
What is the year?	/30	10 for each year error
Total error		
100 – total error		Can be a negative number

From Levin et al. (1979), with permission from Wolters Kluwer Publishers.

determining the length of PTA in individuals with TBI (Shores et al., 1986, 2008; Shores, 1989, 1995; Meares et al., 2011). The WPTAS was designed to assess the ability of individuals who have sustained a TBI to lay down new memories over a 24-hour period (Shores et al., 1986). The WPTAS measures orientation (time, person, place), together with memory for new information from one day to the next (Fig. 12.2). In individuals with mTBI, the A-WPTAS has facilitated the assessment of an individual's orientation and capacity to lay down new information over short periods of time (Ponsford et al., 2004). An individual is considered to have emerged from PTA on the first of 3 consecutive days where a recall score of 12 has been achieved (Tate et al., 2006).

DESCRIPTIVE EPIDEMIOLOGY

Collating and describing epidemiologic TBI data pose some challenges. There are a number of confounding variables in determining the epidemiology of TBI. For instance, there is no universally accepted and systematic manner in which epidemiologic data are collected (Faul et al., 2010). As shown above, there are a variety of TBI definitions, classification criteria, and methods for collecting data that have been used. Incomplete sampling is also an issue that makes it difficult to describe epidemiologic TBI data accurately. For example, there is an absence of data on some individuals suffering an mTBI who do not present to hospital, and those who present but are discharged at the ED (Ratcliff et al., 2014). As such, and in addition to the fact that most cases of mTBI are nonfatal, and individuals may not present to the ED, and also may not have been hospitalized if they did present, historically prevalence rates have not been well documented and estimates are often based on existing disabilities. This is clearly highlighted by the estimated rates of undiagnosed TBIs of 11% (Kraus et al., 1987), 61% (Boswell et al., 2002), and 75% (Kay et al., 1993), and the reported figure of 89 000 undiagnosed TBIs annually (Schootman and Fuortes, 2000).

Incidence rates

TBI is a significant public-health concern. Estimates by the National Institutes of Health (NIH) Consensus Development Panel on Rehabilitation of Persons with TBI reveal that 2.5–6.5 million US citizens live with TBI-related disabilities. It is estimated that approximately 1.5–2 million TBIs occurs annually in the USA. Of the 1.7 million individuals in the USA who are diagnosed with TBI annually, 1 365 000, or 80%, were treated in the ED and released alive (Faul et al., 2010). TBIs represent 1.4% of all visits seen in an ED and 4.8% of the total injuries presented in an ED (Faul et al., 2010). TBIs account for around 275 000 (15.1%) hospital admissions, and 52 000 deaths in the USA each year (Faul et al., 2010). Surveillance data from 11 US states revealed that 74.9% of all hospitalized TBIs were mild (Langlois et al., 2005). Of all injury-related hospitalizations ($n = 1\ 826\ 548$), hospitalizations for TBI constitute 15.1% of the injury burden ($n = 275\ 000$) and 0.7% of all hospitalizations (Faul et al., 2010). This has an enormous social and financial cost, with estimates of the annual financial burden associated with TBI ranging between 9 and 10 billion US dollars.

The incidence of mTBI is approximately 131 cases per 100 000 people, the incidence of moderate TBI is about 15 cases per 100 000 people, and the incidence of severe TBI is approximately 14 cases per 100 000 people (Syed et al., 2007). mTBI is the most common form of TBI, with an annual worldwide incidence estimated at 45 million (Faul et al., 2010). Among US children, 7 400 TBI-related fatalities, over 60 000 hospitalizations, and more than 600 000 ED visits occur annually. There are approximately 56 000 hospitalized children who are discharged

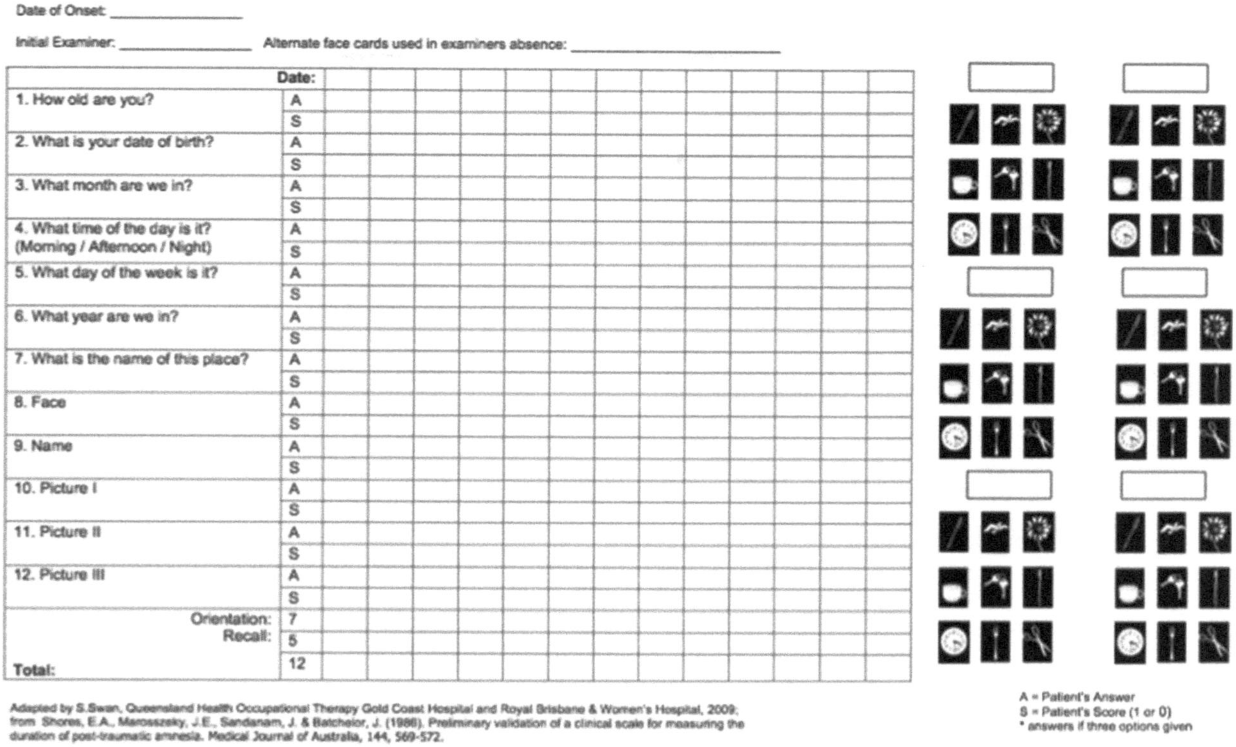

Fig. 12.2. Westmead posttraumatic amnesia (PTA) scale. From Shores EA, Marosszeky JE, Sandanam J and Batchelor J. (1986). Preliminary validation of a clinical scale for measuring the duration of post-traumatic amnesia. Med J Aust 144(11):569–572. © Copyright 1986. The Medical Journal of Australia – reproduced with permission.

home with critical disabilities from TBI, with over 5000 having required intensive inpatient care and rehabilitation (Stanley et al., 2012). TBI prevalence rates are far less commonly calculated and reported in the literature.

Mortality

In the USA approximately 52 000 TBI-related deaths occur each year (Faul et al., 2010). The mortality rate for deaths outside of hospital is approximately 17 per 100 000 hospitalizations; it is approximately 6 per 100 000 people for patients who are hospitalized (Faul et al., 2010). The highest mortality rate (32.8 cases per 100 000 people) is found in persons aged 15–24 years. The mortality rate in patients who are elderly (65 years or older) is about 31.4 individuals per 100 000 people (Langlois et al., 2005).

The leading cause of TBI-related fatalities differed among age ranges; falls was the leading cause of death for individuals older than 64 years, motor vehicle accidents (MVAs) were the leading cause of death among individuals aged 5–24 years, and assaults were the leading cause of death among children aged 0–4 years (Faul et al., 2010). A recent study of TBI in the elderly found that 950 132 TBI-related hospitalizations occurred in adults aged 65 years or older in the USA, along with 107 666 TBI-related deaths between 2000 and 2010 (Haring et al., 2015).

ETIOLOGY

A TBI may occur from any number of causes and may vary according to gender, age, race, and geographic location. Recent TBI data collated from US ED visits, hospitalizations, or deaths from 2006 through 2010 revealed falls as the leading cause of TBI, which accounted for two in every five TBIs. Of individuals over 64 years, 81% of TBIs were a result of a fall, while in children under the age of 15 years, falls accounted for 55% of TBIs. The second leading cause of TBI in the USA from 2006 through 2010 was unintentional blunt trauma (e.g., being struck by an object), which accounted for approximately 15% of TBIs. In children under the age of 15 years, blunt trauma accounted for 25% of TBIs. MVAs were the third leading cause, accounting for approximately 14% of TBIs. The TBI-related death statistics revealed that MVAs accounted for 26% of TBI-related deaths, which was the second leading cause of death for 2006 through 2010. Approximately 10% of all TBIs in the USA from 2006 through 2010 were due to assaults, of which about 75% occurred in individuals aged 15–44 years old (Faul et al., 2010) (Fig. 12.3).

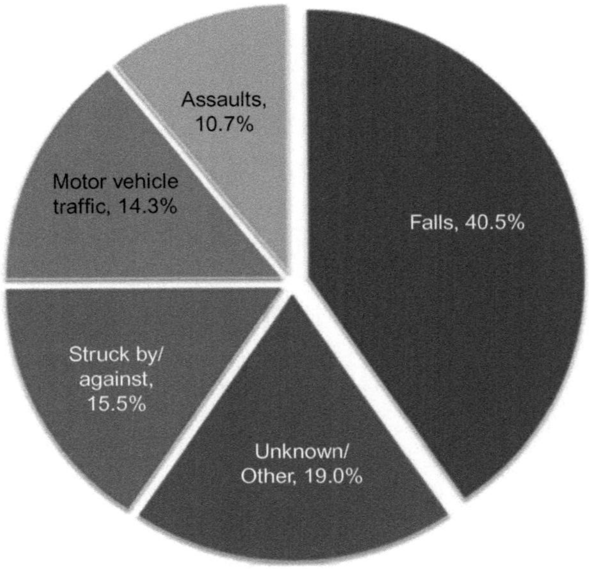

Fig. 12.3. Leading causes of traumatic brain injury. (Reproduced from Faul et al., 2010.)

Sports concussion is another cause of TBI that has attracted considerable attention over the past few decades. Sports concussion is considered to be at the mildest end of the TBI severity continuum, and often involves transient signs and symptoms, and therefore in only rare cases will an athlete present to hospital (McCrory et al., 2013). The Centers for Disease Control estimate that 1.6–3.8 million concussions occur in sports and recreational activities in the USA annually (Langlois et al., 2006).

Risk factors for TBI

In terms of TBI-related mortality, the US 2006–2010 data revealed males had an almost threefold increased risk of TBI-related death than females, and individuals over the age of 64 years had the highest mortality rates. Similarly, the US 2006–2010 data for nonfatal TBI-related injuries revealed that males had a higher rate of both TBI-related hospitalizations and ED visits than females. The highest hospitalization rates were observed among individuals over 64 years, and the highest ED visit rates were observed among children under 5 years. Assaults were the leading cause of TBI-related ED visits for individuals 14–24 years, whereas falls were the leading cause of TBI-related ED visits for all other age groups. The leading cause of TBI-related hospitalizations differed among age ranges; falls was the leading cause of hospitalization for children aged 0–14, and adults over the age of 45 years, whereas MVAs were the leading cause of hospitalization among individuals aged 15–44 years (Faul et al., 2010).

CLINICAL PRESENTATION AND SYMPTOMOLOGY

Symptoms vary greatly depending on the severity of the HI. Cognitive, physical, sensory, somatic, psychologic and psychiatric, or changes in personality may result from TBI depending on the extent of damage and/or disruption to neural circuits. Symptoms may include any of those listed in Table 12.4.

There are numerous post-TBI self-report symptom inventories available to record subjective symptoms and the degree of impact or level of severity each endorsed symptom is having on an individual (for example, Rivermead Post Concussion Symptoms Checklist, Concussion Signs and Symptoms Checklist). Residual signs and symptoms of sport-related concussion and mTBI may include those outlined in Table 12.5. In sport-related concussion, the large majority of athletes self-report resolution of symptoms within 7–10 days, and certainly within 1 month postinjury (McCrea et al., 2003). This pattern of acute disturbance and recovery is remarkably consistent with the pattern of physiologic disturbance and recovery described in neuroscience research (Giza and Hovda, 2001, 2014). In the minority of cases, residual symptom may persist beyond 3 months, at which point an individual meets the criteria for the diagnosis of "postconcussion syndrome" (DSM-IV) or "postconcussional syndrome" (ICD-10-CM). In the data reported from the TRACK-TBI study, 33% of the mTBI

Table 12.4

Traumatic brain injury symptoms

Vomiting	Dizziness and balance problems
Lethargy	Breathing problems
Headache	Slow pulse
Confusion	Difficulty swallowing
Paralysis	Ringing in the ears, or changes in hearing
Coma	Cognitive difficulties
Loss of consciousness	Inappropriate emotional responses
Dilated pupils	Body numbness or tingling
Vision changes (blurred vision or seeing double, not able to tolerate bright light, loss of eye movement, blindness)	Cerebrospinal fluid (which may be clear or blood-tinged) coming out of the ears or nose
Slow breathing rate, with an increase in blood pressure	Speech difficulties (slurred speech, inability to understand and/or articulate words)
Droopy eyelid or facial weakness	Loss of bowel control or bladder control

Table 12.5

Common signs and symptoms of mild traumatic brain injury and sport-related concussion

Cognitive	Physical	Emotional/mood	Sleep disturbance
Difficulty thinking clearly	Headache	Irritability	Sleeping more than usual
Difficulty remembering	Nausea/vomiting	Feeling more emotional	Sleeping less than usual
Difficulty concentrating	Neck pain	Sadness	Trouble falling asleep
Feeling slowed down	"Pressure in the head"	Anxiety	
Feeling like "in a fog"	Balance problems	Nervousness	
"Don't feel right"	Dizziness		
Confusion	Sensitivity to noise		
Drowsiness	Sensitivity to light		
	Blurred vision		
	Fatigue, lacking energy		

Table 12.6

Traumatic brain injury categorization based on neurologic damage

Injury mechanism	Focal	Diffuse
Primary	Skull fracture	Diffuse axonal injury
	Cortical contusion	Petechial hemorrhage
	Focal hemorrhage	Excitotoxicity
	Intracranial hematoma	
	Focal axonal injury	
Secondary	Microvascular injury	
	Hypoxic-ischemic injury	
	Neuroinflammation	
	Hypometabolism	
	Edema and herniation	
	Excitotoxicity	

subjects were functionally impaired (Glasgow Outcome Scale-Extended score ≤ 6) at 3 months postinjury, and 22.4% (of the mTBI subjects available for follow-up) were still below full functional status at 12 months post-injury (McMahon et al., 2014).

The effects of TBI

While TBIs are classified according to the level of severity (i.e., mild, moderate, severe), they can also be conceptualized (or categorized) according to the evidence of neurologic damage present on investigation (Table 12.6).

Primary effects of TBI

Focal injuries

Skull fracture

Skull fracture indicates that the force of impact was substantial and that the individual is at significantly more risk of having a subarachnoid, subdural, or epidural hemorrhage. Fractures of the base of the skull are problematic if they cause injury to nerves, arteries, or other structures. If a fracture extends into the sinuses, there may be leakage of cerebrospinal fluid from the nose or ears. Depressed skull fractures are those in which part of the bone presses on or into the brain. These may require surgical treatment. The damage caused by depressed skull fractures depends upon the region of the brain in which they are located and also upon the possible coexistence of any associated diffuse brain injury (Tseng et al., 2011).

Cortical contusion

TBI-related contusions commonly occur in the anatomic locations where the brain makes contact with the inside of the skull, where irregular bony protuberances are present. These bony protuberances are common in the base of the skull. Such contusions are commonly demonstrated in the inferior aspect of the frontal lobes, the frontal pole, and inferolateral area of the temporal lobes. Contusion can also occur directly below the site of impact (coup contusions) and/or opposite the site of impact (contrecoup contusions) (Graham et al., 1991).

Intraparenchymal hemorrhage

TBI-related hemorrhages occur as a result of vascular damage. Bleeding in the brain is space occupying and may result in delayed clinical symptoms (contingent upon the vessel affected and the rate at which blood is leaking) due to raised intracranial pressure and herniation (Graham et al., 1991).

Subdural hematomas

A subdural hematoma is a collection of blood outside the brain between the dura and the arachnoid layer (i.e., in the subdural space). It is space occupying and, if left unresolved, can result in increased intracranial pressure, leading to LOC (Graham et al., 1991).

Epidural hematomas

An epidural hematoma is similar to a subdural hematoma, but the collection of blood occurs between the dura and the skull (Graham et al., 1991).

Diffuse injuries

Diffuse vascular injury

Capillary shearing resulting in petechial hemorrhages is commonly observed in fatal cases of severe TBI because they are commonly below the sensitivity level of current neuroimaging techniques.

Diffuse axonal injury (DAI)

DAI describes the widespread and extensive lesions in white-matter tracts. While neural tissue is elastic and capable of deforming in response to normal head movement, exposure to rapid acceleration/deceleration (A/D) forces may exceed the maximum elasticity of tissue and result in DAI (Smith et al., 1999).

Excitotoxicity and oxidative stress

Neurotrauma leads to a number of crises that result in a cascade of primary and secondary issues. The abrupt release of neurotransmitters (for example, glutamate) in response to the trauma subsequently results in considerable increases in intracellular calcium, glucose hypermetabolism, kinase activation, and diminished cerebral blood flow (Chen et al., 2008).

Secondary effects of TBI

Vascular injury

A progressive secondary hemorrhage can occur within hours of a TBI and result in expansion of tissue damage, as a result of intracranial pressure but also ischemia, hypoxia, free radical formation, and induction of inflammation (Ghajar, 2000).

Hypoxic-ischemic injury

Hypotension, hypoxia, and ischemia are common secondary events following TBI (Ghajar, 2000).

Neuroinflammation

Following TBI, there are both reparative and degenerative mechanisms that lead to an inflammatory environment. Within minutes of injury, proinflammatory cytokines are released. The proinflammatory cytokine interleukin-1β (IL-1β) is one such cytokine that exacerbates the tissue damage caused by TBI (Helmy et al., 2011). TBI may cause significant damage to vital components to the brain, including the blood–brain barrier. IL-1β causes DNA fragmentation and apoptosis, and, together with tumor necrosis factor-α, may cause damage to the blood–brain barrier and infiltration of leukocytes (Helmy et al., 2011; Ramlackhansingh et al., 2011).

Edema and herniation

Increased intracranial pressure from secondary brain swelling is commonly due to edema and herniation. Edema is typically a result of compromise to the permeability of the blood–brain barrier, which results in altered fluid homeostasis (Marmarou, 2007).

Computed tomography (CT) and outcome of TBI

The introduction of CT dramatically improved the ability to detect intracerebral abnormalities (e.g., contusions, hematomas). The presence of such abnormalities has been used to predict outcome in severe TBI, although they do not appear to add significantly to the predictive value of the clinical variables (e.g., GCS, pupillary responses) used to determine overall outcome in severe TBI. The utility of CT in predicting the outcome of moderate or mild head injuries is even more limited because this modality lacks the sensitivity to detect subtle anatomic lesions. CT is also clinically useful for acute care and formulating decisions around the need for early surgical intervention.

Magnetic resonance imaging (MRI) and outcome of TBI

Structural MRI is more sensitive than CT for detecting abnormalities in the brain after TBI (Fig. 12.4) (Orrison et al., 1994). In mTBI patients with normal head CT at hospital admission, 27% demonstrated abnormal brain MRI (Yuh et al., 2013). This modality is very sensitive for identifying DAI and nonhemorrhagic contusions Yokota et al., 1991; Paterakis et al., 2000). More advanced MR techniques have demonstrated some promise in improving sensitivity for detecting subtle TBI-related abnormalities. Diffusion tensor imaging (DTI) enables the visualization of white-matter tracts (Jeter et al., 2013). Measures are expressed using various indices such as the fractional anisotropy that reflects the directionality of water diffusion or the degree of anisotropy. Because axonal injury is common in TBI, DTI is a promising biomarker. Tractography is a DTI method that allows for other sophisticated analyses of tract integrity (Tournier et al., 2004; Farquharson et al., 2013).

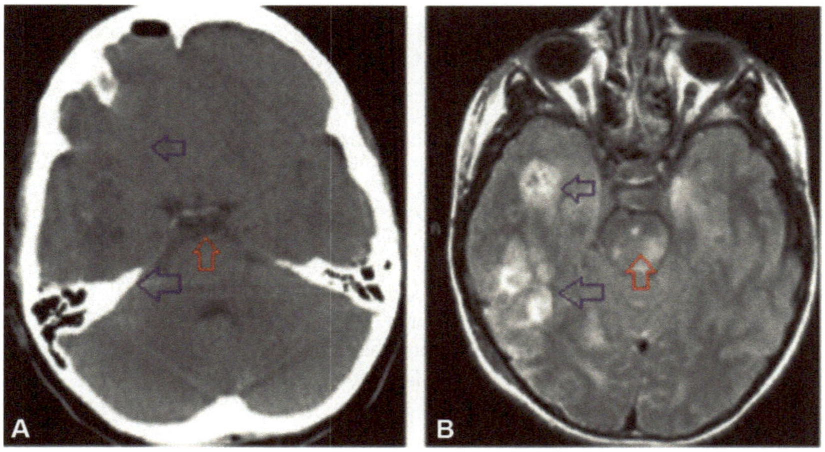

Fig. 12.4. (**A**) Noncontrasted axial computed tomography (CT) scan and (**B**) a fluid-attenuated inversion recovery (FLAIR) magnetic resonance (MR, right) image of a 10-year-old boy 48 hours after he sustained a severe closed head injury. The region of hyperintense signal in the brainstem (red arrows) and the hyperintense signal in the right temporal region (blue arrows) are visible on the MR image, but cannot be detected on the CT scan.

Magnetic resonance spectroscopy allows for measurements of brain metabolites that may be affected by concussion, including *N*-acetylaspartate, a marker of neuronal health; glutathione, an antioxidant that is depleted during oxidative stress, which is common after TBI; glutamate, an excitatory neurotransmitter that is altered by TBI, and *myo*-inosotol, a marker of glial cell activity (Hunter et al., 2012; Gardner et al., 2014a). Susceptibility-weighted imaging is a technique that exploits differences in magnetic susceptibility between tissues, and is sensitive to microhemorrhages after mTBI (Hunter et al., 2012; Shenton et al., 2012). Resting-state functional MRI provides a measurement of the brain's functional connectivity networks. Arterial spin labeling is a method that involves magnetically labeling blood before it enters the brain and is capable of quantifying cerebral blood flow (Kim et al., 2010). As studies report hemodynamic impairment in patients with TBI (Maugans et al., 2012), arterial spin labeling represents a strong TBI biomarker candidate.

Other important mitigating clinical factors

Epidemiologic research has considerably advanced the field's understanding of TBI and factors that may influence outcome. Recovery from TBI is influenced by classes of factors such as individual patient characteristics, social-environmental factors (e.g., family support systems), and barriers to rehabilitation access. In the clinical setting it is important to collect a thorough medical history to determine the extent to which the presenting signs and symptoms may be attributable to the TBI versus other potential premorbid factors. From a cognitive perspective, level of education, work history, a past history of previous TBI, the severity and recovery time, the presence of any learning difficulties and/or disorders, attention deficit disorder and/or hyperactivity, mental health history (e.g., any psychiatric conditions such as anxiety or depression), past and/or present substance use (including prescribed medication, illicit substances, and alcohol use), pain symptoms, and any other neurologic conditions are all important characteristics for the case formulation and interpretation of post-TBI presentation and performance.

Prognostics

The heterogeneity of TBI is one of the most significant challenges associated with predicting individual outcome. There are a number of objective measures of functional outcome that have been developed, the most commonly reported being the Glasgow Outcome Scale (GOS). The GOS (Jennett and Bond, 1975; Teasdale et al., 1998) and the extended version (GOSE) (Wilson et al., 1998) are global scales for functional outcome following TBI that classify an individual's status into one of five (GOS), or one of eight (GOSE) categories (Table 12.7).

The most recent guidelines from the American Association of Neurological Surgeons and the Joint Section on Neurotrauma and Critical Care, pertaining to clinical indicators of outcome in TBI, indicated that the five factors that are most frequently used to determine the prognosis of severe TBI are the GCS score, age, pupillary responses, presence of hypotension, and abnormalities on CT.

The initial GCS score is inversely correlated with the likelihood of death. An admission GCS score of 3 was

Table 12.7

Glasgow Outcome Scale (GOS) and the Extended Glasgow Outcome Scale (GOSE)

Glasgow Outcome Scale			Extended Glasgow Outcome Scale	
Score	Outcome	Definition	Score	Outcome
1	Dead	Dead	1	Dead (D)
2	Persistent vegetative state	Wakefulness without awareness	2	Vegetative state (VS)
3	Severe disability	Conscious but dependent	3	Lower severe disability (SD–)
4	Moderate disability	Independent but disabled	4	Upper severe disability (SD+)
5	Good recovery	Fully integrated into society (may have nondisabling sequelae)	5	Lower moderate disability (MD–)
			6	Upper moderate disability (MD+)
			7	Lower good recovery (GR–)
			8	Upper good recovery (GR+)

Adapted from Jennett and Teasdale (1974).

associated with mortality rates between 65% and 100%, although 7% of patients with an initial GCS of 3 have a good outcome (GOS score of 4 or 5). An initial GCS score of 7 was associated with mortality rates ranging from 15% to 27% (Braakman et al., 1980; Fearnside et al., 1993).

The age of a patient with a severe TBI at presentation is also inversely related to the likelihood of death. Although the age-related mortality rate associated with a severe HI (GCS < 8) falls on a continuum, the critical threshold appears to be around 50 years. The mortality rate of patients with a severe HI over the age of 50 years was 78–84% while it was 28–38% for patients younger than 50 years (Braakman et al., 1980; Narayan et al., 1981).

The absence of bilateral pupillary responses after a severe TBI is also associated with a high mortality rate (91–100%) (Braakman et al., 1980; Cordobés et al., 1981). A weaker predictive factor is the presence of hypotension after a severe TBI. The mortality rates associated with patients who have periods of hypotension range from 42% to 50% compared to 27% or 28% in normotensive patients (Chesnut et al., 1993; Fearnside et al., 1993). In a systematic review of multivariable prognostic models for moderate to severe TBI, the best predictors of outcome were reported to be the GCS, pupil reactivity, and CT (Savola and Hillbom, 2003). In a systematic review of multivariable prognostic models for mTBI, there were no multivariable prognostic models that adequately predicted individual patient outcomes. However, preinjury mental health and early postinjury neuropsychologic functioning were identified as the most robust prognostic factors in the context of multivariable models. In addition, it was found that women and adults with early postinjury anxiety had worse prognoses (Silverberg et al., 2015).

While duration of PTA is used as a measure of TBI severity, it is also a reasonable predictor of longer-term outcomes in children (Briggs et al., 2015), and in adults moderate to severe TBI (Walker et al., 2010). Similarly, time to follow commands has also been demonstrated as a clinically useful injury severity rating for predicting outcome in pediatric TBI (Austin et al., 2013; Davis et al., 2015).

Risk for later-life neurodegenerative disease

There is a growing body of research into the potential longer-term consequences of TBI. Moderate to severe TBI is increasingly recognized as a risk factor for dementia (DeKosky et al., 2013; Nordström et al., 2014; Johnson and Stewart, 2015), and represents one of the strongest identified environmental risk factors (Nordström et al., 2014). Several studies have explored the increased incidence of a variety of dementias associated with TBI of all severity levels (Lehman et al., 2012; Gardner et al., 2014b, 2015), with current evidence suggesting there is a possible "dose" and frequency-dependent association between TBI and risk of dementia (Roozenbeek et al., 2013; Gardner et al., 2014b; Nordström et al., 2014). A hazard ratio for all-cause dementia was reported to be 1.5 for patients with mTBI, and 2.3 for severe TBI, relative to people without TBI; however the overarching finding in this study was that individuals who suffer a moderate to severe TBI over the age of 55 years, or mTBI over the age of 65 years had an increased risk of developing dementia (i.e., a higher risk of dementia is associated with an older age of onset of TBI) (Gardner et al., 2014b). Concern has also been raised following the observation of a strong association between TBI and "young-onset" dementia (Gardner et al., 2014b).

BIOMECHANICS AND PATHOPHYSIOLOGY

Epidemiologic studies have considerably advanced our understanding of the pathogenesis of TBI, particularly in the area of biomechanics. TBI is typically caused by accelerations or decelerations of the head involving both linear and rotational forces (Elson and Ward, 1994). These external forces can result in a brain injury at both the site of contact (coup) and the opposite side of the brain (contrecoup) due to inertial forces (Fig. 12.5).

The first comprehensive theory of TBI, developed by Ommaya and Gennarelli (1974), predicted three important elements related to concussion: (1) that the direction of force can determine the severity of injury; (2) that a continuum of injury exists whereby mild, moderate, and severe brain injuries caused by A/D forces are not random in magnitude but occur on a continuum; and (3) that A/D forces alone are sufficient to cause severe TBI. Rotational acceleration is common during either impact or impulsive head loading. Due to the physical properties and complexity of the human brain, its tissue deforms more readily in response to shear forces compared with other biologic tissues. Importantly, A/D forces differentially affect gray and white matter. Rapid head rotations generate shear forces throughout the brain, and therefore, rotational accelerations have a high potential to cause shear-induced tissue damage (not to be confused with the physical "shearing" or cutting of axons – a common misconception that persists today). Lateral (coronal) plane accelerations have the greatest likelihood for producing damage within the deep internal structures of the human brain (Meaney and Smith, 2011). With the advancement of technology and increased professionalism of sport, telemetry technology has been implemented in helmets of athletes to record impacts sustained by players. Figure 12.6 shows the range of impacts sustained by a 21-year-old collegiate American football defensive

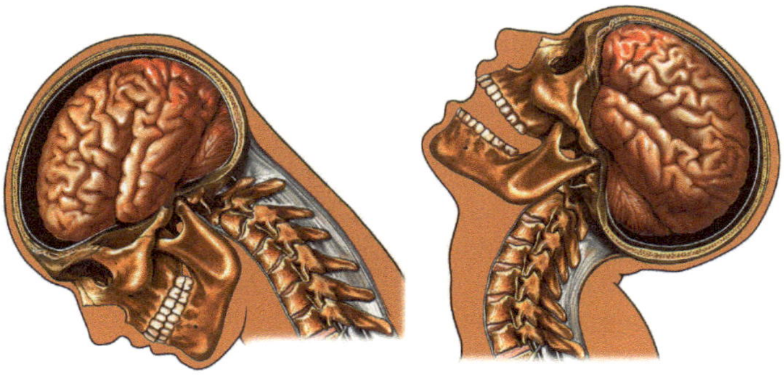

Fig. 12.5. Depiction of (left) the coup and (right) contrecoup injury. The brain impacts the skull in the anterior region as the neck is in extension, and then goes on to impact the skull in the posterior region as the neck is in flexion. This movement can occur in whiplash-type injuries such as those sustained in some motor vehicle accidents. Medical Illustration.

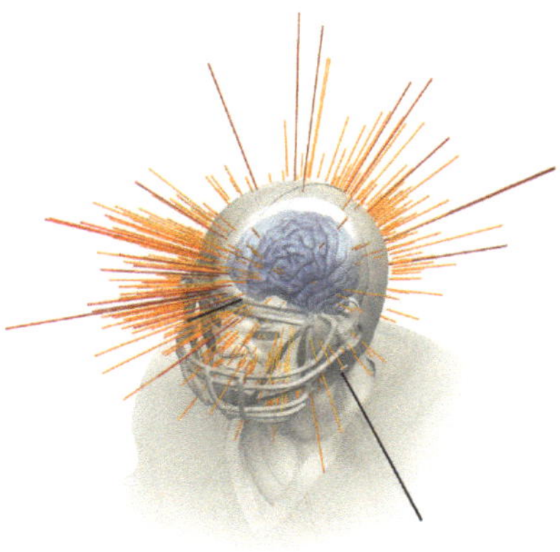

Fig. 12.6. Depiction of the contact made to the helmet of a collegiate-level defensive end during a single season. Yellow rods represent hits below 80 **g**, red rods represent hits above 80 **g**, and the two black rods represent the hits that resulted in a diagnosis of concussion. Credit: Bryan Christie Design. www.bryanchristiedesign.com.

end during football games and practices in a single season. This player sustained 537 hits, of which 417 impacts were recorded at magnitudes of 10 **g** or greater.

In an effort to demonstrate that movement of the brain within the cranial vault may be even more complex than was originally considered, Bayly and colleagues (2005) used MRI to observe that consistent mild accelerations of the head inside the imaging coil generate a remarkable amount of movement of the brain. Three subjects were asked to repeatedly allow the head to fall a short distance downward while lying in the supine position. Results revealed that tethering loads may be borne by the vascular, neural, and dural elements that bind the brain to the base of the skull. These biologic tethers cause the brain to compress and stretch in the anterior–posterior plane followed by rotation backward and upward relative to the skull (Bayly et al., 2005).

When the brain moves within the cranial vault, stretching and compression can affect neuronal cell bodies, axons, and organelles as well as glial cells. The acute neurometabolic and neurochemical changes that can occur in neurons following stretch or compression commence with a disruption of the neuronal cell membrane and axon due to stretching, resulting in ion movement across the plasma membrane, widespread release of neurotransmitters (most importantly, excitatory amino acids such as glutamate), and influx of calcium ions. Further metabolic disruptions are caused by damage to mitochondria (Staal et al., 2007; Wu et al., 2007).

Once an axon is stretched, a pathophysiologic process begins that can lead to structural change and metabolic dysfunction beyond the acute phase. Most studies indicate that it is acute axonal stretch and strain that causes a rapid increase in intracellular calcium. Strain on the axonal membrane causes an abnormal influx of Na^+ through mechanosensitive sodium channels, a reversal of the Na^+-Ca^{2+} exchangers, and activation of voltage-gated Ca^{2+} channels (Wolf et al., 2001). This in turn causes selective proteolysis or a breakdown of Na^+ channels and progressively increasing levels of intra-axonal Ca^{2+} (Iwata et al., 2004). Dramatic elevations in intracellular calcium stores occur quickly, with some estimates being within the first 1–6 seconds following axonal stretch (Staal et al., 2010). The massive influx of Ca^{2+} leads to damage to the axonal cytoskeleton and initiates the pathophysiologic process that follows. It is these pathophysiologic events that lead to the genesis of retraction balls, and eventually, secondary axotomy.

It has long been assumed that axons that have been stretched undergo subsequent axonal disconnection. When disconnection occurs close to the cell body, neuronal death ensues (Giehl et al., 1997; Bonatz et al., 2000), with substantial neuronal atrophy occurring with more distal lesions. New evidence suggests that axonal injury causes (1) persistent neuronal atrophy in neocortex but not neuronal death, and (2) spontaneous structural plasticity (Greer et al., 2011). Equally important is the fact that injured neurons produce an elongation of their proximal disconnection over a 28-day period, consistent with regeneration and reorganization (Greer et al., 2011). Cortical impact models demonstrate that dendrites can also be affected when exposed to mild forces. Specifically, the dendrites of neurons in injured cortex swell in a manner similar to that described for axons, a condition that will obviously impact postsynaptic potential production. Further, repetitive injuries in cultured cells result in reduced neurite (developing dendrites and axons in cultured cells) length and number. In the presence of high extracellular glutamate (such as occurs during the complex neurometabolic and neurochemical cascade that occurs post mTBI), extracellular glutamate induces a decline in GLT-1 and GLAST expression in cultured cortical astrocytes (Lehmann et al., 2009). This, of course, can be linked to glutamate increases in the extracellular space that occurs in a feedforward manner. Furthermore, astrocytes and microglia are important mediators of inflammation following concussion. Inflammation in the central nervous system is driven by the microglia, astrocytes, and peripheral macrophages that are all capable of releasing anti- and proinflammatory cytokines, chemokines, neurotransmitters, and reactive oxygen species. Precise regulation of inflammation is essential because of the potential impact on neurogenesis (Whitney et al., 2009). It may be the case that astrocyte-mediated proinflammatory cytokines such as IL-1β may be involved in increased susceptibility to repetitive injury (Ralay Ranaivo et al., 2011).

The biochemical cascade associated with mTBI has been well described by Giza and Hovda (2001, 2014). Animal models have revealed that neurometabolic cascade of mTBI includes ionic flux and glutamate release, energy crisis, cytoskeletal damage, axonal dysfunction, altered neurotransmission, inflammation, and cell death (for review, see Giza and Hovda, 2001, 2014). mTBI generates an imbalance between oxidant and antioxidant agents (i.e., oxidative stress) that may result in neural dysfunction and, in some circumstances, neural death. Following an mTBI, oxidative stress markers like carbonylated proteins, lipid peroxides, reactive oxygen, and reactive nitrogen species are generated, while antioxidant defense enzymes (such as GSH, ratio GSH/GSSG, GPx, GR, GST, G-6PD, SOD, and CAT) decrease (Rodriguez-Rodriguez et al., 2014).

CLINICAL TRIALS

There have been numerous clinical trials involving pharmacologic agents and other strategies conducted in TBI over the past three decades, with limited success (Maas et al., 2007). During this period, upward of 40 major clinical trials have involved neuroprotective agents that were designed to enhance the morphologic and functional recovery post-TBI that have failed in phase II or phase III evaluation. In general, clinical trials in TBI have demonstrated little clinical benefit, to the extent that most major pharmaceutical companies have dissolved operations and are no longer pursuing research to enhance acute-stage neuroprotection and repair. Despite the significant burden of TBI, a recent pharmaceutical industry report indicated that there were only eight drugs in various stages of clinical trial, with only one in phase III evaluation (Stein et al., 2015). Clinical trials for the treatment of TBI have demonstrated little success. For example the proTECT III clinical trial did not find any outcome benefit of administration of progesterone over placebo in acute TBI patients (Wright et al., 2014); in addition, primary and secondary analysis of the SyNAPSe phase III clinical trial, which was conducted in severe TBI patients, demonstrated no clinical benefit of progesterone (Skolnick et al., 2014). The phase III clinical trial data of the Citicoline Brain Injury Treatment Trial (COBRIT) found that 90 days of citicoline administration to TBI patients did not improve function or cognitive outcomes compared to placebo (Zafonte et al., 2012). Glucocorticoids and magnesium sulfate have been trialed and discontinued due to significantly higher mortality rates in the treatment groups compared to controls.

The reasons why large clinical trials fail are often difficult to elucidate. Typically there is no simple solution, although in the case of clinical trials in TBI it is worth considering the premise upon which many pharmaceutical agents are developed and promoted to trials. The supposition that a successful pharmaceutical agent should target a specific mechanism or signaling pathway is unlikely to be the breakthrough strategy when evaluating such a heterogeneous and complex systemic disorder like TBI (Stein et al., 2015). For this reason, pharmaceutical approaches that target just one aspect of the TBI cascade are unlikely to demonstrate sufficient clinical benefit. Rather, a drug development strategy that includes receptors or combinations of drugs that may act to reduce the multiple factors producing the cascade of secondary injury may be a more beneficial approach (Maas et al., 2007; Stein et al., 2015). Advancing the understanding of the subspecific aspects of the phenotype and targeting a biomarker with a well-defined endophenotype may provide an opportunity for a more targeted approach, rather than the less sophisticated, overly simplistic, and broad approach of trying multiple avenues simultaneously.

Maas and colleagues (2007) eloquently highlight the complexity of clinical trials in TBI, by drawing attention to two important unique elements that distinguish TBI clinical trials from all other fields of medicine. That is, TBI is not one single disease entity but includes a very heterogeneous and complex spectrum of pathologies, ranging from DAI through to focal contusions and extracerebral hematomas. Additionally, TBI patients are usually extremely heterogeneous in relation to clinical severity and baseline prognostic risk (Maas et al., 2007).

While many trials have failed, there remain a number of ongoing clinical trials that demonstrate varying degrees of promise; for example, clinical trials looking at hormonal agents (e.g., progesterone, estrogen), glutamate antagonists, immunomodulators (e.g., statins, cyclosporin), antioxidants, and coagulation modulator therapies (Stein et al., 2015). Time will reveal the extent to which these clinical trials are successful.

Transforming Traumatic Brain Injury Research and Clinical Care (TRACK-TBI)

The NIH National Institute of Neurological Disorders and Stroke (NINDS) funded TRACK-TBI, a multicenter study with the objective of validating the feasibility of implementing a broad range of TBI common data elements (CDEs) in a prospective observational study. Patients were recruited through four study sites, including three acute care sites (San Francisco General Hospital, University of Pittsburgh Medical Center, and University Medical Center Brackenridge in Austin, TX) and a rehabilitation center (Mount Sinai Rehabilitation Center in New York City). Although this study concluded that implementation of the TBI CDEs is feasible, TRACK-TBI researchers identified several gaps related to the outcome-related CDEs for adult TBI. These gaps included the identification of a lack of validated outcome measures for TBI patients living with greater disability. In addition, while the TBI CDE workgroup considered the availability of outcome measures validated among non-English-speaking patients as a factor in recommending specific measures, TRACK-TBI found several measures that could not be administered to non-English-speaking patients. TRACK-TBI has been expanded and is now an ongoing study designed to establish more precise methods for TBI diagnosis and prognosis, improve outcome assessment, and compare the effectiveness and costs of tests, treatments, and services (Yue et al., 2013).

SYNOPSIS

TBI is a common injury across all age ranges and can occur as a result of a range of etiologies (e.g., MVAs, falls, assaults, sports concussion, work-related accidents). No individual is immune from TBI or its potential consequences, which at worst can be fatal (approximately 52 000 TBI-related deaths occur in the USA annually) and at best a full recovery can be achieved with time. TBI severity is categorized as mild, moderate, or severe; approximately 80% of all TBIs fall into the mild category, with 10% each falling into the moderate and severe categories. In the clinical setting, signs and symptoms vary substantially depending on the severity of TBI and whether the individual is in the acute, subacute, or chronic phase post-TBI. Cognitive, physical, sensory, somatic, psychologic, psychiatric, or changes in personality may result from TBI and can persist for any length of time post-TBI. The heterogeneity of TBI is considered one of the most significant barriers to finding effective therapeutic interventions, with most therapeutic treatment methods simply targeting specific symptoms, which can be an effective method in some instances, but not all TBI cases.

HIGHLIGHTS

The TBI neuroepidemiology data have provided important detail and advanced the field's understanding of the etiology pertaining to risk factors for sustaining TBI, which has in turn resulted in preventive management strategies. Mandatory wearing of seatbelts in motor vehicles, campaigns to encourage cyclists to wear helmets, and improved work safety equipment and standards have assisted in reducing the risk of TBI in these specific environments. However, there remain a number of identified risks that require further education and modification in order to curb the current rates of TBI. For example, alcohol is a major factor in many TBIs and often is associated with the leading causes of TBI (i.e., MVAs, falls, and assaults). In addition, TBI is a work-related risk for many contact sports athletes and military personnel, who may potentially be exposed to multiple TBIs during their careers, and for children and adolescent athletes. Identifying at-risk behaviors (or plays in sport-specific context) within specific environments and modifying those behaviors (in sport, this may be achieved through rule modification and sanctions for violations). In addition, identifying at-risk individuals (i.e., those who are more vulnerable to TBIs, protracted recovery, and then have greater vulnerability to subsequent impacts) is another avenue currently being pursued by many researchers around the world. Despite numerous clinical trials evaluating neuroprotective agents and pharmaceutical treatment for TBI, there has been very little clinical benefit to date that may help modify and reduced the risk of concussive injury.

References

Alexander MP (1995). Mild traumatic brain injury: pathophysiology, natural history, and clinical management. Neurology 45 (7): 1253–1260.

Austin CA, Slomine BS, Dematt EJ et al. (2013). Time to follow commands remains the most useful injury severity variable for predicting WeeFIM® scores 1 year after paediatric TBI. Brain Inj 27 (9): 1056–1062.

Bayly PV, Cohen TS, Leister EP et al. (2005). Deformation of the human brain induced by mild acceleration. J Neurotrauma 22 (8): 845–856.

Bonatz H, Röhrig S, Mestres P et al. (2000). An axotomy model for the induction of death of rat and mouse corticospinal neurons in vivo. J Neurosci Methods 100 (1–2): 105–115.

Boswell JE, McErlean M, Verdile VP (2002). Prevalence of traumatic brain injury in an ED population. Am J Emerg Med 20 (3): 177–180.

Bowen A, Chamberlain MA, Tennant A et al. (1999). The persistence of mood disorders following traumatic brain injury: a 1 year follow-up. Brain Inj 13 (7): 547–553.

Braakman R, Gelpke GJ, Habbema JD et al. (1980). Systematic selection of prognostic features in patients with severe head injury. Neurosurgery 6 (4): 362–370.

Briggs R, Brookes N, Tate R et al. (2015). Duration of post-traumatic amnesia as a predictor of functional outcome in school-age children: a systematic review. Dev Med Child Med 57: 618–627.

Centers for Disease Control and Prevention (n.d.). Injury Prevention and Control: Traumatic Brain Injury. Retrieved July 10, 2015, from http://www.cdc.gov/traumaticbraininjury/basics.html.

Chen JK, Johnston KM, Petrides M et al. (2008). Recovery from mild head injury in sports: evidence from serial functional magnetic resonance imaging studies in male athletes. Clin J Sport Med 18 (6): 1101–1119.

Chesnut RM, Marshall LF, Klauber MR et al. (1993). The role of secondary brain injury in determining outcome from severe head injury. J Trauma 34 (2): 216–222.

Cordobés F, Lobato RD, Rivas JJ et al. (1981). Observations on 82 patients with extradural hematoma. Comparison of results before and after the advent of computerized tomography. J Neurosurg 54 (2): 179–186.

Davis KC, Slomine BS, Salorio CF et al. (2015). Time to follow commands and duration of posttraumatic amnesia predict GOS-E Peds scores 1 to 2 years after TBI in children requiring inpatient rehabilitation. J Head Trauma Rehabil 31: E39–E-47.

DeKosky ST, Blennow K, Ikonomovic MD et al. (2013). Acute and chronic traumatic encephalopathies: pathogenesis and biomarkers. Nature Reviews. Neurology 9 (4): 192–200.

Ellenberg JH, Levin HS, Saydjari C (1996). Posttraumatic amnesia as a predictor of outcome after severe closed head injury. Arch Neurol 53 (8): 782–791.

Elson LM, Ward CC (1994). Mechanisms and pathophysiology of mild head injury. Semin Neurol 14 (1): 8–18.

Evans RW (2006). Neurology and Trauma, 2nd edn. Oxford University Press, Oxford.

Farquharson S, Tournier JD, Calamante F et al. (2013). White matter fiber tractography: why we need to move beyond DTI. J Neurosurg 118 (6): 1367–1377.

Faul M, Xu L, Wald MM et al. (2010). Traumatic brain injury in the United States: emergency department visits, hospitalizations, and deaths, Centers for Disease Control and Prevention, Atlanta (GA).

Fearnside MR, Cook RJ, McDougall P et al. (1993). The Westmead Head Injury Project outcome in severe head injury. A comparative analysis of pre-hospital, clinical and CT variables. Br J Neurosurg 7 (3): 267–279.

Gardner A, Iverson GL, Stanwell P (2014a). A systematic review of proton magnetic resonance spectroscopy findings in sport-related concussion. J Neurotrauma 31 (1): 1–18.

Gardner RC, Burke JF, Nettiksimmons J et al. (2014b). Dementia risk after traumatic brain injury vs nonbrain trauma: the role of age and severity. JAMA Neurol 71 (12): 1490–1497.

Gardner RC, Burke JF, Nettiksimmons J et al. (2015). Traumatic brain injury in later life increases risk for Parkinson disease. Ann Neurol 77 (6): 987–995.

Ghajar J (2000). Traumatic brain injury. Lancet 356 (9233): 923–929.

Giehl KM, Schacht CM, Yan Q et al. (1997). GDNF is a trophic factor for adult rat corticospinal neurons and promotes their long-term survival after axotomy in vivo. Eur J Neurosci 9 (11): 2479–2488.

Giza CC, Hovda DA (2001). The Neurometabolic Cascade of Concussion. J Athl Train 36 (3): 228–235.

Giza CC, Hovda DA (2014). The new neurometabolic cascade of concussion. Neurosurgery 75 (Suppl 4): S24–S33.

Graham DI, Gennarelli TA, McIntosh T (1991). Trauma. In: DI Graham, P Lantos (Eds.), Greenfield's Neuropathology, Arnold, London, pp. 823–898.

Greer JE, McGinn MJ, Povlishock JT (2011). Diffuse traumatic axonal injury in the mouse induces atrophy, c-Jun activation, and axonal outgrowth in the axotomized neuronal population. J Neurosci 31 (13): 5089–5105.

Haring RS, Narang K, Canner JK et al. (2015). Traumatic brain injury in the elderly: morbidity and mortality trends and risk factors. J Surg Res 195: 1–9.

Helmy A, De Simoni MG, Guilfoyle MR et al. (2011). Cytokines and innate inflammation in the pathogenesis of human traumatic brain injury. Prog Neurobiol 95 (3): 352–372.

Hunter JV, Wilde EA, Tong KA et al. (2012). Emerging imaging tools for use with traumatic brain injury research. J Neurotrauma 29 (4): 654–671.

Iwata A, Stys PK, Wolf JA et al. (2004). Traumatic axonal injury induces proteolytic cleavage of the voltage-gated sodium channels modulated by tetrodotoxin and protease inhibitors. J Neurosci 24 (19): 4605–4613.

Jackson WT, Novack TA, Dowler RN (1998). Effective serial measurement of cognitive orientation in rehabilitation: the orientation log. Arch Phys Med Rehabil 79 (6): 718–720.

Jennett B, Bond M (1975). Assessment of outcome after severe brain damage. Lancet 1 (7905): 480–484.

Jennett B, Teasdale G (1974). Assessment of coma and impaired consciousness. A practical scale. Lancet 2 (7872): 81–84.

Jeter CB, Hergenroeder GW, Hylin MJ et al. (2013). Biomarkers for the diagnosis and prognosis of mild traumatic brain injury/concussion. J Neurotrauma 30: 657–670.

Johnson VE, Stewart W (2015). Traumatic brain injury: age at injury influences dementia risk after TBI. Nat Rev Neurol 11 (3): 128–130.

Kay T, Harrington DE, Adams R et al. (1993). Definition of mild traumatic brain injury. J Head Trauma Rehabil 8 (1): 86–87.

Khan F, Baguley IJ, Cameron ID (2003). Rehabilitation after traumatic brain injury. Med J Aust 178 (6): 290–295.

Kim J, Whyte J, Patel S et al. (2010). Resting cerebral blood flow alterations in chronic traumatic brain injury: an arterial spin labeling perfusion FMRI study. J Neurotrauma 27 (8): 1399–1411.

Kraus JF, Nourjah P (1988). The epidemiology of mild, uncomplicated brain injury. J Trauma 28 (12): 1637–1643.

Kraus JF, Fife D, Conroy C (1987). Pediatric brain injuries: the nature, clinical course, and early outcomes in a defined United States' population. Pediatrics 79 (4): 501–507.

Langlois JA, Rutland-Brown W, Thomas KE (2005). The incidence of traumatic brain injury among children in the United States: differences by race. J Head Trauma Rehabil 20 (3): 229–238.

Langlois JA, Rutland-Brown W, Wald MM (2006). The epidemiology and impact of traumatic brain injury: a brief overview. J Head Trauma Rehabil 21 (5): 375–378.

Lau BC, Kontos AP, Collins MW et al. (2011). Which on-field signs/symptoms predict protracted recovery from sport-related concussion among high school football players? Am J Sport Med 39 (11): 2311–2318.

Lehman EJ, Hein MJ, Baron SL et al. (2012). Neurodegenerative causes of death among retired National Football League players. Neurology 79 (19): 1970–1974.

Lehmann C, Bette S, Engele J (2009). High extracellular glutamate modulates expression of glutamate transporters and glutamine synthetase in cultured astrocytes. Brain Res 1297: 1–8.

Levin HS, O'Donnell VM, Grossman RG (1979). The Galveston Orientation and Amnesia Test. A practical scale to assess cognition after head injury. J Nerv Ment Dis 167 (11): 675–684.

Maas AI, Marmarou A, Murray GD et al. (2007). Prognosis and clinical trial design in TBI: The IMPACT Study. J Neurotrauma 24 (2): 232–238.

Marmarou A (2007). A review of progress in understanding the pathophysiology and treatment of brain edema. Neurosurg Focus 22 (5): E1.

Maugans TA, Farley C, Altaye M et al. (2012). Pediatric sports-related concussion produces cerebral blood flow alterations. Pediatrics 129: 28–37.

McCrea M, Guskiewicz KM, Marshall SW et al. (2003). Acute effects and recovery time following concussion in collegiate football players: the NCAA Concussion Study. JAMA 290 (19): 2556–2563.

McCrory P, Meeuwisse WH, Aubry M et al. (2013). Consensus statement on concussion in sport: the 4th International Conference on Concussion in Sport held in Zurich, November 2012. Br J Sports Med 47 (5): 250–258.

McMahon P, Hricik A, Yue JK et al. (2014). Symptomatology and functional outcome in mild traumatic brain injury: results from the prospective TRACK-TBI study. J Neurotrauma 31 (1): 26–33.

McMillan TM (2015). Post traumatic amnesia. In: JD Wright (Ed.), International Encyclopedia of Social and

Behavioural Sciences, 2nd edn. Elsevier, Amsterdam, pp. 636–639.

Meaney DF, Smith DH (2011). Biomechanics of concussion. Clin Sports Med 30 (1): 19–31. vii.

Meares S, Shores EA, Taylor AJ et al. (2011). Validation of the Abbreviated Westmead Post-traumatic Amnesia Scale: a brief measure to identify acute cognitive impairment in mild traumatic brain injury. Brain Inj : [BI] 25 (12): 1198–1205.

Nakase-Richardson R, Sherer M, Seel RT et al. (2011). Utility of post-traumatic amnesia in predicting 1-year productivity following traumatic brain injury: comparison of the Russell and Mississippi PTA classification intervals. J Neurol Neurosurg Psychiatry 82 (5): 494–499.

Narayan RK, Greenberg RP, Miller JD et al. (1981). Improved confidence of outcome prediction in severe head injury. A comparative analysis of the clinical examination, multimodality evoked potentials, CT scanning, and intracranial pressure. J Neurotrauma 54 (6): 751–762.

Nordström P, Michaëlsson K, Gustafson Y et al. (2014). Traumatic brain injury and young onset dementia: a nationwide cohort study. Ann Neurol 75 (3): 274–281.

Novack T (2000). The Orientation Log. The Center for Outcome Measurement in Brain Injury, Retrieved July 5, 2015, from, http://www.tbims.org/combi/olog.

Ommaya AK, Gennarelli TA (1974). Cerebral concussion and traumatic unconsciousness. Correlation of experimental and clinical observations of blunt head injuries. Brain 97 (4): 633–654.

Orrison WW, Gentry LR, Stimac GK et al. (1994). Blinded comparison of cranial CT and MR in closed head injury evaluation. AJNR Am J Neuroradiol 15 (2): 351–356.

Paterakis K, Karantanas AH, Komnos A et al. (2000). Outcome of patients with diffuse axonal injury: the significance and prognostic value of MRI in the acute phase. J Trauma 49 (6): 1071–1075.

Ponsford J, Willmott C, Rothwell A et al. (2004). Use of the Westmead PTA scale to monitor recovery after mild head injury. Brain Inj 18 (6): 603–614.

Ralay Ranaivo H, Zunich SM, Choi N et al. (2011). Mild stretch-induced injury increases susceptibility to interleukin-1β-induced release of matrix metalloproteinase-9 from astrocytes. J Neurotrauma 28 (9): 1757–1766.

Ramlackhansingh AF, Brooks DJ, Greenwood RJ et al. (2011). Inflammation after trauma: microglial activation and traumatic brain injury. Ann Neurol 70 (3): 374–383.

Ratcliff JJ, Adeoye O, Lindsell CJ et al. (2014). TRACK-TBI Investigators. Emergency department disposition of the GCS 13-15 traumatic brain injury patient: analysis of the TRACK-TBI study. Am J Emerg Med 32 (8): 844–850.

Ribot T (1881). Les Maladies de la Mémoire [Diseases of Memory], Appleton-Century-Crofts, New York.

Rodriguez-Rodriguez A, Egea-Guerrero JJ, Murillo-Cabezas F et al. (2014). Oxidative stress in traumatic brain injury. Curr Med Chem 21 (10): 1201–1211.

Roozenbeek B, Maas AIR, Menon DK (2013). Changing patterns in the epidemiology of traumatic brain injury. Nature Reviews. Neurology 9 (4): 231–236.

Russell WR, Nathan PW (1946). Traumatic amnesia. Brain 69: 280–300.

Russell WR, Smith A (1961). Post-traumatic amnesia in closed head injury. Arch Neurol 5 (1): 4–17.

Sasser SM, Hunt RC, Faul M (2012). Guidelines for field triage of injured patients: recommendations of the National Expert Panel on Field Triage, 2011. MMWR Recomm Rep. 2012 Jan 13; 61 (RR-1): 1–20.

Savola O, Hillbom M (2003). Early predictors of post-concussion symptoms in patients with mild head injury. Eur J Neurol 10 (2): 175–181.

Schootman M, Fuortes LJ (2000). Ambulatory care for traumatic brain injuries in the US, 1995–1997. Brain Inj 14 (4): 373–381.

Shenton ME, Hamoda HM, Schneiderman JS et al. (2012). A review of magnetic resonance imaging and diffusion tensor imaging findings in mild traumatic brain injury. Brain Imaging Behav 6 (2): 137–192.

Sherer M, Sander AM, Nick TG et al. (2002). Early cognitive status and productivity outcome after traumatic brain injury: findings from the TBI model systems. Arch Phys Med Rehabil 83 (2): 183–192.

Shores EA (1989). Comparison of the Westmead PTA Scale and Glasgow Coma Scale as predictors of neuropsychological outcome following extremely severe blunt head injury. J Neurol Neurosurg Psychiatry 52 (1): 126–127.

Shores EA (1995). Further concurrent validity on the Westmead PTA Scale. Appl Neuropsychol 2 (3–4): 167–169.

Shores EA, Marosszeky JE, Sandanam J et al. (1986). Preliminary validation of a clinical scale for measuring the duration of post-traumatic amnesia. Med J Aust 144 (11): 569–572.

Shores EA, Lammél A, Hullick C et al. (2008). The diagnostic accuracy of the Revised Westmead PTA Scale as an adjunct to the Glasgow Coma Scale in the early identification of cognitive impairment in patients with mild traumatic brain injury. J Neurol Neurosurg Psychiatry 79 (10): 1100–1106.

Silverberg ND, Gardner AJ, Brubacher JR et al. (2015). Systematic review of multivariable prognostic models for mild traumatic brain injury. J Neurotrauma 32 (8): 517–526.

Skolnick BE, Maas AI, Narayan RK et al. (2014). A clinical trial of progesterone for severe traumatic brain injury. N Engl J Med 371 (26): 2467–2476.

Smith DH, Wolf JA, Lusardi TA et al. (1999). High tolerance and delayed elastic response of cultured axons to dynamic stretch injury. J Neurosci 19 (11): 4263–4269.

Staal JA, Dickson TC, Chung RS et al. (2007). Cyclosporin-A treatment attenuates delayed cytoskeletal alterations and secondary axotomy following mild axonal stretch injury. Dev Neurobiol 67 (14): 1831–1842.

Staal JA, Dickson TC, Gasperini R et al. (2010). Initial calcium release from intracellular stores followed by calcium dysregulation is linked to secondary axotomy following transient axonal stretch injury. J Neurotrauma 112 (5): 1147–1155.

Stanley RM, Bonsu BK, Zhao W et al. (2012). US estimates of hospitalized children with severe traumatic brain injury: implications for clinical trials. Pediatrics 129 (1): e24–e30.

Stein DG, Geddews RI, Sribnick EA (2015). Traumatic Brain Injury Part I. In: JC Masdeu, R Gilberto González (Eds.), Handbook of Clinical Neurology. Elsevier, Amsterdam.

Syed AT, Lone NA, Afzai Wani M et al. (2007). Clinical management of patients with minor head injuries. Int J Health Sci 1 (1): 131–140.

Tate RL, Pfaff A, Baguley IJ et al. (2006). A multicentre, randomised trial examining the effect of test procedures measuring emergence from post-traumatic amnesia. J Neurol Neurosurg Psychiatry 77 (7): 841–849.

Teasdale GM, Pettigrew LE, Wilson JT et al. (1998). Analyzing outcome of treatment of severe head injury: a review and update on advancing the use of the Glasgow Outcome Scale. J Neurotrauma 15 (8): 587–597.

Tournier JD, Calamante F, Gadian DG et al. (2004). Direct estimation of the fiber orientation density function from diffusion-weighted MRI data using spherical deconvolution. Neuroimage 23 (3): 1176–1185.

Tseng WC, Shih HM, Su YC et al. (2011). The association between skull bone fractures and outcomes in patients with severe traumatic brain injury. J Trauma 71 (6): 1611–1614.

Walker WC, Ketchum JM, Marwitz JH et al. (2010). A multicentre study on the clinical utility of post-traumatic amnesia duration in predicting global outcome after moderate-sever traumatic brain injury. J Neurol Neurosurg Psychiatry 81 (1): 87–89.

Whitney NP, Eidem TM, Peng H et al. (2009). Inflammation mediates varying effects in neurogenesis: relevance to the pathogenesis of brain injury and neurodegenerative disorders. J Neurochem 108 (6): 1343–1359.

Wilson JTL, Pettigrew LEL, Teasdale GM (1998). Structured interviews for the Glasgow Outcome Scale and the Extended Glasgow Outcome Scale: guidelines for their use. J Neurotrauma 15 (8): 573–585.

Wolf JA, Stys PK, Lusardi T et al. (2001). Traumatic axonal injury induces calcium influx modulated by tetrodotoxin-sensitive sodium channels. J Neurosci 21 (6): 1923–1930.

Wright DW, Yeatts SD, Silbergleit R et al. (2014). Very early administration of progesterone for acute traumatic brain injury. N Engl J Med 371 (26): 2457–2466.

Wu A, Ying Z, Gomez-Pinilla F (2007). Omega-3 fatty acids supplementation restores mechanisms that maintain brain homeostasis in traumatic brain injury. J Neurotrauma 24 (10): 1587–1595.

Yokota H, Kurokawa A, Otsuka T et al. (1991). Significance of magnetic resonance imaging in acute head injury. J Trauma 31 (3): 351–357.

Yue JK, Vassar MJ, Lingsma HF et al. (2013). Transforming research and clinical knowledge in traumatic brain injury pilot: multicenter implementation of common data elements for traumatic brain injury. J Neurotrauma 30 (22): 1831–1844.

Yuh EL, Mukherjee P, Lingsma HF et al. (2013). MRI improves 3-month outcome prediction in mild traumatic brain injury. Ann Neurol 73 (2): 224–235.

Zafonte RD, Bagiella E, Ansel BM et al. (2012). Effect of citicoline on functional and cognitive status among patients with traumatic brain injury: Citicoline Brain Injury Treatment Trial (COBRIT). JAMA 308 (19): 1993–2000.

Zasler ND, Katz DI, Zafonte RD (2007). Clinical continuum of care and natural history. In: Brain injury medicine: principles and practice, Demos Medical Publishing, New York, pp. 3–13.

Handbook of Clinical Neurology, Vol. 138 (3rd series)
Neuroepidemiology
C. Rosano, M.A. Ikram, and M. Ganguli, Editors
http://dx.doi.org/10.1016/B978-0-12-802973-2.00013-6

Chapter 13

The epidemiology of amyotrophic lateral sclerosis

E.O. TALBOTT[1]*, A.M. MALEK[2], AND D. LACOMIS[3]
[1]*Department of Epidemiology, University of Pittsburgh Graduate School of Public Health, Pittsburgh, PA, USA*
[2]*Department of Public Health Sciences, Medical University of South Carolina, Charleston, SC, USA*
[3]*Departments of Neurology and Pathology, University of Pittsburgh Medical Center, Pittsburgh, PA, USA*

Abstract

Amyotrophic lateral sclerosis (ALS) is the most common motor neuron disease in adults and is characterized by neurodegeneration of motor neurons in the brain and spinal cord. The incidence of ALS is approximately 1–2.6 cases per 100 000 persons annually, whereas the prevalence is approximately 6 cases per 100 000. The average age of onset of ALS is currently 58–60 years and the average survival from onset to death is 3–4 years. Between October 19, 2010 and December 31, 2011, there were an estimated 12 187 prevalent cases diagnosed with definite ALS in the USA alone. Sporadic ALS (90–95%) constitutes the large majority of cases, while the remaining 5–10% are hereditary and termed familial ALS. Sporadic ALS is suspected to involve genetic susceptibility to environmental risk factors. The purpose of this review is to present a clinical overview of ALS and provide an epidemiologic summary of personal and environmental risk factors shown to be related to the risk of disease. A discussion of the most recent research initiatives is also included.

DEFINITION OF THE DISEASE AND CLINICAL PRESENTATIONS

Amyotrophic lateral sclerosis (ALS), often referred to as "Lou Gehrig's disease," is a progressive neurodegenerative disorder affecting motor neurons in the brain and spinal cord and is the most common motor neuron disease (MND) in adults. The clinical features reflect the presence and location of upper or lower motor neuron degeneration at a given time. At symptom onset, the process is usually focal and spread usually occurs to contiguous ventral spinal or brainstem regions, but more distant or multifocal spread also occurs (Gargiulo-Monachelli et al., 2012). However, progressive degeneration of the motor neurons in ALS eventually leads to the death of the motor neurons. With voluntary muscle action progressively affected, patients in the later stages of the disease may become totally paralyzed.

About one-third of patients have bulbar onset, and such onset is more common in women. Bulbar (from *medulla* = bulb) functions involve activities of the oropharyngeal muscles. Symptoms include dysarthria, dysphagia (usually for liquids more than solids), difficulty chewing, and hypersalivation. There is usually difficulty holding the mouth closed or pursing the lips. Dysarthria usually precedes dysphagia. Exam findings due to pontine as well as medullary disturbances involving nuclei of cranial nerves VII, IX, X, XI, and XII may include facial, jaw, or tongue weakness with tongue atrophy and fasciculations, reduced palate elevation, and pooling of oral secretions. (Mitsumoto and Pioro, 1998; Kuncl, 2002). Involvement of the spinal accessory nucleus leads to weakness of sternocleidomastoid and sometimes trapezius muscles. The oculomotor nuclei are spared until "end-stage" and brainstem sensory pathways are not affected. When corticobulbar (upper motor neuron) fibers are affected, the findings may include a hyperactive jaw jerk, exaggerated gag reflex, jaw clonus, a snout response, forced yawning, slow incoordinated tongue movements without atrophy or fasciculations, and facial

*Correspondence to: Evelyn O. Talbott, University of Pittsburgh Graduate School of Public Health, 130 DeSoto Street, 526A Crabtree Hall, Pittsburgh PA 15261, USA. Tel: +1-412-624-3074, E-mail: eot1@pitt.edu

weakness. Pseudobulbar affect may also occur and is manifest as inappropriate laughing or crying. A weak cough and reduced ability to clear secretions usually occur in patients with bulbar involvement. Laryngospasm occurs rarely.

Most patients have onset in one limb or two upper or lower limbs in an asymmetric fashion. Onset in the upper limb is more commonly on the dominant side. A hemiparetic pattern (Mills variant) is rare. Occasionally, more than two limbs or limbs and bulbar functions are affected at onset. The major symptom of limb involvement is painless weakness. Lower motor neuron weakness follows a spinal segmental pattern, while upper motor neuron/corticospinal tract weakness involves arm flexors and leg extensors disproportionately. With lower motor neuron weakness, there are often associated fasciculations, but only about 1% present with fasciculations without weakness. Cramps occur commonly, and fatigue may be present. Hyporeflexia and hypotonia occur with pure lower motor neuron involvement, but when there is coexisting upper motor neuron dysfunction affecting the same limb, hyperreflexia and hypertonia occur in association with atrophy and fasciculations. That combination of findings is characteristic of ALS and is considered to be the classic phenotype. In addition, stiffness and clumsiness are usually present with upper motor neuron involvement. Painful flexor spasms sometimes occur.

Lower motor neuron degeneration in the cervical spinal cord leads to arm weakness that may be predominant in hand or shoulder (flail arm) involvement. Motor tasks, such as hand writing, dressing, cutting food, and maintaining hygiene, may become difficult. Involvement of upper motor neuron projections causes spasticity, and Hoffman's sign may be present. Patients with pure lower motor neuron involvement are considered to have the progressive muscular atrophy (PMA) phenotype; however, most of these patients develop some upper motor neuron signs over time and have ALS. The course is somewhat slower than with the classic phenotype, and males outnumber females (Kim et al., 2009). Patients with the PMA phenotype require more extensive evaluations to exclude other disorders such as Kennedy disease, spinal muscular atrophy, multifocal mononeuropathy with conduction block, inclusion body myositis, polyradiculopathy, and other motor neuropathies. Similar to the PMA-ALS presentation, upper motor neuron-dominant ALS also has a slower course than the classic phenotype (Soraru et al., 2010).

Lower motor neuron degeneration in the thoracic spinal cord leads to weakness, atrophy, and fasciculations in trunk and paraspinal muscles. Head drop (chin falling on to the chest) may occur along with a lordotic posture. Thoracic upper motor neuron involvement is usually asymptomatic and findings are more subtle and may include loss of superficial abdominal reflexes.

Lumbosacral lower motor neuron degeneration often causes footdrop with distal weakness, while difficulty with stairs and getting out of a chair are common complaints with proximal weakness. Upper motor neuron leg involvement is often associated with a Babinski sign and a spastic circumducting gait.

Patients with upper-limb or bulbar onset may also have ventilatory (respiratory) involvement; rarely, ALS presents with primary ventilatory dysfunction from diaphragm weakness. Symptoms of ventilatory involvement include orthopnea, poor sleep, morning headaches, daytime hypersomnolence, and shortness of breath (usually first with activity). Tachypnea and use of accessory muscles and paradoxic breathing may be observed later.

Dementia, usually of the frontotemporal type, occurs in about 15% or more, and milder cognitive and behavioral dysfunction occurs in about one-third and in up to 75%. Such affected patients may have a shorter survival than those without neurobehavioral dysfunction (Phukan et al., 2007). There is also evidence of multisystem dysfunction occurring in rare patients, including extrapyramidal and spinocerebellar disturbances as well as sensory and autonomic dysfunction (Oey et al., 2002; Isaacs et al., 2007). Pain is reported in about half of ALS patients. It is likely multifactorial, but it is probably underrecognized and undertreated (Chio et al., 2012).

DIAGNOSIS OF ALS

ALS is mainly a clinical diagnosis, and early uncertainty leads to a mean time from symptom onset to diagnosis of about 1 year (Paganoni et al., 2014). The World Federation of Neurology created the El Escorial criteria based on the presence and distribution of upper and lower motor neuron signs (Brooks, 1994). The addition of electrodiagnostic features led to a revision of the El Escorial criteria in 1998 (Table 13.1) (Brooks et al., 2000). The Awaji Island modifications (de Carvalho et al., 2008) loosened the electromyogram (EMG) criteria and therefore made them equivalent to clinical findings; these criteria may be more sensitive (Chen et al., 2010) without sacrificing specificity (Boekestein et al., 2010). In clinical trials, the revised El Escorial criteria are used predominantly.

A useful diagnostic approach was detailed by Shook and Pioro (2009). Nerve conduction studies are performed to exclude multifocal motor neuropathy and to identify the presence of other neuropathies or sensory involvement as may be seen in Kennedy disease. Needle EMG is used to detect the presence and distribution of

Table 13.1

Revised El Escorial criteria for diagnosis of amyotrophic lateral sclerosis (ALS)

Diagnostic category	Features
Clinically definite ALS	Upper and lower motor neuron signs in bulbar and at least two spinal (lumbosacral, thoracic, or cervical) regions or Upper and lower motor neuron signs in three spinal regions
Clinically probable ALS	Upper and lower motor neuron signs in at least two regions (bulbar or spinal) with some upper motor neuron signs rostral to the lower motor neuron signs
Clinically probable ALS – laboratory-supported*	Clinical evidence of upper and lower motor neuron signs in one body region or of upper motor neuron signs in one region and EMG findings of lower motor neuron involvement in at least two body regions
Clinically possible ALS*	Upper and lower motor neuron signs in only the bulbar or only one spinal region or Upper motor neuron signs in two or more regions or Lower motor neuron signs rostral to upper motor neuron signs

*Other diagnoses are excluded by appropriate neuroimaging and laboratory studies.
EMG, electroymyogram.

fasciculation potentials, positive waves, and fibrillation potentials (from denervation) and enlarged remodeled motor unit potentials consistent with reinnervation. Needle EMG may also detect features of myopathies, such as inclusion body myositis, that could masquerade as a lower motor neuron presentation of ALS. Laboratory screening for paraproteinemia, human immunodeficiency virus (HIV) and other infections, heavy metals, and other metabolic, neoplastic, and paraneoplastic processes are considered partly based on the clinical and EMG features (Shook and Pioro, 2009). Genetic testing for spinal muscular atrophy and Kennedy disease is sometimes undertaken. Imaging, usually magnetic resonance imaging, is performed on the brain, spine, or both to assess for structural processes, such as spinal stenosis or brainstem tumors, that could produce some of the symptoms and signs associated with ALS.

INCIDENCE, PREVALENCE, AND TRENDS IN MORTALITY OVER TIME

The global incidence of ALS is about 1–2.6 cases per 100 000 persons annually, and the prevalence is approximately 6 cases per 100 000 (Mitchell and Borasio, 2007; Govoni et al., 2012). During the seventh decade of life in the USA, however, incidence increases to 5 cases per 100 000 (Hirtz et al., 2007). Further, between October 19, 2010 and December 31, 2011, the National ALS Registry identified 12 187 persons in the USA with a diagnosis of definite ALS using an established algorithm (Mehta et al., 2014). This translated to a prevalence of 3.9 cases of ALS per 100 000 persons (Mehta et al., 2014). Previously, annual incidence and prevalence of ALS were reported as approximately 5600 cases per 100 000 persons and 20 000–30 000 cases, respectively (The ALS Association, 2007). Higher incidence of ALS has been reported among Gulf War veterans, as 20 of the nearly 690 000 young Gulf War veterans were diagnosed with ALS during the postwar period of 8 years (1991–1998) (Haley, 2003). Compared with the expected incidence, the observed incidence was more than twofold higher from 1995 to 1998 (13 vs. 5.72 cases; $p=0.006$) and more than threefold higher in 1998 (5 vs. 1.57 cases; $p=0.02$) (Haley, 2003).

Past studies reported increasing rates of ALS/MND-associated death over the last several decades (Lilienfeld et al., 1989; Sejvar et al., 2005). A more recent investigation found no definitive trend in the ALS/MND-associated death rate for the time period 1999–2009, although a possible decrease was suggested ($p=0.05$) (Mehal et al., 2013). Moreover, the mortality rates for males were reported to decrease over this time period, while there was no change in the rate for females. Age-specific results revealed a slight increase in the death rate for adults 20–49 years of age but a decrease for those 65 and over as well as for males. These more recent estimates are not directly comparable to the death rates reported in earlier studies due to differences between the International Classification of Diseases, 9th edition (ICD-9: World Health Organization, 1975) and ICD-10 (World Health Organization, 1992) codes, which include deaths due to all forms of MND, although most are due to ALS (Mehal et al., 2013).

INVESTIGATED RISK FACTORS FOR ALS

Worldwide, ALS occurs among all racial/ethnic groups. However, ALS is more common in those with a family history of ALS, in males, and in older age (Mitsumoto and Pioro, 1998) Approximately 5–10% of ALS cases are hereditary and termed familial ALS (FALS); the remaining 90–95% of cases are considered sporadic

Table 13.2

Investigated risk factors for amyotrophic lateral sclerosis (ALS)

Risk factor	Classification of risk
Metals (e.g., lead, mercury)[1]	Increased risk
Military service[2]	Increased risk
Genetic factors	Increased risk
Pesticides and insecticides[3]	Increased risk
Electromagnetic fields[4]	Inconsistent support
Physical activity, some sports (e.g., soccer, football)[5]	Possible increased risk
Head injuries/trauma[6]	Equivocal support
Smoking[7]	Increased risk
Glutamate toxicity[8]	Increased risk
Solvent exposure[9]	Inconsistent support
Family history of neurodegenerative disease[10]	Inconsistent support

[1]Campbell et al. (1970), Felmus et al. (1976), Rosati et al. (1977), Conradi et al. (1978), Roelofs-Iverson et al. (1984), Gresham et al. (1986), Provinciali and Giovagnoli (1990), Armon et al. (1991b), Gunnarsson et al. (1992), Chancellor et al. (1993), Strickland et al. (1996), Vinceti et al. (1996, 1997), McGuire et al. (1997), Longnecker et al. (2000), Kamel et al. (2002), Gait et al. (2003), Armon (2004).
[2]Weisskopf et al. (2005), Horner et al. (2008).
[3]McGuire et al. (1997), Kamel et al. (2012), Malek et al. (2012).
[4]Kheifets et al. (2009).
[5]Kurland et al. (1992), Okamoto et al. (2009).
[6]Kurland et al. (1992), Cruz et al. (1999), Piazza et al. (2004), Ince et al. (2011), Armon and Nelson (2012).
[7]Nelson et al. (2000b), Wang et al. (2011).
[8]Eisen et al. (1993a, b), Bristol and Rothstein (1996), Hugon et al. (1996), Asahara et al. (1999), Honig et al. (2000), Ferrarese et al. (2001).
[9]Gunnarsson et al. (1992), Chancellor et al. (1993).
[10]Gunnarsson et al. (1992), Cruz et al. (1999).

ALS (SALS). Men develop ALS at 1.3–1.56 times the rate of women (Hirtz et al., 2007; Mehta et al., 2014). To date, several risk factors have been investigated in relation to ALS, some of which include: smoking, physical activity, environmental and occupational exposures, infectious agents, genetics, nutritional intake, military service, and trauma. Investigated risk factors for ALS are displayed in Table 13.2.

ADVANCES IN GENETIC MARKERS OF RISK FOR ALS

A number of genetic mutations have been identified for FALS and SALS, and have been reviewed by Marangi and Traynor by location and suggested role in ALS (Marangi and Traynor, 2015). Superoxide dismutase 1 (SOD1), which occurs among approximately 12–20% of FALS cases and 1–2% of SALS cases, was the first mutation to be discovered (Rosen, 1993; Chio et al., 2008; Millecamps et al., 2010). Recent identification of C9ORF72 marked an important breakthrough in ALS research as it accounts for about 40% of FALS and 7% of SALS cases (DeJesus-Hernandez et al., 2011; Renton et al., 2011). ALS with frontotemporal dementia (ALS-FTD) is also associated with mutations in C9ORF72 as well as in several other genes. The previously discussed SOD1 protein has been associated with a small proportion of SALS though two mutations, I113T and D90A (Beal et al., 2005). These genes are both considered to be expressed by low penetrance. Because of this, it may be difficult to differentiate environmental factors from genetic factors (Beal et al., 2005). Survival motor neuron (SMN) and neuronal apoptosis inhibitory protein (NAIP) are involved in the development of spinal muscular atrophies and have been studied as possible susceptibility genes. The SMN and NAIP genes were found to be mapped to a modifier gene in an animal model of SOD1 (Armon, 2003a). Other possible genes that may be risk factors for SALS include those coding for the haemocromatosis gene (HFE) and neurofilaments (NEFL, NEFM, NEFH) (Migliore and Coppede, 2009). An increase of copy numbers of SMN genes and the HFE 63D variant was reported to be involved in ALS risk according to a pooled analyis (Migliore and Coppede, 2009). A polymorphism in the dipeptidyl-peptidase 6 gene has also been discovered to be linked to ALS susceptibility (Migliore and Coppede, 2009).

DNA methylation is essential to many cellular processes, such as genomic imprinting and embryonic development; however, methylation errors have been related to carcinogenesis and other outcomes (Hou et al., 2011). An association has been observed between low-dose benzene exposure and DNA methylation. (Bollati et al., 2007) Although the mechanism is not known, a relationship between benzene exposure and acute myelogenous leukemia has been found by previous studies (International Agency for Research on Cancer, 1987). DNA methylation has also been reported following occupational exposure to lead (Li et al., 2013), whereas a slight change in DNA methylation was found among boilermaker welders with exposure to $PM_{2.5}$ (acute or chronic) (Kile et al., 2013; Fan et al., 2014). Further, DNA methylation levels are known to change with age (Horvath, 2013; Hannum et al., 2013; Marioni et al., 2015). With regard to ALS research, DNA methlyation is a new area of interest that is promising given the suscepted genetic susceptibility to certain environmental risk factors. See Figure 13.1 for proposed mechanisms related to ALS neurodegeneration, including DNA methylation.

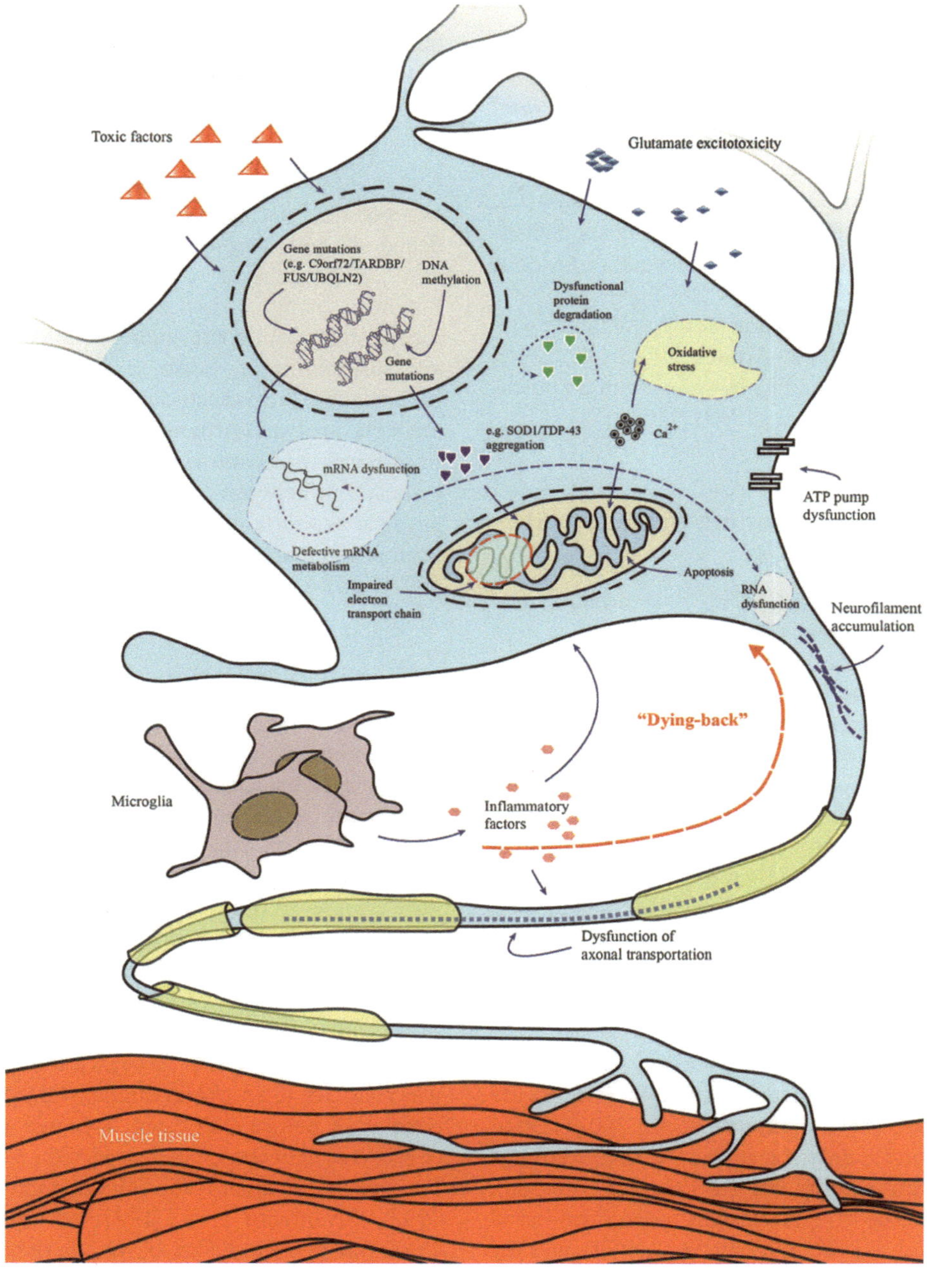

Fig. 13.1. Proposed molecular targets and mechanisms underlying neurodegeneration in amyotrophic lateral sclerosis (ALS). Many of the initial pathologic changes in models of ALS occur in the peripheral motor system, supporting a "dying-back" view of pathogenesis, though a causal primacy of lower motor neuron over upper motor neuron degeneration remains an issue of debate. The transgenic *SOD1* mouse model has been used extensively to dissect the likely pathogenic mechanisms. Many of these illustrated pathways are mechanisms of cell death common to a range of neurologic disorders, whereas more recent genetic discoveries have yet to be elucidated at a molecular level. Pathophysiologic mechanisms involved in ALS might include combinations of glutamate excitoxicity, generation of free radicals, mutant enzymes, as well as disruption of axonal transport processes and mitochondrial dysfunction. Mutations in several ALS-causative genes are related to the formation of intracellular aggregates. Mitochondrial dysfunction, which is associated with increased production of reactive oxygen species and aggregates of SOD1, might induce increased susceptibility to glutamate-mediated excitotoxicity, disturbance in energy production, and apoptosis. Activation of microglia results in secretion of cytokines, resulting in further toxicity. (Reproduced from He et al., 2015, with permission from SAGE Publications.)

PERSONAL RISK FACTORS

Smoking and alcohol consumption

Cigarette smoking has been shown repeatedly in several prospective and retrospective studies to be related to increased risk of ALS (Nelson et al., 2000b; Sutedja et al., 2007; Armon, 2009; Alonso et al., 2010b; Weisskopf et al., 2010), although not all studies, including a meta-analysis, have reported an association (Alonso et al., 2010a; Pamphlett and Ward, 2012).

No association has been found for alcohol intake with ALS (Nelson et al., 2000b; Armon, 2003b); however, one study reported confounding by smoking in the potential relationship (Kamel et al., 1999).

Physical activity including professional sports

ALS gained national attention in 1941 when famed New York Yankees baseball player Lou Gehrig died from the disease, which was subsequently nicknamed Lou Gehrig's Disease in the USA (Lewis et al., 2007). Recent research has suggested a link between frequent strenuous physical activity and onset of disease. Increased rates of ALS death have been reported among professional athletes (Chio, 2005; Lehman et al., 2012) compared to the general population, especially for soccer and football players. However, to date, only a few population-based studies have investigated the potential relationship between vigorous-intensity physical activity or past participation in athletics and the occurrence of ALS, with inconclusive results (Harwood et al., 2009). Case-control studies have provided moderate evidence for a positive association between leisure physical activity and sports participation and incidence of ALS later in life (Strickland et al., 1996; Scarmeas et al., 2002). Other studies have reported no association between the volume of leisure physical activity and ALS (Longstreth et al., 1998; Armon, 2007). The largest population-based case-control study of lifetime physical activity and the risk of ALS was conducted recently by Huisman et al. (2013) in the Netherlands and included cases identified from January 1, 2006 through December 31, 2010. Cases were ascertained from multiple sources. A total of 636 cases and 2166 controls were identified who completed a semistructured questionnaire on lifelong occupations, sport activity, and recreational active hobbies. They used metabolic equivalent scores which were assigned to each activity. ALS patients had significantly higher levels of leisure time physical activity compared with controls (odds ratio (OR) = 1.08, 95% confidence interval (CI) 1.02–1.14, $p = 0.008$) (Huisman et al., 2013). No significant effects were noted in cases compared to controls for vigorous activity such as marathon or triathlon participation or in occupational activity.

Trauma

History of traumatic brain injury has been hypothesized as a potential risk factor for ALS due to increased incidence of ALS among professional athletes (Qureshi et al., 2006; Schmidt et al., 2010; Turner et al., 2010; Sundman et al., 2014). However, results have been inconsistent, with some studies reporting a relationship and others failing to do so, and trauma discounted as a probable risk factor according to a 2003 evidence-based medicine analysis (Armon, 2003b; Lehman et al., 2012). In addition, case-control studies of MND have not found associations between risk for the disorder and history of physical trauma (e.g., fractures), history of polio, polio or tetanus immunization, or surgeries (Chancellor et al., 1993; Cruz et al., 1999).

Electric occupations and magnetic fields

Researchers have examined the potential relationship between electric work and/or electric shocks and ALS. Deapen and Henderson (1986) carried out one of the first case-control studies of 518 ALS cases and 518 controls and found that more cases than controls reported working in occupations with risk of electric exposure (OR = 3.8, 95% CI 1.4–13.0). Electric shocks resulting in unconsciousness occurred more often among cases than controls, although this was borderline statistically significant (OR = 2.8, 95% CI 1.0–9.9). A review by Kheifets et al. (2009) considered 13 larger-scale case-control or cohort studies identified through a PubMed search that was restricted to articles in English. Of these, seven were conducted before 2000 and appeared to yield more consistent evidence of increased risk, although not all findings were significant (Gunnarsson et al., 1991, 1992; Davanipour et al., 1997; Johansen and Olsen, 1998; Savitz et al., 1998a, b). Of the remaining six conducted after 2000, only two clearly showed a significantly increased risk with ORs that did not include one, and one study was borderline significant (Savitz et al., 1998a, b; Noonan, 2002; Feychting et al., 2003; Hakansson et al., 2003; Park et al., 2005; Roosli et al., 2007; Sorahan and Kheifets, 2007). Kheifets et al. (2009) point out in their review that, of the two that found an effect, one may have been confounded by neurotoxic exposure (Hakansson et al., 2003) and the other was restricted to railway workers (Roosli et al., 2007). Overall, of the 13, only three showed increased relative risks (RRs) or ORs that had CIs that did not include one (not significant).

Metals

The relationship between exposure to heavy metals, such as lead, mercury, and selenium, and risk of ALS has been examined by a number of studies. These have included primarily occupational exposures but have also involved exposure to selenium and other metals in the environment at large (Currier and Haerer, 1968; Conradi et al., 1982; Gresham et al., 1986; Durlach et al., 1997; McGuire et al., 1997; Bar-Sela et al., 2001; Kamel et al., 2002). Since the 1970s, lead has been hypothesized to be associated with ALS, with at least 15 studies reporting an association (Campbell et al., 1970; Felmus et al., 1976; Rosati et al., 1977; Conradi et al., 1978; Roelofs-Iverson et al., 1984; Gresham et al., 1986; Armon et al., 1991b; Gunnarsson et al., 1992; Chancellor et al., 1993; Strickland et al., 1996; Vinceti et al., 1996; McGuire et al., 1997; Longnecker et al., 2000; Kamel et al., 2002; Armon, 2004), although several studies have failed to confirm this finding (Gresham et al., 1986; McGuire et al., 1997; Vinceti et al., 1997; Gait et al., 2003). The association between occupational exposure to lead and ALS was examined through a systematic review and meta-analysis of epidemiologic studies (Wang et al., 2014). This analysis included the epidemiologic literature through 2013 and nine studies were chosen representing the largest and most well documented of studies. The risk of developing ALS among individuals with a history of exposure to lead was almost doubled (OR = 1.81; 95% CI 1.39–2.36) on the basis of these investigations that included case-control studies with specific lead exposure information, with no apparent heterogeneity across included studies ($I^2 = 14\%$). The attributable risk of ALS because of exposure to lead was estimated to be 5%. An association has been reported for exposure to selenium and mercury and the development of ALS (Provinciali and Giovagnoli, 1990). However, not all studies have replicated the relationship between exposure to metals and development of ALS (Gait et al., 2003), and some have failed to find an association with specific metals (e.g., mercury, aluminum, cadmium, chromium, and manganese) (McGuire et al., 1997).

Trace metals including selenium, magnesium, calcium, and aluminum have also been studied with regard to ALS development. A case-control study by Nelson et al. in 2000 did not find calcium, lead, copper, or mercury by way of dietary intake to be associated with ALS risk (Nelson et al., 2000a). ALS incident cases ($n = 161$) and 321 age- and sex-matched population controls were included in the study. Aluminum was suspected to play a role in the Western Pacific variant of the disease, but has not been linked to SALS (Noonan, 2002). In 2000, a case-control study of 107 cases and 262 community controls carried out by Longnecker et al. (2000) reported that dietary magnesium was protective for ALS.

It has long been recognized that a high intake of the trace element selenium may be toxic to human and animal health. An Italian cohort study conducted by Vinceti et al. in 1996 examined ALS incidence among 5182 persons exposed to high selenium levels in their water. The standardized incidence ratio was 4.22 (95% CI 1.15–10.80) for the four cases of ALS that developed over the 9-year study period using the remainder of the municipal population as the reference group (Vinceti et al., 1996). Only one case of ALS was actually expected for this population and during the time period specified. An additional cluster of ALS in South Dakota (Kilness and Hichberg, 1977) reported four cases of ALS in a population of 4000 in a region with endemic selenosis in farm animals. In the past, studies in the USA have not found a relationship between areas of high selenium environmental content with ALS death rates (Noonan, 2002). Other studies utilizing blood and tissue samples have produced inconsistent results with regard to selenium exposure and risk of ALS (Noonan, 2002).

The most recent potential selenium-related SALS "cluster" has been the association of early-onset ALS with service during the First Gulf War (August 2, 1990 to July 31, 1991) (Haley, 2003). The crude oil from that region has about 1 gram selenium/liter – enough to have caused a huge aerial dose during the First Gulf War while the oil fires were burning. Since approximately 1 metric ton of selenium was released into the air per day from the oil-well fires in Iraq in 1991 (Al Shatti and Harrington, 1991) and 800 μg/day of selenium is enough to produce toxic symptoms, it is possible selenium (along with intrinsic genetic susceptibility) may have played a role in the veteran ALS cluster. These high levels of environmental selenium in the Gulf suggest that toxic exposures to this compound leading to nervous system injury are at least theoretically possible.

Chemical agents and pesticides

Solvent, agricultural chemical and pesticide exposure have also been evaluated as potential risk factors for ALS development, but findings have been inconsistent (McGuire et al., 1997; Mitchell, 2000; Malek et al., 2014a). A relationship between exposure to solvents (e.g., cleaning solvents and degreasers, polychlorinated biphenyls, and cosmetology occupations) and risk of ALS has also been reported by a number of studies (McGuire et al., 1997; Park et al., 2005; Steenland et al., 2006; Malek et al., 2015), whereas others investigating alcohols or ketones, benzene, styrene, phenols, paints, solvent-based inks or dyes, and adhesives have failed to produce an association (Welp et al., 1996; McGuire et al., 1997; Gait et al., 2003; Park et al., 2005).

Pesticide exposure has been associated with ALS by some studies (McGuire et al., 1997; Burns et al., 2001; Park et al., 2005), including three meta-analyses (Kamel et al., 2012; Malek et al., 2012; Kang et al., 2014). However, upon further examination of exposure to various classes or concentrations of pesticides by two of the same studies, pesticide exposure was refuted as a risk factor for ALS (McGuire et al., 1997; Burns et al., 2001).

Hazardous air pollutants

Exposure to ambient air pollution has been investigated in relation to development of neurologic disorders, including autism, Alzheimer's disease, and Parkinson's disease, although only one study to date has evaluated the potential association with ALS. Exposure to 35 suspected neurotoxicant hazardous air pollutants and risk of ALS was examined by a case-control study involving ALS cases ($n=51$) and hospital- and population-based controls ($n=51$), matched by age, race, and sex (Malek et al., 2015). Table 13.3 presents suspected neurologic toxicant hazardous air pollutants by chemical group. Existing US Environmental Protection Agency National-Scale Air Toxics data were linked by census track to geocoded residential addresses of cases and controls for three time periods, 2005, 2002, and 1999. Similarly to other studies, toxicants were grouped by structural groups (e.g., metals, pesticides, aromatic solvents) and quartiles were used in an attempt to normalize the data (Windham et al., 2006; Roberts et al., 2013; deCastro, 2014). Findings suggested a potential association between exposure to aromatic solvents by place of residence in 2002 (OR 5.03, 95% CI 1.29–19.53) and 1999 (OR 4.27, 95% CI 1.09–16.79) and ALS after adjustment for smoking, education, and other exposure groups (Malek et al., 2015).

Chromium and hexavalent chromium have not been studied extensively as potential risk factors for MND, although the metals were found in air emission data to likely be an environmental exposure (Schulte et al., 1996).

Mechanisms of injury to motor neurons

Motor neurons may be affected through glutamate toxicity, oxidative damage, accumulation of intracellular aggregates, glial cell pathway disturbances, mitochondrial dysfunction, defects in axonal transport, growth factor deficiency, and aberrant RNA metabolism. Disease onset and progression may be influenced by a combination of these various processes. In summary, at least 10 mutations have been associated with SALS. Four oxidative stress biomarkers and two angiogenesis factors have also been linked to SALS development (Bruijn et al., 2004). Lead and mercury toxicity are known to produce neurologic effects (Noonan, 2002). Neuropathy, encephalopathy, and a syndrome similar to MND have been linked with heavy lead exposure (Noonan, 2002). Because of these associations, heavy metals and lead are suspected in the etiology of MND.

Table 13.3

Suspected neurologic toxicants by chemical group (ATSDR, 2000; US EPA, 2003; Cal/EPA, 2005; Windham et al., 2006)

Chemical group				
Metals	Aromatic solvents	Chlorinated solvents	Other hazardous air pollutants	Pesticides
Arsenic	Benzene	Methylene chloride	Hydrazine	Ethylene dibromide
Cadmium	Ethyl benzene	Perchloroethylene	Hexachloroethane	Ethylene dichloride
Lead	Styrene	Trichloroethylene	Allyl chloride	Hexachlorobenzene
Manganese	Toluene	Vinyl chloride	Polychlorinated biphenyls (PCBs)	
Mercury	Xylene	Hexane	Cyanide compounds	
Nickel	2,4-dinitrotoluene	Methyl chloride	Acylamide	
Selenium		Carbon disulfide		
		Ethylene oxide		
		Cresols and cresylic acid		
		Chloroform		
		Carbon tetrachloride		
		1,1,1-trichloroethane		
		1,1,2,2-tetrachloroethane		

Spatial clustering of ALS

Cluster analysis of ALS is of considerable epidemiologic significance since the etiology of ALS is unknown. Many studies reporting localized clusters of ALS have been based on anecdotal observations involving a small number of cases within a neighborhood or family (Melmed and Krieger, 1982; Armon et al., 1991a). They showed no statistically significant evidence for clustering beyond what might be expected to occur by chance. One study describes clustering of what the authors call "conjugal amyotrophic lateral sclerosis" in southeastern France (Corcia et al., 2003). It is based on nine case couples in which both spouses were affected by ALS. The authors did not apply any statistical analysis, but noted that, based on the incidence of ALS in France, the expected number of cases would have been 2.5. In an important large-scale study, the authors examined 1000 cases of ALS distributed throughout Finland who died between June 1985 and December 1995 (Sabel et al., 2003). The investigators used the spatial-scan statistic to examine whether there were significant clusters of the disease at both time of birth and time of death. Two significant, neighboring clusters were identified in southeast and south-central Finland at the time of death. A single significant cluster was identified in southeast Finland at the time of birth, closely matching one of the clusters identified at the time of death. The identification of significant ALS clusters, at both places of birth and death, is etiologically important. The unknown etiology of the disease and the fact that there may be a long latency period between contact with environmental risks and onset mean that the search for clustering may be as important at the time of birth as it is at the time of death. Since this study is based on a relatively large sample of cases, it provides forceful evidence of spatial clustering of ALS. It also shows that, if the cluster analysis is conducted at different stages of the cases' life cycle, different conclusions might be drawn about what potential risk factors might exist.

FUTURE DIRECTIONS

Given the rarity of ALS, future epidemiologic studies and especially gene–environment (interaction studies must continue to be conducted on a large scale or through a consortium. Further, as potential risk factors for SALS are identified and still others replicated, risk modification is becoming a possibility.

The search for additional genetic causes will continue and more mutations will be uncovered, but it is not likely that another common genetic form will be found. However, these genetic discoveries should improve our understanding of disease pathways and lead to more clinical trials, such as ongoing studies of antisense oligonucleotides, and study of the role of RNA toxicity.

As we await better understanding of pathogenesis and environmental triggers for SALS, more global therapies, especially stem cell studies using different types of cells and delivery methods, will continue with goals of trophic support as well as neuronal repopulation. Later, it may also be possible to provide more individualized therapies if the major pathways of neuronal degeneration are identified in a patient or group of patients. Advances in biomarker identification and use of induced pluripotent stem cell analysis might lead to such identification, promoting more targeted clinical drug trials.

In addition, greater use of currently available educational and research resources that exist for scientists and persons with ALS could prove fruitful for future research and collaborations. The National ALS Registry was created in 2010 by the Agency for Toxic Substances and Disease Registry for surveillance and research purposes, and specifically to examine risk factors for ALS as well as to describe the prevalence, incidence, and associated demographic characteristics (Antao and Horton, 2012). Enrollment includes persons with ALS from across the USA, with numbers currently in the thousands. A research mechanism tool is also available for recruitment that links scientists with persons with ALS who expressed interest in participating in research studies upon enrollment. This process has been described elsewhere and more details of the National ALS Registry can be found on their website: https://wwwn.cdc.gov/als/Default.aspx (Malek et al., 2014b). To summarize, the National ALS Registry is a valuable resource for clinical trial and epidemiologic study recruitment on a national level.

The Northeast ALS Consortium (NEALS) is an international group consisting of members or sites that conduct clinical research for ALS and other MNDs. Further, NEALS provides access to a sample repository of specimens, such as extracted DNA, serum, and cerebrospinal fluid, for research purposes. More information is available on the NEALS website: http://www.alsconsortium.org/index.php. The ALS Association and the Muscular Dystrophy Association are two organizations that participate in clinical care at local ALS clinics, in addition to being involved in research and often having local chapters that are open to patients and caregivers as well as scientists.

CONCLUSIONS

After decades of study, the etiology of SALS remains an enigma, although some progress has been made with regard to identification of genetic mutations. Environmental risk factors, such as heavy metals, pesticides, chemical solvents, and others, have been explored as causative agents with intriguing but sometimes inconsistent results. Findings have also been conflicting for the association of personal risk factors, such as excessive physical activity, diet, and head trauma, and development of ALS. While a number of retrospective and prospective studies have been conducted worldwide to investigate potential risk factors for ALS, several past studies have been limited by small sample sizes, biases (e.g., selection and information), and potential misclassification of exposure due to lacking exposure assessments. However, as more genetic research is conducted, it is likely that additional mutations will be found to be associated with SALS.

The future of ALS research is very promising for all involved, including patients and researchers, whether in the area of treatment or prevention. Weight of evidence points to some environmental factors mediating age of onset – in particular, the oxidative stress molecule, which may explain why some develop ALS earlier in life. In conclusion, many studies have been carried out to evaluate risk factors for developing ALS, although future research is needed to further elucidate causes of ALS and to examine potential gene–environment interactions. Specifically, new studies should focus on the use of DNA and methylation biomarkers. This may require large-scale collaboration through an international consortium in order to consider the presence of the gene in question, exposure of individuals throughout their lifetime, and their interaction.

References

Al Shatti AKS, Harrington JM (1991). The Environmental and Health Impact of the Kuwait Oil Fires. In: Proceedings of an International Symposium Held at the University of Birmingham, University of Birmingham.

Alonso A, Logroscino G, Hernan MA (2010a). Smoking and the risk of amyotrophic lateral sclerosis: a systematic review and meta-analysis. J Neurol Neurosurg Psychiatry 81: 1249–1252.

Alonso A, Logroscino G, Jick SS et al. (2010b). Association of smoking with amyotrophic lateral sclerosis risk and survival in men and women: a prospective study. BMC Neurol 10: 6.

Antao VC, Horton DK (2012). The National Amyotrophic Lateral Sclerosis (ALS) Registry. J Environ Health 75: 28–30.

Armon C (2003a). Amyotrophic Lateral Sclerosis. In: LM Nelson, CM Tanner, SK Van Den Eeden (Eds.), Neuroepidemiology. From Principles to Practice, Oxford University Press, Oxford.

Armon C (2003b). An evidence-based medicine approach to the evaluation of the role of exogenous risk factors in sporadic amyotrophic lateral sclerosis. Neuroepidemiology 22: 217–228.

Armon C (2004). Amyotrophic Lateral Sclerosis. In: LM Nelson, CM Tanner, SK Van Den Eeden et al. (Eds.), Neuroepidemiology. From Principles to Practice, Oxford University Press, Oxford.

Armon C (2007). Sports and trauma in amyotrophic lateral sclerosis revisited. J Neurol Sci 262: 45–53.

Armon C (2009). Smoking may be considered an established risk factor for sporadic ALS. Neurology 73: 1693–1698.

Armon C, Daube JR, O'brien PC et al. (1991a). When is an apparent excess of neurologic cases epidemiologically significant? Neurology 41: 1713–1718.

Armon C, Kurland LT, O'brien PC et al. (1991b). Antecedent medical diseases in patients with amyotrophic lateral sclerosis. A population-based case-controlled study in Rochester, Minn, 1925 through 1987. Arch Neurol 48: 283–286.

Armon C, Nelson LM (2012). Is head trauma a risk factor for amyotrophic lateral sclerosis? An evidence based review. Amyotroph Lateral Scler 13: 351–356.

Asahara H, Taniwaki T, Ohyagi Y et al. (1999). Glutamate enhances phosphorylation of neurofilaments in cerebellar granule cell culture. J Neurol Sci 171: 84–87.

Bar-Sela S, Reingold S, Richter ED (2001). Amyotrophic lateral sclerosis in a battery-factory worker exposed to cadmium. Int J Occup Environ Health 7: 109–112.

Beal MF, Lang AE, Ludolph AC (2005). Neurodegenerative Diseases: Neurobiology, Pathogenesis and Therapeutics. Cambridge University Press, Cambridge.

Boekestein WA, Kleine BU, Hageman G et al. (2010). Sensitivity and specificity of the 'Awaji' electrodiagnostic criteria for amyotrophic lateral sclerosis: retrospective comparison of the Awaji and revised El Escorial criteria for ALS. Amyotroph Lateral Scler 11: 497–501.

Bollati V, Baccarelli A, Hou L et al. (2007). Changes in DNA methylation patterns in subjects exposed to low-dose benzene. Cancer Res 67: 876–880.

Brooks BR (1994). El Escorial World Federation of Neurology criteria for the diagnosis of amyotrophic lateral sclerosis. Subcommittee on Motor Neuron Diseases/Amyotrophic Lateral Sclerosis of the World Federation of Neurology Research Group on Neuromuscular Diseases and the El Escorial "Clinical limits of amyotrophic lateral sclerosis" workshop contributors. J Neurol Sci 124 (Suppl): 96–107.

Brooks BR, Miller RG, Swash M et al. (2000). El Escorial revisited: revised criteria for the diagnosis of amyotrophic lateral sclerosis. Amyotroph Lateral Scler Other Motor Neuron Disord 1: 293–299.

Bristol LA, Rothstein JD (1996). Glutamate transporter gene expression in amyotrophic lateral sclerosis motor cortex. Ann Neurol 39: 676–679.

Bruijn LI, Miller TM, Cleveland DW (2004). Unraveling the mechanisms involved in motor neuron degeneration in ALS. Annu Rev Neurosci 27: 723–749.

Burns CJ, Beard KK, Cartmill JB (2001). Mortality in chemical workers potentially exposed to 2,4-dichlorophenoxyacetic acid (2,4-D) 1945–94: an update. Occup Environ Med 58: 24–30.

Campbell AM, Williams ER, Barltrop D (1970). Motor neurone disease and exposure to lead. J Neurol Neurosurg Psychiatry 33: 877–885.

Chancellor AM, Slattery JM, Fraser H et al. (1993). Risk factors for motor neuron disease: a case-control study based on patients from the Scottish Motor Neuron Disease Register. J Neurol Neurosurg Psychiatry 56: 1200–1206.

Chen A, Weimer L, Brannagan 3RD T et al. (2010). Experience with the Awaji Island modifications to the ALS diagnostic criteria. Muscle Nerve 42: 831–832.

Chio A (2005). Mortality trends in ALS: an increasingly intricate puzzle. Lancet Neurol 4: 453–454.

Chio A, Traynor BJ, Lombardo F et al. (2008). Prevalence of SOD1 mutations in the Italian ALS population. Neurology 70: 533–537.

Chio A, Canosa A, Gallo S et al. (2012). Pain in amyotrophic lateral sclerosis: a population-based controlled study. Eur J Neurol 19: 551–555.

Conradi S, Ronnevi LO, Vesterberg O (1978). Increased plasma levels of lead in patients with amyotrophic lateral sclerosis compared with control subjects as determined by flameless atomic absorption spectrophotometry. J Neurol Neurosurg Psychiatry 41: 389–393.

Conradi S, Ronnevi LO, Norris FH (1982). Motor neuron disease and toxic metals. Adv Neurol 36: 201–231.

Corcia P, Jafari-Schluep HF, Lardillier D et al. (2003). A clustering of conjugal amyotrophic lateral sclerosis in southeastern France. Arch Neurol 60: 553–557.

Cruz DC, Nelson LM, Mcguire V et al. (1999). Physical trauma and family history of neurodegenerative diseases in amyotrophic lateral sclerosis: a population-based case-control study. Neuroepidemiology 18: 101–110.

Currier RD, Haerer AF (1968). Amyotrophic lateral sclerosis and metallic toxins. Arch Environ Health 17: 712–719.

Davanipour Z, Sobel E, Bowman JD et al. (1997). Amyotrophic lateral sclerosis and occupational exposure to electromagnetic fields. Bioelectromagnetics 18: 28–35.

De Carvalho M, Dengler R, Eisen A et al. (2008). Electrodiagnostic criteria for diagnosis of ALS. Clin Neurophysiol 119: 497–503.

Deapen DM, Henderson BE (1986). A case-control study of amyotrophic lateral sclerosis. Am J Epidemiol 123: 790–799.

Decastro BR (2014). Acrolein and asthma attack prevalence in a representative sample of the United States adult population 2000-2009. PLoS One 9: e96926.

Dejesus-Hernandez M, Mackenzie IR, Boeve BF et al. (2011). Expanded GGGGCC hexanucleotide repeat in noncoding region of C9ORF72 causes chromosome 9p-linked FTD and ALS. Neuron 72: 245–256.

Durlach J, Bac P, Durlach V et al. (1997). Are age-related neurodegenerative diseases linked with various types of magnesium depletion? Magnes Res 10: 339–353.

Eisen A, Pant B, Stewart H (1993a). Cortical excitability in amyotrophic lateral sclerosis: a clue to pathogenesis. Can J Neurol Sci 20: 11–16.

Eisen A, Stewart H, Schulzer M, Cameron D (1993b). Anti-glutamate therapy in amyotrophic lateral sclerosis: a trial using lamotrigine. Can J Neurol Sci 20: 297–301.

Fan T, Fang SC, Cavallari JM et al. (2014). Heart rate variability and DNA methylation levels are altered after short-term metal fume exposure among occupational welders: a repeated-measures panel study. BMC Public Health 14: 1279.

Felmus MT, Patten BM, Swanke L (1976). Antecedent events in amyotrophic lateral sclerosis. Neurology 26: 167–172.

Ferrarese C, Sala G, Riva R et al. (2001). Decreased platelet glutamate uptake in patients with amyotrophic lateral sclerosis. Neurology 56: 270–272.

Feychting M, Jonsson F, Pedersen NL et al. (2003). Occupational magnetic field exposure and neurodegenerative disease. Epidemiology 14: 413–419. discussion 427–8.

Gait R, Maginnis C, Lewis S et al. (2003). Occupational exposure to metals and solvents and the risk of motor neuron disease. A case-control study. Neuroepidemiology 22: 353–356.

Gargiulo-Monachelli GM, Janota F, Bettini M et al. (2012). Regional spread pattern predicts survival in patients with sporadic amyotrophic lateral sclerosis. Eur J Neurol 19: 834–841.

Govoni V, Cesnik E, Casetta I et al. (2012). Temporal trend of amyotrophic lateral sclerosis incidence in southern Europe: a population study in the health district of Ferrara, Italy. J Neurol 259: 1623–1631.

Gresham LS, Molgaard CA, Golbeck AL et al. (1986). Amyotrophic lateral sclerosis and occupational heavy metal exposure: a case-control study. Neuroepidemiology 5: 29–38.

Gunnarsson LG, Lindberg G, Soderfeldt B et al. (1991). Amyotrophic lateral sclerosis in Sweden in relation to occupation. Acta Neurol Scand 83: 394–398.

Gunnarsson LG, Bodin L, Soderfeldt B et al. (1992). A case-control study of motor neurone disease: its relation to heritability, and occupational exposures, particularly to solvents. Br J Ind Med 49: 791–798.

Hakansson N, Gustavsson P, Johansen C et al. (2003). Neurodegenerative diseases in welders and other workers exposed to high levels of magnetic fields. Epidemiology 14: 420–426. discussion 427–8.

Haley RW (2003). Excess incidence of ALS in young Gulf War veterans. Neurology 61: 750–756.

Hannum G, Guinney J, Zhao L et al. (2013). Genome-wide methylation profiles reveal quantitative views of human aging rates. Mol Cell 49: 359–367.

Harwood CA, Mcdermott CJ, Shaw PJ (2009). Physical activity as an exogenous risk factor in motor neuron disease (MND): a review of the evidence. Amyotroph Lateral Scler 10: 191–204.

He J, Mangelsdorf M, Fan D et al. (2015). Amyotrophic lateral sclerosis genetic studies: from genome-wide association mapping to genome sequencing. Neuroscientist 21: 599–615.

Hirtz D, Thurman DJ, Gwinn-Hardy K et al. (2007). How common are the "common" neurologic disorders? Neurology 68: 326–337.

Horner RD, Grambow SC, Coffman CJ et al. (2008). Amyotrophic lateral sclerosis among 1991 Gulf War veterans: evidence for a time-limited outbreak. Neuroepidemiology 31: 28–32.

Horvath S (2013). DNA methylation age of human tissues and cell types. Genome Biol 14: R115.

Hou L, Zhang X, Tarantini L et al. (2011). Ambient PM exposure and DNA methylation in tumor suppressor genes: a cross-sectional study. Part Fibre Toxicol 8: 25.

Hugon J, Vallat JM, Dumas M (1996). Role of glutamate and excitotoxicity in neurologic diseases. Rev Neurol (Paris) 152: 239–248.

Huisman MH, Seelen M, De Jong SW et al. (2013). Lifetime physical activity and the risk of amyotrophic lateral sclerosis. J Neurol Neurosurg Psychiatry 84: 976–981.

Ince PG, Highley JR, Kirby J et al. (2011). Molecular pathology and genetic advances in amyotrophic lateral sclerosis: an emerging molecular pathway and the significance of glial pathology. Acta Neuropathol 122: 657–671.

International Agency for Research on Cancer (1987). Overall evaluations of carcinogenicity: an updating of IARC monographs ARC Monogr Eval. Carcinog Risks Hum Suppl volumes 1 to 42, 1–44.

Isaacs JD, Dean AF, Shaw CE et al. (2007). Amyotrophic lateral sclerosis with sensory neuropathy: part of a multisystem disorder? J Neurol Neurosurg Psychiatry 78: 750–753.

Johansen C, Olsen JH (1998). Mortality from amyotrophic lateral sclerosis, other chronic disorders, and electric shocks among utility workers. Am J Epidemiol 148: 362–368.

Kamel F, Umbach DM, Munsat TL et al. (1999). Association of cigarette smoking with amyotrophic lateral sclerosis. Neuroepidemiology 18: 194–202.

Kamel F, Umbach DM, Munsat TL et al. (2002). Lead exposure and amyotrophic lateral sclerosis. Epidemiology 13: 311–319.

Kamel F, Umbach DM, Bedlack RS et al. (2012). Pesticide exposure and amyotrophic lateral sclerosis. Neurotoxicology 33: 457–462.

Kang H, Cha ES, Choi GJ et al. (2014). Amyotrophic lateral sclerosis and agricultural environments: a systematic review. J Korean Med Sci 29: 1610–1617.

Kheifets L, Bowman JD, Checkoway H et al. (2009). Future needs of occupational epidemiology of extremely low frequency electric and magnetic fields: review and recommendations. Occup Environ Med 66: 72–80.

Kile ML, Fang S, Baccarelli AA et al. (2013). A panel study of occupational exposure to fine particulate matter and changes in DNA methylation over a single workday and years worked in boilermaker welders. Environ Health 12: 47.

Kilness AW, Hichberg FH (1977). Amyotrophic lateral sclerosis in a high selenium environment. JAMA 237: 2843–2844.

Kim WK, Liu X, Sandner J et al. (2009). Study of 962 patients indicates progressive muscular atrophy is a form of ALS. Neurology 73: 1686–1692.

Kuncl R (Ed.), (2002). Motor Neuron Disease. WB Saunders, London.

Kurland LT, Radhakrishnan K, Smith GE et al. (1992). Mechanical trauma as a risk factor in classic amyotrophic lateral sclerosis: lack of epidemiologic evidence. J Neurol Sci 113: 133–143.

Lehman EJ, Hein MJ, Baron SL et al. (2012). Neurodegenerative causes of death among retired National Football League players. Neurology 79: 1970–1974.

Lewis N, Dollman J, Dale M (2007). Trends in physical activity behaviours and attitudes among South Australian youth between 1985 and 2004. J Sci Med Sport 10: 418–427.

Li C, Yang X, Xu M et al. (2013). Epigenetic marker (LINE-1 promoter) methylation level was associated with occupational lead exposure. Clin Toxicol (Phila) 51: 225–229.

Lilienfeld DE, Chan E, Ehland J et al. (1989). Rising mortality from motoneuron disease in the USA, 1962-84. Lancet 1: 710–713.

Longnecker MP, Kamel F, Umbach DM et al. (2000). Dietary intake of calcium, magnesium and antioxidants in relation to risk of amyotrophic lateral sclerosis. Neuroepidemiology 19: 210–216.

Longstreth WT, Mcguire V, Koepsell TD et al. (1998). Risk of amyotrophic lateral sclerosis and history of physical activity: a population-based case-control study. Arch Neurol 55: 201–206.

Malek AM, Barchowsky A, Bowser B et al. (2012). Pesticide exposure as a risk factor for amyotrophic lateral sclerosis: a meta-analysis of epidemiological studies. Environ Res 117: 112–119.

Malek AM, Barchowsky A, Bowser R et al. (2014a). Environmental and occupational risk factors for amyotrophic lateral sclerosis: a case-control study. Neurodegener Dis 14: 31–38.

Malek AM, Stickler DE, Antao VC et al. (2014b). The National ALS Registry: a recruitment tool for research. Muscle Nerve 50: 830–834.

Malek AM, Barchowsky A, Bowser R et al. (2015). Exposure to hazardous air pollutants and the risk of amyotrophic lateral sclerosis. Environ Pollut 197: 181–186.

Marangi G, Traynor BJ (2015). Genetic causes of amyotrophic lateral sclerosis: new genetic analysis methodologies entailing new opportunities and challenges. Brain Res 1607: 75–93.

Marioni RE, Shah S, Mcrae AF et al. (2015). DNA methylation age of blood predicts all-cause mortality in later life. Genome Biol 16: 25.

Mcguire V, Longstreth Jr WT, Nelson LM et al. (1997). Occupational exposures and amyotrophic lateral sclerosis. A population-based case-control study. Am J Epidemiol 145: 1076–1088.

Mehal JM, Holman RC, Schonberger LB et al. (2013). Amyotrophic lateral sclerosis/motor neuron disease deaths in the United States, 1999–2009. Amyotroph Lateral Scler Frontotemporal Degener 14: 346–352.

Mehta P, Antao V, Kaye W et al. (2014). Prevalence of amyotrophic lateral sclerosis – United States, 2010–2011. MMWR Surveill Summ 63 (Suppl 7): 1–14.

Melmed C, Krieger C (1982). A cluster of amyotrophic lateral sclerosis. Arch Neurol 39: 595–596.

Migliore L, Coppede F (2009). Genetics, environmental factors and the emerging role of epigenetics in neurodegenerative diseases. Mutat Res 667: 82–97.

Millecamps S, Salachas F, Cazeneuve C et al. (2010). SOD1, ANG, VAPB, TARDBP, and FUS mutations in familial amyotrophic lateral sclerosis: genotype-phenotype correlations. J Med Genet 47: 554–560.

Mitchell JD (2000). Amyotrophic lateral sclerosis: toxins and environment. Amyotroph Lateral Scler Other Motor Neuron Disord 1: 235–250.

Mitchell JD, Borasio GD (2007). Amyotrophic lateral sclerosis. Lancet 369: 2031–2041.

Mitsumoto HCD, Pioro EP (Eds.), (1998). Amyotrophic lateral sclerosis, FA Davis, Philadelphia, PA.

Nelson LM, Matkin C, Longstreth Jr WT et al. (2000a). Population-based case-control study of amyotrophic lateral sclerosis in western Washington State. II. Diet. Am J Epidemiol 151: 164–173.

Nelson LM, Mcguire V, Longstreth Jr WT et al. (2000b). Population-based case-control study of amyotrophic lateral sclerosis in western Washington State. I. Cigarette smoking and alcohol consumption. Am J Epidemiol 151: 156–163.

Noonan CW (2002). Motor Neuron Disease/Amyotrophic Lateral Sclerosis: Preliminary Review of Environmental Risk Factors and Mortality in Bexar County, Texas. [Online]. Agency for Toxic Susbtances and Disease Registry (ATSDR), Available: http://www.atsdr.cdc.gov/NEWS/alsreport.html. [Accessed 2/9/09 2009].

Oey PL, Vos PE, Wieneke GH et al. (2002). Subtle involvement of the sympathetic nervous system in amyotrophic lateral sclerosis, Muscle Nerve 25: 402–408.

Okamoto K, Kihira T, Kondo T et al. (2009). Lifestyle factors and risk of amyotrophic lateral sclerosis: a case-control study in Japan. Ann Epidemiol 19: 359–364.

Paganoni S, Macklin EA, Lee A et al. (2014). Diagnostic timelines and delays in diagnosing amyotrophic lateral sclerosis (ALS). Amyotroph Lateral Scler Frontotemporal Degener 15: 453–456.

Pamphlett R, Ward EC (2012). Smoking is not a risk factor for sporadic amyotrophic lateral sclerosis in an Australian population. Neuroepidemiology 38: 106–113.

Park RM, Schulte PA, Bowman JD et al. (2005). Potential occupational risks for neurodegenerative diseases. Am J Ind Med 48: 63–77.

Phukan J, Pender NP, Hardiman O (2007). Cognitive impairment in amyotrophic lateral sclerosis. Lancet Neurol 6: 994–1003.

Piazza O, Siren AL, Ehrenreich H (2004). Soccer, neurotrauma and amyotrophic lateral sclerosis: is there a connection? Curr Med Res Opin 20: 505–508.

Provinciali L, Giovagnoli AR (1990). Antecedent events in amyotrophic lateral sclerosis: do they influence clinical onset and progression? Neuroepidemiology 9: 255–262.

Qureshi MM, Hayden D, Urbinelli L et al. (2006). Analysis of factors that modify susceptibility and rate of progression in amyotrophic lateral sclerosis (ALS). Amyotroph Lateral Scler 7: 173–182.

Renton AE, Majounie E, Waite A et al. (2011). A hexanucleotide repeat expansion in C9ORF72 is the cause of chromosome 9p21-linked ALS-FTD. Neuron 72: 257–268.

Roberts AL, Lyall K, Hart JE et al. (2013). Perinatal air pollutant exposures and autism spectrum disorder in the children of Nurses' Health Study II participants. Environ Health Perspect 121: 978–984.

Roelofs-Iverson RA, Mulder DW, Elveback LR et al. (1984). ALS and heavy metals: a pilot case-control study. Neurology 34: 393–395.

Roosli M, Lortscher M, Egger M et al. (2007). Mortality from neurodegenerative disease and exposure to extremely low-frequency magnetic fields: 31 years of observations on Swiss railway employees. Neuroepidemiology 28: 197–206.

Rosati G, Pinna L, Granieri E et al. (1977). Studies on epidemiological, clinical, and etiological aspects of ALS disease in Sardinia, Southern Italy. Acta Neurol Scand 55: 231–244.

Rosen DR (1993). Mutations in Cu/Zn superoxide dismutase gene are associated with familial amyotrophic lateral sclerosis. Nature 364: 362.

Sabel CE, Boyle PJ, Loytonen M et al. (2003). Spatial clustering of amyotrophic lateral sclerosis in Finland at place of birth and place of death. Am J Epidemiol 157: 898–905.

Savitz DA, Checkoway H, Loomis DP (1998a). Magnetic field exposure and neurodegenerative disease mortality among electric utility workers. Epidemiology 9: 398–404.

Savitz DA, Loomis DP, Tse CK (1998b). Electrical occupations and neurodegenerative disease: analysis of U.S. mortality data. Arch Environ Health 53: 71–74.

Scarmeas N, Shih T, Stern Y et al. (2002). Premorbid weight, body mass, and varsity athletics in ALS. Neurology 59: 773–775.

Schmidt S, Kwee LC, Allen KD et al. (2010). Association of ALS with head injury, cigarette smoking and APOE genotypes. J Neurol Sci 291: 22–29.

Schulte PA, Burnett CA, Boeniger MF et al. (1996). Neurodegenerative diseases: occupational occurrence and potential risk factors, 1982 through 1991. Am J Public Health 86: 1281–1288.

Sejvar JJ, Holman RC, Bresee JS et al. (2005). Amyotrophic lateral sclerosis mortality in the United States, 1979–2001. Neuroepidemiology 25: 144–152.

Shook SJ, Pioro EP (2009). Racing against the clock: recognizing, differentiating, diagnosing, and referring the amyotrophic lateral sclerosis patient. Ann Neurol 65 (Suppl 1): S10–S16.

Sorahan T, kheifets L (2007). Mortality from Alzheimer's, motor neuron and Parkinson's disease in relation to magnetic field exposure: findings from the study of UK electricity generation and transmission workers, 1973–2004. Occup Environ Med 64: 820–826.

Soraru G, Ermani M, Logroscino G et al. (2010). Natural history of upper motor neuron-dominant ALS. Amyotroph Lateral Scler 11: 424–429.

Steenland K, Hein MJ, Cassinelli 2ND RT et al. (2006). Polychlorinated biphenyls and neurodegenerative disease mortality in an occupational cohort. Epidemiology 17: 8–13.

Strickland D, Smith SA, Dolliff G et al. (1996). Amyotrophic lateral sclerosis and occupational history. A pilot case-control study. Arch Neurol 53: 730–733.

Sundman MH, Hall EE, Chen NK (2014). Examining the relationship between head trauma and neurodegenerative disease: A review of epidemiology, pathology and neuroimaging techniques. J Alzheimers Dis Parkinsonism 4.

Sutedja NA, Veldink JH, Fischer K et al. (2007). Lifetime occupation, education, smoking, and risk of ALS. Neurology 69: 1508–1514.

The ALS Association (2007). About ALS. Who Gets ALS [Online]. Available: www.alsa.org [Accessed 02/10/09].

Turner MR, Abisgold J, Yeates DG et al. (2010). Head and other physical trauma requiring hospitalisation is not a significant risk factor in the development of ALS. J Neurol Sci 288: 45–48.

Vinceti M, Guidetti D, Pinotti M et al. (1996). Amyotrophic lateral sclerosis after long-term exposure to drinking water with high selenium content. Epidemiology 7: 529–532.

Vinceti M, Guidetti D, Bergomi M et al. (1997). Lead, cadmium, and selenium in the blood of patients with sporadic amyotrophic lateral sclerosis. Ital J Neurol Sci 18: 87–92.

Wang H, O'Reilly EJ, Weisskopf MG et al. (2011). Smoking and risk of amyotrophic lateral sclerosis: a pooled analysis of 5 prospective cohorts. Arch Neurol 68: 207–213.

Wang MD, Gomes J, Cashman NR et al. (2014). A meta-analysis of observational studies of the association between chronic occupational exposure to lead and amyotrophic lateral sclerosis. J Occup Environ Med 56: 1235–1242.

Weisskopf MG, O'Reilly EJ, McCullough ML et al. (2005). Prospective study of military service and mortality from ALS. Neurology 64: 32–37.

Weisskopf MG, Gallo V, O'reilly EJ et al. (2010). Smoking may be considered an established risk factor for sporadic ALS. Neurology 74: 1927–1928. author reply 1928–9.

Welp E, Kogevinas M, Andersen A et al. (1996). Exposure to styrene and mortality from nervous system diseases and mental disorders. Am J Epidemiol 144: 623–633.

Windham GC, Zhang L, Gunier R et al. (2006). Autism spectrum disorders in relation to distribution of hazardous air pollutants in the San Francisco bay area. Environ Health Perspect 114: 1438–1444.

World Health Organization (1975). International Classification of Diseases, 9th edition. Centers for Disease Control and Prevention, Atlanta, GA.

World Health Organization (1992). International Statistical Classification of Diseases and Related Health Problems, 10th Revision (ICD-10), WHO, Geneva.

Handbook of Clinical Neurology, Vol. 138 (3rd series)
Neuroepidemiology
C. Rosano, M.A. Ikram, and M. Ganguli, Editors
http://dx.doi.org/10.1016/B978-0-12-802973-2.00014-8

Chapter 14

Cerebrovascular disease

M.L.P. PORTEGIES[1]*, P.J. KOUDSTAAL[2], AND M.A. IKRAM[1]

[1]*Department of Epidemiology, Erasmus MC University Medical Center, Rotterdam, The Netherlands*

[2]*Department of Neurology, Erasmus MC University Medical Center, Rotterdam, The Netherlands*

Abstract

With 16.9 million people who suffered a first-ever stroke in 2010 worldwide, stroke is a very common vascular disease. Epidemiologic studies have played an essential role in assessing this burden and in detecting the risk factors for stroke. Primary prevention of these risk factors, primarily hypertension, smoking, diabetes, and atrial fibrillation, has reduced the incidence in high-income countries. However, stroke remains a major cause of death and disability, and therefore research should be continued.

Subarachnoid hemorrhages are less prevalent than strokes but have an even higher risk of death. Similar to stroke, epidemiologic studies identified smoking and hypertension as its most important risk factors, together with excessive alcohol intake.

Although rare, arterial dissections, CADASIL, arteriovenous malformations, venous sinus thrombosis, moyamoya disease, and vasculitis can lead to serious symptoms. The burden and risk factors of those rare diseases are more challenging to assess. Whenever possible, they should be recognized in a timely manner for their increased risk of stroke, but most often they are diagnosed only at the time of stroke.

Some cerebrovascular abnormalities do not result in immediate symptoms. This subclinical cerebrovascular disease includes silent infarcts, white-matter lesions, and microbleeds, and is incidentally found by neuroimaging. These lesions are not innocent, as several epidemiologic studies have associated subclinical cerebrovascular disease with an increased risk of stroke, cognitive decline, dementia, and death.

INTRODUCTION

Stroke is the most common manifestation of cerebrovascular disease. It occurs if an artery supplying the brain is occluded or ruptures. The purpose of the first part of this chapter is to explain how epidemiologic studies contributed to our current knowledge about stroke. This includes the clinical presentation, diagnosis, and treatment, but also the burden, risk factors, and consequences of stroke (see Table 14.1 for an overview). Along with that, we will discuss the characteristics of its equally perilous equivalent, the transient ischemic attack (TIA), and how occurrence of a TIA influences the risk of stroke.

Cerebrovascular diseases are all diseases in which the blood vessels of the brain are primarily involved (Capildeo et al., 1978). Other diseases that fulfill this criterion are subarachnoid hemorrhages, arterial dissections, cerebral autosomal-dominant arteriopathy with subcortical infarcts and leukoencephalopathy (CADASIL), arteriovenous malformations, venous sinus thrombosis, moyamoya disease, and vasculitis. In the second part of this chapter we will discuss their characteristics, occurrence, risk factors, and outcomes, with a main focus on their epidemiology (see Table 14.2 for an overview).

Cerebrovascular diseases often lead to serious symptoms. However, cerebrovascular abnormalities can be present without any acute symptoms as well. Silent brain infarcts, white-matter lesions, and microbleeds are examples of such subclinical cerebrovascular disease. Epidemiologic imaging studies were designed to assess the occurrence of these abnormalities in the general

*Correspondence to: M.L.P. Portegies, MD MSc, Department of Epidemiology, Erasmus MC University Medical Center, PO Box 2040, 3000 CA, Rotterdam, The Netherlands. Tel: +31-10-7043390, E-mail: m.portegies@erasmusmc.nl

Table 14.1

Epidemiology of stroke

	Incidence per 100 000 person-years	Mean 1-month case fatality	Mortality rate per 100 000 person-years	Trends over time (difference between 1990 and 2010)
High-income countries				
All stroke	217 (204–231)	19.8%	61 (57–67)	Incidence reduced by 12% (6% to 17%) Mortality reduced by 37% (31% to 41%)
Ischemic stroke	168 (158–180)	14.3%	40 (38–43)	Incidence reduced by 13% (6% to 18%) Mortality reduced by 37% (19% to 39%)
Hemorrhagic stroke	49 (45–52)	41.0%	20 (19–23)	Incidence reduced by 19% (1% to 15%) Mortality reduced by 38% (32% to 43%)
Low- to middle-income countries				
All stroke	281 (245–322)	26.6%	105 (91–112)	Incidence increased by 12% (–3% to 22%) Mortality reduced by 20% (15% to 30%)
Ischemic stroke	182 (159–207)	16.7%	43 (38–52)	Incidence increased by 6% (–7% to 32%) Mortality reduced by 14 (–2% to 32%)
Hemorrhagic stroke	99 (85–116)	38.7%	62 (53–72)	Incidence increased by 22 (5% to 30%) Mortality reduced by 23 (–3% to 36%)

Data are point estimates (95% confidence interval).
Based on a review and results of the Global Burden of Disease, Injuries and Risk Factors Study (GBD 2010) (Feigin et al., 2009, 2014; Krishnamurthi et al., 2013).

Table 14.2

Epidemiology of cerebrovascular diseases

	Incidence per 100 000 person-years	Age at presentation	Gender predominance	% of cases that present with stroke
Subarachnoid hemorrhage	9	Median 50–60 years	W > M	NA
Arterial dissection	2.6–3.0	Mean ~45 years	W < M	67–77%
CADASIL	1.3–4.6	Peak ~30 years	W = M	~20–50%
Arteriovenous malformation	1.1	Peak 20–40 years	W = M	~50% hemorrhagic
Venous sinus thrombosis	0.5–1.3	Peak 20–50 years	W > M	NA
Moyamoya disease	0.35–0.94 in Japan and China 0.05–0.17 in the USA	Peak 1–14 years and 31–59 years	W > M	50–75% ischemic 10–40% hemorrhagic
PCNSV	0.24	Peak 37–59 years	W = M	30–50%

W, women; M, men; CADASIL, cerebral autosomal-dominant arteriopathy with subcortical infarcts and leukoencephalopathy; PCNSV, primary central nervous system vasculitis; NA, not applicable, since both disorders are considered to be a subtype of stroke.
Cerebrovascular diseases are presented in descending order of frequency.

population, the risk factors, and outcomes. The results of those studies will be described in the last part of this chapter.

STROKE AND TRANSIENT ISCHEMIC ATTACK

Definition

The presentation of a stroke and TIA is identical in the acute phase of the event. A stroke is characterized by rapidly developing clinical signs of focal or global disturbance of cerebral function, lasting more than 24 hours or leading to death with no apparent nonvascular cause (Hatano, 1976; WHO MONICA Project, 1988). This can be a consequence of bleeding (hemorrhagic stroke) or ischemia (ischemic stroke). A TIA is also characterized by neurologic dysfunction, but is only caused by cerebral or retinal ischemia and lasts less than 24 hours, usually less than 1 hour, without evidence of acute infarction (Albers et al., 2002; Easton et al., 2009).

The methods to distinguish strokes from TIAs have changed with improved neuroimaging techniques.

Before these techniques were available, it was thought that events shorter than 24 hours did not leave permanent damage in the form of an infarct. However, more recent studies using diffusion magnetic resonance imaging (MRI) showed that in 30–50% of events lasting less than 24 hours, ischemia was present (Ovbiagele et al., 2003; Easton et al., 2009), making the distinction based on duration of symptoms less reliable. A current proposal is therefore to distinguish TIA from ischemic stroke by the absence or presence of ischemia on neuroimaging (Easton et al., 2009). Practically, it is important to realize that the use of this new definition will lead to changes in estimated incidence and prevalence of stroke and TIA (Ovbiagele et al., 2003). Interpreting trends over time will be problematic if different definitions are used.

Clinical presentation

The symptoms of a stroke or TIA depend on the location of the brain damage. An ischemic stroke or TIA in the cerebral hemispheres leads to contralateral weakness or sensory loss, sometimes accompanied by visual field defects or conjugate gaze deviation towards the brain lesion (Bamford et al., 1991; Lewandowski et al., 2008; Muir, 2013). Lesions in the dominant (mostly left-sided) hemisphere can also lead to language disorders (aphasia) and lesions in the nondominant (right) hemisphere to neglect, extinction, or anosognosia (Orfei et al., 2007; Muir, 2013; Nielsen et al., 2013; Vossel et al., 2013). A stroke or TIA in the posterior circulation is characterized by an isolated homonymous visual field defect, ipsilateral cranial nerve palsy with contralateral weakness and/or sensory loss, ipsilateral ataxia, or vertigo (Bamford et al., 1991; Lewandowski et al., 2008; Muir, 2013). Transient retinal ischemia is characterized by transient blindness (amaurosis fugax) (Lewandowski et al., 2008).

Lacunar stroke is a subtype of ischemic stroke. This is a small lesion in the deep gray matter or white matter that leads to distinct patterns of neurologic dysfunction. Examples are a pure motor stroke, a pure sensory stroke, a sensorimotor stroke, or ataxic hemiparesis (Bamford et al., 1991; Bailey et al., 2012).

Most symptoms of hemorrhagic stroke are similar to those of ischemic stroke. Supratentorial lesions can lead to contralateral weakness and sensory loss, aphasia, neglect, gaze deviation, and hemianopia. Infratentorial lesions can lead to ataxia, nystagmus, and dysmetria (Qureshi et al., 2001). Additional symptoms are a decreased consciousness, headache, and/or vomiting. This is a consequence of the increased intracranial pressure. Meningeal irritation with neck stiffness can occur if blood spreads to the ventricles (Qureshi et al., 2001).

Diagnosis

Stroke requires rapid assessment and treatment. People have to call the emergency services quickly and therefore need to recognize the symptoms. The ability to name a stroke symptom varies across studies, ranging from 18% to 94%. Older adults have more difficulty naming stroke symptoms and are less likely to call the emergency services, even though strokes are most common in this group (Jones et al., 2010). The stroke itself might also complicate its recognition. Symptoms of strokes in the right cerebral hemisphere like neglect or anosognosia are difficult to identify. Therefore right-sided strokes can be missed (Foerch et al., 2005; Vossel et al., 2013; Portegies et al., 2015). Whether and how campaigns can increase the awareness of symptoms is still uncertain (Payne et al., 2010; Dombrowski et al., 2013; Rasura et al., 2014).

When stroke patients arrive in the hospital, they need a rapid diagnosis. The time window for thrombolysis, the treatment of ischemic stroke, is limited. Hemorrhagic strokes or stroke mimics (brain abscess, tumor, or other space-occupying lesions) should be excluded for thrombolysis (Mair and Wardlaw, 2014). Neuroimaging helps to distinguish between an ischemic and hemorrhagic stroke. The usual imaging technique in the acute stroke phase is the plain computed tomography (CT), because it is fast, widely available, allows unstable patients to be scanned, and it is accurate for the identification of hemorrhage and stroke mimics (Mair and Wardlaw, 2014). However, CT scans only detect 40–70% of infarcts in the first 6 hours after a stroke (Muir et al., 2006). MRI scans are better for detecting infarcts, which can give greater confidence for the start of thrombolysis (Brazzelli et al., 2009). Diffusion-weighted imaging MRI scans (MRI-DWI) are abnormal within minutes after stroke onset and show even minor ischemic lesions (Muir et al., 2006).

New imaging modalities are CT or MRI perfusion and CT or MRI angiography (CTA or MRA) (Mair and Wardlaw, 2014). Perfusion scans can more adequately show the penumbra, that, is the amount of tissue that might be saved by thrombolysis. Angiography can show which vessels are occluded by a thrombus and might be useful to select patients in which thrombolysis would be most effective (Mair and Wardlaw, 2014). However, more research into the usefulness of these techniques is necessary before they can be used in the acute stroke phase. Until then, rapid imaging is most important, either by using CT or MRI (Mair and Wardlaw, 2014).

Epidemiology

Stroke is one of the most important global health problems; it is the second leading cause of death and the third leading cause of morbidity worldwide (Lozano et al., 2012; Murray et al., 2012). To identify possibilities for prevention of this major health problem, studies need to ascertain its occurrence across the world, its high-risk populations, and risk factors. Furthermore, to explore the effect of preventive options and therapy, studies need to monitor changes in incidence, prevalence, outcome, and risk factors over years. Epidemiologic studies were designed to provide this information (Ovbiagele and Nguyen-Huynh, 2011). Since the mid-1980s there have been several attempts to create guidelines for studies that measure incidence, prevalence, and outcome (Malmgren et al., 1987; Bonita et al., 1995; Sudlow and Warlow, 1996; Feigin and Hoorn, 2004). The main characteristics of ideal studies are a prospective, population-based design, a standardized definition of stroke (Hatano, 1976; WHO MONICA Project, 1988) (World Health Organization criteria), a distinction between first-ever and recurrent stroke, and use of multiple sources of information (Malmgren et al., 1987; Bonita et al., 1995; Sudlow and Warlow, 1996; Feigin and Hoorn, 2004).

In developing countries, studies fulfilling all those criteria are not available (Sajjad et al., 2013). Therefore, it remains difficult to compare studies worldwide. Still, epidemiology has brought many insights into the occurrence, outcomes, risk factors, and prognostic factors of stroke and trends over time (see Table 14.1 for an overview). The following paragraphs discuss this further.

Stroke prevalence and incidence

Prevalence and incidence of stroke vary widely around the world. Estimates from the Global Burden of Diseases, Injuries, and Risk Factors study 2010 showed that the total prevalence of stroke in 2010 was 33 million worldwide, of which 16.9 million people suffered a first-ever stroke (Global Burden of Disease Stroke Expert Group, 2012; Feigin et al., 2014). The prevalence was on average 502 per 100 000 people, but varied between countries from 82 stroke survivors to 1187 stroke survivors per 100 000 people. Incidence was on average 258 per 100 000 person-years, but varied from 60 to 504 stroke cases per 100 000 person-years. These variations were partly due to the fact that incidences in low- to middle-income countries were 23% higher than in high-income countries (Feigin et al., 2014). However, even between countries of a smaller region such as Europe, incidence varies strongly, from 210 cases per 100 000 person-years to 600 cases per 100 000 person-years. This might be a consequence of environmental factors, risk factors, or access to healthcare (Bejot et al., 2007; Feigin et al., 2014).

Other factors that affect the occurrence of stroke are age, sex, and ethnicity. Stroke incidence increases rapidly with age (Asplund et al., 2009; Wieberdink et al., 2012; Koton et al., 2014; Meschia et al., 2014). With 1-year increase in age the risk of ischemic stroke and intracerebral hemorrhage increases to 9% in men and 10% in women (Asplund et al., 2009; Meschia et al., 2014). Children can also suffer from stroke, with an annual incidence of 4.6–6.4 per 100 000 children. Approximately half of those strokes are hemorrhagic (Mozaffarian et al., 2015). Men have slightly higher age-adjusted stroke incidences than women (5.3 per 1000 person-years in men versus 5.1 per 1000 person-years in women) (Carandang et al., 2006). However, above the age of 75, the stroke rates in women equal or exceed those of men (Reeves et al., 2008). This, in combination with the longer life expectancy of women, leads to more strokes in women than in men (Reeves et al., 2008). The Framingham Heart Study showed that women aged 55–75 years have a 20–21% lifetime risk of stroke, compared to 14–17% in men (Seshadri et al., 2006). Furthermore, incidence varies with ethnicity. Mexican Americans and blacks have higher stroke rates than non-Hispanic whites, especially at younger ages (Howard et al., 2011; Morgenstern et al., 2013).

Studies into stroke occurrence often combine ischemic and hemorrhagic stroke. However, most strokes are ischemic (70–90%), followed by intracerebral hemorrhages (10–20%) and subarachnoid hemorrhages (1–10%) (Feigin et al., 2009; Mozaffarian et al., 2015). Incidences per subtype are described in Table 14.1 (Krishnamurthi et al., 2013).

Trends in stroke prevalence and incidence over years

The incidence of stroke in high-income countries is declining (Table 14.1) (Carandang et al., 2006; Feigin et al., 2009; Koton et al., 2014). This might be a consequence of better preventive options, particularly of blood pressure control and smoking cessation. In the past four decades, stroke incidence in these countries has decreased by 42%. By comparison, low- to middle-income countries saw a doubling of the stroke risk (Feigin et al., 2009). These changes were mainly between 1970 and 1990, but still continue (Feigin et al., 2009, 2014).

Besides the development of the country, differences in decline also appear to depend on age and sex. Some studies found an increasing incidence in younger people aged 20–54 years, especially after 2005 (Kissela et al., 2012; Rosengren et al., 2013; Mozaffarian et al., 2015). Other studies found indications for a stronger decrease in men compared to women (Appelros et al., 2009; Wieberdink et al., 2012).

Consequences of stroke

Stroke explains 1 in 20 deaths in the USA (Mozaffarian et al., 2015). The early case-fatality (21 days to 1 month) is approximately 20% and varies by stroke subtype: 14.3% for ischemic stroke, 41.0% for hemorrhagic stroke, and 30.0% for subarachnoid hemorrhage (Feigin et al., 2009). Over the last 40 years this case fatality has reduced, but not significantly (Feigin et al., 2009). Mortality rates of stroke vary between 27 and 264 cases per 100 000 person-years and reduced by 37% in high-income countries and 20% in low-income countries between 1990 and 2010 (Table 14.1) (Feigin et al., 2014).

Disability further explains the high burden of stroke. In the USA, stroke is the most important cause of long-term disability (Centers for Disease Control and Prevention, 2007; Mozaffarian et al., 2015). In a study that examined the discharge location, more than half of patients were discharged to rehabilitation or skilled nursing facilities because of their disabilities. Only 45% of stroke patients returned home after their hospital admission and 32% of them needed home healthcare services (Mozaffarian et al., 2015). Women have a higher risk of disability than men (Mozaffarian et al., 2015).

Survivors of stroke are at high risk for another stroke and other adverse events. The risk of a stroke recurrence is approximately 3.1% at 30 days, 11.1% at 1 year, 26.4% at 5 years, and 39.2% at 10 years after the initial stroke (Mohan et al., 2011). However, the heterogeneity between studies is large (Mohan et al., 2011). Stroke survivors are also at higher risk of dementia. In population-based studies, which excluded prestroke dementia, this risk was 7.4% (Pendlebury and Rothwell, 2009). This risk is highest directly after stroke. In older stroke patients aged >75 years, this risk might even be up to 24% (Allan et al., 2011). Additionally, stroke survivors are at higher risk of depression, with a prevalence of 29%, which remains stable up to 10 years after stroke (Ayerbe et al., 2013).

TIA prevalence and incidence

Estimating TIA occurrence is challenging. People with TIA do not always recognize their symptoms as something serious and then do not consult a physician (Ovbiagele and Nguyen-Huynh, 2011). These attacks are therefore missed in the prevalence or incidence values. Furthermore, TIA might now be classified differently compared to in the past due to increased use of neuroimaging for the diagnosis.

Studies in the USA from 1985 to 2000 reported incidence rates for TIA ranging from 37 to 107 per 100 000 person-years (Ovbiagele et al., 2003). A more recent study in Italy reported a similar incidence rate of 52 per 100 000 person-years (Cancelli et al., 2011). Incidence increases with age. Men, blacks, and Mexican Americans have higher incidences of TIA compared to women and non-Hispanic whites (Mozaffarian et al., 2015).

Risk of stroke after TIA

TIA are serious because of their high short-term risk of a stroke. Two studies meta-analyzed the early risk of stroke after a TIA and reported that the risk of stroke is 3.1% at 2 days, 5.2% at 7 days, 8.0% at 30 days, and 9.2% at 90 days after a TIA (Giles and Rothwell, 2007; Wu et al., 2007). Furthermore, in 23% of people with stroke, this stroke was preceded by a TIA (Rothwell and Warlow, 2005). To detect people who are at highest risk of stroke, the ABCD score was developed (Rothwell et al., 2005; Johnston et al., 2007). This score assigns points to patient characteristics:

- Age (60 years or older = 1 point)
- Blood pressure (>140 mmHg systolic and/or ≥90 mmHg diastolic = 1 point)
- Clinical features (unilateral weakness = 2 points, speech disturbance without weakness = 1 point, other = 0 points)
- Duration of symptoms in minutes (≥60 = 2 points, 10–59 = 1 point, <10 = 0 points).

Later it was extended with diabetes (1 point), after which it was called the ABCD2 score (Johnston et al., 2007). In secondary care, prediction might be improved by adding preclinical or imaging information (Merwick et al., 2010). The purpose of the ABCD2 tool was to distinguish between people who need urgent assessment (score ≥ 4) and those in whom assessment can wait (score < 4). However, a recent meta-analysis questions the value of the ABCD2 score for this, because of low specificity (Wardlaw et al., 2015). Some people who need urgent treatment are still missed, namely people with carotid stenosis or atrial fibrillation (Wardlaw et al., 2015). Therefore, further research in the use of screening tools for the risk of stroke after TIA is warranted.

Pathophysiology

Pathophysiology of ischemic stroke and TIA

An ischemic stroke occurs if an artery to the brain becomes occluded, which leads to brain tissue death and neurologic deficits (Prabhakaran et al., 2015). This can be a consequence of atherosclerosis, cardioembolism, or small-vessel disease (Mohr et al., 1997). For a TIA these causes are similar, but consequences are different because the occlusion is reversible and no permanent damage occurs (Easton et al., 2009). The diagnostic workup and treatment of TIA and ischemic stroke are similar, however.

Atherosclerosis is thought to be an inflammatory process of the vascular wall in which plaques are formed that might rupture and cause thrombosis, embolism, or significant stenosis (Berliner et al., 1995). Atrial fibrillation, valvular heart disease, and other cardiac conditions can lead to thrombi that embolize to the brain (Mohr et al., 1997).

Lacunar strokes are generally the consequence of small-vessel disease (Pantoni, 2010). This is a slowly developing process affecting small arterioles, capillaries, and sometimes venules of the brain and is mostly the consequence of arteriosclerosis (Pantoni, 2010; Wardlaw et al., 2013a). Vessel walls and perivascular brain tissue are damaged by infiltration of inflammatory cells and lipid deposits in arteriolar walls (Wardlaw et al., 2013a). Other mechanisms, such as forming of emboli, hypoperfusion, endothelial failure, and impaired autoregulation, might also play a role (Wardlaw et al., 2013a; Fanning et al., 2014a). Another frequently used term for this is hypertensive small-vessel disease, because it is associated with age, diabetes, and particularly hypertension (Mohr et al., 1997; Pantoni, 2010). Cerebral amyloid angiopathy is another form of small-vessel disease. It results from accumulation of the amyloid protein in the walls of small arteries and arterioles. As a consequence, the arteries become dilated and disrupted, which can lead to both hemorrhagic and ischemic changes in the brain (Pantoni, 2010; Viswanathan and Greenberg, 2011).

Identifying the cause of stroke can lead to specific treatments, such as anticoagulants for strokes based on cardioembolism, or carotid endarterectomy for strokes based on carotid stenosis (Ay et al., 2014). Classification scores have therefore been developed to distinguish clinically between the different causes of ischemic stroke, i.e., the TOAST criteria (Adams et al., 1993) and the causative classification system (Ay et al., 2005, 2007).

Pathophysiology of hemorrhagic stroke

A hemorrhagic stroke occurs if brain vessels rupture. Most hemorrhagic strokes are a consequence of hypertensive small-vessel disease or cerebral amyloid angiopathy, as described in the previous paragraph (Donnan et al., 2008). These hemorrhages often occur in the cerebral hemispheres, basal ganglia, thalamus, and brainstem (Qureshi et al., 2009). Other hemorrhagic strokes can be due to vascular malformations (cavernous angiomas or arteriovenous malformations), infarcts with secondary hemorrhage, or use of oral anticoagulant therapy (Steiner et al., 2006; Donnan et al., 2008; Qureshi et al., 2009).

Risk factors and stroke prevention

Epidemiologic studies have identified many risk factors of stroke. Knowledge about risk factors and the natural history of stroke can guide the start of clinical trials. A hospital-based case-control study among 22 countries showed that 90% of strokes could potentially be prevented with optimal therapy of the risk factors that are currently known (O'Donnell et al., 2010). However, in a longitudinal population-based study, this percentage was only 51% (Bos et al., 2014), suggesting that the search for new etiologic risk factors still needs to be continued. The next paragraphs describe currently known risk factors, results of prevention trials based on those risk factors, and possibilities for future research.

Nonmodifiable risk factors

Not all risk factors can be avoided or treated. For instance, people at higher age, with low birth weight, or with an African American or Hispanic background are at higher risk of stroke (Meschia et al., 2014; Mozaffarian et al., 2015). However, identifying groups of people who are at higher risk of stroke might help to target therapies (Ovbiagele and Nguyen-Huynh, 2011).

A positive parental history of stroke increases the risk of stroke, suggesting a role for genetic factors (Flossmann et al., 2004; Seshadri et al., 2010). Until now some genetic variants for blood pressure, myocardial infarction, and atrial fibrillation have been related to stroke, based on findings from genetic risk scores and genomewide association studies (Ikram et al., 2009; Falcone et al., 2014; Meschia et al., 2014).

Hypertension

Hypertension is the most important modifiable risk factor for both ischemic and hemorrhagic stroke, with an attributable risk of approximately 35% (O'Donnell et al., 2010; Bos et al., 2014). A meta-analysis found a twofold increase in stroke death rate per 20 mmHg increase in blood pressure during middle age (Lewington et al., 2002). Guidelines for treatment are targeted to people with blood pressure higher than 140/90 or 150/90 mmHg (James et al., 2014; Meschia et al., 2014), but prehypertension is also associated with a higher risk of stroke (Huang et al., 2014). Furthermore, focusing on blood pressure levels may not capture all information that blood pressure might give. For instance, increased variability of blood pressure also leads to an increased risk of stroke (Rothwell et al., 2010). The importance of hypertension as a risk factor for stroke is further highlighted by the effect of treatment detected by clinical trials. A large meta-analysis of clinical trials showed that with

10 mmHg reduction in systolic blood pressure, the risk of stroke decreased with 41% (Law et al., 2009).

Cholesterol

Some studies found an increased risk of stroke with higher total cholesterol, whereas others actually found no or inverse associations (Meschia et al., 2014; Mozaffarian et al., 2015). This paradox may be explained by an increased risk of ischemic stroke in people with high cholesterol and an increased risk of hemorrhagic stroke in people with low cholesterol (Wang et al., 2013; Meschia et al., 2014).

Despite those inconsistencies, treatment with lipid-lowering statins does have a clear beneficial effect. A meta-analysis of randomized controlled trials of statins showed that with each 1 mmol/L decrease in low-density lipoprotein cholesterol, the risk of stroke reduced by 21%, while it did not increase the risk of hemorrhagic stroke (Amarenco and Labreuche, 2009).

Cardiac disease

Cardiac disease can cause stroke through distant embolism. Cardiac diseases that increase the risk of stroke are atrial fibrillation, myocardial infarction, cardiomyopathy, valvular disease, patent foramen ovale, atrial septum aneurysms, cardiac tumors, and aortic atherosclerosis (Meschia et al., 2014).

Atrial fibrillation is the most common cardiac cause of ischemic stroke (O'Donnell et al., 2010). It increases the risk of stroke fivefold and becomes increasingly important with higher age, with an attributable risk of 4.6–6.7% at the age of 59–79 increasing to 20.2–36.2% at the age of 80–89 (Bjorck et al., 2013; Romero and Wolf, 2013). Importantly, the risk of stroke due to atrial fibrillation might even be underestimated because people with asymptomatic atrial fibrillation are missed (Meschia et al., 2014; Keach et al., 2015).

To evaluate the risk of stroke and hemorrhagic complications of anticoagulant therapy, several risk scores have been developed. They are described in detail in previous articles (Lip and Lane, 2015; Apostolakis et al., 2012). The $CHADS_2$ and later the CHA_2DS_2-VASc are used to identify people at high risk of stroke (Lip and Lane, 2015). Oral anticoagulants are recommended in people with a CHA_2DS_2-VASc score of ≥ 2. Men with a CHA_2DS_2-VASc score of 0 and women with a score of 1 do not need any antithrombotic therapy (Meschia et al., 2014; Lip and Lane, 2015). Other scores evaluate bleeding risk (HAS-BLED, HEMOR-R_2HAGES). A high bleeding risk is defined as a HAS-BLED score of ≥ 3 or a HEMORR$_2$HAGES score of ≥ 4, but this is only used to identify patients who need more frequent monitoring (Apostolakis et al., 2012; Lip and Lane, 2015).

Vitamin K antagonists are the traditional anticoagulants started to prevent stroke. They are associated with a 37% reduced risk of stroke, compared to treatment with antiplatelet drugs (Hart et al., 2007). Four clinical trials examined the effect of the new oral anticoagulants (NOAC) for the prevention of stroke after atrial fibrillation (Connolly et al., 2009, 2011; Granger et al., 2011; Patel et al., 2011). NOACs directly inhibit thrombin and factor Xa and can be administered in a standard dose with more predictable effects than the vitamin K antagonists (Alberts et al., 2012; Dentali et al., 2012). Clinical trials showed that NOACs had a similar effect on the risk of stroke and systemic embolism compared to vitamin K antagonists, and a lower risk of intracerebral hemorrhage and mortality (Alberts et al., 2012; Verheugt and Granger, 2015).

Diabetes

Diabetes increases ischemic stroke risk by 2.3 times and hemorrhagic stroke risk by 1.8 times (Emerging Risk Factors Collaboration, 2010). The stroke risk further increases with the duration of diabetes (Banerjee et al., 2012), but is most prominent below the age of 65 (Khoury et al., 2013; Meschia et al., 2014; Mozaffarian et al., 2015). Diabetes is a chronic disease with no known cure. Several randomized trials found that intensive glucose control did not have a different effect on the risk of stroke compared to the usual less intensive therapy (Luitse et al., 2012). However, lowering of blood pressure below 130/80 mmHg and treatment with statins do prevent a substantial number of strokes in patients with diabetes (Luitse et al., 2012; Meschia et al., 2014).

Obesity

A large meta-analysis showed that obesity increases the risk of stroke 1.64 times and overweight 1.22 times (Strazzullo et al., 2010). Overweight and obesity are not associated with the risk of hemorrhagic stroke (Strazzullo et al., 2010). The effect of weight reduction on the risk of stroke is not clear, but it does have a beneficial effect on related risk factors such as diabetes and blood pressure (Meschia et al., 2014).

Smoking

Current smoking leads to 1.58 times increased risk of stroke and past smoking to a 1.17 times increased risk. The risk becomes even greater if people smoke more frequently and decreases with time after smoking cessation (Mons et al., 2015). Even secondhand smoking confers a higher risk of stroke (Mozaffarian et al., 2015).

Associations are particularly present for ischemic stroke. The effects on hemorrhagic stroke are less clear (Meschia et al., 2014).

Other

Other factors that could lead to an increased risk of stroke are chronic kidney disease, nutrition, sleep apnea, psychosocial factors like anxiety, distress, and depression, migraine, high alcohol consumption, sleep disordered breathing, hyperhomocysteinemia, and hypercoagulability (Meschia et al., 2014).

Potential

In the future, promising factors might be found in the field of epigenomics, proteomics, or genetics (Hankey, 2006; Ay et al., 2014). Furthermore, the attributable risk of established risk factors might be stronger than is currently detected. In the field of hypertension it has been shown that blood pressure variability might add to the risk of stroke (Rothwell et al., 2010), similar to the amount of abdominal fat as a measure for obesity (Suk et al., 2003). Moreover, there might be more subclinical cardiac disorders that lead to stroke. Many rhythm disorders go undetected. For example, in studies with extensive rhythm monitoring, paroxysmal atrial fibrillation was found in 10–20% of patients (Hart et al., 2014). Likewise, subclinical cardiac dysfunction has been related to stroke (de Bruijn et al., 2015). Additionally, atherosclerosis of the cerebral arteries may play a larger role in the risk of stroke than currently estimated. Namely, intracranial carotid artery stenosis appeared to be a strong risk factor for stroke (Bos et al., 2012). Future studies should determine whether associations between those risk factors and stroke are causal, whether they are modifiable, and whether elimination or treatment of these factors could reduce the risk of stroke.

Tools that combine risk factors to detect persons at high risk of stroke might also be more promising in the future with knowledge of an increasing amount of risk factors. Until now, several prediction tools have been developed, e.g. the Framingham Stroke Risk Score, Qstroke, and recently the Stroke Riskometer (Wolf et al., 1991; Hippisley-Cox et al., 2013; Parmar et al., 2015). Those prediction tools have a poor performance, but the Stroke Riskometer will be further developed using a large amount of data, which might lead to a better prediction of stroke (Parmar et al., 2015).

Treatment

The main goal of acute ischemic stroke treatment is to restore the blood supply to the brain as early as possible. Randomized controlled trials into the effect of intravenous recombinant tissue plasminogen activator (rtPA) showed a better functional outcome after rtPA, and although there was also a higher risk of bleeding, the benefits outweighed the harms if therapy was started within 3 hours. In later trials a beneficial effect within 4.5 hours after the start of symptoms was found (Lees et al., 2010), possibly even until 6 hours (Wardlaw et al., 2012, 2014). Currently, the recommendation is to treat eligible patients presenting within 4.5 hours after the start of their symptoms (Jauch et al., 2013; Prabhakaran et al., 2015).

The effect of rtPA appears to be small for proximal arterial occlusions (Prabhakaran et al., 2015). Therefore, catheter-based devices have been developed to directly remove the clot. The catheter is guided from the femoral artery up to the occluded artery in the brain, then thrombectomy is performed. Early trials using coils or aspiration devices did not have a beneficial effect (Broderick et al., 2013; Ciccone et al., 2013). The recent MRCLEAN trial, however, did find a better functional outcome after treating proximal occlusions with a stent retriever (Berkhemer et al., 2015). Other clinical trials with stent retrievers showed similar results (Campbell et al., 2015; Goyal et al., 2015; Saver et al., 2015). This has led to the recommendation that patients should receive endovascular therapy with a stent retriever if they fulfill the criteria (Powers et al., 2015).

SUBARACHNOID HEMORRHAGE

Definition

A subarachnoid hemorrhage is a bleeding in the subarachnoid space. This is the consequence of a ruptured aneurysm in 85% of cases (van Gijn et al., 2007). In other patients, the subarachnoid hemorrhage can be a consequence of arteriovenous malformations, dural fistulae, dural sinus thrombosis, intracranial artery dissection, mycotic aneurysms, bleeding diseases, or drugs (Ferro et al., 2008). A relatively innocent subarachnoid hemorrhage is the perimesencephalic bleeding. In these cases the blood is confined to the cisterns around the midbrain (van Gijn et al., 2007). The cause is unknown, but the outcome is good.

Clinical presentation

Patients with a subarachnoid hemorrhage usually present with a sudden headache that starts within a second, diffuse, and heavier than they ever experienced. Two-thirds of these patients have decreased consciousness on admission and seizures occur in 1 in 14 patients (van Gijn et al., 2007). Other symptoms that might occur are vomiting and focal disturbances such as motor defects, aphasia, ptosis, third-nerve palsy, visual troubles, and amnesia (van Gijn et al., 2007; Ferro et al., 2008). Neck stiffness is a common finding on physical examination due to the inflammatory response to the blood in the subarachnoid space (van Gijn et al., 2007). Furthermore, fundoscopy can show intraocular hemorrhages due to the high cerebrospinal fluid pressure (van Gijn et al., 2007; Ferro et al., 2008). The diagnosis is usually made by CT scanning. In 95% of cases blood is visible in the subarachnoid space. If the CT is performed within 6 hours and normal, no further investigations are needed. Otherwise, a lumbar puncture may show metabolites of hemoglobin in another 3% of patients (van Gijn et al., 2007).

Epidemiology

The mean incidence of subarachnoid hemorrhages per 100 000 person-years is 9, but increases to 22.7 in Japan and 19.7 in Finland (de Rooij et al., 2007). Incidence increases further with age. Starting at age 55 women have a 1.6 times higher chance of subarachnoid hemorrhage than men (Steiner et al., 2013). Although subarachnoid hemorrhages occur far less frequently than ischemic strokes, their consequences are serious. They occur first of all at younger ages, with a median age of onset between 50 and 60 years (Steiner et al., 2013). They also lead to a high mortality. Better treatment has led to a 17% increased chance of surviving, which is now 65%, but 12% of patients still die before they reach the hospital (Rinkel and Algra, 2011). In total, the proportion of potential life lost from subarachnoid hemorrhage is similar to that of stroke (Rinkel and Algra, 2011). This high risk of dying is partly due to the risk of rebleeding. This risk is 50% within the first 6 months (Steiner et al., 2013). Finally, patients often have long-term problems, such as risk of a new subarachnoid hemorrhage, cognitive problems, depression, and problems with getting back to work (Rinkel and Algra, 2011).

These serious consequences stimulated epidemiologic studies to search for modifiable risk factors. Many of those studies were limited by small numbers, but a meta-analysis of these studies showed that hypertension, smoking, and excess alcohol consumption are the most important risk factors for subarachnoid hemorrhage (Feigin et al., 2005; Andreasen et al., 2013). Additionally, population-based imaging and autopsy studies into asymptomatic aneurysms provided information about risk factors for rupture. Aneurysms develop during life under the influence of vessel wall injury. Weakness of the vessel wall leads to an outpouching, or aneurysm (Andreasen et al., 2013). Asymptomatic aneurysms occur in 3% of the population and more often in women, in people with autosomal-dominant polycystic kidney disease or people with a positive family history for aneurysms or subarachnoid hemorrhages (Vlak et al., 2011). The risk of rupture of an asymptomatic aneurysm is 1.4% at 1 year up to 3.4% at 5 years (Greving et al., 2014). This is determined by age, presence of hypertension, history of a subarachnoid hemorrhage, aneurysm size and location, and geographic region (Greving et al., 2014).

Treatment

The goal of treatment after a subarachnoid hemorrhage is to stop the bleeding and prevent rebleeding by occluding the aneurysm. The options are surgical clipping or endovascular coiling. To determine which of these options is best, the International Subarachnoid Aneurysm Trial was initiated. Although it was conducted in a highly selected population, it showed that endovascular coiling gave a better functional outcome compared to surgical clipping up to 18 years after the aneurysm (Molyneux et al., 2002, 2015). Other studies did not find any difference in mortality or morbidity between clipping and coiling (O'Kelly et al., 2010; Steiner et al., 2013). It seems that the preferred treatment depends on the patient and the aneurysm. Clipping is preferred for aneurysms with a wide neck, branching vessels out of the aneurysm sack, middle cerebral artery aneurysms, or patients with intracerebral hematoma. Coiling is preferred for aneurysms of the basilar artery, or elderly patients (Steiner et al., 2013). The timing of treatment is also an important consideration. A meta-analysis showed that treatment within 72 hours leads to a better outcome than later treatment (de Gans et al., 2002).

ARTERIAL DISSECTIONS

Clinical presentation

Although arterial dissections are relatively uncommon, they are an important cause of stroke in young and middle-aged adults (Debette, 2014). Tearing or stretching of the carotid or vertebral arterial wall can lead to a dissection and bleeding within the layers of the arterial wall. Almost all patients present with a headache or pain in the neck. A dissection between the middle (tunica media) and inner layer (tunica intima) of the artery may result in a stroke due to compression of the arterial lumen, or due to the formation of thrombi if the dissection tears through the luminal wall (Schievink, 2001; Caplan, 2008). A stroke or TIA occurs in 67–77% of patients, making a dissection one of the most common causes of stroke in patients younger than 45 years (Caplan, 2008; Lyrer and Engelter, 2010; Debette, 2014). Dissections between the tunica media and the outer wall (tunica adventitia) may lead to compression of external structures. For instance, cervical carotid artery dissections may also lead to Horner syndrome, tinnitus, and cranial nerve palsy due to stretching of the sympathetic fibers or cranial nerves. Furthermore, those dissections can lead to forming of a pseudoaneurysm and subarachnoid hemorrhages if the intracranial part of the vessel ruptures (Schievink, 2001; Caplan, 2008; Debette, 2014). Radiologic features of a dissection are a long tapered stenosis, a tapered occlusion, or a dissecting aneurysm (Lee et al., 2006; Debette and Leys, 2009).

Epidemiology

The extracranial part of the arteries is most vulnerable for dissections (Debette and Leys, 2009; Debette et al., 2015), with an incidence varying from 2.6 to 3.0 per 100 000 persons per year and a mean age of 45.8 years (Lee et al., 2006; Debette and Leys, 2009). There might be a predominance in men, but this is inconsistent within studies (Debette and Leys, 2009). Incidence of intracranial dissections is unknown, but will be lower (Debette et al., 2015). Extracranial dissections are 1.7 times more common in the carotid artery compared to the vertebral artery (Lee et al., 2006; Debette and Leys, 2009), whereas intracranial dissections are more common in the vertebral artery (Debette et al., 2015).

Arterial dissections seem to be a multifactorial disease (Rubinstein et al., 2005). To gain power to detect risk factors in this rare disease, the Cervical Artery Dissections and Ischemic Stroke Patients (CADISP) consortium was established, including patients from 20 centers in nine countries (Debette et al., 2011). A finding of this consortium is that in 40% of patients a prior cervical trauma has been described. This might be a severe blunt or penetrating trauma, but can also be cervical manipulation therapy, or extreme neck movements during sports injuries or injuries of recreational activities (Engelter et al., 2013). Other risk factors are infections, migraine, or hyperhomocysteinemia (Rubinstein et al., 2005; Debette, 2014). It is hypothesized that people with dissections have an underlying genetic predisposition towards arterial wall weakness, which could make them more vulnerable to trauma or infections, but this needs further exploration (Debette and Leys, 2009; Debette, 2014). Vascular risk factors and atherosclerosis are thought to be less important for arterial dissections compared to other causes of stroke (Rubinstein et al., 2005; Debette, 2014).

Treatment

Treatment consists of antiplatelets or anticoagulants, the goal being to prevent ischemia. No randomized controlled trials are available testing the optimal treatment for dissections, but recent meta-analyses of available studies did not show any difference for treatment with anticoagulant versus antiplatelet therapy (Lyrer and Engelter, 2010; Chowdhury et al., 2015).

CADASIL

Clinical presentation

A hereditary form of cerebrovascular disease is CADASIL. It is the consequence of a mutation in the Notch homolog 3 (*NOTCH3*) gene (Joutel et al., 1996). Due to the autosomal inheritance it occurs mainly in families. The clinical presentation is characterized by presence of migraine, occurrence of TIA and stroke, mood disturbances, apathy, and cognitive deterioration (Chabriat et al., 2009; Tikka et al., 2014). Migraine often starts around the age of 30 years; 20–40% of patients have migraine with aura. Strokes also occur at young age, 20–50% present with stroke, but eventually 60–85% of patients suffers a stroke, with a mean age of 49 years (Chabriat et al., 2009; Adib-Samii et al., 2010; Tikka et al., 2014). Strokes are mostly lacunar, so that CADASIL accounts for 2% of lacunar strokes in those younger than 65 (Chabriat et al., 2009).

MRI images in patients with CADASIL are characterized by the presence of leukoencephalopathy, subcortical infarcts, and microbleeds. The diagnosis is confirmed using mutational screening or skin biopsy (Chabriat et al., 2009; Tikka et al., 2014).

Epidemiology

Due to the rarity of the disease, it is difficult to study the occurrence of CADASIL. Some studies estimated the prevalence based on the amount of subjects diagnosed in a specific geographic area. This prevalence was estimated to be 1.3/100 000 in northeast England (Narayan et al., 2012), 4.1/100 000 in Italy (Bianchi et al., 2015), and 4.6/100 000 in Scotland (Moreton et al., 2014).

Since CADASIL is the consequence of a genetic defect in the *NOTCH3* gene, the presence of this mutation is an essential risk factor. Cardiovascular risk factors such as hypertension, smoking, cholesterol, diabetes, and hyperhomocysteinemia may further influence the expression of the disease (Singhal et al., 2004; Adib-Samii et al., 2010; Tikka et al., 2014).

CADASIL is very severe. Eventually patients will be demented, bedridden, and mute, within a mean of 25 years after the diagnosis (Opherk et al., 2004; Chabriat et al., 2009).

Treatment

The best treatment would be to prevent the expression of the mutated *NOTCH3* allele. However, this is not yet possible (Tikka et al., 2014). The currently available treatment is symptomatic. People with migraine can be treated with acetazolamide. Ischemic events can be prevented by use of antiplatelet drugs and treatment of vascular risk factors (Chabriat et al., 2009; Tikka et al., 2014).

ARTERIOVENOUS MALFORMATIONS

Clinical presentation

Arteriovenous malformations consist of a tangled web of abnormal vessels. This vascular network, or nidus, lacks a capillary bed, so arteries are directly connected to the veins, which results in arteriovenous shunting (Choi and Mohr, 2005). Arteriovenous malformations can rupture, leading to a hemorrhagic stroke. Approximately 35–81% of patients present with an intracerebral hemorrhage (Choi and Mohr, 2005). Other presentations can be epilepsy or a focal neurologic deficit (Hofmeister et al., 2000; Al-Shahi and Warlow, 2001), but many arteriovenous malformations are found as incidental findings (Morris et al., 2009). Although angiography is the gold standard, they can be visualized on CT, CTA, MRI, or MRA (Friedlander, 2007).

Epidemiology

Since arteriovenous malformations are often silent, their occurrence is difficult to assess. A retrospective population-based study estimated that arteriovenous malformations occur in approximately 18 per 100 000 adults (Al-Shahi et al., 2002). The incidence is approximately 1 per 100 000 adults and is similar in men and women (Al-Shahi and Warlow, 2001; Choi and Mohr, 2005; Friedlander, 2007). Silent arteriovenous malformations might be missed, but the incidence has not increased over the last years despite increased use of neuroimaging (Gabriel et al., 2010). The annual risk of rupture is 2–4% in all, and 1–3% in unruptured patients, and prospective population-based studies showed that about 50% of patients present with an intracerebral hemorrhage (Choi and Mohr, 2005; Kim et al., 2014). Consequently, the incidence of an intracerebral hemorrhage due to an arteriovenous malformation is 0.55 per 100 000 person-years (Choi and Mohr, 2005; Friedlander, 2007). Since these hemorrhages occur mostly between the age of 20 and 40 years, arteriovenous malformations account for a third of intracerebral hemorrhages in young adults (Al-Shahi and Warlow, 2001).

Several cohort studies followed people with an arteriovenous malformation to detect risks and risk factors for rupture (Al-Shahi et al., 2003; Halim et al., 2004; Stapf et al., 2006; Kim et al., 2007). Results of these studies were combined in an individual patient data meta-analysis, which showed that people with a hemorrhagic presentation or older age had the highest risk of rupture (Kim et al., 2014). Women and people with an arteriovenous malformation with exclusively deep venous drainage may additionally be at high risk, but further research into these factors is necessary (Kim et al., 2014).

The prognosis of a bleeding from an arteriovenous malformation is better than that of a subarachnoid hemorrhage or other intracerebral hemorrhage, with a case fatality of 1–1.5% per year (Al-Shahi and Stapf, 2005).

Treatment

Many surgical treatments are developed to obliterate the arteriovenous malformation and prevent rupture. These include microsurgical resection, stereotactic radiosurgery, and endovascular embolization (van Beijnum et al., 2011; Al-Shahi Salman et al., 2014; Mohr et al., 2014). However, observational studies detected a high risk of complications after those treatments (Wedderburn et al., 2008; van Beijnum et al., 2011; Al-Shahi Salman et al., 2014) and a recent randomized controlled trial found that medical management was even superior to surgical treatment (Mohr et al., 2014). Whether the beneficial effect will remain after longer follow-up is unclear (Mohr et al., 2014). This needs to be examined in further research.

VENOUS SINUS THROMBOSIS

Clinical presentation

An occluded artery leads to a stroke, but the symptoms of patients with an occluded venous sinus are less specific. Symptoms can occur acutely, subacutely, and chronically. Thrombosis of the venous sinus obstructs drainage of the cerebrospinal fluid, leading to an increased intracranial pressure and headache (Stam, 2005). Other common symptoms are seizures, focal neurologic deficits, altered consciousness, and papilledema (Stam, 2005; Bousser and Ferro, 2007; Ferro and Canhao, 2014). The thrombus can be detected using neuroimaging. Signs on a CT scan are the cord sign (due to a thrombosed cortical or deep vein), dense triangle sign (due to a clot inside the sinus), and empty delta sign (showing the clot after contrast injection). MRI in combination with MR venography (MRV) is currently the best method to show the thrombus, which will appear as an abnormal signal in a sinus on MRI and a corresponding absence of flow on MRV (Stam, 2005; Ferro and Canhao, 2014).

Epidemiology

Venous sinus thrombosis is the cause of stroke in 0.5% of all strokes and can occur at all ages (Bousser and Ferro, 2007). It even occurs in children, particularly in neonates, with an incidence of 0.67 per 100 000 per year (deVeber et al., 2001). In adults, the incidence was estimated as being around 0.5 per 100 000 per year (Bousser and Ferro, 2007). However, a more recent population-based study found a higher incidence of 1.32 per 100 000 per year, which might be a consequence of improved imaging techniques (Coutinho et al., 2012). Venous sinus thrombosis occurs most in people aged 20–50 years and in women, probably because women have more risk factors than men (Coutinho et al., 2012; Coutinho, 2015).

Risk factors often have to do with blood stasis, endothelial damage, or hypercoagulability (Saposnik et al., 2011). They include prothrombotic conditions, pregnancy and puerperium, use of oral contraceptives, cancer, infections, head trauma, and inflammatory diseases (Bousser and Ferro, 2007; Saposnik et al., 2011; Ferro and Canhao, 2014).

In adults, the prognosis is rather good. A meta-analysis showed that 91.6% survived the end of the follow-up period. At a follow-up of 3–6 months, 87.2% of patients had a good outcome with a complete or partial recovery. Only 8.7% had a poor outcome with permanent neurologic deficits (Dentali et al., 2006). The large International Study on Cerebral Vein and Dural Sinus Thrombosis (ISCVT) observed a higher risk of poor outcome in patients older than 37 years, in men, and in patients with coma, mental status disorder, intracranial hemorrhage on admission, thrombosis of the deep cerebral venous system, central nervous system infection, or cancer (Ferro et al., 2004). The outcome in children is relatively good, but neonates have a poor prognosis (Grunt et al., 2010).

Treatment

The treatment goal is to recanalize the occluded sinus or vein, prevent growth of the thrombus, prevent pulmonary embolism, and treat the prothrombotic state (Coutinho et al., 2011; Ferro and Canhao, 2014). A Cochrane review meta-analyzed the results of two small randomized controlled trials about the effect of heparin compared to placebo (Einhaupl et al., 1991; de Bruijn and Stam, 1999; Coutinho et al., 2011). Heparin had a nonsignificant beneficial effect on mortality and dependency. Since observational studies also observed a favorable outcome for patients treated with heparin, the current recommendation is to treat patients with heparin (Ferro et al., 2004; Saposnik et al., 2011).

MOYAMOYA DISEASE

Clinical presentation

Moyamoya disease is a progressive disease leading to stenosis or occlusion of the distal internal carotid artery or its proximal branches and forms a fine network of fragile collateral vessels. This collateral network is visible on angiography resembling a "puff of smoke," or *moyamoya* in Japanese (Kuroda and Houkin, 2008; Scott and Smith, 2009). Due to the stenosed vessels, 50–75% of patients present with an ischemic stroke or TIA in the territory of the internal carotid artery, leading to dysarthria, aphasia, or hemiparesis. Others present with a hemorrhagic stroke (10–40%), as a consequence of rupture of collateral vessels, or headache due to the dilated collateral vessels (Kuroda and Houkin, 2008; Scott and Smith, 2009). The diagnosis is made based on cerebral angiography. This displays the typical collateral pathways (Kuroda and Houkin, 2008; Scott and Smith, 2009).

Epidemiology

Moyamoya is most common in Asian countries. Incidence ranges from 0.35 to 0.94 per 100 000 person-years in Japan and China and 0.05 to 0.17 per 100 000 person-years in the USA (Kleinloog et al., 2012). Most studies found two age peaks, one in childhood and one in early adulthood. Women are more often affected than men (Kleinloog et al., 2012).

The risk factors of the disease are largely unclear. There may be a genetic predisposition, since 10–15% of cases in Japan are familial (Kleinloog et al., 2012; Kronenburg et al., 2014). Other factors that might play a role are vessel wall stress and repair factors, and autoimmune responses (Kronenburg et al., 2014).

Few studies have investigated the prognosis of Moyamoya disease. In children, the prognosis was poor after conservative treatment due to intellectual or motor impairment (Kuroda and Houkin, 2008). Many patients will also develop recurrent ischemic strokes (65%) or hemorrhages (30–65%) (Kuroda and Houkin, 2008).

Treatment

Treatment may prevent (recurrent) strokes. For this purpose, antiplatelets are often prescribed, but their effect is unknown (Kronenburg et al., 2014). Another option is surgical treatment, in which collateral pathways are provided by making a bypass (Kuroda and Houkin, 2008; Scott and Smith, 2009; Pandey and Steinberg, 2011; Kronenburg et al., 2014). Several observational studies found a beneficial effect on the prevention of ischemic stroke (Fung et al., 2005; Pandey and Steinberg, 2011). The sole randomized controlled trial in moyamoya examined the effect of recanalization on hemorrhagic stroke and found a marginal preventive effect against rebleeding (Miyamoto et al., 2014).

CEREBRAL VASCULITIS

Clinical presentation

Cerebral vasculitis is a rare disorder that is difficult to recognize. It is characterized by inflammation and necrosis of blood vessel walls (Fieschi et al., 1998; Berlit, 2010) and can occur in combination with a systemic inflammatory or infectious disease, or isolated in the brain (primary central nervous system vasculitis (PCNSV)) (Berlit and Kraemer, 2014; Hajj-Ali and Calabrese, 2014). The most frequent symptom is headache, followed by an altered cognition, personality changes, and stroke (Salvarani et al., 2012; Berlit and Kraemer, 2014; Hajj-Ali and Calabrese, 2014; Vera-Lastra et al., 2015). Stroke and TIA occur in 30–50% of patients (Hajj-Ali et al., 2011). People rarely have cranial neuropathies, insults, ataxia, or coma (Hajj-Ali and Calabrese, 2014). Fatigue, weight loss, fever, myalgia, arthralgia, and neuropathy occur more often in people with systemic vasculitis (Salvarani et al., 2012; Hajj-Ali and Calabrese, 2014; Vera-Lastra et al., 2015). Laboratory tests might show elevated acute-phase proteins, but these are usually normal in people with PCNSV. Cerebrospinal fluid often shows mild pleiocytosis and elevated protein. MRI scans can reveal new and older ischemic lesions and sometimes hemorrhages. Angiographic findings are stenoses, arterial occlusions, and aneurysms, bilaterally and diffuse in the brain. The gold standard is a biopsy of the brain and meninges, but this is an invasive procedure (Salvarani et al., 2012; Berlit and Kraemer, 2014).

Epidemiology

Cerebral vasculitis secondary to a systemic inflammatory or infectious disease is most common. For instance, about 20% of patients with polyarteritis nodosa, 30% with Behçet, and 22–33.6% with Wegener granulomatosis will develop central nervous system involvement (Berlit, 2010; Berlit and Kraemer, 2014). Other secondary causes can be systemic lupus erythematosus, dermatomyositis, sarcoidosis, and infections such as varicella-zoster virus and hepatitis C (Hajj-Ali and Calabrese, 2014). Primary central nervous system vasculitis is very rare. Therefore, most information is based on case reports and case series. One retrospective study collected 101 cases, later extended to 163 cases, and was able to study the epidemiology in more detail. The incidence rate appeared to be 2.4 per million person-years (Salvarani et al., 2007, 2015). Primary central nervous system vasculitis occurs over a wide age range and is even described in children. Most patients are between 37 and 59 years, with a median age of onset of about 50 years. The occurrence is similar in men and women (Salvarani et al., 2007, 2012, 2015; Cellucci and Benseler, 2010).

The pathophysiology of primary central nervous system vasculitis is still unclear. Infectious agents might be the trigger, since it is known that infectious diseases can lead to vasculitis (Salvarani et al., 2012).

Treatment

Due to the rarity of the disease, the aspecific symptoms, and because the definite diagnosis is only possible with an invasive procedure, namely a biopsy, there is often a delay in the diagnosis and treatment of cerebral vasculitis (Vera-Lastra et al., 2015). The type of treatment depends on whether there is an underlying systemic disease, but often includes corticosteroids and cyclophosphamide (Berlit, 2010; Salvarani et al., 2012). Response to treatment is usually strong, with improvement of the symptoms due to active inflammation (headache and cognitive symptoms) (Lucke and Hajj-Ali, 2014). People with primary central nervous system vasculitis have higher relapse-free survival than people with secondary central nervous system vasculitis (Vera-Lastra et al., 2015).

SUBCLINICAL CEREBROVASCULAR DISEASE

Clinical presentation

Subclinical cerebrovascular disease is incidentally found by population imaging studies or other reasons for imaging, because it does not lead to symptoms (Vernooij et al., 2007). Silent infarcts, white-matter lesions, and microbleeds are all forms of subclinical cerebrovascular disease and are often the consequence of cerebral small-vessel disease (Pantoni, 2010; Wardlaw et al., 2013a). Population-based studies have allowed us to determine incidence, risk factors, and prognosis of those findings, showing that they are not as innocent as they seem. This will be illustrated in the following paragraphs.

Epidemiology

Silent infarcts

Silent infarcts are characterized as infarcts on CT or MRI without a history of a corresponding stroke or TIA (Vermeer et al., 2002, 2007; Fanning et al., 2014a). Infarcts are visible on CT or MRI as focal lesions, greater than 3 mm, with the same signal intensity as cerebrospinal fluid (Vermeer et al., 2007). They are better visible on MRI, where they appear hyperintense on T2 and hypointense on T1 (Vermeer et al., 2007; Fanning et al., 2014a).

Due to different criteria for the detection of infarcts on neuroimaging and divergences in the collection of strokes, prevalence of silent infarcts varies widely within studies (Fanning et al., 2014a, b). Population-based studies are least likely to contain a selective population (Fanning et al., 2014b), but even in those studies the prevalence of silent infarcts differs from 5.7% to 49%. Most estimates fall in the range of 10–20%, with an increasing prevalence with age (Lee et al., 2000; Vermeer et al., 2002, 2007; Fanning et al., 2014b). Incidence is even more difficult to assess, because follow-up MRI scans are needed for its assessment, but seems to vary from 1.9% to 3.7% (Longstreth et al., 2002; Vermeer et al., 2003a; Fanning et al., 2014b).

Most silent infarcts are lacunar infarcts, suggesting that they are a consequence of small-vessel disease (Fanning et al., 2014a). Age and hypertension are its most important risk factors (Fanning et al., 2014b).

Silent infarcts by definition do not lead to symptoms. However, a more detailed assessment does often reveal subtle neurologic deficits, such as cognitive dysfunction (Vermeer et al., 2007; Saini et al., 2012; Fanning et al., 2014a). Furthermore, silent infarcts do contribute to an adverse prognosis, with a two to three times increased risk of dementia (Vermeer et al., 2003c; Liebetrau et al., 2004) and a two to four times increased risk of stroke (Bernick et al., 2001; Vermeer et al., 2003b; Bokura et al., 2006).

White-matter lesions

White-matter lesions are best visible on T2 and fluid-attenuated inversion recovery (FLAIR)-weighted MRI sequences, where they generally appear as symmetric hyperintensities (Wardlaw et al., 2013b). They can occur in the cerebral hemispheres and basal ganglia and occasionally in the brainstem and cerebellum (Wardlaw et al., 2013a). They are highly frequent in the elderly population. In people aged 60–70 years the prevalence is about 87%, whereas the prevalence at the age of 80–90 years even approaches 100% (de Leeuw et al., 2001; Launer, 2004). It is unclear whether occurrence differs by sex (Launer, 2004).

The development of white-matter lesions seems to be a gradual process in which small-vessel disease plays an important role (Pantoni, 2002; de Groot et al., 2013). This is suggested by studies that found associations of white-matter lesions with vascular risk factors, ischemic lesions, and changes in the blood–brain barrier (Pantoni, 2002; Debette and Markus, 2010). Chronic hypoperfusion and disruption of the blood–brain barrier might lead to leakage of plasma in the white matter (Debette and Markus, 2010). Hypertension is a main risk factor for white-matter lesions, suggesting that hypertensive small-vessel disease could be the cause, but cerebral amyloid angiopathy is also associated with white-matter lesions (Pantoni, 2010). Other potential risk factors are atherosclerosis, endothelial dysfunction, and diabetes (Ovbiagele and Saver, 2006).

The small-vessel disease background already suggests that presence of white-matter lesions could be a risk factor for disease. This is confirmed by a meta-analysis that found a three times increased risk of stroke and dementia, and a two times increased risk of death in people with white-matter lesions (Debette and Markus, 2010).

Microbleeds

Cerebral microbleeds are focal lesions characterized by hemosiderin deposits. This is most likely a consequence of leakage of blood cells out of small brain vessels. These deposits cause inhomogeneity in the magnetic field of the MRI images because they are supermagnetic. This can be visualized on T2*-weighted gradient-recalled echo MRI images, where microbleeds appear as black or hypointense lesions (Greenberg et al., 2009).

The prevalence of microbleeds varies between 10% and 25% (Jeerakathil et al., 2004; Sveinbjornsdottir et al., 2008; Greenberg et al., 2009; Poels et al., 2010) and increases with age (Vernooij et al., 2008; Poels et al., 2010). Few studies examined incidence of microbleeds, which varied from 10% over 3 years in the population-based Rotterdam Study to 12% in 2 years in a memory clinic population (Goos et al., 2010; Poels et al., 2011).

The most robust risk factors for microbleeds appear to be age and hypertension (Greenberg et al., 2009). Furthermore, lacunar infarcts and white-matter lesions have been identified as risk factors (Goos et al., 2010; Poels et al., 2010). Risk factors seem to differ according to microbleed location (Poels et al., 2010), which could be a function of the pathophysiologic mechanism (Greenberg et al., 2009). Microbleeds can be the consequence of hypertensive small-vessel disease or cerebral amyloid angiopathy (Fazekas et al., 1999; Greenberg et al., 2009; Yates et al., 2014). APOE ϵ4, which is related to amyloid angiopathy, is more related to lobar microbleeds, whereas lacunar infarcts, white-matter lesions, and previous microbleeds are more related to deep microbleeds (Poels et al., 2010, 2011).

Microbleeds can be a precursor for large bleedings. Stroke patients, but also stroke-free elderly, are at increased risk of hemorrhagic stroke if they have microbleeds (Charidimou et al., 2013; Akoudad et al., 2015). However, this is not the only possible adverse event. People with microbleeds are also at higher risk of ischemic stroke (Charidimou et al., 2013; Akoudad et al., 2015), cognitive decline (Poels et al., 2012; Lei et al., 2013), dementia (Qiu et al., 2010; Miwa et al., 2014), and death

(Akoudad et al., 2013; Benedictus et al., 2015). The location of the microbleed might influence the risk of adverse events. Lobar microbleeds might be more related to hemorrhagic events, because of their relation with amyloid angiopathy, whereas deep microbleeds might be more associated with ischemic events and cardiovascular death (Akoudad et al., 2013, 2015).

Treatment

The question is whether subclinical cerebrovascular disease should be treated. As shown in the previous paragraphs, silent infarcts, white-matter lesions, and microbleeds are all related to a higher risk of adverse outcomes. However, to initiate an appropriate treatment, knowledge about the pathophysiology is necessary and the pathophysiology of small-vessel disease is still largely unclear (Wardlaw et al., 2013a). Hypertension is an important risk factor, but trial results on the effect of antihypertensive treatment against progression of white-matter lesions are conflicting (Dufouil et al., 2005; Weber et al., 2012). One review suggested a role of endothelial failure with increased permeability of the blood–brain barrier. Leakage of plasma components might lead to edema and tissue damage (Wardlaw et al., 2013a). It has also been suggested that silent infarcts, white-matter lesions, and microbleeds are only the tip of the iceberg of subclinical vascular damage. Other changes on MRI, such as disrupted microvascular integrity measured by diffusion tensor imaging, might precede these abnormalities and might be an earlier marker of subclinical cerebrovascular disease (de Groot et al., 2013). Until the pathophysiology is clear and has provided new insights for further trials, vascular risk factor control is the best available approach to reduce the consequences of subclinical cerebrovascular disease (Wardlaw et al., 2013a).

CONCLUSIONS

Epidemiology has brought many insights in the field of stroke. Over the last four decades, an increasing number of countries have collected information about stroke prevalence and incidence, which has made it possible to assess the global burden of stroke. Additionally, epidemiologic studies identified many potentially modifiable risk factors of stroke. Their preventive effect was further explored in clinical trials, which formed the basis for the prevention of stroke as it currently is. In high-income countries, prevention has led to a large reduction in stroke incidence. However, stroke still is the second leading cause of death, which suggests that current preventive options are insufficient and continuous stroke research remains necessary. Novel risk factors might be found in the field of genetics, especially when these studies focus more on specific etiologic subtypes. Furthermore, the understanding of the relation between existing factors and stroke may be improved. The risk of stroke is often based on a single measurement of a risk factor, while trajectories of risk factors, or response to treatment may provide additional information. Moreover, the additive value of stroke risk factors to stroke prediction needs to be further explored. Until now, many subclinical markers of large-vessel, small-vessel, and heart disease have been related to stroke. Future studies are necessary to determine whether such factors add to existing stroke prediction scores, whether screening is cost-effective, and what treatment can be added to high-risk populations.

Subarachnoid hemorrhages, arterial dissections, arteriovenous malformations, venous sinus thrombosis, CADASIL, vasculitis, and moyamoya disease are rarer. Cohort studies are often too small for the detection of rare diseases, which makes the assessment of occurrence and detection of risk factors of these diseases more challenging. Still, many studies attempted to assess the incidence of those diseases and meta-analyses of these studies provided more reliable estimates. Since the consequences of these diseases are serious, due to either the high risk of dying or of stroke, detection of risk factors is important. Some risk factors are already detected, but the search for new factors is often hampered by low power. Promising results are therefore to be expected from studies that combine data of several studies in a consortium or in individual patient data meta-analyses.

Epidemiologic imaging studies have revealed the prevalence and consequences of subclinical cerebrovascular disease. Studies with multiple follow-up scans are needed to inform about the incidence and development of those abnormalities. This will help explain more of its natural history, pathophysiology, risk factors, and consequences, which is the first step in discovering how those subclinical abnormalities can be handled best. Furthermore, new imaging techniques such as diffusion tensor imaging might add to our knowledge of subclinical vascular disease.

HIGHLIGHTS

Many epidemiologic studies examined occurrence and risk factors for stroke. Prevention of currently known risk factors has reduced the incidence and mortality of stroke, but since stroke remains the second cause of mortality worldwide, more research remains necessary. This could be through identification of new risk factors, but stroke prediction studies will also remain important to detect people at high risk of stroke, who might benefit from more aggressive treatment.

Epidemiologic studies also estimated the incidence of the rarer cerebrovascular diseases. Some risk factors have been identified, but this may be further improved using patient data of multiple studies through meta-analyses or the formation of consortia.

Finally, imaging studies have shown the impact of subclinical cerebrovascular disease. In coming years, imaging studies with follow-up scans will be important, since they can inform about the incidence and development of subclinical vascular disease and thereby better inform on the total burden of cerebrovascular disease.

References

Adams Jr HP, Bendixen BH, Kappelle LJ et al. (1993). Classification of subtype of acute ischemic stroke. Definitions for use in a multicenter clinical trial. TOAST. Trial of Org 10172 in Acute Stroke Treatment. Stroke 24: 35–41.

Adib-Samii P, Brice G, Martin RJ et al. (2010). Clinical spectrum of CADASIL and the effect of cardiovascular risk factors on phenotype: study in 200 consecutively recruited individuals. Stroke 41: 630–634.

Akoudad S, Ikram MA, Koudstaal PJ et al. (2013). Cerebral microbleeds and the risk of mortality in the general population. Eur J Epidemiol 28: 815–821.

Akoudad S, Portegies ML, Koudstaal PJ et al. (2015). Cerebral microbleeds are associated with an increased risk of stroke: the Rotterdam study. Circulation 132: 509–516.

Albers GW, Caplan LR, Easton JD et al. (2002). Transient ischemic attack – proposal for a new definition. N Engl J Med 347: 1713–1716.

Alberts MJ, Eikelboom JW, Hankey GJ (2012). Antithrombotic therapy for stroke prevention in non-valvular atrial fibrillation. Lancet Neurol 11: 1066–1081.

Allan LM, Rowan EN, Firbank MJ et al. (2011). Long term incidence of dementia, predictors of mortality and pathological diagnosis in older stroke survivors. Brain 134: 3716–3727.

Al-Shahi Salman R, White PM, Counsell CE et al. (2014). Outcome after conservative management or intervention for unruptured brain arteriovenous malformations. JAMA 311: 1661–1669.

Al-Shahi R, Stapf C (2005). The prognosis and treatment of arteriovenous malformations of the brain. Pract Neurol 5: 194–205.

Al-Shahi R, Warlow C (2001). A systematic review of the frequency and prognosis of arteriovenous malformations of the brain in adults. Brain 124: 1900–1926.

Al-Shahi R, Fang JS, Lewis SC et al. (2002). Prevalence of adults with brain arteriovenous malformations: a community based study in Scotland using capture-recapture analysis. J Neurol Neurosurg Psychiatry 73: 547–551.

Al-Shahi R, Bhattacharya JJ, Currie DG et al. (2003). Prospective, population-based detection of intracranial vascular malformations in adults: the Scottish Intracranial Vascular Malformation Study (SIVMS). Stroke 34: 1163–1169.

Amarenco P, Labreuche J (2009). Lipid management in the prevention of stroke: review and updated meta-analysis of statins for stroke prevention. Lancet Neurol 8: 453–463.

Andreasen TH, Bartek Jr J, Andresen M et al. (2013). Modifiable risk factors for aneurysmal subarachnoid hemorrhage. Stroke 44: 3607–3612.

Apostolakis S, Lane DA, Guo Y et al. (2012). Performance of the HEMORR(2)HAGES, ATRIA, and HAS-BLED bleeding risk-prediction scores in patients with atrial fibrillation undergoing anticoagulation: the AMADEUS (evaluating the use of SR34006 compared to warfarin or acenocoumarol in patients with atrial fibrillation) study. J Am Coll Cardiol 60: 861–867.

Appelros P, Stegmayr B, Terent A (2009). Sex differences in stroke epidemiology: a systematic review. Stroke 40: 1082–1090.

Asplund K, Karvanen J, Giampaoli S et al. (2009). Relative risks for stroke by age, sex, and population based on follow-up of 18 European populations in the MORGAM Project. Stroke 40: 2319–2326.

Ay H, Furie KL, Singhal A et al. (2005). An evidence-based causative classification system for acute ischemic stroke. Ann Neurol 58: 688–697.

Ay H, Benner T, Arsava EM et al. (2007). A computerized algorithm for etiologic classification of ischemic stroke: the Causative Classification of Stroke System. Stroke 38: 2979–2984.

Ay H, Arsava EM, Andsberg G et al. (2014). Pathogenic ischemic stroke phenotypes in the NINDS-stroke genetics network. Stroke 45: 3589–3596.

Ayerbe L, Ayis S, Wolfe CD et al. (2013). Natural history, predictors and outcomes of depression after stroke: systematic review and meta-analysis. Br J Psychiatry 202: 14–21.

Bailey EL, Smith C, Sudlow CL et al. (2012). Pathology of lacunar ischemic stroke in humans – a systematic review. Brain Pathol 22: 583–591.

Bamford J, Sandercock P, Dennis M et al. (1991). Classification and natural history of clinically identifiable subtypes of cerebral infarction. Lancet 337: 1521–1526.

Banerjee C, Moon YP, Paik MC et al. (2012). Duration of diabetes and risk of ischemic stroke: the Northern Manhattan Study. Stroke 43: 1212–1217.

Bejot Y, Benatru I, Rouaud O et al. (2007). Epidemiology of stroke in Europe: geographic and environmental differences. J Neurol Sci 262: 85–88.

Benedictus MR, Prins ND, Goos JD et al. (2015). Microbleeds, mortality, and stroke in Alzheimer disease: the MISTRAL study. JAMA Neurol 72: 539–545.

Berkhemer OA, Fransen PS, Beumer D et al. (2015). A randomized trial of intraarterial treatment for acute ischemic stroke. N Engl J Med 372: 11–20.

Berliner JA, Navab M, Fogelman AM et al. (1995). Atherosclerosis: basic mechanisms. Oxidation, inflammation, and genetics. Circulation 91: 2488–2496.

Berlit P (2010). Diagnosis and treatment of cerebral vasculitis. Ther Adv Neurol Disord 3: 29–42.

Berlit P, Kraemer M (2014). Cerebral vasculitis in adults: what are the steps in order to establish the diagnosis? Red flags and pitfalls. Clin Exp Immunol 175: 419–424.

Bernick C, Kuller L, Dulberg C et al. (2001). Silent MRI infarcts and the risk of future stroke: the cardiovascular health study. Neurology 57: 1222–1229.

Bianchi S, Zicari E, Carluccio A et al. (2015). CADASIL in central Italy: a retrospective clinical and genetic study in 229 patients. J Neurol 262: 134–141.

Bjorck S, Palaszewski B, Friberg L et al. (2013). Atrial fibrillation, stroke risk, and warfarin therapy revisited: a population-based study. Stroke 44: 3103–3108.

Bokura H, Kobayashi S, Yamaguchi S et al. (2006). Silent brain infarction and subcortical white matter lesions increase the risk of stroke and mortality: a prospective cohort study. J Stroke Cerebrovasc Dis 15: 57–63.

Bonita R, Broad JB, Anderson NE et al. (1995). Approaches to the problems of measuring the incidence of stroke: the Auckland Stroke Study, 1991–1992. Int J Epidemiol 24: 535–542.

Bos D, van der Rijk MJ, Geeraedts TE et al. (2012). Intracranial carotid artery atherosclerosis: prevalence and risk factors in the general population. Stroke 43: 1878–1884.

Bos MJ, Koudstaal PJ, Hofman A et al. (2014). Modifiable etiological factors and the burden of stroke from the Rotterdam study: a population-based cohort study. PLoS Med 11: e1001634.

Bousser MG, Ferro JM (2007). Cerebral venous thrombosis: an update. Lancet Neurol 6: 162–170.

Brazzelli M, Sandercock PA, Chappell FM et al. (2009). Magnetic resonance imaging versus computed tomography for detection of acute vascular lesions in patients presenting with stroke symptoms. Cochrane Database Syst Rev: CD007424.

Broderick JP, Palesch YY, Demchuk AM et al. (2013). Endovascular therapy after intravenous t-PA versus t-PA alone for stroke. N Engl J Med 368: 893–903.

Campbell BC, Mitchell PJ, Kleinig TJ et al. (2015). Endovascular therapy for ischemic stroke with perfusion-imaging selection. N Engl J Med 372: 1009–1018.

Cancelli I, Janes F, Gigli GL et al. (2011). Incidence of transient ischemic attack and early stroke risk: validation of the ABCD2 score in an Italian population-based study. Stroke 42: 2751–2757.

Capildeo R, Haberman S, Rose FC (1978). The definition and classification of stroke. A new approach. Q J Med 47: 177–196.

Caplan LR (2008). Dissections of brain-supplying arteries. Nat Clin Pract Neurol 4: 34–42.

Carandang R, Seshadri S, Beiser A et al. (2006). Trends in incidence, lifetime risk, severity, and 30-day mortality of stroke over the past 50 years. JAMA 296: 2939–2946.

Cellucci T, Benseler SM (2010). Central nervous system vasculitis in children. Curr Opin Rheumatol 22: 590–597.

Centers for Disease Control and Prevention (2007). Outpatient rehabilitation among stroke survivors – 21 States and the District of Columbia, 2005. MMWR Morb Mortal Wkly Rep 56: 504–507.

Chabriat H, Joutel A, Dichgans M et al. (2009). Cadasil. Lancet Neurol 8: 643–653.

Charidimou A, Kakar P, Fox Z et al. (2013). Cerebral microbleeds and recurrent stroke risk: systematic review and meta-analysis of prospective ischemic stroke and transient ischemic attack cohorts. Stroke 44: 995–1001.

Choi JH, Mohr JP (2005). Brain arteriovenous malformations in adults. Lancet Neurol 4: 299–308.

Chowdhury MM, Sabbagh CN, Jackson D et al. (2015). Antithrombotic treatment for acute extracranial carotid artery dissections: a meta-analysis. Eur J Vasc Endovasc Surg 50: 148–156.

Ciccone A, Valvassori L, Nichelatti M et al. (2013). Endovascular treatment for acute ischemic stroke. N Engl J Med 368: 904–913.

Connolly SJ, Ezekowitz MD, Yusuf S et al. (2009). Dabigatran versus warfarin in patients with atrial fibrillation. N Engl J Med 361: 1139–1151.

Connolly SJ, Eikelboom J, Joyner C et al. (2011). Apixaban in patients with atrial fibrillation. N Engl J Med 364: 806–817.

Coutinho JM (2015). Cerebral venous thrombosis. J Thromb Haemost 13 (Suppl 1): S238–S244.

Coutinho J, de Bruijn SF, Deveber G et al. (2011). Anticoagulation for cerebral venous sinus thrombosis. Cochrane Database Syst Rev: CD002005.

Coutinho JM, Zuurbier SM, Aramideh M et al. (2012). The incidence of cerebral venous thrombosis: a cross-sectional study. Stroke 43: 3375–3377.

de Bruijn SF, Stam J (1999). Randomized, placebo-controlled trial of anticoagulant treatment with low-molecular-weight heparin for cerebral sinus thrombosis. Stroke 30: 484–488.

de Bruijn RF, Portegies ML, Leening MJ et al. (2015). Subclinical cardiac dysfunction increases the risk of stroke and dementia: the Rotterdam Study. Neurology 84: 833–840.

de Gans K, Nieuwkamp DJ, Rinkel GJ et al. (2002). Timing of aneurysm surgery in subarachnoid hemorrhage: a systematic review of the literature. Neurosurgery 50: 336–340. discussion 340–332.

de Groot M, Verhaaren BF, de Boer R et al. (2013). Changes in normal-appearing white matter precede development of white matter lesions. Stroke 44: 1037–1042.

de Leeuw FE, de Groot JC, Achten E et al. (2001). Prevalence of cerebral white matter lesions in elderly people: a population based magnetic resonance imaging study. The Rotterdam Scan Study. J Neurol Neurosurg Psychiatry 70: 9–14.

de Rooij NK, Linn FH, van der Plas JA et al. (2007). Incidence of subarachnoid haemorrhage: a systematic review with emphasis on region, age, gender and time trends. J Neurol Neurosurg Psychiatry 78: 1365–1372.

Debette S (2014). Pathophysiology and risk factors of cervical artery dissection: what have we learnt from large hospital-based cohorts? Curr Opin Neurol 27: 20–28.

Debette S, Leys D (2009). Cervical-artery dissections: predisposing factors, diagnosis, and outcome. Lancet Neurol 8: 668–678.

Debette S, Markus HS (2010). The clinical importance of white matter hyperintensities on brain magnetic resonance imaging: systematic review and meta-analysis. BMJ 341: c3666.

Debette S, Metso T, Pezzini A et al. (2011). Association of vascular risk factors with cervical artery dissection and ischemic stroke in young adults. Circulation 123: 1537–1544.

Debette S, Compter A, Labeyrie MA et al. (2015). Epidemiology, pathophysiology, diagnosis, and management of intracranial artery dissection. Lancet Neurol 14: 640–654.

Dentali F, Gianni M, Crowther MA et al. (2006). Natural history of cerebral vein thrombosis: a systematic review. Blood 108: 1129–1134.

Dentali F, Riva N, Crowther M et al. (2012). Efficacy and safety of the novel oral anticoagulants in atrial fibrillation: a systematic review and meta-analysis of the literature. Circulation 126: 2381–2391.

deVeber G, Andrew M, Adams C et al. (2001). Cerebral sinovenous thrombosis in children. N Engl J Med 345: 417–423.

Dombrowski SU, Mackintosh JE, Sniehotta FF et al. (2013). The impact of the UK 'Act FAST' stroke awareness campaign: content analysis of patients, witness and primary care clinicians' perceptions. BMC Public Health 13: 915.

Donnan GA, Fisher M, Macleod M et al. (2008). Stroke. Lancet 371: 1612–1623.

Dufouil C, Chalmers J, Coskun O et al. (2005). Effects of blood pressure lowering on cerebral white matter hyperintensities in patients with stroke: the PROGRESS (Perindopril Protection Against Recurrent Stroke Study) magnetic resonance imaging substudy. Circulation 112: 1644–1650.

Easton JD, Saver JL, Albers GW et al. (2009). Definition and evaluation of transient ischemic attack: a scientific statement for healthcare professionals from the American Heart Association/American Stroke Association Stroke Council; Council on Cardiovascular Surgery and Anesthesia; Council on Cardiovascular Radiology and Intervention; Council on Cardiovascular Nursing; and the Interdisciplinary Council on Peripheral Vascular Disease. The American Academy of Neurology affirms the value of this statement as an educational tool for neurologists. Stroke 40: 2276–2293.

Einhaupl KM, Villringer A, Meister W et al. (1991). Heparin treatment in sinus venous thrombosis. Lancet 338: 597–600.

Emerging Risk Factors Collaboration (2010). Diabetes mellitus, fasting blood glucose concentration, and risk of vascular disease: a collaborative meta-analysis of 102 prospective studies. Lancet 375: 2215–2222.

Engelter ST, Grond-Ginsbach C, Metso TM et al. (2013). Cervical artery dissection: trauma and other potential mechanical trigger events. Neurology 80: 1950–1957.

Falcone GJ, Malik R, Dichgans M et al. (2014). Current concepts and clinical applications of stroke genetics. Lancet Neurol 13: 405–418.

Fanning JP, Wesley AJ, Wong AA et al. (2014a). Emerging spectra of silent brain infarction. Stroke 45: 3461–3471.

Fanning JP, Wong AA, Fraser JF (2014b). The epidemiology of silent brain infarction: a systematic review of population-based cohorts. BMC Med 12: 119.

Fazekas F, Kleinert R, Roob G et al. (1999). Histopathologic analysis of foci of signal loss on gradient-echo T2*-weighted MR images in patients with spontaneous intracerebral hemorrhage: evidence of microangiopathy-related microbleeds. AJNR Am J Neuroradiol 20: 637–642.

Feigin V, Hoorn SV (2004). How to study stroke incidence. Lancet 363: 1920.

Feigin VL, Rinkel GJ, Lawes CM et al. (2005). Risk factors for subarachnoid hemorrhage: an updated systematic review of epidemiological studies. Stroke 36: 2773–2780.

Feigin VL, Lawes CM, Bennett DA et al. (2009). Worldwide stroke incidence and early case fatality reported in 56 population-based studies: a systematic review. Lancet Neurol 8: 355–369.

Feigin VL, Forouzanfar MH, Krishnamurthi R et al. (2014). Global and regional burden of stroke during 1990–2010: findings from the Global Burden of Disease Study 2010. Lancet 383: 245–254.

Ferro JM, Canhao P (2014). Cerebral venous sinus thrombosis: update on diagnosis and management. Curr Cardiol Rep 16: 523.

Ferro JM, Canhao P, Stam J et al. (2004). Prognosis of cerebral vein and dural sinus thrombosis: results of the International Study on Cerebral Vein and Dural Sinus Thrombosis (ISCVT). Stroke 35: 664–670.

Ferro JM, Canhao P, Peralta R (2008). Update on subarachnoid haemorrhage. J Neurol 255: 465–479.

Fieschi C, Rasura M, Anzini A et al. (1998). Central nervous system vasculitis. J Neurol Sci 153: 159–171.

Flossmann E, Schulz UG, Rothwell PM (2004). Systematic review of methods and results of studies of the genetic epidemiology of ischemic stroke. Stroke 35: 212–227.

Foerch C, Misselwitz B, Sitzer M et al. (2005). Difference in recognition of right and left hemispheric stroke. Lancet 366: 392–393.

Friedlander RM (2007). Clinical practice. Arteriovenous malformations of the brain. N Engl J Med 356: 2704–2712.

Fung LW, Thompson D, Ganesan V (2005). Revascularisation surgery for paediatric moyamoya: a review of the literature. Childs Nerv Syst 21: 358–364.

Gabriel RA, Kim H, Sidney S et al. (2010). Ten-year detection rate of brain arteriovenous malformations in a large, multiethnic, defined population. Stroke 41: 21–26.

Giles MF, Rothwell PM (2007). Risk of stroke early after transient ischaemic attack: a systematic review and meta-analysis. Lancet Neurol 6: 1063–1072.

Global Burden of Disease Stroke Expert Group (2012). Methodology of the global and regional burden of stroke study. Neuroepidemiology 38: 30–40.

Goos JD, Henneman WJ, Sluimer JD et al. (2010). Incidence of cerebral microbleeds: a longitudinal study in a memory clinic population. Neurology 74: 1954–1960.

Goyal M, Demchuk AM, Menon BK et al. (2015). Randomized assessment of rapid endovascular treatment of ischemic stroke. N Engl J Med 372: 1019–1030.

Granger CB, Alexander JH, McMurray JJ et al. (2011). Apixaban versus warfarin in patients with atrial fibrillation. N Engl J Med 365: 981–992.

Greenberg SM, Vernooij MW, Cordonnier C et al. (2009). Cerebral microbleeds: a guide to detection and interpretation. Lancet Neurol 8: 165–174.

Greving JP, Wermer MJ, Brown Jr RD et al. (2014). Development of the PHASES score for prediction of risk of rupture of intracranial aneurysms: a pooled analysis of six prospective cohort studies. Lancet Neurol 13: 59–66.

Grunt S, Wingeier K, Wehrli E et al. (2010). Cerebral sinus venous thrombosis in Swiss children. Dev Med Child Neurol 52: 1145–1150.

Hajj-Ali RA, Calabrese LH (2014). Diagnosis and classification of central nervous system vasculitis. J Autoimmun 48–49: 149–152.

Hajj-Ali RA, Singhal AB, Benseler S et al. (2011). Primary angiitis of the CNS. Lancet Neurol 10: 561–572.

Halim AX, Johnston SC, Singh V et al. (2004). Longitudinal risk of intracranial hemorrhage in patients with arteriovenous malformation of the brain within a defined population. Stroke 35: 1697–1702.

Hankey GJ (2006). Potential new risk factors for ischemic stroke: what is their potential? Stroke 37: 2181–2188.

Hart RG, Pearce LA, Aguilar MI (2007). Meta-analysis: antithrombotic therapy to prevent stroke in patients who have nonvalvular atrial fibrillation. Ann Intern Med 146: 857–867.

Hart RG, Diener HC, Coutts SB et al. (2014). Embolic strokes of undetermined source: the case for a new clinical construct. Lancet Neurol 13: 429–438.

Hatano S (1976). Experience from a multicentre stroke register: a preliminary report. Bull World Health Organ 54: 541–553.

Hippisley-Cox J, Coupland C, Brindle P (2013). Derivation and validation of QStroke score for predicting risk of ischaemic stroke in primary care and comparison with other risk scores: a prospective open cohort study. BMJ 346: f2573.

Hofmeister C, Stapf C, Hartmann A et al. (2000). Demographic, morphological, and clinical characteristics of 1289 patients with brain arteriovenous malformation. Stroke 31: 1307–1310.

Howard VJ, Kleindorfer DO, Judd SE et al. (2011). Disparities in stroke incidence contributing to disparities in stroke mortality. Ann Neurol 69: 619–627.

Huang Y, Cai X, Li Y et al. (2014). Prehypertension and the risk of stroke: a meta-analysis. Neurology 82: 1153–1161.

Ikram MA, Seshadri S, Bis JC et al. (2009). Genomewide association studies of stroke. N Engl J Med 360: 1718–1728.

James PA, Oparil S, Carter BL et al. (2014). 2014 evidence-based guideline for the management of high blood pressure in adults: report from the panel members appointed to the Eighth Joint National Committee (JNC 8). JAMA 311: 507–520.

Jauch EC, Saver JL, Adams Jr HP et al. (2013). Guidelines for the early management of patients with acute ischemic stroke: a guideline for healthcare professionals from the American Heart Association/American Stroke Association. Stroke 44: 870–947.

Jeerakathil T, Wolf PA, Beiser A et al. (2004). Cerebral microbleeds: prevalence and associations with cardiovascular risk factors in the Framingham Study. Stroke 35: 1831–1835.

Johnston SC, Rothwell PM, Nguyen-Huynh MN et al. (2007). Validation and refinement of scores to predict very early stroke risk after transient ischaemic attack. Lancet 369: 283–292.

Jones SP, Jenkinson AJ, Leathley MJ et al. (2010). Stroke knowledge and awareness: an integrative review of the evidence. Age Ageing 39: 11–22.

Joutel A, Corpechot C, Ducros A et al. (1996). Notch3 mutations in CADASIL, a hereditary adult-onset condition causing stroke and dementia. Nature 383: 707–710.

Keach JW, Bradley SM, Turakhia MP et al. (2015). Early detection of occult atrial fibrillation and stroke prevention. Heart 101: 1097–1102.

Khoury JC, Kleindorfer D, Alwell K et al. (2013). Diabetes mellitus: a risk factor for ischemic stroke in a large biracial population. Stroke 44: 1500–1504.

Kim H, Sidney S, McCulloch CE et al. (2007). Racial/Ethnic differences in longitudinal risk of intracranial hemorrhage in brain arteriovenous malformation patients. Stroke 38: 2430–2437.

Kim H, Al-Shahi Salman R, McCulloch CE et al. (2014). Untreated brain arteriovenous malformation: patient-level meta-analysis of hemorrhage predictors. Neurology 83: 590–597.

Kissela BM, Khoury JC, Alwell K et al. (2012). Age at stroke: temporal trends in stroke incidence in a large, biracial population. Neurology 79: 1781–1787.

Kleinloog R, Regli L, Rinkel GJ et al. (2012). Regional differences in incidence and patient characteristics of moyamoya disease: a systematic review. J Neurol Neurosurg Psychiatry 83: 531–536.

Koton S, Schneider AL, Rosamond WD et al. (2014). Stroke incidence and mortality trends in US communities, 1987 to 2011. JAMA 312: 259–268.

Krishnamurthi RV, Feigin VL, Forouzanfar MH et al. (2013). Global and regional burden of first-ever ischaemic and haemorrhagic stroke during 1990–2010: findings from the Global Burden of Disease Study 2010. Lancet Glob Health 1: e259–e281.

Kronenburg A, Braun KP, van der Zwan A et al. (2014). Recent advances in moyamoya disease: pathophysiology and treatment. Curr Neurol Neurosci Rep 14: 423.

Kuroda S, Houkin K (2008). Moyamoya disease: current concepts and future perspectives. Lancet Neurol 7: 1056–1066.

Launer LJ (2004). Epidemiology of white matter lesions. Top Magn Reson Imaging 15: 365–367.

Law MR, Morris JK, Wald NJ (2009). Use of blood pressure lowering drugs in the prevention of cardiovascular disease: meta-analysis of 147 randomised trials in the context of expectations from prospective epidemiological studies. BMJ 338: b1665.

Lee SC, Park SJ, Ki HK et al. (2000). Prevalence and risk factors of silent cerebral infarction in apparently normal adults. Hypertension 36: 73–77.

Lee VH, Brown Jr RD, Mandrekar JN et al. (2006). Incidence and outcome of cervical artery dissection: a population-based study. Neurology 67: 1809–1812.

Lees KR, Bluhmki E, von Kummer R et al. (2010). Time to treatment with intravenous alteplase and outcome in stroke: an updated pooled analysis of ECASS, ATLANTIS, NINDS, and EPITHET trials. Lancet 375: 1695–1703.

Lei C, Lin S, Tao W et al. (2013). Association between cerebral microbleeds and cognitive function: a systematic review. J Neurol Neurosurg Psychiatry 84: 693–697.

Lewandowski CA, Rao CP, Silver B (2008). Transient ischemic attack: definitions and clinical presentations. Ann Emerg Med 52: S7–S16.

Lewington S, Clarke R, Qizilbash N et al. (2002). Age-specific relevance of usual blood pressure to vascular mortality: a meta-analysis of individual data for one million adults in 61 prospective studies. Lancet 360: 1903–1913.

Liebetrau M, Steen B, Hamann GF et al. (2004). Silent and symptomatic infarcts on cranial computerized tomography in relation to dementia and mortality: a population-based study in 85-year-old subjects. Stroke 35: 1816–1820.

Lip GY, Lane DA (2015). Stroke prevention in atrial fibrillation: a systematic review. JAMA 313: 1950–1962.

Longstreth Jr WT, Dulberg C, Manolio TA et al. (2002). Incidence, manifestations, and predictors of brain infarcts defined by serial cranial magnetic resonance imaging in the elderly: the Cardiovascular Health Study. Stroke 33: 2376–2382.

Lozano R, Naghavi M, Foreman K et al. (2012). Global and regional mortality from 235 causes of death for 20 age groups in 1990 and 2010: a systematic analysis for the Global Burden of Disease Study 2010. Lancet 380: 2095–2128.

Lucke M, Hajj-Ali RA (2014). Advances in primary angiitis of the central nervous system. Curr Cardiol Rep 16: 533.

Luitse MJ, Biessels GJ, Rutten GE et al. (2012). Diabetes, hyperglycaemia, and acute ischaemic stroke. Lancet Neurol 11: 261–271.

Lyrer P, Engelter S (2010). Antithrombotic drugs for carotid artery dissection. Cochrane Database Syst Rev: CD000255.

Mair G, Wardlaw JM (2014). Imaging of acute stroke prior to treatment: current practice and evolving techniques. Br J Radiol 87: 20140216.

Malmgren R, Warlow C, Bamford J et al. (1987). Geographical and secular trends in stroke incidence. Lancet 2: 1196–1200.

Merwick A, Albers GW, Amarenco P et al. (2010). Addition of brain and carotid imaging to the ABCD(2) score to identify patients at early risk of stroke after transient ischaemic attack: a multicentre observational study. Lancet Neurol 9: 1060–1069.

Meschia JF, Bushnell C, Boden-Albala B et al. (2014). Guidelines for the primary prevention of stroke: a statement for healthcare professionals from the American Heart Association/American Stroke Association. Stroke 45: 3754–3832.

Miwa K, Tanaka M, Okazaki S et al. (2014). Multiple or mixed cerebral microbleeds and dementia in patients with vascular risk factors. Neurology 83: 646–653.

Miyamoto S, Yoshimoto T, Hashimoto N et al. (2014). Effects of extracranial-intracranial bypass for patients with hemorrhagic moyamoya disease: results of the Japan Adult Moyamoya Trial. Stroke 45: 1415–1421.

Mohan KM, Wolfe CD, Rudd AG et al. (2011). Risk and cumulative risk of stroke recurrence: a systematic review and meta-analysis. Stroke 42: 1489–1494.

Mohr JP, Albers GW, Amarenco P et al. (1997). American Heart Association Prevention Conference. IV Prevention and Rehabilitation of Stroke. Etiology of stroke Stroke 28: 1501–1506.

Mohr JP, Parides MK, Stapf C et al. (2014). Medical management with or without interventional therapy for unruptured brain arteriovenous malformations (ARUBA): a multicentre, non-blinded, randomised trial. Lancet 383: 614–621.

Molyneux A, Kerr R, Stratton I et al. (2002). International Subarachnoid Aneurysm Trial (ISAT) of neurosurgical clipping versus endovascular coiling in 2143 patients with ruptured intracranial aneurysms: a randomised trial. Lancet 360: 1267–1274.

Molyneux AJ, Birks J, Clarke A et al. (2015). The durability of endovascular coiling versus neurosurgical clipping of ruptured cerebral aneurysms: 18 year follow-up of the UK cohort of the International Subarachnoid Aneurysm Trial (ISAT). Lancet 385: 691–697.

Mons U, Muezzinler A, Gellert C et al. (2015). Impact of smoking and smoking cessation on cardiovascular events and mortality among older adults: meta-analysis of individual participant data from prospective cohort studies of the CHANCES consortium. BMJ 350: h1551.

Moreton FC, Razvi SS, Davidson R et al. (2014). Changing clinical patterns and increasing prevalence in CADASIL. Acta Neurol Scand 130: 197–203.

Morgenstern LB, Smith MA, Sanchez BN et al. (2013). Persistent ischemic stroke disparities despite declining incidence in Mexican Americans. Ann Neurol 74: 778–785.

Morris Z, Whiteley WN, Longstreth Jr WT et al. (2009). Incidental findings on brain magnetic resonance imaging: systematic review and meta-analysis. BMJ 339: b3016.

Mozaffarian D, Benjamin EJ, Go AS et al. (2015). Heart disease and stroke statistics – 2015 update: a report from the American Heart Association. Circulation 131: e29–e322.

Muir KW (2013). Stroke. Medicine 41: 169–174.

Muir KW, Buchan A, von Kummer R et al. (2006). Imaging of acute stroke. Lancet Neurol 5: 755–768.

Murray CJ, Vos T, Lozano R et al. (2012). Disability-adjusted life years (DALYs) for 291 diseases and injuries in 21 regions, 1990–2010: a systematic analysis for the

Global Burden of Disease Study 2010. Lancet 380: 2197–2223.

Narayan SK, Gorman G, Kalaria RN et al. (2012). The minimum prevalence of CADASIL in northeast England. Neurology 78: 1025–1027.

Nielsen JA, Zielinski BA, Ferguson MA et al. (2013). An evaluation of the left-brain vs. right-brain hypothesis with resting state functional connectivity magnetic resonance imaging. 8: e71275.

O'Donnell MJ, Xavier D, Liu L et al. (2010). Risk factors for ischaemic and intracerebral haemorrhagic stroke in 22 countries (the INTERSTROKE study): a case-control study. Lancet 376: 112–123.

O'Kelly CJ, Kulkarni AV, Austin PC et al. (2010). The impact of therapeutic modality on outcomes following repair of ruptured intracranial aneurysms: an administrative data analysis. Clinical article. J Neurosurg 113: 795–801.

Opherk C, Peters N, Herzog J et al. (2004). Long-term prognosis and causes of death in CADASIL: a retrospective study in 411 patients. Brain 127: 2533–2539.

Orfei MD, Robinson RG, Prigatano GP et al. (2007). Anosognosia for hemiplegia after stroke is a multifaceted phenomenon: a systematic review of the literature. Brain 130: 3075–3090.

Ovbiagele B, Nguyen-Huynh MN (2011). Stroke epidemiology: advancing our understanding of disease mechanism and therapy. Neurotherapeutics 8: 319–329.

Ovbiagele B, Saver JL (2006). Cerebral white matter hyperintensities on MRI: current concepts and therapeutic implications. Cerebrovasc Dis 22: 83–90.

Ovbiagele B, Kidwell CS, Saver JL (2003). Epidemiological impact in the United States of a tissue-based definition of transient ischemic attack. Stroke 34: 919–924.

Pandey P, Steinberg GK (2011). Neurosurgical advances in the treatment of moyamoya disease. Stroke 42: 3304–3310.

Pantoni L (2002). Pathophysiology of age-related cerebral white matter changes. Cerebrovasc Dis 13 (Suppl 2): 7–10.

Pantoni L (2010). Cerebral small vessel disease: from pathogenesis and clinical characteristics to therapeutic challenges. Lancet Neurol 9: 689–701.

Parmar P, Krishnamurthi R, Ikram MA et al. (2015). The Stroke Riskometer(TM) App: validation of a data collection tool and stroke risk predictor. Int J Stroke 10: 231–244.

Patel MR, Mahaffey KW, Garg J et al. (2011). Rivaroxaban versus warfarin in nonvalvular atrial fibrillation. N Engl J Med 365: 883–891.

Payne GH, Fang J, Fogle CC et al. (2010). Stroke awareness: surveillance, educational campaigns, and public health practice. J Public Health Manag Pract 16: 345–358.

Pendlebury ST, Rothwell PM (2009). Prevalence, incidence, and factors associated with pre-stroke and post-stroke dementia: a systematic review and meta-analysis. Lancet Neurol 8: 1006–1018.

Poels MM, Vernooij MW, Ikram MA et al. (2010). Prevalence and risk factors of cerebral microbleeds: an update of the Rotterdam scan study. Stroke 41: S103–S106.

Poels MM, Ikram MA, van der Lugt A et al. (2011). Incidence of cerebral microbleeds in the general population: the Rotterdam Scan Study. Stroke 42: 656–661.

Poels MM, Ikram MA, van der Lugt A et al. (2012). Cerebral microbleeds are associated with worse cognitive function: the Rotterdam Scan Study. Neurology 78: 326–333.

Portegies ML, Selwaness M, Hofman A et al. (2015). Left-sided strokes are more often recognized than right-sided strokes: the Rotterdam study. Stroke 46: 252–254.

Powers WJ, Derdeyn CP, Biller J et al. (2015). 2015 AHA/ASA Focused Update of the 2013 Guidelines for the Early Management of Patients With Acute Ischemic Stroke Regarding Endovascular Treatment: A Guideline for Healthcare Professionals From the American Heart Association/American Stroke Association. Stroke. http://dx.doi.org/10.1161/STR.0000000000000074.

Prabhakaran S, Ruff I, Bernstein RA (2015). Acute stroke intervention: a systematic review. JAMA 313: 1451–1462.

Qiu C, Cotch MF, Sigurdsson S et al. (2010). Cerebral microbleeds, retinopathy, and dementia: the AGES-Reykjavik Study. Neurology 75: 2221–2228.

Qureshi AI, Tuhrim S, Broderick JP et al. (2001). Spontaneous intracerebral hemorrhage. N Engl J Med 344: 1450–1460.

Qureshi AI, Mendelow AD, Hanley DF (2009). Intracerebral haemorrhage. Lancet 373: 1632–1644.

Rasura M, Baldereschi M, Di Carlo A et al. (2014). Effectiveness of public stroke educational interventions: a review. Eur J Neurol 21: 11–20.

Reeves MJ, Bushnell CD, Howard G et al. (2008). Sex differences in stroke: epidemiology, clinical presentation, medical care, and outcomes. Lancet Neurol 7: 915–926.

Rinkel GJ, Algra A (2011). Long-term outcomes of patients with aneurysmal subarachnoid haemorrhage. Lancet Neurol 10: 349–356.

Romero JR, Wolf PA (2013). Epidemiology of stroke: legacy of the Framingham Heart Study. Glob Heart 8: 67–75.

Rosengren A, Giang KW, Lappas G et al. (2013). Twenty-four-year trends in the incidence of ischemic stroke in Sweden from 1987 to 2010. Stroke 44: 2388–2393.

Rothwell PM, Warlow CP (2005). Timing of TIAs preceding stroke: time window for prevention is very short. Neurology 64: 817–820.

Rothwell PM, Giles MF, Flossmann E et al. (2005). A simple score (ABCD) to identify individuals at high early risk of stroke after transient ischaemic attack. Lancet 366: 29–36.

Rothwell PM, Howard SC, Dolan E et al. (2010). Prognostic significance of visit-to-visit variability, maximum systolic blood pressure, and episodic hypertension. Lancet 375: 895–905.

Rubinstein SM, Peerdeman SM, van Tulder MW et al. (2005). A systematic review of the risk factors for cervical artery dissection. Stroke 36: 1575–1580.

Saini M, Ikram K, Hilal S et al. (2012). Silent stroke: not listened to rather than silent. Stroke 43: 3102–3104.

Sajjad A, Chowdhury R, Felix JF et al. (2013). A systematic evaluation of stroke surveillance studies in low- and middle-income countries. Neurology 80: 677–684.

Salvarani C, Brown Jr RD, Calamia KT et al. (2007). Primary central nervous system vasculitis: analysis of 101 patients. Ann Neurol 62: 442–451.

Salvarani C, Brown Jr RD, Hunder GG (2012). Adult primary central nervous system vasculitis. Lancet 380: 767–777.

Salvarani C, Brown Jr RD, Christianson T et al. (2015). An update of the Mayo Clinic cohort of patients with adult primary central nervous system vasculitis: description of 163 patients. Medicine (Baltimore) 94: e738.

Saposnik G, Barinagarrementeria F, Brown Jr RD et al. (2011). Diagnosis and management of cerebral venous thrombosis: a statement for healthcare professionals from the American Heart Association/American Stroke Association. Stroke 42: 1158–1192.

Saver JL, Goyal M, Bonafe A et al. (2015). Stent-retriever thrombectomy after intravenous t-PA vs. t-PA alone in stroke. N Engl J Med 372: 2285–2295.

Schievink WI (2001). Spontaneous dissection of the carotid and vertebral arteries. N Engl J Med 344: 898–906.

Scott RM, Smith ER (2009). Moyamoya disease and moyamoya syndrome. N Engl J Med 360: 1226–1237.

Seshadri S, Beiser A, Kelly-Hayes M et al. (2006). The lifetime risk of stroke: estimates from the Framingham Study. Stroke 37: 345–350.

Seshadri S, Beiser A, Pikula A et al. (2010). Parental occurrence of stroke and risk of stroke in their children: the Framingham study. Circulation 121: 1304–1312.

Singhal S, Bevan S, Barrick T et al. (2004). The influence of genetic and cardiovascular risk factors on the CADASIL phenotype. Brain 127: 2031–2038.

Stam J (2005). Thrombosis of the cerebral veins and sinuses. N Engl J Med 352: 1791–1798.

Stapf C, Mast H, Sciacca RR et al. (2006). Predictors of hemorrhage in patients with untreated brain arteriovenous malformation. Neurology 66: 1350–1355.

Steiner T, Rosand J, Diringer M (2006). Intracerebral hemorrhage associated with oral anticoagulant therapy: current practices and unresolved questions. Stroke 37: 256–262.

Steiner T, Juvela S, Unterberg A et al. (2013). European Stroke Organization guidelines for the management of intracranial aneurysms and subarachnoid haemorrhage. Cerebrovasc Dis 35: 93–112.

Strazzullo P, D'Elia L, Cairella G et al. (2010). Excess body weight and incidence of stroke: meta-analysis of prospective studies with 2 million participants. Stroke 41: e418–e426.

Sudlow CL, Warlow CP (1996). Comparing stroke incidence worldwide: what makes studies comparable? Stroke 27: 550–558.

Suk SH, Sacco RL, Boden-Albala B et al. (2003). Abdominal obesity and risk of ischemic stroke: the Northern Manhattan Stroke Study. Stroke 34: 1586–1592.

Sveinbjornsdottir S, Sigurdsson S, Aspelund T et al. (2008). Cerebral microbleeds in the population based AGES-Reykjavik study: prevalence and location. J Neurol Neurosurg Psychiatry 79: 1002–1006.

Tikka S, Baumann M, Siitonen M et al. (2014). CADASIL and CARASIL. Brain Pathol 24: 525–544.

van Beijnum J, van der Worp HB, Buis DR et al. (2011). Treatment of brain arteriovenous malformations: a systematic review and meta-analysis. JAMA 306: 2011–2019.

van Gijn J, Kerr RS, Rinkel GJ (2007). Subarachnoid haemorrhage. Lancet 369: 306–318.

Vera-Lastra O, Sepulveda-Delgado J, Cruz-Dominguez Mdel P et al. (2015). Primary and secondary central nervous system vasculitis: clinical manifestations, laboratory findings, neuroimaging, and treatment analysis. Clin Rheumatol 34: 729–738.

Verheugt FW, Granger CB (2015). Oral anticoagulants for stroke prevention in atrial fibrillation: current status, special situations, and unmet needs. Lancet 386: 303–310.

Vermeer SE, Koudstaal PJ, Oudkerk M et al. (2002). Prevalence and risk factors of silent brain infarcts in the population-based Rotterdam Scan Study. Stroke 33: 21–25.

Vermeer SE, Den Heijer T, Koudstaal PJ et al. (2003a). Incidence and risk factors of silent brain infarcts in the population-based Rotterdam Scan Study. Stroke 34: 392–396.

Vermeer SE, Hollander M, van Dijk EJ et al. (2003b). Silent brain infarcts and white matter lesions increase stroke risk in the general population: the Rotterdam Scan Study. Stroke 34: 1126–1129.

Vermeer SE, Prins ND, den Heijer T et al. (2003c). Silent brain infarcts and the risk of dementia and cognitive decline. N Engl J Med 348: 1215–1222.

Vermeer SE, Longstreth Jr WT, Koudstaal PJ (2007). Silent brain infarcts: a systematic review. Lancet Neurol 6: 611–619.

Vernooij MW, Ikram MA, Tanghe HL et al. (2007). Incidental findings on brain MRI in the general population. N Engl J Med 357: 1821–1828.

Vernooij MW, van der Lugt A, Ikram MA et al. (2008). Prevalence and risk factors of cerebral microbleeds: the Rotterdam Scan Study. Neurology 70: 1208–1214.

Viswanathan A, Greenberg SM (2011). Cerebral amyloid angiopathy in the elderly. Ann Neurol 70: 871–880.

Vlak MH, Algra A, Brandenburg R et al. (2011). Prevalence of unruptured intracranial aneurysms, with emphasis on sex, age, comorbidity, country, and time period: a systematic review and meta-analysis. Lancet Neurol 10: 626–636.

Vossel S, Weiss PH, Eschenbeck P et al. (2013). Anosognosia, neglect, extinction and lesion site predict impairment of daily living after right-hemispheric stroke. Cortex 49: 1782–1789.

Wang X, Dong Y, Qi X et al. (2013). Cholesterol levels and risk of hemorrhagic stroke: a systematic review and meta-analysis. Stroke 44: 1833–1839.

Wardlaw JM, Murray V, Berge E et al. (2012). Recombinant tissue plasminogen activator for acute ischaemic stroke: an updated systematic review and meta-analysis. Lancet 379: 2364–2372.

Wardlaw JM, Smith C, Dichgans M (2013a). Mechanisms of sporadic cerebral small vessel disease: insights from neuroimaging. Lancet Neurol 12: 483–497.

Wardlaw JM, Smith EE, Biessels GJ et al. (2013b). Neuroimaging standards for research into small vessel disease and its contribution to ageing and neurodegeneration. Lancet Neurol 12: 822–838.

Wardlaw JM, Murray V, Berge E et al. (2014). Thrombolysis for acute ischaemic stroke. Cochrane Database Syst Rev 7: CD000213.

Wardlaw JM, Brazzelli M, Chappell FM et al. (2015). ABCD2 score and secondary stroke prevention: meta-analysis and effect per 1,000 patients triaged. Neurology 85: 373–380.

Weber R, Weimar C, Blatchford J et al. (2012). Telmisartan on top of antihypertensive treatment does not prevent progression of cerebral white matter lesions in the prevention regimen for effectively avoiding second strokes (PRoFESS) MRI substudy. Stroke 43: 2336–2342.

Wedderburn CJ, van Beijnum J, Bhattacharya JJ et al. (2008). Outcome after interventional or conservative management of unruptured brain arteriovenous malformations: a prospective, population-based cohort study. Lancet Neurol 7: 223–230.

WHO MONICA Project PI (1988). The World Health Organization MONICA Project (monitoring trends and determinants in cardiovascular disease): a major international collaboration. WHO MONICA Project Principal Investigators. J Clin Epidemiol 41: 105–114.

Wieberdink RG, Ikram MA, Hofman A et al. (2012). Trends in stroke incidence rates and stroke risk factors in Rotterdam, the Netherlands from 1990 to 2008. Eur J Epidemiol 27: 287–295.

Wolf PA, D'Agostino RB, Belanger AJ et al. (1991). Probability of stroke: a risk profile from the Framingham Study. Stroke 22: 312–318.

Wu CM, McLaughlin K, Lorenzetti DL et al. (2007). Early risk of stroke after transient ischemic attack: a systematic review and meta-analysis. Arch Intern Med 167: 2417–2422.

Yates PA, Villemagne VL, Ellis KA et al. (2014). Cerebral microbleeds: a review of clinical, genetic, and neuroimaging associations. Front Neurol 4: 205.

Handbook of Clinical Neurology, Vol. 138 (3rd series)
Neuroepidemiology
C. Rosano, M.A. Ikram, and M. Ganguli, Editors
http://dx.doi.org/10.1016/B978-0-12-802973-2.00015-X

Chapter 15

Peripheral neuropathies

R. HANEWINCKEL[1], M.A. IKRAM[1]*, AND P.A. VAN DOORN[2]
[1]*Department of Epidemiology, Erasmus MC University Medical Center, Rotterdam, The Netherlands*
[2]*Department of Neurology, Erasmus University Medical Center, Rotterdam, The Netherlands*

Abstract

Peripheral neuropathies are diseases of the peripheral nervous system that can be divided into mononeuropathies, multifocal neuropathies, and polyneuropathies. Symptoms usually include numbness and paresthesia. These symptoms are often accompanied by weakness and can be painful. Polyneuropathies can be divided into axonal and demyelinating forms, which is important for diagnostic reasons. Most peripheral neuropathies develop over months or years, but some are rapidly progressive. Some patients only suffer from mild, unilateral, slowly progressive tingling in the fingers due to median nerve compression in the wrist (carpal tunnel syndrome), while other patients can be tetraplegic, with respiratory insufficiency within 1–2 days due to Guillain–Barré syndrome. Carpal tunnel syndrome, with a prevalence of 5% and incidence of 1–2 per 1000 person-years, is the most common mononeuropathy. Population-based data for chronic polyneuropathy are relatively scarce. Prevalence is estimated at 1% and increases to 7% in persons over 65 years of age. Incidence is approximately 1 per 1000 person-years. Immune-mediated polyneuropathies like Guillain–Barré syndrome and chronic inflammatory demyelinating polyradiculoneuropathy are rare diseases, with an annual incidence of approximately 1–2 and 0.2–0.5 per 100 000 persons respectively. Most peripheral neuropathies are more prevalent in older adults and in men, except for carpal tunnel syndrome, which is more common in women. Diabetes is a common cause of peripheral neuropathy and is associated with both mono- and polyneuropathies. Among the group of chronic polyneuropathies, in about 20–25% no direct cause can be found. These are slowly progressive axonal polyneuropathies.

INTRODUCTION

Peripheral neuropathies are common neurologic diseases and are often encountered in hospitals as well as outpatient clinics and general practitioners' offices. Peripheral neuropathy encompasses a wide spectrum of clinical syndromes, which can be divided according to anatomic region and distribution across the peripheral nervous system. A first distinction that can be made is the involvement of one peripheral nerve, in case of mononeuropathies, or multiple peripheral nerves, in case of multifocal neuropathies and polyneuropathies. These three main categories can be subdivided into smaller groups, making distinctions according to cause of the disease (compressive or noncompressive), time course of the disease (chronic or acute), or type of neuropathy (axonal or demyelinating neuropathy). Despite the high frequency of occurrence of peripheral neuropathies, there is much less epidemiologic data available than for diseases of the central nervous system, such as stroke, dementia, or Parkinson's disease. In this chapter we discuss the clinical phenotype, prevalence, incidence, risk factors, and treatment of peripheral neuropathies, with a focus on the most common and the most studied variants (Table 15.1 and Fig. 15.1). The chapter is divided into four parts. In the first part we discuss mononeuropathies. Multifocal neuropathies are discussed in the second part. Acute and chronic polyneuropathies are discussed in the last two parts of this chapter.

*Correspondence to: M.A. Ikram, MD PhD, Department of Epidemiology, Erasmus University Medical Center, PO Box 2040, 3000 CA, Rotterdam, The Netherlands. Tel: +31-10-7043489, E-mail: m.a.ikram@erasmusmc.nl

Table 15.1

Epidemiologic features of peripheral neuropathies

	Prevalence per 100 000	Incidence* per 100 000 person-years	Gender predominance	Age
Mononeuropathy				
Carpal tunnel syndrome	~5000 (5%)	100–376	F>M	Peak 50–60 years
Ulnar neuropathy	7–14	20	M>F	Peak 50–70 years
Radial neuropathy	ND	2	M>F	ND
Peroneal neuropathy	19	17	M>F	ND
Meralgia paraesthetica	71	11–43	F=M	Peak 50–60 years
Cranial nerves				
Facial nerve (Bell's) palsy	28–143	12–40	F=M	Peak 15–45 years
Sixth-nerve palsy	ND	10	M>F	Increasing with age
Fourth-nerve palsy	ND	6	M>F	Increasing with age
Third-nerve palsy	ND	4	M>F	Increasing with age
Multifocal neuropathies				
Mononeuritis multiplex	ND	ND	ND	ND
Multifocal motor neuropathy	1	ND	M>F	Peaks 20–50 years
Acute polyneuropathy				
Guillain–Barré syndrome	ND	1–2	M>F	Increasing with age
Chronic polyneuropathy	1000 (1%)	77	M>F	Increasing with age
Diabetic polyneuropathy	~200–600	23–54	CR	Increasing with age
Idiopathic polyneuropathy	~200	32	M>F	Increasing with age
Hereditary polyneuropathy	40	ND	ND	–
CIDP	1–8	0.2–0.5	M>F	Increasing with age
Toxic neuropathies	~100	ND	M>F	Increasing with age

*Prevalence data for nonchronic diseases like carpal tunnel syndrome, Bell's palsy, and Guillain–Barré syndrome is less informative. Incidence of these diseases is important from an epidemiologic perspective. These rates are mostly derived from registered new cases in a certain time period. ND, no data available; CR, conflicting results; CIDP, chronic inflammatory demyelinating polyradiculoneuropathy.

MONONEUROPATHIES

Definition and clinical presentation

Mononeuropathy refers to the condition in which only one single peripheral nerve is involved. Typically, patients complain of numbness, prickling or tingling sensations, and pain in an area specific to a single peripheral nerve. Occasionally, pain may radiate into other regions (especially in carpal tunnel syndrome (CTS)). Aside from these sensory complaints, mononeuropathy can also lead to muscle weakness, for example, weakness of intrinsic hand muscles in case of an ulnar neuropathy, wrist drop in case of radial neuropathy, or foot drop in case of a peroneal neuropathy. Mononeuropathy can be caused by internal (e.g., by soft-tissue masses such as ganglia and lipomas) or external compression, entrapment, or trauma of a peripheral nerve. Trauma often leads to an acute lesion of the nerve, while compression often leads to a more gradual onset of injury. Other causes include metabolic and endocrine disorders (diabetes mellitus, hypothyroidism), inflammation (neuralgic amyotrophy, vasculitis), infection (Lyme disease, leprosy, human immunodeficiency virus (HIV), Epstein–Barr virus), and drugs and toxins.

Frequency and risk factors

The most common mononeuropathies are median nerve compression at the wrist (CTS), ulnar nerve involvement near the elbow, peroneal nerve compression at the fibular head, and neuropathy of the lateral femoral cutaneous nerve of the thigh (meralgia paraesthetica). Neuropathy can also occur in cranial nerves. The most commonly seen mononeuropathy of the cranial nerves is Bell's palsy, an idiopathic neuropathy of the facial nerve that leads to numbness and paresis or paralysis of one side of the face. Mononeuropathy of ocular nerves, leading to diplopia, is also occasionally encountered in neurologic clinics. Other cranial nerve neuropathies are less common. Most important causes of cranial nerve neuropathies include trauma, vascular diseases, and compression by neoplasms or carcinomatous metastasis within the cerebrospinal fluid. Discussing all mononeuropathies goes beyond the scope of this chapter. Moreover,

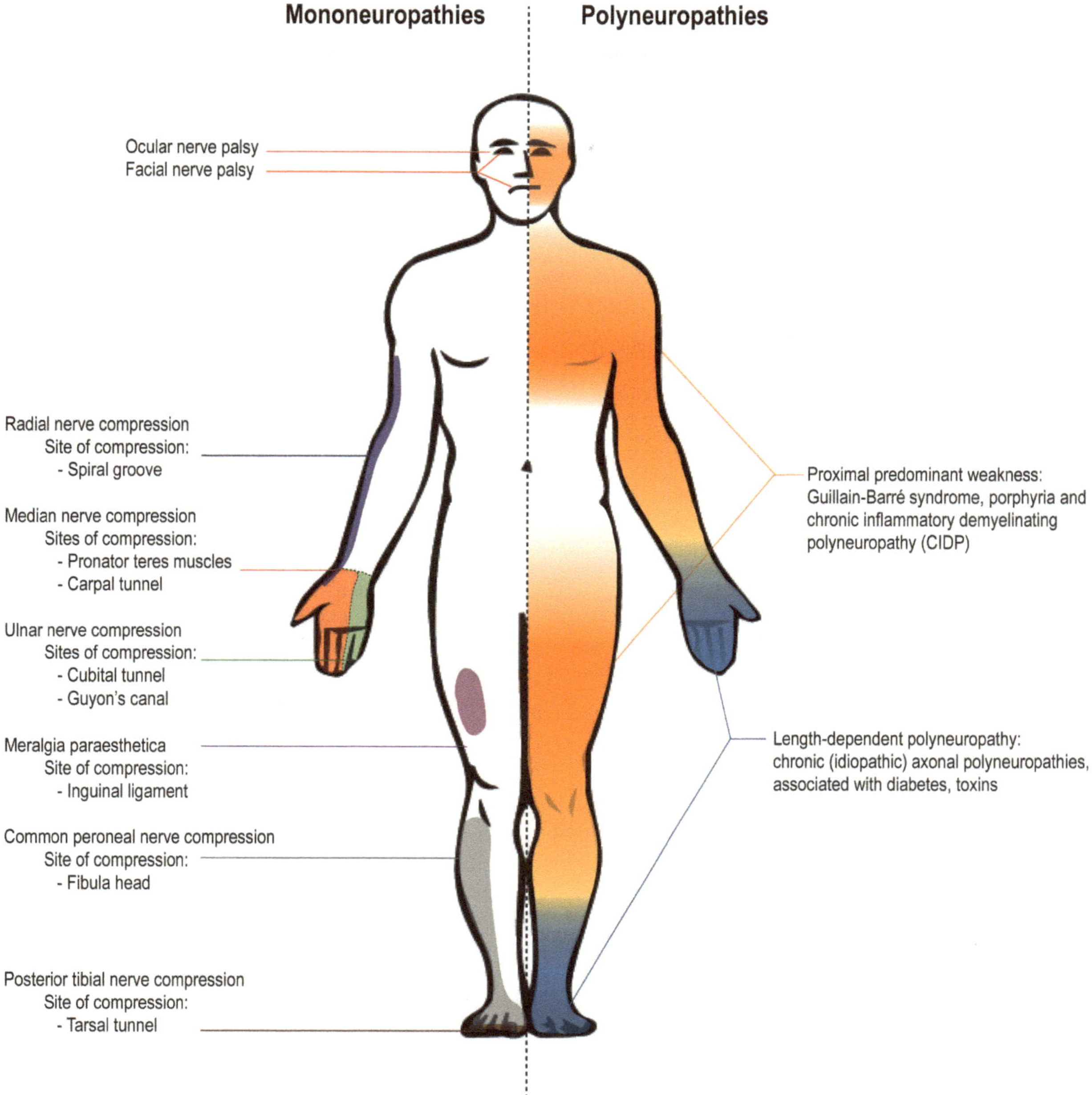

Fig. 15.1. Overview of peripheral neuropathies.

epidemiologic data on less frequent or rare variants are scarce. In the next paragraphs of this chapter, we will therefore focus on the most common mononeuropathies.

Carpal tunnel syndrome

A consensus panel proposed diagnostic criteria for CTS, designed for epidemiologic studies (Rempel et al., 1998). These criteria state that the most accurate diagnosis of CTS can be made based on a combination of symptoms and abnormal median nerve conduction studies. Neurologic examination adds little diagnostic value to these criteria, but is important for the differential diagnosis. Studies that report on the epidemiology of CTS using these criteria show that CTS is a common disorder, with an age-adjusted prevalence around 5% in the general population (de Krom et al., 1992; Atroshi et al., 1999; Middleton and Anakwe, 2014) and an incidence of 100–376 per 100 000 person-years. These numbers vary by country and employed methodology (differences in data sources, retrospective or prospective data collection, different CTS definitions) (Stevens et al., 1988; Nordstrom et al., 1998; Mondelli et al., 2002; Bland and Rudolfer, 2003; Latinovic et al., 2006; Gelfman et al., 2009). Incidence seems to have increased in the last decades of the 20th century, but this could also (partly) indicate better awareness and diagnostic procedures (more sensitive electrophysiologic techniques leading to detection of milder cases) to detect CTS (Mondelli et al., 2002; Gelfman et al., 2009). CTS occurs more frequently in females than in males, with a ratio of 2–4:1, and often occurs bilaterally (Atroshi et al., 1999; Mondelli et al., 2002; Bland and Rudolfer, 2003; Latinovic et al., 2006; Middleton and Anakwe, 2014). Incidence peaks in both sexes during the sixth decade, with another peak in the eighth decade, predominantly in men (Mondelli et al., 2002; Bland and Rudolfer, 2003).

Several risk factors for CTS have been identified. There is a higher incidence of CTS in workers than in nonworkers, especially if work involves manual labor (Nordstrom et al., 1997; Atroshi et al., 1999; Roquelaure et al., 2008). This can probably be attributed to the use of power tools or machinery, hand/wrist activities with a high degree of repetition, and working with vibrating tools (Nordstrom et al., 1997; Palmer et al., 2007; Gelfman et al., 2009; Barcenilla et al., 2012; Dale et al., 2015). An association with computer use has not been established (Mediouni et al., 2014). Other factors that are associated with a higher incidence of CTS are a positive family history, a high body mass index (BMI), pregnancy, and having certain musculoskeletal and endocrine diseases, such as osteoarthritis, rheumatoid arthritis, diabetes mellitus, and hypothyroidism (Nordstrom et al., 1997; van Dijk et al., 2003; Geoghegan et al., 2004; Padua et al., 2010; Ibrahim et al., 2012). However, screening for these associated disorders is generally not recommended, as most patients are often already diagnosed with these disorders. CTS rarely is the presenting symptom of an underlying disease (van Dijk et al., 2003; de Rijk et al., 2007).

CTS is recognized as an important cause of work-related disability and is associated with considerable social and economic costs (Atroshi et al., 1999; Mondelli et al., 2002; Gelfman et al., 2009; Ibrahim et al., 2012).

Ulnar neuropathy

Entrapment of the ulnar nerve usually occurs within the cubital tunnel at the elbow, where the nerve is subject to mechanic stretch and compression (cubital tunnel syndrome), but can also occur in the wrist (Guyon's canal syndrome) (Robertson and Saratsiotis, 2005; Assmus et al., 2011). Although ulnar neuropathy is considered the second most common mononeuropathy of the upper extremity, there is not much population-based epidemiologic data available. Prevalence of ulnar neuropathy in two door-to-door survey studies, one from India and one from Egypt, was 14 and 7 per 100 000 people respectively (Bharucha et al., 1991; Kandil et al., 2012). The first report on incidence of ulnar nerve neuropathy (at the elbow) originates from 2005 (Mondelli et al., 2005). Over a 5-year period, medical records of electromyography laboratories and discharge diagnoses of local hospitals were checked retrospectively to obtain cases with ulnar nerve neuropathy. Cases were selected if they had symptoms or signs consistent with ulnar nerve neuropathy and slowing of motor conduction velocity across the elbow, or surgical evidence of nerve compression at the elbow. In a population of 251 930 people, 311 cases were identified, yielding a crude incidence of 24.7 cases per 100 000 person-years and a standardized incidence of 20.0 per 100 000 person-years. Most new cases occurred between the age of 50 and 70 years. Crude incidence was almost twice as high in males than in females (32.7 compared to 17.2, ratio 1.9:1). Subsequent studies found similar numbers (Latinovic et al., 2006).

This male predominance has been found in other studies as well (Richardson et al., 2001; Latinovic et al., 2006). Perhaps men are more prone to develop ulnar neuropathy at the elbow because the ulnar nerve and the blood vessels supplying it are less well protected by subcutaneous fat at the tubercle of the coronoid process than in females (Contreras et al., 1998). Another explanation for this difference can be a different exposure to (work-related) risk factors. Ulnar neuropathy may result from some manual labour activities, such as repetitively holding a tool in certain position, forceful work, and some specific professions or activities, but there is not much evidence attributing ulnar neuropathy to work (Descatha et al., 2004; Mondelli et al., 2005; Svendsen et al., 2012). Resting or pressing the elbow upon an armchair or desk while working on a computer, or at the window frame when driving a car and prolonged periods of flexion of the elbow, for example, after immobilization of an arm following a fracture, sleeping with arms in flexed position, or extensive phone use, can also increase the risk of developing neuropathy of the ulnar nerve (Robertson and Saratsiotis, 2005; Juratli et al., 2010). Ulnar nerve dysfunction has been observed to occur quite often as a complication of surgical treatment of distal humerus fractures (Wiggers et al., 2012) and of procedures performed on anesthetized patients in general, likely due to the position in which the patient is placed during the operation. Mainly very thin and obese males with prolonged hospitalization are at risk of developing neuropathy after surgery (Warner et al., 1994). Other persons at risk are patients receiving hemodialysis (Nardin et al., 2005), diabetics (Rota et al., 2014), bedridden patients (Assmus et al., 2011), cyclists (ulnar neuropathy at the wrist, often referred to as cyclist's palsy or handlebar palsy) (Patterson et al., 2003), and possibly smokers (Frost et al., 2013).

Peroneal neuropathy

Peroneal neuropathy is often considered as the most common mononeuropathy of the lower extremities (Aprile et al., 2005), but there is not much evidence to support this statement. In fact, we identified only one study reporting the prevalence of peroneal neuropathy. In this door-to-door study, prevalence was 19 per 100 000 people (Kandil et al., 2012). No information about incidence of this disorder has been published. One study investigated the relative risk of cancer on peroneal neuropathy and, although not reported, from their data it is possible to calculate incidence rates. The authors reported 372 newly diagnosed patients with peroneal neuropathy during a 5-year study period. With a catchment population of 433 142, this would yield a crude incidence rate of 17.2 per 100 000 person-years (Koehler et al., 1997). Risk factors for this neuropathy are mainly described in case series and include trauma, surgery, weight loss, habitual leg crossing (especially in the presence of weight loss), other prolonged postures (for example, squatting), prolonged hospitalization, underlying neuropathies, and diabetes (Sotaniemi, 1984; Katirji and Wilbourn, 1988; Cruz-Martinez et al., 2000; Aprile et al., 2005; Stewart, 2008). Peroneal neuropathy seems to be more common in males than in females (Katirji and Wilbourn, 1988; Aprile et al., 2005).

Meralgia paraesthetica

In a door-to-door study from India, prevalence of meralgia was 71 per 100 000 people (Bharucha et al., 1991). Three studies reported the incidence of meralgia paraesthetica (van Slobbe et al., 2004; Latinovic et al., 2006; Parisi et al., 2011). In the first study, performed in general practices in the Netherlands, an incidence rate of 43 per 100 000 person-years was reported, with a slightly higher incidence in females (50 per 100 000 person-years) than in males (35 per 100 000 person-years) and a peak in incidence between the age of 41 and 60 years (van Slobbe et al., 2004). This medical records-based study also found that people suffering from other mononeuropathies, such as CTS, and pregnant women were at higher risk for the development of meralgia paraesthetica, while no association was found for documented diabetes mellitus, osteoarthritis, and overweight. In a similar study, performed in primary care in the UK, an incidence rate of 11 per 100 000 was found for males and of 13 per 100 000 for females. A similar peak in incidence was found for people between 55 and 64 years of age (Latinovic et al., 2006). Risk factors were not investigated in this study. The third study was performed in the USA (Parisi et al., 2011). In Olmsted County medical records of all healthcare are available for epidemiologic research. In a 10-year study period, records of nearly the whole population of approximately 106 000 inhabitants were investigated. The overall age- and sex-adjusted incidence was 33 per 100 000. Incidence did not differ significantly between males and females. In this study, the highest incidence was also observed for people between 55 and 64 years. Meralgia paraesthetica was associated with a higher BMI and presence of diabetes. The association with BMI was also found in another study (Mondelli et al., 2007).

Overall, meralgia paraesthetica is a quite common mononeuropathy of the lower extremities. There is no consensus whether incidence among genders differs, but BMI and diabetes seem to be associated with this mononeuropathy. Other possible risk factors, based on early case series or case reports, include tight trousers, carrying heavy wallets, seat belt use, pregnancy, and surgery (Chhuttani et al., 1966; Boyce, 1984; Nahabedian and Dellon, 1995).

Cranial nerves: facial nerve neuropathy

Facial nerve neuropathy results in rapid onset of usually unilateral paresis or paralysis of the musculature of the face, which can be accompanied by postauricular pain, altered facial sensation, taste disorders, and tear flow (caused by reduced blinking of the eyes). About two-thirds of all facial mononeuropathies are of idiopathic origin, known as Bell's palsy. The remainder are caused by varicella-zoster infection (Ramsay Hunt syndrome), Lyme disease, trauma, tumors, vascular lesions, congenital factors, Guillain–Barré syndrome (GBS: usually bilateral), diabetes, and other systemic disorders or less frequent causes (Peitersen, 2002; Eviston et al., 2015). Bell's palsy was found in 28 and 143 per 100 000 people in the previously mentioned door-to-door studies in Egypt and India (Bharucha et al., 1991; Kandil et al., 2012). Incidence of Bell's palsy varies between 11.5 and 40.2 per 100 000 person-years (Eviston et al., 2015). Some studies suggested a seasonal variation in incidence, but a large study of 2570 patients with facial neuropathy (1701 patients with Bell's palsy) revealed no such trend (Peitersen, 2002). There is no difference in incidence between males and females. There is a peak in incidence between the age of 15 and 45 years.

Without treatment, the majority of patients experience initial signs of recovery within 3 weeks. Patients with such early remission are more likely to achieve normal facial nerve function than patients in whom recovery only starts after a few months. Chances of complete recovery decrease significantly with higher age and are lower when accompanying symptoms are present.

In case of an incomplete Bell's palsy, 94% of patients will fully recover, compared to 61% in case of a complete Bell's palsy. Patients who do not fully recover can have a residual paresis, contractures, or synkinesis. Patients with a facial palsy due to herpes zoster infection or diabetes mellitus have a worse prognosis than patients with a Bell's palsy (Peitersen, 2002).

Cranial nerves: ocular nerve neuropathy

Palsies of the third, fourth, and sixth cranial nerve (the oculomotor, trochlear, and abducens nerve) are reported in 3.5, 6.3, and 10.4 per 100 000 person-years respectively. Incidence for these cranial nerve palsies is higher for males than for females and increases with age, reaching a peak in the seventh decade of life (Martinez-Thompson et al., 2014). The higher incidence in older people can probably be explained by a higher prevalence of cardiovascular risk factors, mainly hypertension and diabetes, which have been associated with these cranial nerve palsies (Patel et al., 2005; Martinez-Thompson et al., 2014). The higher incidence in males may be due to a higher incidence of head trauma in males, which also is an important cause of these cranial nerve palsies (Martinez-Thompson et al., 2014). Other causes include compression by intracranial neoplasms or aneurysms, infections (meningitis, encephalitis, syphilis) or by lesions located in the central nervous system, thereby interfering with the cell body of the specific cranial nerve, in case of stroke or multiple sclerosis. The ocular nerves, as well as other cranial nerves, can also be paralyzed by (variants of) GBS, which we will discuss later. In about a quarter to a third of patients with a palsy of the ocular nerves no cause can be established (Richards et al., 1992; Tiffin et al., 1996).

Treatment and randomized controlled trials

Carpal tunnel syndrome

Treatment of CTS depends on the severity, cause, and duration of symptoms. When symptoms are triggered by certain (work-related) factors or diseases (such as thyroid dysfunction and rheumatoid arthritis), avoiding these triggers (changing work environment) or treating the underlying disease can be helpful. When this approach fails, or when there is no specific trigger, other treatment options include a conservative approach (local injection with corticosteroids or wrist splinting during the night) or, when symptoms are severe or when other options have failed, a surgical approach, where the pressure on the median nerve is released by cutting the transverse carpal ligament (Middleton and Anakwe, 2014). Although conservative methods provide relief of symptoms (Marshall et al., 2007; Page et al., 2012; Atroshi et al., 2013), several randomized controlled trials (RCTs) showed that surgical treatment results in better outcomes than treatment with wrist splinting (Gerritsen et al., 2002), steroid injections (Hui et al., 2005), or other nonsurgical treatments (Jarvik et al., 2009). Especially in very severe cases or in patients with long-standing complaints, surgery probably is the best option.

Ulnar neuropathy

Treatment for ulnar neuropathy at the elbow can be conservative (ergonomic advice, physical therapy, splints) or surgical. There is insufficient evidence to determine which treatment is superior (Caliandro et al., 2012).

Peroneal neuropathy

In general, prognosis of this neuropathy is good, especially when there is no underlying disease such as diabetes. Spontaneous improvement usually occurs within 6 months of onset of symptoms. Rehabilitation may help recovery (Aprile et al., 2009; Bsteh et al., 2013).

Meralgia paraesthetica

Several treatments have been suggested for meralgia paraesthetica, although the disease often improves without any intervention. Suggested therapies include steroid injections and surgery, but no trials have been performed to investigate their efficacy (Khalil et al., 2012).

Cranial nerves: facial nerve neuropathy

Several treatments have been proposed for Bell's palsy, including thermal heating or cooling, electrotherapy, facial exercises, massages, and surgery, but there is no evidence that shows a beneficial effect of these treatments (Peitersen, 2002). Until 2007, trials investigating the effect of corticosteroids have shown disappointing results, mainly due to suboptimal quality and small patient groups included in these trials. Hereafter, conclusive evidence was published that showed a beneficial effect of early treatment with steroids in contrast to placebo or antiviral agents (Sullivan et al., 2007; Engstrom et al., 2008). Adding antiviral agents to steroids is not beneficial and is therefore not recommended. An important part of the treatment of facial nerve palsy constitutes protection of the eye against prolonged drying and irritation that can occur due to dysfunction of the orbicularis oculi muscle.

MULTIFOCAL NEUROPATHIES

Definition and clinical presentation

In this part we describe multifocal neuropathies, diseases that affect multiple nerves, although not in a symmetric distribution as in most polyneuropathies. There are different types of multifocal neuropathies. In the following paragraphs we will discuss mononeuritis multiplex, or multiple mononeuropathy, and multifocal motor neuropathy (MMN).

Mononeuritis multiplex or multiple mononeuropathy

Mononeuritis multiplex refers to a group of disorders that generally result in a painful, asymmetric, sensorimotor neuropathy, where at least two nerves in separate body parts are affected. With progression of the condition more nerves may become affected, which in some cases may finally lead to a more or less symmetric distribution of symptoms, resembling polyneuropathy (England and Asbury, 2004). Mononeuritis multiplex often is the clinical phenotype of an underlying vasculitis (Collins, 2012).

Multifocal motor neuropathy

MMN is an immune-mediated demyelinating disorder of the peripheral nervous system that affects motor nerves. MMN leads to slowly progressive, asymmetric, predominantly distal muscle weakness and atrophy that commonly starts in the arms. Weakness can also occur distally in the legs, but during the course of the disease weakness of the arms will be most prominent (Van Asseldonk et al., 2005; Vlam et al., 2012). Since fasciculations can also occur, the clinical picture of MMN may initially resemble that of motor neuron disease, which is a much more severe and often fatal disorder. Differentiation is extremely important, because MMN is a treatable disorder with a much better prognosis. The course and distribution of symptoms and signs, together with nerve conduction studies and the presence of anti-GM1 antibodies, can help to distinguish MMN from other disorders (Van Asseldonk et al., 2005; Vlam et al., 2012).

Frequency and risk factors

Mononeuritis multiplex or multiple mononeuropathy

Data on prevalence or incidence of mononeuritis multiplex in general are not available, probably due to the heterogeneity of this entity and its causes. Mononeuritis multiplex is often associated with primary systemic vasculitis (polyarteritis nodosa, Churg–Strauss disease, Wegener granulomatosis), and can also be a manifestation of connective tissue diseases such as systemic lupus erythematosus, Sjögren syndrome, and rheumatoid arthritis (secondary systemic vasculitis). Systemic vasculitides have an annual incidence of 14 per 100 000 adults (Collins, 2012). In primary vasculitis, neuropathy is very common, occurring in up to 75% of patients, and mononeuritis multiplex is the classic initial presentation (Gorson, 2007). In secondary systemic vasculitis neuropathy is less common and the neuropathy more often follows a distal symmetric distribution, but there is large variation in the clinical presentation, especially in Sjögren syndrome (Gorson, 2007; Grantz, 2010a). Mononeuritis multiplex is also described in association with diabetes (discussed in more detail in the section on chronic neuropathies), sarcoidosis, and certain infectious diseases, such as HIV, hepatitis, and Lyme disease (Kelkar and Parry, 2003; England and Asbury, 2004; Gorson, 2007; Grantz, 2010a; Grantz and Huan, 2010; Singhal et al., 2015). Patients may also have a nonsystemic vasculitis with only peripheral nerve involvement.

Multifocal motor neuropathy

Epidemiologic characteristics of MMN have not been studied in detail. Only a few studies reported on the prevalence of MMN. In the Netherlands prevalence was reported to be 0.6 per 100 000 persons (Cats et al., 2010) and in the UK 0.5 per 100 000 (Mahdi-Rogers and Hughes, 2014), but these number may be underestimated. Incidence of MMN is unknown. MMN is more common in males than in females, with a ratio of 2.7:1. The mean age of onset is around 40 years, and the majority are between 20 and 50 years of age at onset of the disease (Nobile-Orazio, 2001; Cats et al., 2010).

Treatment and randomized controlled trials

Mononeuritis multiplex or multiple mononeuropathy

The course of disease in vasculitis can be very rapidly progressive and requires rapid proper diagnostic procedures and adequate treatment. Treatment depends on the cause of the neuropathy, but usually includes corticosteroids and immunosuppressive agents (Gorson, 2007).

Multifocal motor neuropathy

Patients with MMN do not benefit from steroids or plasma exchange and may even worsen after receiving these treatments (Van Asseldonk et al., 2005; Joint Task Force of the European Federation of Neurological Sciences/Peripheral Nerve Society, 2010; Vlam et al., 2012; Umapathi et al., 2015). Nonrandomized trials and RCTs showed an improvement of muscle strength in about 80% of patients treated with intravenous immunoglobulins (IVIg) compared to only 4% of patients receiving placebo, without serious side-effects (van Schaik et al., 2005). Other immunosuppressive or immunomodulatory agents have not been studied in RCTs, and there are insufficient data to conclude anything about the effect of these treatments. Some uncontrolled studies showed a beneficial effect of cyclophosphamide, but this treatment has several (severe) adverse effects, especially in patients requiring long-term treatment (Van Asseldonk et al., 2005; Joint Task Force of the European Federation of Neurological Societies/Peripheral Nerve Society, 2010; Umapathi et al., 2015). Less toxic drugs that can be considered for clinical trials include azathioprine, cyclosporine, methotrexate, and rituximab (Vlam et al., 2012).

ACUTE POLYNEUROPATHIES

Definition and clinical presentation

Polyneuropathies can be divided according to the clinical course (acute versus chronic) or according to the type of neuropathy (axonal versus demyelinating neuropathies: Fig. 15.2). In this chapter, polyneuropathies are divided according to the clinical course. In the following section of this chapter we discuss the most important acute polyneuropathies.

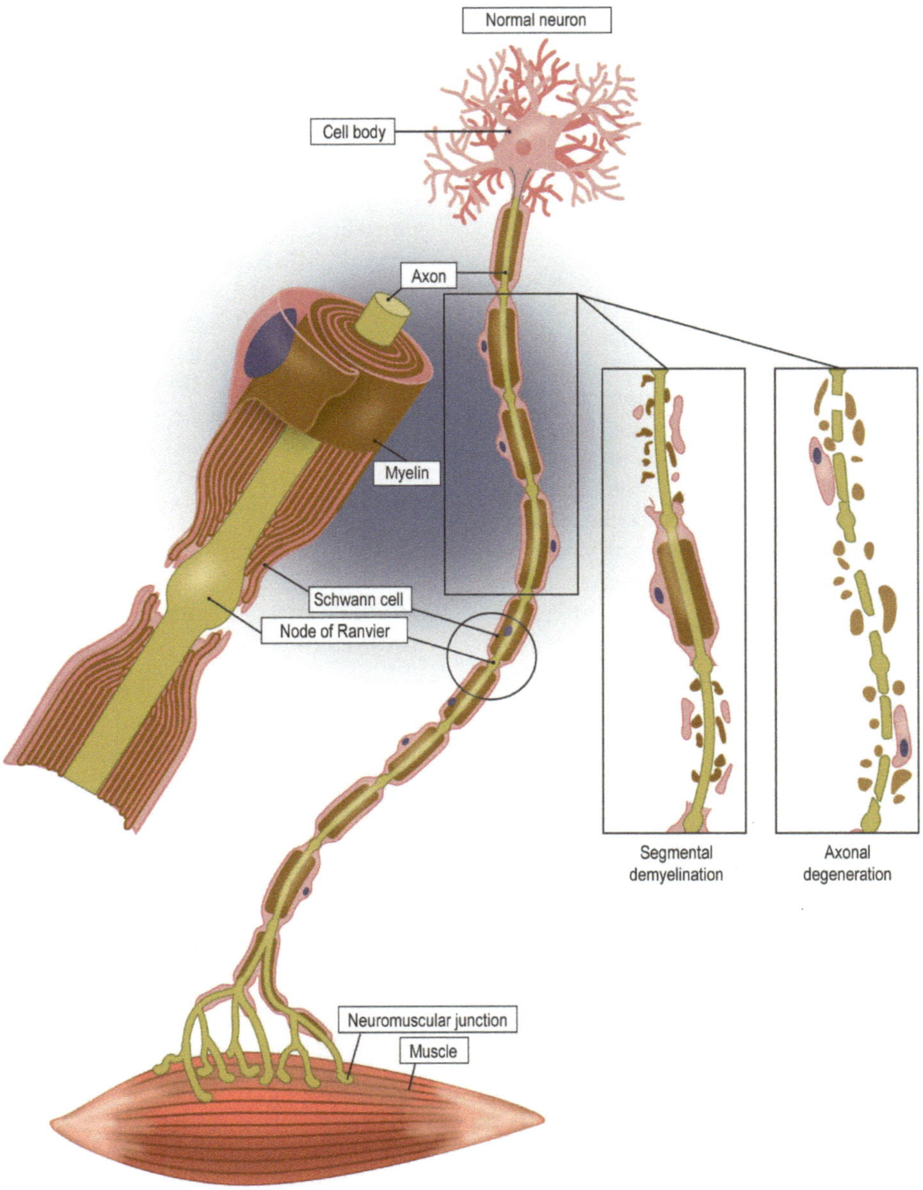

Fig. 15.2. Axonal and demyelinating polyneuropathy. Schematic illustration of a normal (motor) neuron. The cell body of the neuron is located in the spinal cord and its fiber travels through a peripheral nerve to a muscle. The fiber consists of an axon, which is surrounded by a layer of myelin, the myelin sheath. Damage to these neurons (neuropathy) can occur in two forms: demyelination (destruction of the myelin sheath) and axonal degeneration (destruction of the axon). (Modified from van Doorn, 2007.)

GBS is a potentially life-threatening disorder, and probably the most frequent and best example of an acute polyneuropathy. Acute polyneuropathies that may mimic GBS in its clinical course are polyneuropathies associated with severe vitamin B_1 deficiency (beriberi), toxic agents (especially heavy metals), paraneoplastic syndromes, vasculitis, Lyme disease, and porphyria (Levin, 2004; van den Berg et al., 2014). Chronic inflammatory demyelinating polyradiculoneuropathy (CIDP) can also have an acute onset (A-CIDP). Features of CIDP will be discussed in the section on chronic neuropathies. Mononeuritis multiplex, the main presenting phenotype of vasculitis, has been discussed earlier in this chapter. The following paragraphs outline the characteristics of GBS, and neuropathies that are associated with paraneoplastic syndromes, porphyria, lead intoxication, and severe vitamin B_1 deficiency.

Guillain–Barré syndrome

GBS is a heterogeneous immune-mediated polyradiculoneuropathy that consists of several variants. Typically, GBS presents as an acute monophasic flaccid paralysis, provoked by a preceding infection, characterized by rapidly progressive fairly symmetric weakness of the extremities and areflexia, that reaches its maximum within 4 weeks (Fig. 15.3). Sensory symptoms, ataxia, pain, and cranial nerve deficits often also occur. Respiratory insufficiency requiring mechanic ventilation occurs in about 20–25% of GBS patients. Four to 6 weeks after onset of symptoms, GBS patients enter a plateau phase that can last from several days to up to 6 months or even more, before initial recovery occurs (van den Berg et al., 2014).

Paraneoplastic neuropathy

Paraneoplastic disorders refer to disorders that are immune-mediated, nonmetastatic complications of cancer (Darnell and Posner, 2003; Hoftberger et al., 2015). An antineuronal immune response, triggered by the expression of neuronal proteins by a tumor, results in the symptoms and signs of paraneoplastic neurologic disorders that may be the initial presentation of the tumor. Cerebellar degeneration and limbic encephalitis are the most common paraneoplastic disorders of the central nervous system. Sensory neuronopathy and Lambert–Eaton myasthenic syndrome are among the most common presentations in the peripheral nervous system (Giometto et al., 2010; Gozzard et al., 2015). Sensory neuronopathy classically presents with a subacute, asymmetric onset of numbness, pain, ataxia, and loss of proprioception in affected areas (especially the arms), that rapidly progresses in weeks. Motor weakness can also be present and may be a manifestation of lower motor neuron degeneration (Graus et al., 2004; Koike et al., 2011; Graus and Dalmau, 2013; Gozzard et al., 2015).

Other acute polyneuropathies

Vitamin B_1 deficiency in combination with chronic alcohol use generally leads to a chronic, sensory-predominant polyneuropathy. However, an acute form

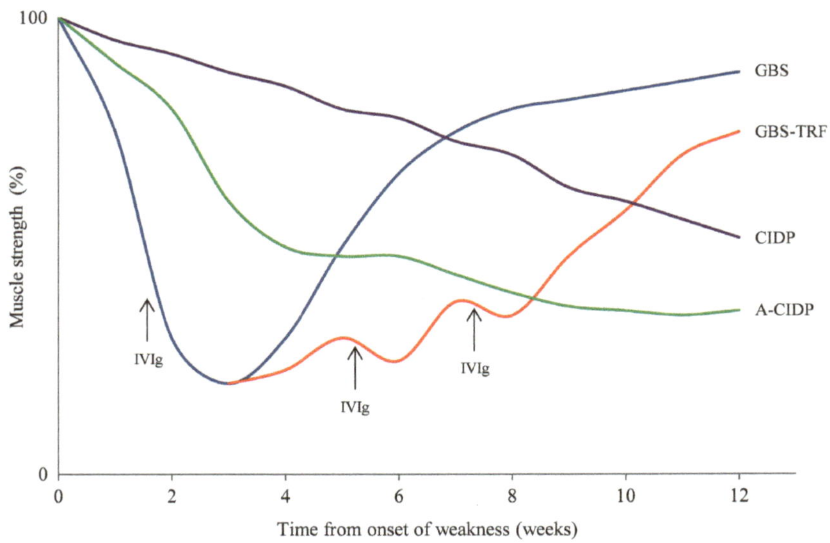

Fig. 15.3. Clinical course of weakness of important (motor-predominant) polyneuropathies. GBS, Guillain–Barré syndrome; GBS-TRF, Guillain–Barré syndrome, treatment-related fluctuation; CIDP, chronic inflammatory demyelinating polyradiculoneuropathy; A-CIDP, acute-onset chronic inflammatory demyelinating polyradiculoneuropathy; IVIg, intravenous immunoglobulin. (Modified from van Doorn et al., 2008.)

with severe weakness, also known as dry beriberi, has also been described (Koike et al., 2003). Recognition is important to prevent associated heart failure (wet beriberi), Wernicke's encephalopathy, and related long-term effects. Administration of intravenous thiamine usually leads to improvement, but residual deficits are common.

Porphyria is another disorder that can mimic GBS. Porphyria consists of a group of metabolic disorders characterized by abnormal functioning of heme biosynthesis enzymes, leading to accumulation of heme precursors (Puy et al., 2010). Porphyria can present with life-threatening attacks, which are characterized by severe abdominal pain, nausea, and vomiting. Seizures and psychiatric symptoms such as anxiety, confusion, and paranoia are common. Severe motor axonal neuropathy resembling GBS can also occur, especially when porphyrinogenic drugs (drugs that induce cytochrome P450 enzymes, e.g., rifampicin, estrogen, antiepileptic drugs) are used during an attack. After an attack, motor weakness will gradually recover, but is often incomplete and may result in a chronic neuropathy. Treatment with hemin will not reverse neuropathy, but may halt its progression (Puy et al., 2010; Simon and Herkes, 2011).

Lead intoxication may present with a neuropathy that mimics the neuropathy seen in porphyria, but classically presents with early and severe weakness of wrist and finger extensors (Thomson and Parry, 2006). History usually reveals high-intensity exposure to lead and the diagnosis can be confirmed by documenting elevated lead levels in serum (Simon and Herkes, 2011). Chronic exposure to lead, for example, by inhalation of lead-polluted air (e.g., during industrial work (battery manufacturing, smelting), or due to use of lead-based fuel), can also result in a chronic, sensory polyneuropathy. This chronic neuropathy shares less pathogenic mechanisms with porphyria than the (sub)acute form, but is more likely a result of a direct neurotoxic effect of lead (Rubens et al., 2001).

Frequency and risk factors

Guillain–Barré syndrome

GBS is a rare disease. Incidence of GBS is reported to be 1–2 per 100 000 person-years and affects males more than females (ratio 1.5:1). Incidence is lower in children and increases with age (Kuwabara and Yuki, 2013; van den Berg et al., 2014). Acute inflammatory demyelinating polyneuropathy (AIDP), a sensorimotor polyneuropathy often accompanied with cranial nerve deficits and autonomic dysfunction, and acute motor axonal neuropathy (AMAN), a pure motor neuropathy where cranial nerves are rarely affected, are the most common forms of GBS. Less common is acute motor and sensory axonal neuropathy (AMSAN), which is similar to AMAN but with sensory deficits. Other variants of GBS include the Miller Fisher syndrome, which is characterized by ophthalmoplegia, ataxia, and areflexia, the pharyngeal-cervical-brachial variant, where weakness is most prominent in oropharyngeal, facial, and neck muscles, and the paraparetic variant of GBS. In European countries and North America, AIDP is the most common subtype, while the axonal form AMAN is more common in Asia and Central and South America. These differences may be explained by different exposure to certain pathogens, but also to differences in genetic polymorphisms between individuals (Kuwabara and Yuki, 2013; van den Berg et al., 2014).

The immune system plays a key role in the pathogenesis of GBS. Two-thirds of all GBS patients report having experienced symptoms attributable to a respiratory or gastrointestinal tract infection before the onset of weakness. *Campylobacter jejuni* is the most commonly identified pathogen as cause for this antecedent infection, especially in the AMAN and Miller Fisher subtypes. Other associated pathogens include cytomegalovirus, Epstein–Barr virus, hepatitis E virus, influenza A virus, *Mycoplasma pneumoniae*, and *Haemophilus influenzae* (van den Berg et al., 2014). After an infection, the immune system produces antibodies against these pathogens that may cross-react with specific gangliosides at nerve membranes. This molecular mimicry, in combination with complement activation, ultimately leads to nerve dysfunction and symptoms of GBS. Several of these antiganglioside antibodies have been identified in association with specific subtypes of GBS (anti-GM1a, anti-GM1b, anti-GD1a, and anti-GalNAc-GD1a in AMAN, and especially anti-GQ1b in Miller Fisher syndrome) (van den Berg et al., 2014).

To further study and describe important epidemiologic features of GBS and to identify clinical predictors and biomarkers of the disease course and outcome in individual patients, the International GBS Outcome Study (IGOS) was launched. IGOS is a large, observational, collaborative study in more than 15 countries worldwide. Results of this study can be expected in the coming years.

Prognosis of GBS is quite variable and depends on the severity of the disease, preceding infection, comorbidity, and the age of the patient. There are two prognostic models that can be used in clinical practice: modified Erasmus GBS Outcome Score (mEGOS), administered 1 week after admission to predict outcome after 6 months (Walgaard et al., 2011), and Erasmus GBS Respiratory Insufficiency Score (EGRIS), administered at hospital admission to predict the chance that a patient will need artificial ventilation within 2 weeks (Walgaard et al., 2010). GBS Mortality ranges from 3% to 7% and, although current treatment options accelerate recovery and improve outcome at 1 year, they have not improved mortality (van den Berg et al., 2014).

Paraneoplastic neuropathy

Sensory neuronopathy is one of the most commonly encountered paraneoplastic syndromes, accounting for 24.3% of the 979 registered patients in the Paraneoplastic Neurologic Syndrome Euronetwork Database (Giometto et al., 2010). Autonomic neuropathies and sensorimotor poly(radiculo)neuropathies are less common. The prevalence and incidence of these disorders in the general population are unknown. Small-cell lung carcinoma is the most common tumor associated with paraneoplastic disorders and accounts for more than 80% of tumor types associated with sensory neuronopathies (Giometto et al., 2010; Graus and Dalmau, 2013; Gozzard et al., 2015). Other tumors that are associated with paraneoplastic disorders are breast cancer, ovary cancer, thymoma, and hematologic malignancies.

Paraneoplastic antibodies to Hu, SOX2, CRMP5, amphiphysin, voltage-gated calcium channel, and *N*-methyl-D-aspartate receptor have been associated with neuronopathies and neuropathies. Sensory neuronopathy is mainly associated with the occurrence of Hu antibodies and axonal and demyelinating polyneuropathy can be a manifestation of a paraneoplastic syndrome associated with CRMP5 antibodies (Hoftberger et al., 2015). Polyneuropathy that resembles GBS and vasculitic neuropathy may also occur, especially in association with lymphomas (Giometto et al., 2010; Koike et al., 2011; Graus and Dalmau, 2013).

Other acute polyneuropathies

Not much is known about the prevalence and incidence of GBS mimics.

Dry beriberi is rare in Western populations; exact numbers are not available. Chronic alcoholics and obese people undergoing gastric surgery are at an increased risk of developing this nutritional polyneuropathy (Juhasz-Pocsine et al., 2007; Koike et al., 2008; Kazemi et al., 2010).

As discussed, porphyria can also present with a GBS-like phenotype. The most common form of porphyria that is associated with neurologic symptoms is acute intermittent porphyria, which affects approximately 1 in 75 000 people. Acute porphyria attacks have an annual incidence of 1–2 per 100 000 persons, occur more often in females than in males, and are most frequent in the third decade. Presentation is mostly nonspecific, and neuropathy occurs in 10–40% of episodes (Puy et al., 2010; Simon and Herkes, 2011). The prevalence and incidence of acute lead-induced neuropathy are unknown.

Treatment and randomized controlled trials

Guillain–Barré syndrome

Regardless of the subtype, most patients are treated with IVIg or plasma exchange together with multidisciplinary supportive medical care. Whether specific subtypes of GBS require specific treatment remains to be investigated. Most RCTs have been performed in populations where AIDP is the predominant subtype. These trials showed that both plasma exchange and IVIg are (equally) effective treatments (Raphael et al., 2012; Hughes et al., 2014; van den Berg et al., 2014). Treatment with corticosteroids is not proven effective as standalone treatment, nor as additional treatment (Hughes and van Doorn, 2012). Very few other treatments have been formally tested in GBS and those that have been studied did not show a beneficial effect (Hughes et al., 2013). There are limited ongoing controlled trials from which results are expected in the coming years. One randomized placebo-controlled trial is investigating the effect of second immunoglobulin course in patients with a predicted poor prognosis (SID-GBS trial), and one trial is investigating the effect of eculizumab, a humanized monoclonal antibody that inhibits the complement cascade (ICA-GBS trial). The international second-dose IVIg study (I-SID GBS) is part of the observational IGOS. The IGOS can serve as an infrastructure for the development and conduct of future GBS treatment studies.

Paraneoplastic neuropathy

The most important aspect of the treatment of paraneoplastic neuronopathy is treating the underlying tumor. Other therapies that have been suggested to treat paraneoplastic neuronopathies include corticosteroids, IVIg, plasma exchange, and more aggressive immunotherapies such as rituximab, cyclophosphamide, tacrolimus, and cyclosporine, but the effect of these treatments is unclear and RCTs are lacking (Giometto et al., 2012; Graus and Dalmau, 2013).

CHRONIC POLYNEUROPATHIES

Definition and clinical presentation

Chronic polyneuropathy generally is a more generalized, often length-dependent disease of peripheral nerves. This indicates that most symptoms are present in arms and legs, and distally more severe than proximally. Onset is gradual and progression is slow. Chronic axonal polyneuropathies often lead to more or less symmetric complaints of numbness, paresthesia, and pain in the lower legs or feet, sometimes accompanied with distal muscle weakness. Chronic demyelinating polyneuropathies are often more proximal and generally lead to more muscle weakness (Fig. 15.1). Polyneuropathy has a great variety of causes. Discussing all these causes goes beyond the scope of this chapter. In this last part of the chapter we discuss the clinical phenotype and epidemiology of what are probably the most frequent chronic polyneuropathies in the Western world.

Diabetic neuropathy

The best-known risk factor for chronic axonal polyneuropathy is diabetes. Patients with diabetes have an increased risk of developing mononeuropathies, radiculopathies, plexopathies, and length-dependent polyneuropathy, which is the most common form (Callaghan et al., 2012a). Polyneuropathy associated with diabetes often is a painful, sensorimotor neuropathy with a slowly progressive course, although rapidly progressive painful neuropathies can also occur, especially just after the start of insulin treatment.

Chronic inflammatory demyelinating polyradiculoneuropathy

CIDP is an example of a chronic demyelinating polyneuropathy. CIDP is an immune-mediated polyneuropathy characterized by monophasic, relapsing-remitting or slowly progressive sensory and motor symptoms that develop over a period of more than 8 weeks (Vallat et al., 2010; Van den Bergh et al., 2010). Several variants of CIDP have been described. These include pure motor, pure sensory, or sensory-predominant variants. Distal acquired demyelinating symmetric neuropathy (DADS) can sometimes be considered as a variant of CIDP. DADS often runs a more slowly progressive course. When DADS is associated with an IgM paraprotein in combination with anti-MAG antibodies, the diagnosis is not compatible with CIDP (Van den Bergh et al., 2010). Multifocal acquired demyelinating sensory and motor neuropathy (MADSAM), also known as the Lewis–Sumner syndrome, is a multifocal variety of CIDP (Mathey et al., 2015). Distinguishing CIDP from GBS can be difficult, especially when a patient with GBS has multiple treatment-related fluctuations (secondary deterioration after initial improvement following treatment). About 16% of patients have an acute onset of CIDP (A-CIDP) (McCombe et al., 1987). Patients with an A-CIDP have a nadir within 8 weeks, which is followed by a chronic relapsing-remitting or progressive course (Fig. 15.3). A-CIDP is more likely than GBS with treatment-related fluctuations when patients initially diagnosed as GBS have more than two deteriorations or when a deterioration occurs after 8 weeks of onset of first symptoms (Ruts et al., 2010).

Paraproteinemic neuropathy

Paraproteinemic neuropathies refer to neuropathies that are associated with the presence of a clone of plasma cells that produce monoclonal proteins, also called M-proteins. The proliferation of plasma cells can range from benign, low-grade monoclonal gammopathy of undetermined significance (MGUS) to malignant disorders, such as multiple myeloma, Waldenström macroglubulinemia, amyloidosis, and POEMS (polyneuropathy, organomegaly, endocrinopathy, M-protein, and skin changes) syndrome. The clinical phenotype of neuropathy can be diverse and depends on the specific protein. Patients with an IgM MGUS-associated neuropathy, especially when having anti-MAG antibodies, present with slowly progressive distal, predominantly sensory polyneuropathy (DADS; see section on CIDP, above), often accompanied with tremor and sensory ataxia. IgM M-proteins can also result in neuropathy with ataxia, ophthalmoplegia, dysphagia, and dysarthria, known as CANOMAD (chronic ataxic neuropathy with ophthalmoplegia, M-protein, cold agglutinins and disialosyl antibodies). Neuropathy in POEMS syndrome is a subacute, rapidly progressive and very painful, sensorimotor demyelinating and axonal polyneuropathy, whereas neuropathy associated with multiple myeloma typically presents with a mild, sensory-dominant, distal, axonal polyneuropathy (Mathey et al., 2015; Raheja et al., 2015).

Hereditary polyneuropathies

Hereditary polyneuropathies, generally known as Charcot–Marie–Tooth (CMT) disease, present with a variety of subtypes with differences in severity, clinical course, electrodiagnostic features, and gene mutations. Over 80 causative genes have been identified (Fridman et al., 2015). Mutations occur in genes encoding for proteins involved in myelin architecture, axonal transport, cytoskeletal formation, endosomal vesicular transport, and microtubule interaction (Klein et al., 2013). Most hereditary neuropathies present with length-dependent weakness, atrophy, and sensory deficit, without marked positive symptoms like paresthesia and burning pain (hereditary motor and sensory neuropathies). High arches, hammer toes, and inverted-champagne-bottle appearance of the legs are common features of inherited neuropathies. Subtypes with mainly sensory and autonomic involvement (hereditary sensory and autonomic neuropathy), only sensory involvement (hereditary sensory neuropathy), and only motor involvement (distal hereditary motor neuropathy) have also been described (Klein et al., 2013). Algorithms, mainly based on clinical phenotype and electrophysiology, have been published to guide genetic testing for the diagnosis of the CMT subtype (Saporta et al., 2011).

Toxic and nutritional polyneuropathies

Chemotherapy often induces a polyneuropathy. Cancer is an increasing public health concern, not only because of the worldwide increasing incidence (Bray et al., 2012), but also due to an increased survival of cancer patients and associated long-term consequences of treatment (Quaresma et al., 2015). Chemotherapy-induced neuropathy is a predominantly sensory polyneuropathy which may be accompanied by pain, autonomic disturbances, and motor deficit. This polyneuropathy develops in almost 70% of patients treated with chemotherapy during or within 1 month after the end of the therapy. After 6 months 30% of treated patients still suffer from this complication (Seretny et al., 2014). Several single-nucleotide polymorphisms have been associated with an increased risk of developing neuropathy after chemotherapy. Other risk factors may be pre-existing neuropathy, smoking, and a decreased creatinine clearance (Seretny et al., 2014). Aside from chemotherapy, several other drugs have neurotoxic adverse effects (e.g., amiodarone, nitrous oxide, isoniazid, metronidazole, and antiretroviral drugs) (Manji, 2011).

Nutritional deficiencies are often associated with alcohol overuse and bariatric surgery. Thiamine deficiency in patients without alcoholism may lead to the clinical picture of beriberi with a motor-predominant neuropathy, while alcoholism without thiamine deficiency is mainly sensory, affecting predominantly small fibers. Polyneuropathy is described in 25–66% of chronic alcoholics. Mixed small- and large-fiber polyneuropathies are most frequently found (Mellion et al., 2011). Vitamin B_{12} deficiency can lead to subacute combined degeneration of the spinal cord (posterior columns and the corticospinal tracts), which results in sensory ataxia, and loss of proprioception and vibration sense. In addition, an axonal sensorimotor polyneuropathy may occur. Vitamin B_{12} deficiency is associated with bariatric surgery, but can also be a manifestation of pernicious anemia, gastritis, or Crohn disease.

Frequency and risk factors

There are few high-quality data available about the prevalence and incidence of chronic polyneuropathy. A recent review article summarized studies investigating the epidemiology of chronic polyneuropathy (Hanewinckel et al., 2016). Most of the 29 identified studies included in this review had methodologic limitations, mainly caused by the absence of a gold-standard test for the diagnosis of polyneuropathy. Consequently, polyneuropathy definitions and assessment protocols across studies varied substantially. Some studies defined polyneuropathy as the presence of only one symptom or sign, while others used a combination of symptoms and signs, sometimes combined into a composite score. Very few epidemiologic studies have incorporated nerve conduction studies into the study protocol. The combination of symptoms, signs, and nerve conduction studies provides the most accurate diagnosis of polyneuropathy (England et al., 2005).

Currently, the most reliable estimate of the prevalence of chronic polyneuropathy is based on the results of the Italian Longitudinal Study on Aging (ILSA), a population-based cohort study that focuses on chronic diseases in the elderly (age range 65–84 years). In this study, participants with a positive screening (a questionnaire and a brief neurologic examination) are subjected to a more thorough neurologic examination performed by a neurologist to confirm the diagnosis of polyneuropathy. Polyneuropathy was present in 334 of 4478 participants, yielding a prevalence of 7.5% (age-adjusted prevalence of 7.0%, standardized to the 1992 Italian population). Diagnosis was based on clinical judgment; nerve conduction studies were not routinely performed. A few years after the initial screening, 2845 were rescreened with the same procedure. This resulted in an adjusted incidence rate of 7.9 per 1000 person-years (Baldereschi et al., 2007). Prevalence of chronic polyneuropathy across all age categories is estimated at 1% and incidence at approximately 0.7 per 1000 person-years (Visser et al., 2015; Hanewinckel et al., 2016).

Over 100 potential causes and risk factors for chronic polyneuropathy have been identified (Grantz, 2010a, b; Grantz and Huan, 2010). Information about the frequency of these causes is mainly derived from hospital-based studies, which probably resulted in biased estimates due to selection and referral bias. These studies divided chronic neuropathies according to the most likely cause of the disease. Although this approach is not entirely correct, since multiple causes can be present in one individual and polyneuropathy probably results from a multifactorial process where the presence of merely one risk factor is not sufficient for the development of neuropathy (e.g., not all diabetics have neuropathy), these studies give insight into the most important associated risk factors (Fig. 15.4). Diabetes is the most important risk factor for polyneuropathy, accounting for approximately one-third of all chronic polyneuropathies. Toxic (alcohol and medication), hereditary, and immune-mediated (CIDP, paraneoplastic, and paraproteinemic neuropathy) factors are other important causes. In a quarter to a third of polyneuropathy patients no associated disease or risk factor can be found. These patients are often designated as having chronic idiopathic axonal polyneuropathy (CIAP). The incidence of CIAP is estimated at 31.6 per 100 000 person-years (Visser et al., 2015).

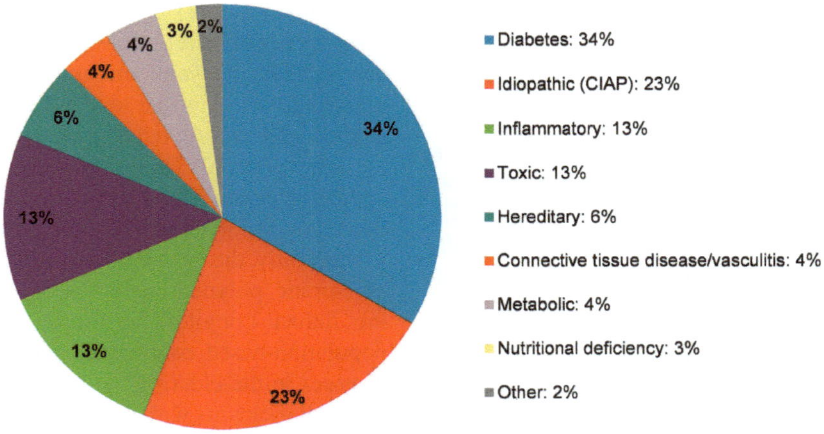

Fig. 15.4. Common associated diseases and risk factors for chronic polyneuropathy in hospital settings. Distribution of causes, based on nine hospital-based studies, total 2613 patients (George and Twomey, 1986; Lin et al., 1993; Johannsen et al., 2001; Mygland and Monstad, 2001; Rosenberg et al., 2001; Verghese et al., 2001; Vrancken et al., 2006; Rudolph and Farbu, 2007; Visser et al., 2015). Inflammatory neuropathies here include Guillain–Barré syndrome, chronic inflammatory demyelinating polyradiculoneuropathy, and neuropathies associated with paraproteins, paraneoplastic antibodies, and infections (human immunodeficiency virus, Lyme disease, leprosy). Toxic neuropathies include alcohol- and drug-related neuropathy. Metabolic neuropathies include neuropathy related to thyroid dysfunction, renal failure, and liver disease. CIAP, chronic idiopathic axonal polyneuropathy.

Subtypes of chronic polyneuropathy that have been studied in more detail are discussed in the following paragraphs.

Diabetic neuropathy

Prevalence of diabetes is around 6–7% worldwide and 14% in the USA, and these numbers are expected to rise the next decades (Nolan et al., 2011; Menke et al., 2015). Up to 50% of persons with diabetes will ultimately develop polyneuropathy during the course of the disease (Callaghan et al., 2012a). Prevalence of diabetic polyneuropathy in the ILSA (age 65–84 years) was 3%. Incidence of polyneuropathy in diabetics was 32.2 per 1000 person-years and 5.8 per 1000 person-years in nondiabetics, yielding a rate ratio of 5.6 (Baldereschi et al., 2007). Incidence of diabetic polyneuropathy across all ages in the general population is estimated at 23–54 per 100 000 person-years (MacDonald et al., 2000; Visser et al., 2015). Overall, diabetes probably accounts for approximately one-third to one-half of all polyneuropathy cases (Hanewinckel et al., 2016). Diabetics who also have other cardiovascular risk factors, such as dyslipidemia, hypertension, and obesity, are at an increased risk of developing polyneuropathy, especially in type 2 diabetes (Tesfaye et al., 2005).

Chronic inflammatory demyelinating polyradiculoneuropathy

Prevalence of CIDP varies between 1 and 7.7 per 100 000 people and the annual incidence of CIDP is estimated at 0.2–0.5 per 100 000 people. Males are more often affected than females (Lunn et al., 1999; McLeod et al., 1999; Mygland and Monstad, 2001; Chio et al., 2007; Iijima et al., 2008; Mahdi-Rogers and Hughes, 2014). These numbers are often considered as minimum prevalence, as they are thought to be underestimations of the true prevalence. However, a recent report from a single American center pointed to the fact that diagnosing CIDP can be a challenge. It showed a high degree of misclassification of the diagnosis: in a group of 59 patients diagnosed as CIDP, up to 47% failed to meet the European Federation of Neurological Societies/Peripheral Nerve Society criteria for CIDP, and likely had another diagnosis. Most common alternative diagnoses were diabetic neuropathy, motor neuron disease, and, very surprisingly, also fibromyalgia (Allen and Lewis, 2015). This may suggest a lower prevalence of CIDP than previously reported, but since evaluation was limited to only one referral centre, other studies are required (Gorson and Gooch, 2015). About half of all cases with CIDP present with the typical sensorimotor phenotype. Lewis–Sumner syndrome, acute-onset CIDP, DADS and motor predominant CIDP are less common, each accounting for 5–15% of cases (Mathey et al., 2015).

Paraproteinemic neuropathy

M-proteins are present in 1% of the general population, and this number increases to over 5% for persons older than 70 years of age. MGUS is the most common presentation, accounting for approximately two-thirds of all monoclonal proteins. The most common M-protein in the general population is IgG (60–70%), followed by IgM (18%) and IgA (11%) (Raheja et al., 2015).

Neuropathies are mainly associated with IgM M-proteins. Up to 50% of persons with IgM paraproteins have neuropathy, while this is only in 5% of persons with IgG and in 15% of persons with IgA. IgM-associated neuropathies mostly present with DADS, or with a phenotype similar to CIDP. Around 2% of patients with an IgM paraproteinemic neuropathy present as CANOMAD (Mathey et al., 2015; Raheja et al., 2015). IgG and IgA MGUS-related neuropathies are less common, and it is uncertain whether the presence of these proteins is related to the neuropathy, or is merely coincidental. POEMS syndrome is even more uncommon. POEMS syndrome is usually associated with IgA or IgG plasma cell neoplasms. Multiple myeloma is mainly associated with IgG paraproteins (Raheja et al., 2015). Exact prevalence numbers of M-protein-related neuropathies in the general population are not available.

Hereditary polyneuropathies

CMT disease has a prevalence of 40 per 100 000 individuals (Fridman et al., 2015). The most common clinical form is CMT1, which is an autosomal-dominantly inherited, demyelinating polyneuropathy, most often caused by a duplication in the *PMP22* gene (CMT1A). Other common mutations are mutations in *GJB1* (CMT1X, X-linked), *MPZ* (CMT1B), *MFN2* (CMT2A, dominantly inherited axonal neuropathy) and *PMP22* deletion (hereditary neuropathy with liability to pressure palsies). Together, these five mutations account for almost 90% of all genetically confirmed CMT cases. Genetic confirmation is possible in around 60% of all patients with clinically diagnosed CMT (in 91% of patients with CMT1, compared to only 43% of cases with CMT2) (Fridman et al., 2015).

Treatment and randomized controlled trials

In general, most chronic axonal neuropathies can only be treated symptomatically. Antidepressant (amitriptyline, nortriptyline, duloxetine) and anticonvulsant (gabapentin, pregabalin, valproate, carbamazepine) drugs have shown to be effective to reduce neuropathic pain (Finnerup et al., 2005). Demyelinating neuropathies are often more responsive to treatment. Treatment and RCTs for specific subtypes of polyneuropathy are discussed in this section.

Diabetic neuropathy

Enhanced glucose control to prevent the development of neuropathy is especially helpful in patients with type 1 diabetes, but may also reduce the incidence of neuropathy in type 2 diabetes. However, this approach may lead to severe hypoglycemic episodes, so caution is warranted (Callaghan et al., 2012b). Optimal control of other metabolic factors, like blood pressure, obesity, and dyslipidemia, may also lower the risk of development or progression of polyneuropathy, but there is insufficient evidence to establish a causal relationship (Tesfaye et al., 2005; Callaghan et al., 2012a). Several trials investigated the effect of aldose reductase inhibitors, a drug that inhibits the enzyme necessary for the polyol pathway. A meta-analysis of these studies, however, showed no beneficial effect of this treatment (Chalk et al., 2007).

Chronic inflammatory demyelinating polyradiculoneuropathy

CIDP is a treatable chronic polyneuropathy. Treatment consists of corticosteroids, plasma exchange, or IVIg. Multiple case series showed improvement after corticosteroid therapy, but there are no randomized placebo controlled trials that investigate the effect of corticosteroids. One – not blinded – trial compared prednisone with no treatment and found a beneficial effect of prednisone on impairment and measures of nerve conduction, but this study had several limitations, resulting in a high risk of bias (Dyck et al., 1982). Another trial compared oral prednisolone with oral dexamethasone pulse treatment and found that the effects of these treatments were equal (van Schaik et al., 2010; Hughes and Mehndiratta, 2015). There have been several RCTs that assessed the effect of IVIg compared with that of placebo, plasma exchange, or corticosteroids. These studies showed that immunoglobulins lead to significant improvement in disability. Immunoglobulins are more effective than placebo, and equally effective as plasma exchange and (oral or intravenous) corticosteroids (Mehndiratta and Hughes, 2012; Nobile-Orazio et al., 2012; Eftimov et al., 2013). One open study investigated the effect of IVIg in 281 patients with CIDP: 76% improved after IVIg. Of the nonresponders, around 60% improved after a second-line treatment (steroids or plasma exchange), and even 75% improved after failure of the first two proven effective treatments when they were given the third proven effective treatment for CIDP (Kuitwaard et al., 2015). Long-term effect has only been shown in one study investigating IVIg (Hughes et al., 2008). More trials are needed to confirm this long-term effect and to determine the long-term effect of the other effective treatments. Observational studies suggested an effect of several immunomodulatory and immunosuppressive treatments, other than steroids, immunoglobulins, and plasma exchange. However, in the randomized trials that have been performed, azathioprine, interferon beta-1a, and (a relatively low dose of) methotrexate were not effective, although potential small beneficial effects could not be shown due to the small sample size of the performed trials (Mahdi-Rogers et al., 2013).

Paraproteinemic neuropathy

Treatment depends on the cause and subtype. Immunotherapy (as for CIDP) is effective for DADS without presence of a paraprotein, but response is usually poor in patients who have a paraprotein. For MGUS IVIg and rituximab may be beneficial, but results are inconclusive. Corticosteroids are not effective (Raheja et al., 2015). In malignant diseases like POEMS, multiple myeloma, and amyloidosis, the underlying disease should be treated, but this may not improve or control the neuropathy.

SUMMARY AND CONCLUSIONS

During the last decades, a lot of research has been performed to elucidate pathophysiologic mechanisms of peripheral nerve disorders. This has led to a better understanding of several diseases, especially for immune-mediated, hereditary, and diabetic polyneuropathies. Prevalence and incidence of most disorders have been described. However, although peroneal mononeuropathy and chronic polyneuropathy are among the most common peripheral nervous system diseases, well-designed population-based studies investigating these diseases are lacking. Moreover, in every 1 in 4 patients with chronic polyneuropathy no cause can be established. The absence of a uniform gold standard for the diagnosis of polyneuropathy makes field studies difficult. These studies are necessary, especially to determine the prevalence, incidence, and risk factors for chronic axonal polyneuropathies. Studies that use medical databases or registries are useful for diseases that follow a detrimental course with predominantly motor weakness (e.g., GBS, CIDP, MMN), since these patients are often seen and diagnosed by neurologists practicing in larger hospitals. Prevalence and incidence of chronic polyneuropathies can be severely underestimated with this approach, since a substantial proportion may be undiagnosed or not referred for additional investigations. For this purpose, cohort studies that collect information on potential risk factors preferentially should incorporate screening for polyneuropathy. For rare diseases, international collaboration studies like the IGOS are very important to investigate differences in regional subtypes, patient characteristics, and potential new therapies.

References

Allen JA, Lewis RA (2015). CIDP diagnostic pitfalls and perception of treatment benefit. Neurology 85: 498–504.

Aprile I, Caliandro P, La Torre G et al. (2005). Multicenter study of peroneal mononeuropathy: clinical, neurophysiologic, and quality of life assessment. J Peripher Nerv Syst 10: 259–268.

Aprile I, Tonali P, Caliandro P et al. (2009). Italian multicentre study of peroneal mononeuropathy: multiperspective follow-up. Neurol Sci 30: 37–44.

Assmus H, Antoniadis G, Bischoff C et al. (2011). Cubital tunnel syndrome – a review and management guidelines. Cent Eur Neurosurg 72: 90–98.

Atroshi I, Gummesson C, Johnsson R et al. (1999). Prevalence of carpal tunnel syndrome in a general population. JAMA 282: 153–158.

Atroshi I, Flondell M, Hofer M et al. (2013). Methylprednisolone injections for the carpal tunnel syndrome: a randomized, placebo-controlled trial. Ann Intern Med 159: 309–317.

Baldereschi M, Inzitari M, Di Carlo A et al. (2007). Epidemiology of distal symmetrical neuropathies in the Italian elderly. Neurology 68: 1460–1467.

Barcenilla A, March LM, Chen JS et al. (2012). Carpal tunnel syndrome and its relationship to occupation: a meta-analysis. Rheumatology (Oxford) 51: 250–261.

Bharucha NE, Bharucha AE, Bharucha EP (1991). Prevalence of peripheral neuropathy in the Parsi community of Bombay. Neurology 41: 1315–1317.

Bland JD, Rudolfer SM (2003). Clinical surveillance of carpal tunnel syndrome in two areas of the United Kingdom, 1991–2001. J Neurol Neurosurg Psychiatry 74: 1674–1679.

Boyce JR (1984). Meralgia paresthetica and tight trousers. JAMA 251: 1553.

Bray F, Jemal A, Grey N et al. (2012). Global cancer transitions according to the Human Development Index (2008–2030): a population-based study. Lancet Oncol 13: 790–801.

Bsteh G, Wanschitz JV, Gruber H et al. (2013). Prognosis and prognostic factors in non-traumatic acute-onset compressive mononeuropathies – radial and peroneal mononeuropathies. Eur J Neurol 20: 981–985.

Caliandro P, La Torre G, Padua R et al. (2012). Treatment for ulnar neuropathy at the elbow. Cochrane Database Syst Rev 7. CD006839.

Callaghan BC, Cheng HT, Stables CL et al. (2012a). Diabetic neuropathy: clinical manifestations and current treatments. Lancet Neurol 11: 521–534.

Callaghan BC, Little AA, Feldman EL et al. (2012b). Enhanced glucose control for preventing and treating diabetic neuropathy. Cochrane Database Syst Rev 6. CD007543.

Cats EA, van der Pol WL, Piepers S et al. (2010). Correlates of outcome and response to IVIg in 88 patients with multifocal motor neuropathy. Neurology 75: 818–825.

Chalk C, Benstead TJ, Moore F (2007). Aldose reductase inhibitors for the treatment of diabetic polyneuropathy. Cochrane Database Syst Rev CD004572.

Chhuttani PN, Chawla LS, Sharma TD (1966). Meralgia paraesthetica. Acta Neurol Scand 42: 483–490.

Chio A, Cocito D, Bottacchi E et al. (2007). Idiopathic chronic inflammatory demyelinating polyneuropathy: an epidemiological study in Italy. J Neurol Neurosurg Psychiatry 78: 1349–1353.

Collins MP (2012). The vasculitic neuropathies: an update. Curr Opin Neurol 25: 573–585.

Contreras MG, Warner MA, Charboneau WJ et al. (1998). Anatomy of the ulnar nerve at the elbow: potential relationship of acute ulnar neuropathy to gender differences. Clin Anat 11: 372–378.

Cruz-Martinez A, Arpa J, Palau F (2000). Peroneal neuropathy after weight loss. J Peripher Nerv Syst 5: 101–105.

Dale AM, Zeringue A, Harris-Adamson C et al. (2015). General population job exposure matrix applied to a pooled study of prevalent carpal tunnel syndrome. Am J Epidemiol 181: 431–439.

Darnell RB, Posner JB (2003). Paraneoplastic syndromes involving the nervous system. N Engl J Med 349: 1543–1554.

de Krom MC, Knipschild PG, Kester AD et al. (1992). Carpal tunnel syndrome: prevalence in the general population. J Clin Epidemiol 45: 373–376.

de Rijk MC, Vermeij FH, Suntjens M et al. (2007). Does a carpal tunnel syndrome predict an underlying disease? J Neurol Neurosurg Psychiatry 78: 635–637.

Descatha A, Leclerc A, Chastang JF et al. (2004). Incidence of ulnar nerve entrapment at the elbow in repetitive work. Scand J Work Environ Health 30: 234–240.

Dyck PJ, O'Brien PC, Oviatt KF et al. (1982). Prednisone improves chronic inflammatory demyelinating polyradiculoneuropathy more than no treatment. Ann Neurol 11: 136–141.

Eftimov F, Winer JB, Vermeulen M et al. (2013). Intravenous immunoglobulin for chronic inflammatory demyelinating polyradiculoneuropathy. Cochrane Database Syst Rev 12. CD001797.

England JD, Asbury AK (2004). Peripheral neuropathy. Lancet 363: 2151–2161.

England JD, Gronseth GS, Franklin G et al. (2005). Distal symmetric polyneuropathy: a definition for clinical research: report of the American Academy of Neurology, the American Association of Electrodiagnostic Medicine, and the American Academy of Physical Medicine and Rehabilitation. Neurology 64: 199–207.

Engstrom M, Berg T, Stjernquist-Desatnik A et al. (2008). Prednisolone and valaciclovir in Bell's palsy: a randomised, double-blind, placebo-controlled, multicentre trial. Lancet Neurol 7: 993–1000.

Eviston TJ, Croxson GR, Kennedy PG et al. (2015). Bell's palsy: aetiology, clinical features and multidisciplinary care. J Neurol Neurosurg Psychiatry 86: 1356–1361.

Finnerup NB, Otto M, McQuay HJ et al. (2005). Algorithm for neuropathic pain treatment: an evidence based proposal. Pain 118: 289–305.

Fridman V, Bundy B, Reilly MM et al. (2015). CMT subtypes and disease burden in patients enrolled in the Inherited Neuropathies Consortium natural history study: a cross-sectional analysis. J Neurol Neurosurg Psychiatry 86: 873–878.

Frost P, Johnsen B, Fuglsang-Frederiksen A et al. (2013). Lifestyle risk factors for ulnar neuropathy and ulnar neuropathy-like symptoms. Muscle Nerve 48: 507–515.

Gelfman R, Melton 3rd LJ, Yawn BP et al. (2009). Long-term trends in carpal tunnel syndrome. Neurology 72: 33–41.

Geoghegan JM, Clark DI, Bainbridge LC et al. (2004). Risk factors in carpal tunnel syndrome. J Hand Surg Br 29: 315–320.

George J, Twomey JA (1986). Causes of polyneuropathy in the elderly. Age Ageing 15: 247–249.

Gerritsen AA, de Vet HC, Scholten RJ et al. (2002). Splinting vs surgery in the treatment of carpal tunnel syndrome: a randomized controlled trial. JAMA 288: 1245–1251.

Giometto B, Grisold W, Vitaliani R et al. (2010). Paraneoplastic neurologic syndrome in the PNS Euronetwork database: a European study from 20 centers. Arch Neurol 67: 330–335.

Giometto B, Vitaliani R, Lindeck-Pozza E et al. (2012). Treatment for paraneoplastic neuropathies. Cochrane Database Syst Rev 12. CD007625.

Gorson KC (2007). Vasculitic neuropathies: an update. Neurologist 13: 12–19.

Gorson KC, Gooch CL (2015). The (mis)diagnosis of CIDP: The high price of missing the mark. Neurology 85: 488–489.

Gozzard P, Woodhall M, Chapman C et al. (2015). Paraneoplastic neurologic disorders in small cell lung carcinoma: a prospective study. Neurology 85: 235–239.

Grantz M (2010a). Unusual peripheral neuropathies. Part II: intrinsic reactive causes. Semin Neurol 30: 396–404.

Grantz M (2010b). Unusual peripheral neuropathies. Part III: intrinsic inherited causes. Semin Neurol 30: 405–415.

Grantz M, Huan MC (2010). Unusual peripheral neuropathies. Part I: extrinsic causes. Semin Neurol 30: 387–395.

Graus F, Dalmau J (2013). Paraneoplastic neuropathies. Curr Opin Neurol 26: 489–495.

Graus F, Delattre JY, Antoine JC et al. (2004). Recommended diagnostic criteria for paraneoplastic neurological syndromes. J Neurol Neurosurg Psychiatry 75: 1135–1140.

Hanewinckel R, Van Oijen M, Ikram MA et al. (2016). The epidemiology and risk factors of chronic polyneuropathy. Eur J Epidemiol 31: 5–20.

Hoftberger R, Rosenfeld MR, Dalmau J (2015). Update on neurological paraneoplastic syndromes. Curr Opin Oncol 27: 489–495.

Hughes RA, Mehndiratta MM (2015). Corticosteroids for chronic inflammatory demyelinating polyradiculoneuropathy. Cochrane Database Syst Rev 1. CD002062.

Hughes RA, van Doorn PA (2012). Corticosteroids for Guillain–Barré syndrome. Cochrane Database Syst Rev 8. CD001446.

Hughes RA, Donofrio P, Bril V et al. (2008). Intravenous immune globulin (10% caprylate-chromatography purified) for the treatment of chronic inflammatory demyelinating polyradiculoneuropathy (ICE study): a randomised placebo-controlled trial. Lancet Neurol 7: 136–144.

Hughes RA, Pritchard J, Hadden RD (2013). Pharmacological treatment other than corticosteroids, intravenous immunoglobulin and plasma exchange for Guillain–Barré syndrome. Cochrane Database Syst Rev 2. CD008630.

Hughes RA, Swan AV, van Doorn PA (2014). Intravenous immunoglobulin for Guillain–Barré syndrome. Cochrane Database Syst Rev 9. CD002063.

Hui AC, Wong S, Leung CH et al. (2005). A randomized controlled trial of surgery vs steroid injection for carpal tunnel syndrome. Neurology 64: 2074–2078.

Ibrahim I, Khan WS, Goddard N et al. (2012). Carpal tunnel syndrome: a review of the recent literature. Open Orthop J 6: 69–76.

Iijima M, Koike H, Hattori N et al. (2008). Prevalence and incidence rates of chronic inflammatory demyelinating polyneuropathy in the Japanese population. J Neurol Neurosurg Psychiatry 79: 1040–1043.

Jarvik JG, Comstock BA, Kliot M et al. (2009). Surgery versus non-surgical therapy for carpal tunnel syndrome: a randomised parallel-group trial. Lancet 374: 1074–1081.

Johannsen L, Smith T, Havsager AM et al. (2001). Evaluation of patients with symptoms suggestive of chronic polyneuropathy. J Clin Neuromuscul Dis 3: 47–52.

Joint Task Force of the European Federation of Neurological Societies/Peripheral Nerve Society (2010). European Federation of Neurological Societies/Peripheral Nerve Society guideline on management of multifocal motor neuropathy. Report of a joint task force of the European Federation of Neurological Societies and the Peripheral Nerve Society – first revision. J Peripher Nerv Syst 15: 295–301.

Juhasz-Pocsine K, Rudnicki SA, Archer RL et al. (2007). Neurologic complications of gastric bypass surgery for morbid obesity. Neurology 68: 1843–1850.

Juratli SM, Nayan M, Fulton-Kehoe D et al. (2010). A population-based study of ulnar neuropathy at the elbow in Washington State workers' compensation. Am J Ind Med 53: 1242–1251.

Kandil MR, Darwish ES, Khedr EM et al. (2012). A community-based epidemiological study of peripheral neuropathies in Assiut, Egypt. Neurol Res 34: 960–966.

Katirji MB, Wilbourn AJ (1988). Common peroneal mononeuropathy: a clinical and electrophysiologic study of 116 lesions. Neurology 38: 1723–1728.

Kazemi A, Frazier T, Cave M (2010). Micronutrient-related neurologic complications following bariatric surgery. Curr Gastroenterol Rep 12: 288–295.

Kelkar P, Parry GJ (2003). Mononeuritis multiplex in diabetes mellitus: evidence for underlying immune pathogenesis. J Neurol Neurosurg Psychiatry 74: 803–806.

Khalil N, Nicotra A, Rakowicz W (2012). Treatment for meralgia paraesthetica. Cochrane Database Syst Rev 12. CD004159.

Klein CJ, Duan X, Shy ME (2013). Inherited neuropathies: clinical overview and update. Muscle Nerve 48: 604–622.

Koehler PJ, Buscher M, Rozeman CA et al. (1997). Peroneal nerve neuropathy in cancer patients: a paraneoplastic syndrome? J Neurol 244: 328–332.

Koike H, Iijima M, Sugiura M et al. (2003). Alcoholic neuropathy is clinicopathologically distinct from thiamine-deficiency neuropathy. Ann Neurol 54: 19–29.

Koike H, Ito S, Morozumi S et al. (2008). Rapidly developing weakness mimicking Guillain-Barre syndrome in beriberi neuropathy: two case reports. Nutrition 24: 776–780.

Koike H, Tanaka F, Sobue G (2011). Paraneoplastic neuropathy: wide-ranging clinicopathological manifestations. Curr Opin Neurol 24: 504–510.

Kuitwaard K, Hahn AF, Vermeulen M et al. (2015). Intravenous immunoglobulin response in treatment-naive chronic inflammatory demyelinating polyradiculoneuropathy. J Neurol Neurosurg Psychiatry 86: 1331–1336.

Kuwabara S, Yuki N (2013). Axonal Guillain–Barré syndrome: concepts and controversies. Lancet Neurol 12: 1180–1188.

Latinovic R, Gulliford MC, Hughes RA (2006). Incidence of common compressive neuropathies in primary care. J Neurol Neurosurg Psychiatry 77: 263–265.

Levin KH (2004). Variants and mimics of Guillain–Barré syndrome. Neurologist 10: 61–74.

Lin KP, Kwan SY, Chen SY et al. (1993). Generalized neuropathy in Taiwan: an etiologic survey. Neuroepidemiology 12: 257–261.

Lunn MP, Manji H, Choudhary PP et al. (1999). Chronic inflammatory demyelinating polyradiculoneuropathy: a prevalence study in south east England. J Neurol Neurosurg Psychiatry 66: 677–680.

MacDonald BK, Cockerell OC, Sander JW et al. (2000). The incidence and lifetime prevalence of neurological disorders in a prospective community-based study in the UK. Brain 123 (Pt 4): 665–676.

Mahdi-Rogers M, Hughes RA (2014). Epidemiology of chronic inflammatory neuropathies in southeast England. Eur J Neurol 21: 28–33.

Mahdi-Rogers M, van Doorn PA, Hughes RA (2013). Immunomodulatory treatment other than corticosteroids, immunoglobulin and plasma exchange for chronic inflammatory demyelinating polyradiculoneuropathy. Cochrane Database Syst Rev 6. CD003280.

Manji H (2011). Toxic neuropathy. Curr Opin Neurol 24: 484–490.

Marshall S, Tardif G, Ashworth N (2007). Local corticosteroid injection for carpal tunnel syndrome. Cochrane Database Syst Rev. CD001554.

Martinez-Thompson JM, Diehl NN, Holmes JM et al. (2014). Incidence, types, and lifetime risk of adult-onset strabismus. Ophthalmology 121: 877–882.

Mathey EK, Park SB, Hughes RA et al. (2015). Chronic inflammatory demyelinating polyradiculoneuropathy: from pathology to phenotype. J Neurol Neurosurg Psychiatry 90: 951–962.

McCombe PA, Pollard JD, McLeod JG (1987). Chronic inflammatory demyelinating polyradiculoneuropathy. A clinical and electrophysiological study of 92 cases. Brain 110 (Pt 6): 1617–1630.

McLeod JG, Pollard JD, Macaskill P et al. (1999). Prevalence of chronic inflammatory demyelinating polyneuropathy in New South Wales, Australia. Ann Neurol 46: 910–913.

Mediouni Z, de Roquemaurel A, Dumontier C et al. (2014). Is carpal tunnel syndrome related to computer exposure at work? A review and meta-analysis. J Occup Environ Med 56: 204–208.

Mehndiratta MM, Hughes RA (2012). Plasma exchange for chronic inflammatory demyelinating polyradiculoneuropathy. Cochrane Database Syst Rev 9. CD003906.

Mellion M, Gilchrist JM, de la Monte S (2011). Alcohol-related peripheral neuropathy: nutritional, toxic, or both? Muscle Nerve 43: 309–316.

Menke A, Casagrande S, Geiss L et al. (2015). Prevalence of and trends in diabetes among adults in the United States, 1988–2012. JAMA 314: 1021–1029.

Middleton SD, Anakwe RE (2014). Carpal tunnel syndrome. BMJ 349: g6437.

Mondelli M, Giannini F, Giacchi M (2002). Carpal tunnel syndrome incidence in a general population. Neurology 58: 289–294.

Mondelli M, Giannini F, Ballerini M et al. (2005). Incidence of ulnar neuropathy at the elbow in the province of Siena (Italy). J Neurol Sci 234: 5–10.

Mondelli M, Rossi S, Romano C (2007). Body mass index in meralgia paresthetica: a case-control study. Acta Neurol Scand 116: 118–123.

Mygland A, Monstad P (2001). Chronic polyneuropathies in Vest-Agder, Norway. Eur J Neurol 8: 157–165.

Nahabedian MY, Dellon AL (1995). Meralgia paresthetica: etiology, diagnosis, and outcome of surgical decompression. Ann Plast Surg 35: 590–594.

Nardin R, Chapman KM, Raynor EM (2005). Prevalence of ulnar neuropathy in patients receiving hemodialysis. Arch Neurol 62: 271–275.

Nobile-Orazio E (2001). Multifocal motor neuropathy. J Neuroimmunol 115: 4–18.

Nobile-Orazio E, Cocito D, Jann S et al. (2012). Intravenous immunoglobulin versus intravenous methylprednisolone for chronic inflammatory demyelinating polyradiculoneuropathy: a randomised controlled trial. Lancet Neurol 11: 493–502.

Nolan CJ, Damm P, Prentki M (2011). Type 2 diabetes across generations: from pathophysiology to prevention and management. Lancet 378: 169–181.

Nordstrom DL, Vierkant RA, DeStefano F et al. (1997). Risk factors for carpal tunnel syndrome in a general population. Occup Environ Med 54: 734–740.

Nordstrom DL, DeStefano F, Vierkant RA et al. (1998). Incidence of diagnosed carpal tunnel syndrome in a general population. Epidemiology 9: 342–345.

Padua L, Di Pasquale A, Pazzaglia C et al. (2010). Systematic review of pregnancy-related carpal tunnel syndrome. Muscle Nerve 42: 697–702.

Page MJ, Massy-Westropp N, O'Connor D et al. (2012). Splinting for carpal tunnel syndrome. Cochrane Database Syst Rev 7. CD010003.

Palmer KT, Harris EC, Coggon D (2007). Carpal tunnel syndrome and its relation to occupation: a systematic literature review. Occup Med (Lond) 57: 57–66.

Parisi TJ, Mandrekar J, Dyck PJ et al. (2011). Meralgia paresthetica: relation to obesity, advanced age, and diabetes mellitus. Neurology 77: 1538–1542.

Patel SV, Holmes JM, Hodge DO et al. (2005). Diabetes and hypertension in isolated sixth nerve palsy: a population-based study. Ophthalmology 112: 760–763.

Patterson JM, Jaggars MM, Boyer MI (2003). Ulnar and median nerve palsy in long-distance cyclists. A prospective study. Am J Sports Med 31: 585–589.

Peitersen E (2002). Bell's palsy: the spontaneous course of 2,500 peripheral facial nerve palsies of different etiologies. Acta Otolaryngol Suppl: 4–30.

Puy H, Gouya L, Deybach JC (2010). Porphyrias. Lancet 375: 924–937.

Quaresma M, Coleman MP, Rachet B (2015). 40-year trends in an index of survival for all cancers combined and survival adjusted for age and sex for each cancer in England and Wales, 1971–2011: a population-based study. Lancet 385: 1206–1218.

Raheja D, Specht C, Simmons Z (2015). Paraproteinemic neuropathies. Muscle Nerve 51: 1–13.

Raphael JC, Chevret S, Hughes RA et al. (2012). Plasma exchange for Guillain–Barré syndrome. Cochrane Database Syst Rev 7. CD001798.

Rempel D, Evanoff B, Amadio PC et al. (1998). Consensus criteria for the classification of carpal tunnel syndrome in epidemiologic studies. Am J Public Health 88: 1447–1451.

Richards BW, Jones Jr FR, Younge BR (1992). Causes and prognosis in 4,278 cases of paralysis of the oculomotor, trochlear, and abducens cranial nerves. Am J Ophthalmol 113: 489–496.

Richardson JK, Green DF, Jamieson SC et al. (2001). Gender, body mass and age as risk factors for ulnar mononeuropathy at the elbow. Muscle Nerve 24: 551–554.

Robertson C, Saratsiotis J (2005). A review of compressive ulnar neuropathy at the elbow. J Manipulative Physiol Ther 28: 345.

Roquelaure Y, Ha C, Pelier-Cady MC et al. (2008). Work increases the incidence of carpal tunnel syndrome in the general population. Muscle Nerve 37: 477–482.

Rosenberg NR, Portegies P, de Visser M et al. (2001). Diagnostic investigation of patients with chronic polyneuropathy: evaluation of a clinical guideline. J Neurol Neurosurg Psychiatry 71: 205–209.

Rota E, Zavaroni D, Parietti L et al. (2014). Ulnar entrapment neuropathy in patients with type 2 diabetes mellitus: an electrodiagnostic study. Diabetes Res Clin Pract 104: 73–78.

Rubens O, Logina I, Kravale I et al. (2001). Peripheral neuropathy in chronic occupational inorganic lead exposure: a clinical and electrophysiological study. J Neurol Neurosurg Psychiatry 71: 200–204.

Rudolph T, Farbu E (2007). Hospital-referred polyneuropathies – causes, prevalences, clinical- and neurophysiological findings. Eur J Neurol 14: 603–608.

Ruts L, Drenthen J, Jacobs BC et al. (2010). Distinguishing acute-onset CIDP from fluctuating Guillain–Barré syndrome: a prospective study. Neurology 74: 1680–1686.

Saporta AS, Sottile SL, Miller LJ et al. (2011). Charcot–Marie–Tooth disease subtypes and genetic testing strategies. Ann Neurol 69: 22–33.

Seretny M, Currie GL, Sena ES et al. (2014). Incidence, prevalence, and predictors of chemotherapy-induced peripheral neuropathy: a systematic review and meta-analysis. Pain 155: 2461–2470.

Simon NG, Herkes GK (2011). The neurologic manifestations of the acute porphyrias. J Clin Neurosci 18: 1147–1153.

Singhal NS, Irodenko VS, Margeta M et al. (2015). Sarcoid polyneuropathy masquerading as chronic inflammatory demyelinating polyneuropathy. Muscle Nerve 52: 664–668.

Sotaniemi KA (1984). Slimmer's paralysis–peroneal neuropathy during weight reduction. J Neurol Neurosurg Psychiatry 47: 564–566.

Stevens JC, Sun S, Beard CM et al. (1988). Carpal tunnel syndrome in Rochester, Minnesota, 1961 to 1980. Neurology 38: 134–138.

Stewart JD (2008). Foot drop: where, why and what to do? Pract Neurol 8: 158–169.

Sullivan FM, Swan IR, Donnan PT et al. (2007). Early treatment with prednisolone or acyclovir in Bell's palsy. N Engl J Med 357: 1598–1607.

Svendsen SW, Johnsen B, Fuglsang-Frederiksen A et al. (2012). Ulnar neuropathy and ulnar neuropathy-like symptoms in relation to biomechanical exposures assessed by a job exposure matrix: a triple case-referent study. Occup Environ Med 69: 773–780.

Tesfaye S, Chaturvedi N, Eaton SE et al. (2005). Vascular risk factors and diabetic neuropathy. N Engl J Med 352: 341–350.

Thomson RM, Parry GJ (2006). Neuropathies associated with excessive exposure to lead. Muscle Nerve 33: 732–741.

Tiffin PA, MacEwen CJ, Craig EA et al. (1996). Acquired palsy of the oculomotor, trochlear and abducens nerves. Eye (Lond) 10 (Pt 3): 377–384.

Umapathi T, Hughes RA, Nobile-Orazio E et al. (2015). Immunosuppressant and immunomodulatory treatments for multifocal motor neuropathy. Cochrane Database Syst Rev 3. CD003217.

Vallat JM, Sommer C, Magy L (2010). Chronic inflammatory demyelinating polyradiculoneuropathy: diagnostic and therapeutic challenges for a treatable condition. Lancet Neurol 9: 402–412.

Van Asseldonk JT, Franssen H, Van den Berg-Vos RM et al. (2005). Multifocal motor neuropathy. Lancet Neurol 4: 309–319.

van den Berg B, Walgaard C, Drenthen J et al. (2014). Guillain–Barré syndrome: pathogenesis, diagnosis, treatment and prognosis. Nat Rev Neurol 10: 469–482.

Van den Bergh PY, Hadden RD, Bouche P et al. (2010). European Federation of Neurological Societies/Peripheral Nerve Society guideline on management of chronic inflammatory demyelinating polyradiculoneuropathy: report of a joint task force of the European Federation of Neurological Societies and the Peripheral Nerve Society – first revision. Eur J Neurol 17: 356–363.

van Dijk MA, Reitsma JB, Fischer JC et al. (2003). Indications for requesting laboratory tests for concurrent diseases in patients with carpal tunnel syndrome: a systematic review. Clin Chem 49: 1437–1444.

van Doorn PA (2007). [Guideline on polyneuropathy.] Richtlijn 'Polyneuropathie'. Ned Tijdschr Geneeskd 151: 1566–1573.

van Doorn PA, Ruts L, Jacobs BC (2008). Clinical features, pathogenesis, and treatment of Guillain-Barre syndrome. Lancet Neurol 7: 939–950.

van Schaik IN, van den Berg LH, de Haan R et al. (2005). Intravenous immunoglobulin for multifocal motor neuropathy. Cochrane Database Syst Rev CD004429.

van Schaik IN, Eftimov F, van Doorn PA et al. (2010). Pulsed high-dose dexamethasone versus standard prednisolone treatment for chronic inflammatory demyelinating polyradiculoneuropathy (PREDICT study): a double-blind, randomised, controlled trial. Lancet Neurol 9: 245–253.

van Slobbe AM, Bohnen AM, Bernsen RM et al. (2004). Incidence rates and determinants in meralgia paresthetica in general practice. J Neurol 251: 294–297.

Verghese J, Bieri PL, Gellido C et al. (2001). Peripheral neuropathy in young-old and old-old patients. Muscle Nerve 24: 1476–1481.

Visser NA, Notermans NC, Linssen RS et al. (2015). Incidence of polyneuropathy in Utrecht, the Netherlands. Neurology 84: 259–264.

Vlam L, van der Pol WL, Cats EA et al. (2012). Multifocal motor neuropathy: diagnosis, pathogenesis and treatment strategies. Nat Rev Neurol 8: 48–58.

Vrancken AF, Kalmijn S, Buskens E et al. (2006). Feasibility and cost efficiency of a diagnostic guideline for chronic polyneuropathy: a prospective implementation study. J Neurol Neurosurg Psychiatry 77: 397–401.

Walgaard C, Lingsma HF, Ruts L et al. (2010). Prediction of respiratory insufficiency in Guillain-Barre syndrome. Ann Neurol 67: 781–787.

Walgaard C, Lingsma HF, Ruts L et al. (2011). Early recognition of poor prognosis in Guillain–Barré syndrome. Neurology 76: 968–975.

Warner MA, Warner ME, Martin JT (1994). Ulnar neuropathy. Incidence, outcome, and risk factors in sedated or anesthetized patients. Anesthesiology 81: 1332–1340.

Wiggers JK, Brouwer KM, Helmerhorst GT et al. (2012). Predictors of diagnosis of ulnar neuropathy after surgically treated distal humerus fractures. J Hand Surg Am 37: 1168–1172.

Handbook of Clinical Neurology, Vol. 138 (3rd series)
Neuroepidemiology
C. Rosano, M.A. Ikram, and M. Ganguli, Editors
http://dx.doi.org/10.1016/B978-0-12-802973-2.00016-1

Chapter 16

Migraine

K.R. PECK, Y.L. JOHNSON, AND T.A. SMITHERMAN*
Department of Psychology, University of Mississippi, Oxford, MS, USA

Abstract

This chapter presents an overview of migraine epidemiology and mechanisms. Migraine is a common and disabling neurologic disorder characterized by episodic attacks of severe head pain and other symptoms, including interference with activity, nausea, and sensitivity to light and sound. A number of risk factors for migraine onset and progression have been identified, including the presence of comorbid disorders and overuse of acute headache medications. Though the pathophysiology of migraine is complex and incompletely understood, advances in genetics research and clinical trials methodology offer promise for better understanding the underlying pathophysiologic mechanisms. These advances presently center on genome-wide studies, development of antibodies targeting calcitonin gene-related peptide, and understanding the psychologic mechanisms that underlie the efficacy of some interventions. Studies of both pharmacologic and behavioral interventions for migraine and its common comorbidities also offer promise for understanding the neuroepidemiologic mechanisms of migraine. Clinical trials relevant to these mechanisms are reviewed, and methodologic considerations for future trials are discussed.

EPIDEMIOLOGY

Diagnosis

The *International Classification of Headache Disorders* (3rd edition, beta version) delineates diagnostic criteria for migraine and other headache disorders (Headache Classification Committee of the International Headache Society, 2013). Diagnosis of migraine requires that one has experienced at least five attacks fulfilling three conditions: (1) attacks lasting 4–72 hours; (2) pain that is characterized by two or more of the following: unilateral location, pulsating quality, moderate or severe intensity, and aggravation by or causing avoidance of physical activity; (3) nausea, vomiting, or both photophobia (sensitivity to light) and phonophobia (sensitivity to sound) during the attack. As with all primary headache disorders (e.g., tension-type headache, cluster headache), the symptoms must not be attributed to another disorder, in which case a "secondary" headache disorder takes precedent. Approximately 30% of migraineurs experience auras, which are temporary distortions in vision (or less commonly, sensation or speech) that precede the head pain and typically resolve within 1 hour. In light of the accompanying symptoms of migraine, migraine is recognized as a chronic neurologic disease with intermittent attack frequency.

Prevalence and impact

According to data from the 2010 Global Burden of Disease study, migraine is the third most common medical condition in the world (Vos et al., 2012). Migraine is most prevalent in Europe (15%) and least prevalent in Africa (5%: Stovner et al., 2007). Data from the American Migraine Prevalence and Prevention Study indicate that annual prevalence in the USA is 12% (Lipton et al., 2007). Migraine affects three times as many women as men (Lipton et al., 2007; Smitherman et al., 2013). For both sexes, prevalence is highest during the peak productive years of life – between ages 30 and 39 – and is lowest among those aged 60 and older. Migraine prevalence is also higher among Caucasians than African Americans, and highest in individuals of lower income (Lipton et al.,

*Correspondence to: Todd A. Smitherman, PhD., 207 Peabody, University of Mississippi Psychology Department, University MS 38677, USA. Tel: +1-662-915-7383, E-mail: tasmithe@olemiss.edu

2007; Smitherman et al., 2013). Onset most commonly occurs during late adolescence or early adulthood, with 75% of migraineurs experiencing onset of migraine before age 35 (Stewart et al., 2013). Approximately one-third of migraineurs have never sought treatment from a physician, and 40% of those who have received consultation go undiagnosed (Lipton et al., 1998).

The majority of migraineurs (63%) experience 1–4 days with migraine per month (Lipton et al., 2007). Individuals also experience substantial impairment in functioning on days with headache, such as restricted activity and interference with obligations at home or work. Over half report that their migraine attacks cause significant impairment in daily activities or require bed rest, and 1 in every 5 migraineurs experience moderate or greater disability as quantified by migraine-specific disability measures. Migraine imposes substantial financial costs to society as a result of missed work or reduced productivity, such that 1 in every 4 individuals with migraine misses at least 1 day of work or school every 3 months as a result (Lipton et al., 2007). A recent survey of over 2700 migraineurs showed that 50% report that headaches negatively impact their professional advancement (Malone et al., 2015), indicating that the personal impact of migraine extends well beyond its direct effects on attendance. Globally, migraine is the eighth leading cause for years lived with disability, accounting for the most years lived with disability of any neurologic disorder, surpassing epilepsy, Alzheimer's disease, and Parkinson's disease (Vos et al., 2012).

Risk factors and migraine progression

In addition to demographic variables, several other factors contribute to an individual's likelihood of developing migraine. The majority of migraineurs have a family history of migraine, suggesting that genetic predisposition also plays a major role. A Danish population-based twin study found that pairwise concordance rates for migraine with aura were significantly higher in monozygotic (34%) than dizygotic twins (12%: Ulrich et al., 2001). These and other family and twin studies indicate that variants and polymorphisms of certain genes may be responsible for some forms of migraine, but the influences for the common forms of migraine are likely polygenetic and still poorly understood (Ducros et al., 2002).

Modifiable risk factors are those related to the individual's environment and lifestyle that portend increased risk for developing migraine, for headache progression, or that precipitate individual headache attacks (i.e., "triggers"). While the majority of migraineurs eventually see a decrease in frequency and some will become headache-free within months or years, a small percentage experience a gradual increase in frequency of attacks over time. Approximately 3% of headache sufferers experience headache "chronification," or progression from episodic migraine (<15 headache days per month) to chronic migraine (≥15 headache days per month; Scher et al., 2003). This increase in frequency reflects further sensitization of an already-hypersensitive central nervous system both within and between attacks (i.e., central sensitization). Emerging research on chronification indicates that overuse of acute headache medications (>2 days/week; Bigal et al., 2008) and onset or exacerbation of comorbid medical and psychiatric disorders are potent risk factors (Bigal and Lipton, 2006a; Scher et al., 2008).

Regarding triggers, over 75% of migraineurs report experiencing specific migraine triggers, the most common of which are stress, missing meals, too much or too little sleep, and menstruation (in women: Kelman, 2007). Odors, neck pain, light, and alcohol use also are reported, though less frequently. No single factor acts as a trigger for all migraineurs, and it is rare for a reported trigger to always provoke an attack upon exposure. Understanding the causal role of various migraine triggers is complicated by a preponderance of retrospective versus prospective studies, the latter of which require considerable experimental control over the individual, trigger manipulation, and other potential triggers (Turner et al., 2013).

Comorbid disorders

Compared to those without headache, individuals with migraine are at increased risk for experiencing chronic medical (odds ratio (OR) = 2.1), other pain (OR = 3.3), and psychiatric (OR = 3.1) conditions (Saunders et al., 2008). Risk is greatest for those with chronic versus episodic attack frequency (Buse et al., 2013). The following is an overview of common comorbid disorders; descriptions of potential mechanisms are discussed later in this chapter.

Regarding medical comorbidities, migraine is strongly associated with high blood pressure, irritable-bowel syndrome, ulcers, and both hearing and vision problems (Saunders et al., 2008). Chronic pain conditions such as arthritis, neck pain, and back pain affect nearly 60% of migraineurs, compared to 36% of headache-free individuals (Saunders et al., 2008). Obesity is positively associated with migraine frequency, such that overweight and obese individuals are, respectively, at three and five times greater odds of developing chronic headache relative to individuals in the normal weight range (Scher et al., 2003). Obesity thus plays a significant role in migraine chronification, but does not appear to be a strong risk factor for increased frequency of tension-type headache (Bigal and Lipton, 2006b). An abundance of data also indicates a twofold risk of ischemic stroke among individuals who have migraine with aura (Schürks et al., 2009), which is potentially attributable to shared mechanistic underpinnings (Mawet et al., 2015). This risk appears to be highest among women, smokers, and those who use oral contraceptives. Relatedly, migraine has been implicated as a risk factor for structural changes in the brain. A recent meta-analysis found that migraine with aura was associated with increased odds of white-matter abnormalities (OR = 1.68), though migraine without aura was not associated with increased risk (Bashir et al., 2015). The clinical significance of these abnormalities, however, is unclear, and large longitudinal studies are needed.

Sleep disturbances (i.e., too much or too little sleep) and sleep-disordered breathing (i.e., sleep apnea) are common among headache sufferers. Sleep disorders can cause initial onset of migraine, aggravate an existing headache diagnosis, and/or be maintained by the presence of migraine (Rains and Poceta, 2006). Specifically, insomnia is the most prevalent sleep disorder among treatment-seeking migraineurs, with a prevalence three times higher than that of the general population (Kelman and Rains, 2005).

The most common psychiatric comorbidities among migraineurs are depression (OR = 2.2–4.0) and anxiety disorders (OR = 3.5–5.3: Hamelsky and Lipton, 2006; Jette et al., 2008). Lifetime prevalence ranges from 21% to 32% for major depressive disorder, and 50% for anxiety disorders (e.g., panic disorder, generalized anxiety disorder, specific phobia; Breslau, 1998). Both depression and panic disorder appear to have bidirectional relationships with migraine, indicating that psychiatric disorders do not always develop in response to living with migraine. Exposure to traumatic life events has been linked to development of migraine but does not appear to be as strongly related to migraine as posttraumatic stress disorder *per se* (Smitherman and Kolivas, 2013). Other psychiatric comorbidities include bipolar spectrum disorders (Fornaro et al., 2015), personality disorders (Rothrock et al., 2007), and attempted suicide (Breslau et al., 2012), although fewer studies have been conducted on these conditions. Migraine does not appear to be strongly associated with alcohol dependence (Jette et al., 2008), perhaps in part because alcohol is a trigger for many with migraine and thus often avoided.

These comorbid disorders are problematic in that they compound the already significant burden and impact of migraine (Saunders et al., 2008) and may complicate treatment or worsen prognosis. Longitudinal data suggest that the presence of one or more psychiatric disorders during adolescence or childhood is predictive of unchanging or worsening headache later in life (Guidetti et al., 1998). Those with multiple comorbid psychiatric disorders are more likely to have significant disability, restricted activity, or impaired quality of life compared to individuals with only one or none of these conditions (Jette et al., 2008). The literature suggests that individuals with chronic daily headache and who frequently use acute headache medications are at highest risk of psychiatric comorbidity (Juang et al., 2000; Radat et al., 2005), underscoring both the close relationship between migraine and other behavioral health problems and the need to screen for these comorbidities and behaviors clinically (Hamelsky and Lipton, 2006).

MECHANISMS OF MIGRAINE AND MIGRAINE TREATMENT

Pathophysiologic mechanisms

The pathophysiology of migraine is complex and remains incompletely understood. Contrary to the previous vascular theory of migraine, which held that migraine resulted from constriction and dilation of blood vessels innervating the head, migraine is now recognized as resulting fundamentally from a hypersensitive central nervous system that has difficulty properly modulating pain (Goadsby et al., 2009). In the current neurovascular model, the vascular changes that occur are recognized as secondary phenomena. The primary components of a migraine attack involve interactions between the brainstem, the cortex, and the trigeminovascular system. The brainstem is involved in descending modulation of pain, neuronal inhibition that traverses the cortex (cortical spreading depression) is the recognized cause of migraine aura, and resulting sensitization and activation of trigeminal afferents are the sources of pain. (Whether spreading depression occurs in migraine without aura is unclear but has been postulated.) Culminating from this sequence is a release of neuropeptides, dilation of meningeal blood vessels, neurogenic inflammation, and both within attacks and over time central sensitization manifests. The pathophysiologic influences reviewed herein thus do not operate in isolation, but instead influence a cascade of interacting events within the nervous system. Four of the most central concepts in migraine pathophysiology are discussed below.

GENETICS

Migraine runs in families and has a strong genetic component, the best exemplar of which is familial hemiplegic migraine, a rare autosomal-dominant subtype of migraine with aura that includes motor weakness and that presents in a first- or second-degree relative. Three genetic mutations corresponding to three variants of familial hemiplegic migraine have been identified, all of which code for ion-channel transport: the *CACNA1A* gene on chromosome 19, the *SCN1A* gene on chromosome 2, and the *ATP1A2* gene on chromosome 1 (Headache Classification Committee of the International Headache Society, 2013).

Precise genetic contributions to the more common forms of migraine, however, have not been firmly established, owing in large part to the varied phenotypes, triggers, treatment responses, and comorbidities of migraine that point to multigenetic influences. Candidate gene association studies have targeted those implicated in metabolism of serotonin and dopamine, neurogenic inflammation via tumor necrosis factor, sex hormones, and vascular oxidative stress. A meta-analysis concluded that the *MTHFR* 677TC genotype was associated with increased likelihood for migraine with aura only (pooled OR = 1.48) and was specific to non-Caucasians (Schürks et al., 2010). Otherwise candidate gene studies have produced negative or mixed results (Schürks, 2012), and coinciding with advances in gene-mapping technologies, an emphasis on genomewide approaches has emerged.

Linkage analyses endeavor to detect the chromosomal loci of migraine genes, and indeed, several loci have been linked to migraine broadly and separately to migraine with aura versus migraine without aura (Wessman et al., 2007; Schürks, 2012). These studies have identified different loci without much consistency, precluding definitive conclusions but also implying some degree of genetic heterogeneity. Genomewide association studies (GWAS) attempt to identify single-nucleotide polymorphisms associated with a disease by comparing cohorts (one with and one without the disease) across the genome. Though small in number, existing GWAS have implicated genetic variants impacting glutamate and pain pathways (Schürks, 2012). A 2013 meta-analysis identified four single-nucleotide polymorphisms (rs10166942, rs11172113, rs2078371, and rs2651899) significantly associated with migraine (with and without aura) after appropriate statistical correction (Esserlind et al., 2013). Likely future studies will identify additional loci and better clarify the involvement of previously identified genetic variants.

Significant methodologic challenges are inherent in GWAS. One pragmatic challenge is significant cost that typically requires substantial external funding against a background of strikingly low public funding for migraine research (Shapiro and Goadbsy, 2007). Others include the lack of an established migraine biomarker that necessitates adherence to phenotype-based diagnostic criteria via latent class analysis or trait component analysis, gender and population risk differences, and a need for adequate sample size and proper statistical correction for the numerous genetic variants assessed (Wessman et al., 2007). The impact of these challenges is particularly relevant for study replication. With increased public funding and analytic procedures, genomewide studies ultimately may spur significant revisions to the current phenotypic classification of migraine.

Neuronal hypersensitivity

Similar to other pain conditions, manifestations of neuronal hypersensitivity are evident in migraine, including hyperalgesia and allodynia (pain from a typically nonnoxious stimulus) that typically manifests cutaneously (e.g., while brushing one's hair, showering, touching the face). Allodynia results from sensitization of central nociceptive neurons (Burstein et al., 2014) and is most typically evident during migraine attacks, particularly among those with high-frequency headache. A series of experiments by Burstein and colleagues (2000) showed that cutaneous allodynia occurs commonly during migraine attacks and interferes with response to triptan therapy, such that triptan efficacy is contingent upon administration prior to development of allodynia (Burstein and Jakubowski, 2004; Burstein et al., 2004). These findings in part underlie the clinical recommendation that effective acute treatment requires early administration of triptans (i.e., within an hour of pain onset and before central sensitization develops).

The most well-established preventive medications for episodic migraine are specific tricyclic antidepressants (e.g., amitriptyline), beta-blockers (e.g., propranolol), and anticonvulsants (valproic acid, topiramate; Silberstein et al., 2012). Their mechanisms of action for migraine are complex but involve, among others, central nervous system modulation via the serotonergic system and sympathetic inhibition via gamma-aminobutyric acid enhancement. Peripheral hypersensitivity also occurs in migraine, principally through sensitization of meningeal nociceptors, and preventive migraine medications may also act peripherally. Botulinum toxin A, for instance, selectively inhibits C-meningeal nociceptors relevant for mechanical pain (Burstein et al., 2014). Botulinum toxin A is administered peripherally through a series of injections to the head/neck and has US Food and Drug Administration approval for the prophylactic treatment of chronic migraine (Blumenfeld et al., 2010), though the clinical significance of its superiority to placebo injections appears somewhat modest (Jackson et al., 2012).

Frequent use of acute headache medications (e.g., triptans, analgesics) perpetuates neuronal hypersensitivity and can lead to the development of medication overuse headache (MOH). MOH is diagnosed when an individual has chronic headache frequency in conjunction with overuse of acute headache medications, which for most agents is defined as use on ≥10 days per month on average for more than 3 months (Headache Classification Committee of the International Headache Society, 2013). All acute headache agents can produce MOH, but opioids are the most common culprit, and both the American Headache Society and American Academy of Neurology explicitly recommend against using opioids (or butalbital) for migraine except as a last resort (Langer-Gould et al., 2013; Loder et al., 2013). Chronic opioid administration not only results in a gradual loss of analgesic efficacy but produces increased excitatory neurotransmission at the dorsal horn and nucleus caudalis and increased expression of calcitonin gene-related peptide (CGRP) in trigeminal primary afferent neurons (Gardell et al., 2002; De Felice and Porreca, 2009). Chronic administration of nonnarcotic analgesics or triptans produces changes in serotonin metabolism and results in serotonin depletion and further impairment of pain inhibition (Srikiatkhachorn et al., 2000; Ayzenberg et al., 2008). Frequent use of acute migraine agents thus serves to further sensitize an already-delicate nervous system and set the stage for progressive increases in headache frequency. As a result, withdrawal from overused acute medications, sometimes in an inpatient setting, is the standard of care for initial treatment of MOH, often in combination with initiation of an appropriate migraine preventive.

Serotonergic availability

Central serotonergic dysfunction appears to play a role in migraine pathogenesis, but this contention is largely based on circumstantial evidence, including the high comorbidity with depression and the efficacy of low-dose tricyclic antidepressants for migraine prevention (Silberstein et al., 2012). (Notably, however, selective serotonin reuptake inhibitors (SSRIs) and serotonin–norepinephrine reuptake inhibitors (SNRIs) do not have established efficacy for migraine prevention; Banzi et al., 2015.) Low serotonergic bioavailability contributes to risk for cortical spreading depression and sensitization of trigeminovascular pathways, and plasma serotonin concentrations are increased during migraine attacks but lowered interictally (Hamel, 2007).

The migraine-specific acute treatments of choice are $5\text{-HT}_{1B/1D}$ agonists (i.e., triptans); ergotamine derivatives likewise increase serotonergic availability, but with lower selective affinity. The efficacy of triptans is well established across numerous clinical trials (Ferrari et al., 2002; Marmura et al., 2015) and is presumably related to their action of constricting cerebral blood vessels and tempering trigeminovascular pain pathways. However, the most direct evidence of serotonergic involvement comes from positron emission topography tracer studies showing increased brain synthesis of serotonin in migraine without aura (Chugani et al., 1999) that is reduced after triptan administration (Sakai et al., 2008, 2014). Though presently few in number, studies such as these that explore *in vivo* metabolism of serotonin in the brain hold great promise for advancing our understanding of this neurotransmitter in migraine.

CGRP

CGRP causes vasodilation and neurogenic inflammation when released at trigeminal sensory neurons, and at present CGRP is the most-researched target for migraine drug development. Initial efforts in developing CGRP antagonists as acute migraine treatments showed promise but were abandoned after a large multicenter randomized controlled trial (RCT) was terminated because 13 patients taking telcagepant developed significant liver aminotransferase elevations (Ho et al., 2014). Since then, drug development efforts have centered on humanized monoclonal antibodies against CGRP and its receptors. Long half-lives allow for periodic intravenous injections and thus use as preventive migraine agents, potentially with less risk for hepatic toxicity than acute CGRP antagonists and less cardiovascular risk than the triptans (Karsan and Goadsby, 2015).

Many headache specialists believe that these anti-CGRP antibodies may be the next major pharmacologic breakthrough in migraine treatment. Several have been developed and are in testing phases by numerous drug companies, and recent phase II trial results are encouraging. In one trial, 218 episodic migraine patients (with 4–14 headache days/month) across 35 sites were randomized to 150 mg of LY2951742 or placebo subcutaneous injections every 2 weeks for 12 weeks (Dodick et al., 2014a). LY2951742 significantly outperformed placebo on the primary endpoint of changes in monthly headache frequency from baseline at week 12 (4.2 vs. 3.0 days/month reduction). These frequency reductions correspond to a 62.5% decrease for the CGRP group and a 42.3% decrease for the placebo group. Improvements in secondary endpoints of migraine days and migraine attacks were also superior for LY2951742, and the treatment was well tolerated. In another trial, 174 patients with episodic migraine (5–14 days/month) across 26 sites were randomized to a single intravenous injection of 1000 mg ALD403 or placebo and followed for 24 weeks (Dodick et al., 2014b). ALD403 yielded significantly greater reductions in the primary endpoint of monthly migraine days than placebo at 5–8 weeks, though again the magnitude of this effect was rather modest (1.0 fewer migraine days/month; 5.6 fewer days vs. 4.6 fewer days for placebo).

Both of these trials found that the CGRP antagonists were generally well tolerated and appeared safe, but they were not large enough to detect rare serious adverse events or long enough to establish long-term efficacy or safety. Both also had high placebo response rates, which is not uncommon in preventive migraine trials but does call into question the clinical significance of the reductions in headache frequency versus placebo in individuals with episodic migraine at these frequencies. Two other recent phase II clinical trials have shown efficacy versus placebo for yet another anti-CGRP monoclonal antibody, TEV-48125, among patients with "high-frequency episodic migraine" (8–14 migraine days/month; Bigal et al., 2015a) and chronic migraine (Bigal et al., 2015b). These phase II trials set the stage for forthcoming phase III trials designed to address limitations inherent to prior trials by assessing efficacy and safety at varying doses and longer intervals. Moreover, they help establish a mechanistic role for CGRP in migraine pathogenesis.

Psychologic mechanisms

To the extent that their co-occurrence is driven at least in part by shared mechanisms, the study of conditions commonly comorbid with migraine may help further our understanding of migraine pathology.

Insomnia

Sleep disruptions are commonly associated with migraine, with 38% of migraineurs sleeping 6 hours per night on average (Kelman and Rains, 2005). Headache frequency is associated with sleep problems such that individuals with chronic migraine sleep less and experience more difficulty falling and staying asleep than individuals with episodic migraine. Nonetheless, it is unclear whether increased headache frequency is the cause or result of disrupted sleep. The co-occurrence of sleep disturbances and migraine is likely the product of shared pathophysiology. Physiologic mechanisms include serotonergic influences and decreased concentrations of melatonin, both of which are affected by fluctuations in circadian biorhythms as a function of hypothalamic activity (Holland, 2014). In conjunction with physiologic underpinnings, common behaviors for coping with migraine can disrupt sleep. These include napping in a dark room and taking either sedating medications or caffeine-containing compounds; frequent engagement in these behaviors may perpetuate nighttime sleep disruptions and disturb circadian pathways (Ong and Park, 2012).

Of the sleep disorders comorbid with migraine, insomnia is the most common. Insomnia is defined as difficulty falling or staying asleep, early-morning awakening, or nonrestorative sleep, all of which contribute to impairment in daytime functioning. Behavioral interventions designed to eliminate sleep-interfering behaviors and reassociate the bed with sleep have well-established efficacy for insomnia (Morgenthaler et al., 2006) that rivals pharmacotherapy (Jacobs et al., 2004). Two small RCTs have examined the efficacy of behavioral insomnia treatments for individuals with chronic migraine and comorbid insomnia. In both, adults with chronic migraine and comorbid insomnia were randomly assigned to receive either behavioral treatment (e.g., stimulus control, sleep restriction, visualization techniques) or "sham" behavioral instructions to change presumably irrelevant lifestyle behaviors (e.g., training in and daily practice of skills pertaining to consistent food/liquid intake, range-of-motion exercises, acupressure). In one, migraineurs who received a single session of behavioral insomnia treatment obtained significantly greater reductions in headache frequency and severity than the control group, and 49% of the treatment group reverted from chronic to episodic migraine (Calhoun and Ford, 2007). In the other, individuals receiving three sessions of behavioral insomnia treatment were 60% less likely than controls to experience headache at 6-week follow-up and evidenced improved sleep time and sleep efficiency as quantified by actigraphy (Smitherman et al., 2016).

Preliminary evidence from these small trials thus highlights the role of insomnia-related sleep behaviors as a maintaining factor in chronic migraine and provides some evidence that treating insomnia in turn improves migraine, even if migraine itself is not directly targeted. Nonetheless, the control treatment in one of these trials also obtained nontrivial reductions in migraine frequency (Smitherman et al., 2016), highlighting the continued difficulty in administering truly inert pseudotherapy control "treatments." This limitation is in part a function of the inability to double-blind almost all behavioral trials, as patients are aware of their treatment group because they must practice the skills they are taught and the therapists are aware of what skills they are teaching. A related difficulty is that any behavioral intervention involves interpersonal contact with an empathic provider. As such, pragmatic challenges surround efforts to determine whether observed reductions in headache frequency are attributable to changes in sleep behaviors or to nonspecific factors common to the interpersonal context of the intervention. Nevertheless, these trials convey the value of targeting comorbid insomnia as a means of reducing migraine, but larger studies including comparisons with established migraine preventive medications are needed.

Obesity

As with insomnia, multiple mechanisms may explain the previously described association between obesity and migraine. Cytokines that regulate weight through their effects on appetite and metabolism may promote an inflammatory process common to both conditions. Additionally, lifestyle factors such as stress, decreased physical activity, and the consumption of a high-fat, high-calorie diet are risk factors for both obesity and migraine (Bond et al., 2011a).

Despite strong empiric support for the association between obesity and migraine, few studies have examined the impact of weight loss on migraine. Two independent studies followed patients with migraine who received bariatric surgery to determine whether weight loss following bariatric surgery was associated with improvements in migraine (Bond et al., 2011b; Novack et al., 2011). In both studies, patients reported significant reductions in headache and headache-related disability after weight loss surgery. Patients also reported using fewer acute migraine medications.

In an effort to examine the effects of behavioral treatment of obesity on migraine, 135 obese adolescents with migraine participated in a 12-month treatment program that included dietary education, aerobic training, and cognitive-behavioral therapy (CBT). Significant decreases in weight, waist circumference, headache frequency, headache severity, and acute migraine medication use were observed 6 months after the start of the treatment program and maintained over a 12-month period. Lower body mass index (BMI) at baseline and greater change in BMI from baseline to 12 months were associated with greater decreases in headache frequency, intensity, and medication intake (Verrotti et al., 2013).

Reductions in body weight and headache following surgical and behavioral weight loss interventions indicate that obesity and migraine are closely linked and suggest that weight loss may be incorporated as a sort of "prophylactic" treatment for individuals with migraine. Despite these promising results, previous studies are limited by the absence of a control group and relatively small sample sizes. Moreover, Verrotti and colleagues (2013) failed to measure weight control behaviors; therefore, it is unclear whether improvements in migraine following behavioral weight loss treatment were attributable to weight loss or to alterations in specific behaviors such as dietary intake and exercise. Bond and colleagues (2013) are attempting to address these methodologic shortcomings in a National Institutes of Health-funded RCT designed to examine whether weight loss achieved via behavioral treatment is effective in reducing headache frequency in adults with migraine.

Affective disorders

Proposed mechanisms for the established bidirectional relationship between migraine and affective disorders (e.g., depression, anxiety) include serotonergic and dopaminergic dysfunctions, ovarian hormone fluctuations (for women), hypothalamic–pituitary–adrenal axis dysregulation, and sensitization of both sensory and affective neural pathways (Baskin and Smitherman, 2009). Mood and anxiety disorders are relevant clinically because they are assumed to predict a poor response to migraine treatment; however, few studies have addressed this question directly. A recent exploratory RCT evaluated the efficacy of CBT for comorbid headache and depression (Martin et al., 2015). Patients with headache and major depressive disorder were randomized to receive either 12 weekly sessions of CBT that addressed both headache and depression (e.g., relaxation training, social skills training, cognitive restructuring, relaxation training, pain and trigger management) + routine primary care or routine primary care alone. Compared to routine primary care alone, CBT + routine primary care produced significantly greater reductions in daily headache intensity and scores on two measures of depression from baseline to posttreatment and at 4-month follow-up.

Seng and Holroyd (2012) evaluated 177 participants with migraine in order to determine whether the presence of a mood or anxiety disorder was associated with poorer response to pharmacologic and behavioral migraine treatments. Contrary to historic assumptions, participants diagnosed with mood and/or anxiety disorders reported larger reductions in migraine frequency and disability than participants who did not receive either diagnosis. Considered together, these findings suggest that preventive migraine treatments hold promise for improving symptoms of comorbid mood and anxiety disorders. Nonetheless, additional research is needed that examines the efficacy of pharmacologic treatments, as many trials of preventive pharmacologic migraine treatments exclude individuals with comorbid anxiety or depression, including trials of antidepressants. Although the tricyclic antidepressant amitriptyline is efficacious for migraine prevention (Silberstein et al., 2012), the SSRIs and SNRIs lack efficacy for migraine (Banzi et al., 2015). As a result, it is unclear how individuals with comorbid migraine and affective disorders respond to preventive pharmacologic migraine treatments.

Behavioral and cognitive-behavioral treatments for migraine

Behavioral and established preventive drug treatments for migraine produce comparable reductions in migraine frequency (35–50% reduction on average; Penzien et al., 2002; Rains et al., 2005; Andrasik, 2007). Behavioral therapy and CBT for migraine teach patients to regulate their migraine-related physiology, behaviors, and cognitions via relaxation training, thermal or electromyographic biofeedback, and stress management training, with the ultimate goal of preventing and managing future migraine attacks. Not only are these interventions efficacious in reducing migraine when administered alone, results of two large RCTs indicate that combining preventive drug treatment with behavioral migraine treatment is superior to either modality alone.

In the first, Holroyd and colleagues (2010) randomly assigned 232 adults with migraine to receive one of four preventive treatments: beta-blocker, matched placebo, behavioral migraine management plus placebo, or behavioral migraine management plus beta-blocker. All patients received optimized acute treatment, including triptans. The combination of behavioral migraine management and beta-blocker but neither active treatment alone decreased headache frequency in comparison to optimized acute treatment. In the second, Powers and colleagues (2013) randomly assigned 135 children and adolescents to receive CBT plus amitriptyline or basic headache education plus amitriptyline. After 20 weeks, patients who received CBT plus amitriptyline endorsed greater reductions in days with headache and headache-related disability in comparison to those in the headache education group. These well-conducted RCTs provide compelling evidence that the combination of behavioral and pharmacologic treatments is efficacious in reducing migraine in children, adolescents, and adults.

Mechanisms of behavioral treatments

Despite a large body of well-controlled research documenting the effectiveness of behavioral and cognitive-behavioral treatments for migraine, the precise mechanisms of change are not well established. Proposed mechanisms include vascular changes, increases in self-efficacy, and reductions in pain-related anxiety.

Although physiologic changes are the expressed target of interventions such as biofeedback and relaxation training, the vascular theory of migraine has long since been discredited. Psychologic variables may instead underlie behavioral migraine treatments. This was shown eloquently in a classic study that examined electromyographic (EMG) biofeedback treatment for headache (Holroyd et al., 1984). Regardless of actual changes in muscle tension, participants in the study who believed that they were highly successful in altering muscle tension showed significantly greater improvements in headache activity in comparison to those who believed they were less successful. As such, the effectiveness of EMG biofeedback was concluded to be attributable to changes in self-efficacy rather than to changes in muscle tension. Headache self-efficacy refers to confidence in one's ability to utilize behavioral skills to prevent and manage the occurrence of headache. Studies conducted with tension-type headache patients indicate that EMG biofeedback and CBT are successful in treating headache in part because they increase headache self-efficacy. Seng and Holroyd (2010) found that individuals who received behavioral treatment demonstrated large increases in headache self-efficacy that endured over a 12-month period. Increases in self-efficacy following behavioral treatment were significantly larger than increases in self-efficacy following migraine drug therapy, which did not differ from placebo.

Behavioral treatments for migraine also may be effective in part due to their effects on anxiety and fear related to pain. The Fear-Avoidance Model of pain posits that excessive fear of pain and pain stimuli prompts unwarranted avoidance behavior, which over time promotes increased sensitivity to pain, reduced exposure to reinforcing activities, and social isolation (Lethem et al., 1983; Vlaeyen and Linton, 2000, 2012). A recent study of headache sufferers confirmed that fear of pain was positively associated with headache severity, frequency, and disability and partly mediated the relationship between headache severity and disability (Black et al., 2015). Despite these promising findings, no studies to date have examined fear of pain as a mediator of migraine treatment outcomes or evaluated the utility of directly targeting fear of pain in migraine.

CONCLUSION

Migraine is an extremely prevalent and disabling neurologic disorder that manifests as periodic attacks of severe head pain and is accompanied by associated symptoms of interference with activity, nausea/vomiting, and sensitivity to light and sound. Despite its high prevalence and impact, migraine remains underfunded at the national level, though many recent advances have been made. One such advance is the identification of both demographic and modifiable risk factors for migraine onset and progression, the latter of which includes common comorbid disorders and overuse of acute headache medications.

In parallel with these advances, understanding of migraine pathophysiology has improved but remains incomplete. Migraine is recognized as a recurrent, often chronic condition with strong genetic influences, though the precise genetic underpinnings have been identified only for familial hemiplegic migraine, a rare form of migraine with aura. Existing pathophysiology studies with rodents and humans have confirmed that migraine is fundamentally a disorder involving a hypersensitive central nervous system. This neuronal hypersensitivity manifests both peripherally and centrally and is modulated by both triptans and well-established migraine preventives. Dysfunctional serotonergic transmission and availability have been implicated largely through indirect lines of evidence. At present, however, most novel drug development efforts center on anti-CGRP monoclonal antibodies, many of which are now entering phase III clinical trials. Clinical trials of these and other pharmacologic interventions have highlighted some of the many mechanisms that underlie migraine.

Beyond these reductionist approaches to migraine, other lines of research pertinent to epidemiologic mechanisms center on exploring common migraine comorbidities and psychologic variables that predict response to treatment. Foremost among these comorbidities are insomnia, obesity, depression, and anxiety, and small trials targeting these comorbidities in treatment to assess effects on migraine have begun to appear and inform understanding of shared pathophysiology. Other treatment studies have confirmed the efficacy of behavioral (nonpharmacologic) migraine interventions, particularly in combination with pharmacotherapy. The mechanism of action of these behavioral interventions is not firmly established, but current thinking is that these treatments work principally through enhancing perceptions of self-efficacy and implementation of lifestyle skills that help prevent migraine attacks. These studies do not imply that migraine is a psychologic condition, but instead that it is amenable to behavior change interventions, as are so many other chronic medical conditions.

Despite these advances, major challenges remain in research on migraine, particularly experimental epidemiology. These include the lack of a reliable biomarker that predicates reliance on phenotypic classification and the need for increased funding for further GWAS. Studies on CGRP are promising but at present limited by high placebo response rates, though forthcoming phase III trials should elucidate efficacy at varying doses and over longer intervals. Studies of migraine comorbidities and their treatment have been mostly nonexperimental in nature, and existing clinical trials have used small samples and thus merit replication in larger designs. Behavioral trials are by nature limited by the inability of proper double blinding and the difficulty in establishing a truly inert behavioral placebo; these issues complicate establishment of mechanisms of action. Studies that satisfactorily address these issues will continue to advance our understanding of migraine epidemiology and mechanisms.

References

Andrasik F (2007). What does the evidence show? Efficacy of behavioral treatments for recurrent headaches in adults. Neurol Sci 28: S70–S77.

Ayzenberg I, Oberman M, Leineweber K et al. (2008). Increased activity of serotonin uptake in platelets in medication overuse headache following regular intake of analgesics and triptans. J Headache Pain 9: 109–112.

Banzi R, Cusi C, Randazzo C et al. (2015). Selective serotonin re-uptake inhibitors (SSRIs) and serotonin-norepinephrine reuptake inhibitors (SNRIs) for the prevention of migraine in adults (Review). Cochrane Database Syst Rev 4. CD002919.

Bashir A, Lipton RB, Ashina S et al. (2015). Migraine and structural changes in the brain. Neurology 81: 1260–1268.

Baskin SM, Smitherman TA (2009). Migraine and psychiatric disorders: comorbidities, mechanisms, and clinical applications. Neurol Sci 30: S61–S65.

Bigal M, Lipton R (2006a). Modifiable risk factors for migraine progression. Headache 46: 1334–1343.

Bigal ME, Lipton RB (2006b). Obesity is a risk factor for transformed migraine but not chronic tension-type headache. Neurology 67: 252–257.

Bigal M, Serrano D, Buse D et al. (2008). Acute migraine medications and evolution from episodic to chronic migraine: a longitudinal population-based study. Headache 48: 1157–1168.

Bigal ME, Dodick DW, Rapoport AM et al. (2015a). Safety, tolerability, and efficacy of TEV-48125 for preventive treatment of high-frequency episodic migraine: a multicenter, randomised, double-blind, placebo-controlled, phase 2b study. Lancet Neurol. Retrieved from, http://dx.doi.org/10.1016/S1474-4422(15)00249-5.

Bigal ME, Edvinsson L, Rapoport AM et al. (2015b). Safety, tolerability, and efficacy of TEV-48125 for preventive treatment of chronic migraine: a multicenter, randomised, double-blind, placebo-controlled, phase 2b study. Lancet Neurol. Retrieved from, http://dx.doi.org/10.1016/S1474-4422(15)00245-8.

Black AK, Fulwiler JC, Smitherman TA (2015). The role of fear of pain in headache. Headache 55: 669–679.

Blumenfeld A, Silberstein SD, Dodick DW et al. (2010). Method of injection of onabotulinumtoxinA for chronic migraine: a safe, well-tolerated, and effective treatment paradigm based on the PREEMPT clinical program. Headache 50: 1406–1418.

Bond DS, Roth J, Nash JM et al. (2011a). Migraine and obesity: epidemiology, possible mechanisms and the potential role of weight loss treatment. Obes Rev 12: e362–e371.

Bond DS, Vithiananthan S, Nash JM et al. (2011b). Improvement of migraine headaches in severely obese patients after bariatric surgery. Neurology 76: 1135–1138.

Bond DS, O'Leary KC, Thomas JG et al. (2013). Can weight loss improve migraine headaches in obese women? Rationale and design of the Women's Health and Migraine (WHAM) randomized controlled trial. Contemp Clin Trials 35: 133–144.

Breslau N (1998). Psychiatric comorbidity in migraine. Cephalalgia 18 (Supp. 22): 56–58.

Breslau N, Schultz L, Lipton R et al. (2012). Migraine headaches and suicide attempt. Headache 52: 723–731.

Burstein R, Jakubowski M (2004). Analgesic triptan action in an animal model of intracranial pain: a race against central sensitization. Ann Neurol 55: 27–36.

Burstein R, Yarnitsky D, Goor-Arheh I et al. (2000). An association between migraine and cutaneous allodynia. Ann Neurol 47: 614–624.

Burstein R, Collins B, Jakubowski M (2004). Defeating migraine pain with triptans: a race against the development of cutaneous allodynia. Ann Neurol 55: 19–26.

Burstein R, Zhang X, Levy D et al. (2014). Selective inhibition of meningeal nociceptors by botulinum neurotoxin type A: therapeutic implications for migraine and other pains. Cephalalgia 34: 853–869.

Buse D, Silberstein S, Manack A et al. (2013). Psychiatric comorbidities of episodic and chronic migraine. J Neurol 260: 1960.

Calhoun AH, Ford S (2007). Behavioral sleep modification may revert transformed migraine to episodic migraine. Headache 47: 1178–1183.

Chugani DC, Niimura K, Chaturvedi S et al. (1999). Increased brain serotonin synthesis in migraine. Neurology 53: 1473–1479.

De Felice M, Porreca F (2009). Opiate-induced persistent pronociceptive trigeminal neural adaptations: potential relevance to opiate-induced medication overuse headache. Cephalalgia 29: 1277–1284.

Dodick DW, Goadsby PJ, Spierings ELH et al. (2014a). Safety and efficacy of LY2951742, a monoclonal antibody to calcitonin gene-related peptide, for the prevention of migraine: a phase 2, randomised, double-blind, placebo-controlled study. Lancet Neurol 13: 885–892.

Dodick DW, Goadsby PJ, Silberstein SD et al. (2014b). Safety and efficacy of ALD403, an antibody to calcitonin gene-related peptide, for the prevention of frequent episodic migraine: a randomised, double-blind, placebo-controlled, exploratory phase 2 trial. Lancet Neurol 13: 1100–1107.

Ducros A, Tournier-Lasserve E, Bousser M (2002). Review: the genetics of migraine. Lancet 1: 285–293.

Esserlind AL, Christensen AF, Le H et al. (2013). Replication and meta-analysis of common variants identifies a genome-wide significant locus in migraine. Eur J Neurol 20: 765–772.

Ferrari MD, Goadsby PJ, Roon KI et al. (2002). Triptans (serotonin, 5-HT1B/1D agonists) in migraine: detailed results and methods of a meta-analysis of 53 trials. Cephalalgia 22: 633–658.

Fornaro M, De Berardis D, De Pasquale C et al. (2015). Prevalence and clinical features associated to bipolar disorder–migraine comorbidity: a systematic review. Compr Psychiat 56: 1–16.

Gardell LR, Wang R, Burgess SE et al. (2002). Sustained morphine exposure induces a spinal dynorphin-dependent enhancement of excitatory transmitter release from primary afferent fibers. J Neurosci 22: 6747–6755.

Goadsby PJ, Charbit AR, Andreou AP et al. (2009). Neurobiology of migraine. Neuroscience 161: 327–341.

Guidetti V, Galli F, Fabrizi P et al. (1998). Headache and psychiatric comorbidity: clinical aspects and outcome in an 8-year follow-up study. Cephalalgia 18: 455–462.

Hamel E (2007). Serotonin and migraine: biology and clinical implications. Cephalalgia 27: 1295–1300.

Hamelsky S, Lipton R (2006). Psychiatric comorbidity of migraine. Headache 46: 1327–1333.

Headache Classification Committee of the International Headache Society (2013). The International Classification of Headache Disorders, 3rd edn (beta version). Cephalalgia 33: 629–808.

Ho TW, Conner KM, Zhang Y et al. (2014). Randomized controlled trial of the CGRP receptor antagonist telcagepant for migraine prevention. Neurology 83: 958–966.

Holland PR (2014). Headache and sleep: shared pathophysiological mechanisms. Cephalalgia 34: 725–744.

Holroyd KA, Cottrell CK, O'Donnell FJ et al. (2010). Effect of preventive (beta blocker) treatment, behavioural migraine management, or their combination on outcomes of optimised acute treatment in frequent migraine: randomised controlled trial. BMJ 341: c4871.

Holroyd KA, Penzien DB, Hursey KG et al. (1984). Change mechanisms in EMG biofeedback training: cognitive changes underlying improvements in tension headache. J Consul Clin Psycol 52: 1039–1053.

Jackson JL, Kuriyama A, Hayashino Y (2012). Botulinum toxin A for prophylactic treatment of migraine and tension headaches in adults. JAMA 307: 1736–1745.

Jacobs GD, Pace-Shott EF, Stickgold R et al. (2004). Cognitive behavior therapy and pharmacotherapy for insomnia: a randomized controlled trial and direct comparison. Arch Intern Med 164: 1888–1896.

Jette N, Patten S, Williams J et al. (2008). Comorbidity of migraine and psychiatric disorders – a national population-based study. Headache 48: 501–516.

Juang KD, Wang SJ, Fuh JL et al. (2000). Comorbidity of depressive and anxiety disorders in chronic daily headache and its subtypes. Headache 40: 818–823.

Karsan N, Goadsby PJ (2015). CGRP mechanism antagonists and migraine management. Curr Neurol Neurosci Rep 15: 25.

Kelman L (2007). The triggers or precipitants of the acute migraine attack. Cephalalgia 27: 394–402.

Kelman L, Rains JC (2005). Headache and sleep: examination of sleep patterns and complaints in a large clinical sample of migraineurs. Headache 45: 904–910.

Langer-Gould AM, Anderson WE, Cohen AB et al. (2013). The American Academy of Neurology's top five choosing wisely recommendations. Neurology 81: 1004–1011.

Lethem J, Slade PD, Troup JD et al. (1983). Outline of a Fear-Avoidance Model of exaggerated pain perception – I. Behav Res Ther 21: 401–408.

Lipton RB, Stewart WF, Simon D (1998). Medical consultation for migraine: results from the American Migraine Study. Headache 38: 87–96.

Lipton RB, Bigal ME, Diamond M et al. (2007). Migraine prevalence, disease burden, and the need for preventive therapy. Neurology 68: 343–349.

Loder E, Weizenbaum E, Frishberg B et al. (2013). Choosing wisely in headache medicine: The American Headache Society's list of five things physicians and patients should question. Headache 53: 1651–1659.

Malone CD, Bhowmick A, Wachholtz AB (2015). Migraine: treatments, comorbidities, and quality of life, in the USA. J Pain Res 8: 537–547.

Marmura MJ, Silberstein SD, Schwedt TJ (2015). The acute treatment of migraine in adults: The American Headache

Society evidence assessment of migraine pharmacotherapies. Headache 55: 3–20.

Martin PR, Aiello R, Gilson K et al. (2015). Cognitive behavior therapy for comorbid migraine and/or tension-type headache and major depressive disorder: an exploratory randomized controlled trial. Behav Res Ther 73: 8–18.

Mawet J, Kurth T, Avata C (2015). Migraine and stroke: in search of shared mechanisms. Cephalalgia 35: 165–181.

Morgenthaler T, Kramer M, Alessi C et al. (2006). Practice parameters for the psychological and behavioral treatment of insomnia: an update. An American Academy of Sleep Medicine report. Sleep 29: 1415–1419.

Novack V, Fuchs L, Lantsberg L et al. (2011). Changes in headache frequency in premenopausal women with migraine after bariatric surgery: a case series. Cephalalgia 31: 1336–1342.

Ong JC, Park M (2012). Chronic headaches and insomnia: working toward a biobehavioral model. Cephalalgia 32: 1059–1070.

Penzien D, Rains J, Andrasik F (2002). Behavioral management of recurrent headache: three decades of experience and empiricism. Appl Psychophysiol Biofeedback 27: 163–181.

Powers SW, Kashikar-Zuck S, Allen JR et al. (2013). Cognitive behavioral therapy plus amitriptyline for chronic migraine in children and adolescents: a randomized clinical trial. JAMA 310: 2622–2630.

Radat F, Creac'h C, Swendsen J et al. (2005). Psychiatric comorbidity in the evolution from migraine to medication overuse headache. Cephalalgia 25: 519–522.

Rains J, Poceta J (2006). Headache and sleep disorders: review and clinical implications for headache management. Headache 46: 1344–1363.

Rains JC, Penzien DB, McCrory DC et al. (2005). Behavioral headache treatment: history, review of the empirical literature, and methodological critique. Headache 45: S92–S109.

Rothrock J, Lopez I, Zweilfer R et al. (2007). Borderline personality disorder and migraine. Headache 47: 22–26.

Sakai Y, Dobson C, Diksic M et al. (2008). Sumatriptan normalizes the migraine attack-related increase in brain serotonin synthesis. Neurology 70: 431–439.

Sakai Y, Nishikawa M, Diksic M et al. (2014). α-[^{11}C] methyl-L tryptophan-PET as a surrogate for interictal cerebral serotonin synthesis in migraine without aura. Cephalalgia 34: 165–173.

Saunders K, Merikangas K, Low N et al. (2008). Impact of comorbidity on headache-related disability. Neurology 70: 538–547.

Scher A, Stewart W, Ricci J et al. (2003). Factors associated with the onset and remission of chronic daily headache in a population-based study. Pain 106: 81–89.

Scher A, Midgette L, Lipton R (2008). Risk factors for headache chronification. Headache 48: 16–25.

Schürks M (2012). Genetics of migraine in the age of genome-wide association studies. J Headache Pain 13: 1–9.

Schürks M, Rist P, Bigal M et al. (2009). Migraine and cardiovascular disease: systematic review and meta-analysis. Br Med J 339: b3914.

Schürks M, Rist PM, Kurth T (2010). *MTHFR* 677C > T and *ACE* D/I polymorphisms in migraine: a systematic review and meta-analysis. Headache 50: 588–599.

Seng EK, Holroyd KA (2010). Dynamics of changes in self-efficacy and locus of control expectancies in the behavioral and drug treatment of severe migraine. Ann Behav Med 40: 235–247.

Seng EK, Holroyd KA (2012). Psychiatric comorbidity and response to preventative therapy in the treatment of severe migraine trial. Cephalalgia 32: 390–400.

Shapiro RE, Goadbsy PJ (2007). The long drought: the dearth of public funding for headache research. Cephalalgia 27: 991–994.

Silberstein SD, Holland S, Freitag F et al. (2012). Evidence-based guideline update: pharmacologic treatment for episodic migraine prevention in adults: report of the Quality Standards Subcommittee of the American Academy of Neurology and the American Headache Society. Neurology 78: 1337–1345.

Smitherman TA, Kolivas ED (2013). Trauma exposure versus posttraumatic stress disorder: relative associations with migraine. Headache 53: 775–786.

Smitherman TA, Burch R, Sheikh H et al. (2013). The prevalence, impact, and treatment of migraine and severe headaches in the United States: a review of statistics from national surveillance studies. Headache 53: 427–436.

Smitherman TA, Walters AB, Davis RE et al. (2016). Randomized controlled pilot trial of behavioral insomnia treatment for chronic migraine with comorbid insomnia. Headache 56: 276–291.

Srikiatkhachorn A, Tarasub N, Govitrapong P (2000). Effect of chronic analgesic exposure on the central serotonin system: a possible mechanism of analgesic abuse headache. Headache 40: 343–350.

Stewart WF, Roy J, Lipton RB (2013). Migraine prevalence, socioeconomic status, and social causation. Neurology 81: 948–955.

Stovner LJ, Hagen K, Jensen R et al. (2007). The global burden of headache: a documentation of headache prevalence and disability worldwide. Cephalalgia 27: 193–210.

Turner DP, Smitherman TA, Martin VT et al. (2013). Causality and headache triggers. Headache 53: 628–635.

Ulrich VB, Gervil M, Kyvik KO et al. (2001). Evidence of a genetic factor in migraine with aura: a population-based Danish twin study. Eur J Neurol 8: 33–37.

Verrotti A, Agostinelli S, D'Egidio C et al. (2013). Impact of a weight loss program on migraine in obese adolescents. Eur J Neurol 20: 394–397.

Vlaeyen JWS, Linton SJ (2000). Fear-avoidance and its consequences in chronic musculoskeletal pain: a state of the art. Pain 85: 317–332.

Vlaeyen JWS, Linton SJ (2012). Fear-avoidance model of chronic musculoskeletal pain: 12 years on. Pain 153: 1144–1147.

Vos T, Flaxman A, Naghavi et al. (2012). Years lived with disability (YLDs) for 1160 sequelae of 289 diseases and injuries 1990–2010: a systematic analysis for the Global Burden of Disease Study 2010. Lancet 380: 2163–2196.

Wessman M, Terwindt GM, Kaunisto MA et al. (2007). Migraine: a complex genetic disorder. Lancet Neurol 6: 521–532.

Section 3

Neurocognitive consequences of conditions originating outside the central nervous system

Handbook of Clinical Neurology, Vol. 138 (3rd series)
Neuroepidemiology
C. Rosano, M.A. Ikram, and M. Ganguli, Editors
http://dx.doi.org/10.1016/B978-0-12-802973-2.00017-3

Chapter 17

Neuroepidemiology of cancer and treatment-related neurocognitive dysfunction in adult-onset cancer patients and survivors

P. JEAN-PIERRE[1,2,3]* AND B.C. MCDONALD[3,4]

[1]*Cancer Neurocognitive Translational Research Laboratory, University of Notre Dame, Notre Dame, IN, USA*

[2]*Department of Psychology, University of Notre Dame, Notre Dame, IN, USA*

[3]*Indiana University Melvin and Bren Simon Cancer Center, Indianapolis, IN, USA*

[4]*Center for Neuroimaging, Department of Radiology and Imaging Sciences, Indiana University School of Medicine, Indianapolis, IN, USA*

Abstract

Cancer and treatment-related neurocognitive dysfunction (CRND) – impairments in aspects of cognition commonly including attention and memory, information-processing speed, and executive functioning – can negatively affect patients' and survivors' participation in routine activities of daily living and overall quality of life. CRND can be enduring, and varies in severity level. The epidemiology of CRND is not yet clearly established; reported incidence has ranged from 17% to 75% among patients and survivors of non-central nervous system malignancies, including breast, prostate, cervical, and colorectal cancers. Progress in the development of strategies for assessing and treating CRND has been delayed by limitations in the knowledge of the precise etiology of this adverse condition, as well as the lack of sufficiently sensitive and reliable methods to determine its presence and quantify its severity. In this chapter, we present a brief description of the concept of CRND and its clinical presentation, and discuss the descriptive epidemiology, pathophysiology, risk factors, and availability of treatment interventions. This neuroepidemiologic perspective provides a framework for characterizing CRND, determining its etiology, and understanding its negative effects on routine activities of daily living, to help support the development and testing of reliable interventions to treat this deleterious condition for patients and survivors.

INTRODUCTION

Cancer is a well-known, incapacitating, and potentially deadly disease affecting many individuals and their families across all sociodemographic spectra (e.g., age, sex, race, and ethnicity, education level, occupation and income, social class, spirituality, faith and religion, and culture) in diverse communities throughout the world. Cancer is the second leading cause of death worldwide at all income levels. In 2012, the worldwide incidence of cancer – excluding nonmelanoma skin cancers – was estimated at 14.1 million. The corresponding estimate for worldwide cancer-related deaths was 8.2 million in 2012 (American Cancer Society, 2015). Global and national estimates of cancer incidence and mortality are well documented (American Cancer Society, 2014, 2015; Stewart and Wild, 2014; Ferlay et al., 2015; National Cancer Institute, 2016). However, the incidence and severity of cancer and treatment-related biobehavioral adverse effects are generally not well known. For example, cancer and treatment-related cognitive impairments (e.g., problems in cognitive processes and

*Correspondence to: Pascal Jean-Pierre, PhD, MPH, Cancer Neurocognitive Translational Research Laboratory, Cancer Control and Survivorship Research Program, University of Notre Dame, Notre Dame IN 46566, USA. E-mail: PJeanPierre@nd.edu

behavioral outcomes such as attention and memory), the clinical course of this adverse condition, reliable methods for assessing its presence and severity, and possible effective interventions to help mitigate its impact on cancer patients' and survivors' psychosocial functioning and overall quality of life are not well characterized.

Malignant tumors and their treatments are associated with adverse effects that can negatively impact patients and survivors on multiple biobehavioral (e.g., fatigue, pain, peripheral neuropathy, nausea and emesis, and sleep impairments), neurocognitive (e.g., impairments in attention and memory, information-processing speed, and executive functioning), psychologic (e.g., anxiety, depression, self-perception, coping strategies, and life adjustment), and sociorelational (e.g., familial and interpersonal relationships) dimensions (van Dam et al., 1998; Wefel et al., 2004; Falleti et al., 2005; Vardy and Tannock, 2007; Dietrich et al., 2008; Luckett et al., 2011; Ganz, 2012; Myers, 2012; Cleeland et al., 2013; Sales et al., 2014; Faller et al., 2015; Kenzik et al., 2015; Kurita and Sjogren, 2015; Mackenzie, 2015; Saita et al., 2015). Relative to the data available regarding cancer mortality, however, cancer and treatment-related biobehavioral, neurocognitive, psychologic, and sociorelational sequelae are not well understood. Consequently, systematic efforts and strategies to alleviate the burden and suffering of these adverse conditions are still lacking. Difficulties understanding and treating adverse effects of cancer and its treatments are particularly salient when dealing with neurocognitive sequelae of noncentral nervous system (CNS) cancers, where the underlying etiology is yet to be fully characterized.

Brain cancers (e.g., astrocytoma, glioblastoma, glioma, medulloblastoma, and meningioma) and to some degree spinal cord cancers (including intramedullary and extramedullary tumors) and related treatment (e.g., neurosurgery, CNS-targeted radiotherapy, intrathecal chemotherapy) commonly cause neurocognitive dysfunction as a result of direct insult to CNS tissues, and patients are routinely counseled regarding the likelihood of these effects (Behin and Delattre, 2004; Schiff et al., 2015). In contrast, the detrimental cognitive sequelae resulting from non-CNS cancers (e.g., breast, colorectal, cervical, and prostate carcinomas) and their treatment would not be as readily expected given the lack of direct insult to the CNS. While awareness of these potential effects has increased dramatically in recent years, there remain important limitations in our understanding of the mechanisms underlying these changes, and few available resources with regard to prevention and treatment.

In the present chapter, we examine the effects of adult-onset non-CNS malignancies and their various treatments on neurocognitive functioning from a neuroepidemiologic perspective. Specifically, we present a brief overview of the concept of cancer and treatment-related neurocognitive dysfunction (CRND), its clinical presentation (e.g., presenting symptoms and complaints), descriptive epidemiology (e.g., incidence, prevalence, and mortality), pathophysiology (e.g., functional changes that normally accompany CRND), factors that enable its development and maintenance, and ongoing efforts (e.g., clinical trials of behavioral and pharmacologic interventions) to mitigate this detrimental neurocognitive ailment for cancer patients and survivors throughout the cancer diagnosis, care, and survivorship continuum.

BRIEF OVERVIEW OF CRND

A growing number of patients and survivors of non-CNS malignancies, including breast, colorectal, cervical, and prostate cancers, report experiencing neurocognitive dysfunction (e.g., difficulties paying attention, remembering information, and following routine conversations) before, during, and many years after completion of cancer treatment with cytotoxic chemotherapy, radiotherapy, hormonal therapy, immunotherapy, stem cell transplant, or combined treatment modalities (Wieneke and Dienst, 1995; van Dam et al., 1998; Schagen et al., 1999; Brezden et al., 2000; Ahles et al., 2002; Paraska and Bender, 2003; Saykin et al., 2003; Tchen et al., 2003; Tannock et al., 2004; Wefel et al., 2004; Falleti et al., 2005). CRND, commonly referred to as "chemobrain," can be a transient phenomenon that resolves after treatment completion, but for some individuals may be an enduring and incapacitating condition that can interfere with patients' and survivors' normal cognitive operations, participation in routine activities of daily living, ability to engage in economically gainful employment, involvement in regular social and interpersonal relationships, and normal psychosocial functioning. It should be distinctly noted, however, that the use of the term "chemobrain" – to denote the neurocognitive impairments related to cancer and its treatments – is somewhat of a misnomer (Hurria et al., 2007). Impairments in cognitive processes and behavioral outcomes related to non-CNS malignant tumors and their treatments have been linked to disease management modalities (e.g., radiotherapy, immunotherapy, and hormonal therapy) other than cytotoxic chemotherapy, and studies showing poorer cognitive performance in cancer patients relative to controls prior to adjuvant treatment suggest that patient vulnerability factors or aspects of the disease process itself may negatively affect cognition. Nonetheless, the term chemobrain is very popular, even while understanding of the actual causes of these symptoms is rapidly growing. In this chapter we will use the term CRND to refer to the damaging effects of cancer and its treatments on key cognitive domains of attention (e.g., focused or sustained attention, distractibility, and selective attention), memory (e.g., immediate, delayed, verbal, visual, short-term, or working memory), information-processing speed (e.g., cognitive data manipulation time), and executive functioning (e.g., ability to engage in planning, organizing, problem solving, abstract thinking, mental flexibility, and monitoring and correcting one's own behavior), which have significant psychosocial, economic, emotional, interpersonal, relational, and overall quality-of-life implications for cancer patients and survivors.

CRND generally involves mild to moderate impairments in important neurocognitive domains of attention, memory, information-processing speed, and executive functioning, and can impact the wellbeing and overall quality of life of patients and survivors throughout their cancer diagnosis, treatment, and survivorship continuum (Brezden et al., 2000; Ahles et al., 2002; Saykin et al., 2003; Tannock et al., 2004). Incidence of CRND has been reported and/or documented over the past few decades, with symptoms that are qualitatively different than the common cognitive declines generally associated with normal aging (van Dam et al., 1998; Schagen et al., 1999; Falleti et al., 2005), and researchers have reported higher incidence of neurocognitive deficits or declines in cancer patients and survivors as compared with matched healthy individuals without a history of cancer or cancer treatment (Schagen et al., 1999; Brezden et al., 2000).

The notion that cancer and its treatments can impact brain structures and negatively influence cognitive functioning has been suggested since at least the late 1950s. In 1958, Reitan published results of a study on the validity of the Trail Making Test as an indicator of organic brain damage that included 27 patients with brain cancer in a total sample of 200 patients with verified brain damage and 84 patients without clinical evidence of brain damage. In 1969, Gottschalk and his colleagues published a paper on the effects of total and half-body irradiation on cognitive and emotional processes. A recent literature search of the terms "cancer and cognitive function" using www.PubMed.gov revealed a total of 7336 peer-reviewed papers. Of the manuscripts, eight were published in the 1950s, 54 were published in the 1960s, 149 were published in the 1970s, 278 were published in the 1980s, 744 were published in the 1990s, 2725 were published in the 2000s, and 3384 were published from 2010 up to February 17, 2016 (including 78 manuscripts published in 2016). It is very likely that literature searches using MeSH terms and other search engines will reveal a parallel trend and exponential growth in peer-reviewed publications on CRND. The rapidly increasing interest in and publications about CRND underscore the fact that oncology researchers and clinicians are becoming more cognizant of the negative impact of cancer and its treatments on the brain and its functions.

Despite an increasing recognition of the impact of CRND on cancer survivorship and quality of life, studies that systematically describe the underlying biologic mechanisms involved in CRND are still needed. Future systematic studies that are adequately conceptualized and implemented will inform the development and testing of reliable assessment methods (e.g., neuropsychologic, biochemical, neuroimaging, and integrated approaches) and effective treatment interventions (e.g., behavioral, pharmacologic, or combined) for CRND.

CLINICAL PRESENTATION OF CRND

Individuals struggling with CRND often present with subjective complaints of difficulties paying attention, concentrating, remembering, following conversations, and multitasking, as well as a general sense of declining linguistic and arithmetic capabilities (Boykoff et al., 2009). The presence of CRND has been documented anecdotally and through several single-site studies. Reported severity of CRND generally varies across different individuals. Some cancer patients and survivors have reported very subtle to mild cognitive difficulties (e.g., increased efforts in mentation that do not markedly impact their participation in routine activities of daily functioning and overall quality of life), whereas others have reported more moderate to severe cognitive complications that interfere with their psychosocial activities, ability to maintain gainful employment, routine cognitive operations, and overall quality of life (Boykoff et al., 2009). The severity of reported CRND can be influenced by other factors, including genetics and psychologic predispositions to cognitive impairments, the defining biologic characteristics of their cancer, the treatment regimen, dosage, and associated toxicity level, brain structural changes and atrophy, and patients' and survivors' knowledge about cancer and cancer treatment-related adverse effects. Irrespective of its reported severity, CRND can be an enduring condition, and has been detected in cancer survivors many decades after completion of their cancer treatment (Ahles and Saykin, 2001; Ahles et al., 2002; Koppelmans et al., 2012). It is important to note that the great majority of work in this area to date has been conducted in breast cancer. In recent years, however, there have been studies demonstrating similar findings in other non-CNS cancer populations, including lung, renal cell, head and neck, lymphoma, ovarian, testicular, colorectal, prostate, and mixed cancer samples.

DESCRIPTIVE EPIDEMIOLOGY OF CRND

Understanding the epidemiology or distribution and determinants of CRND is essential to the development, testing, and implementation of reliable interventions to prevent or treat this adverse condition for cancer patients and survivors. Prior studies have provided insights into the incidence and possible magnitude of CRND (Wefel et al., 2004; Jean-Pierre et al., 2012). Most studies in this area, however, have important limitations that need to be carefully addressed in order to facilitate the compilation of more accurate and meaningful epidemiologic data on CRND. For instance, most studies on CRND generally involve small sample size, which is associated with low statistical power and can undermine the reliability of the findings (e.g., low probability of detecting a true effect, low positive predictive value; exaggerated estimates of the magnitude of a true effect when one is detected). Another limitation of most CRND studies involves cross-sectional study design. Despite their contributions to public health organization (e.g., estimation of incidence and prevalence, as well as risk factors of outcomes of interest in specific populations), cross-sectional studies only provide snapshots in time and might be difficult to replicate. Cross-sectional studies can also be influenced by selection bias, and posit many challenges to causal inference. Additional limitations with CRND studies include a lack of pretreatment baseline data and scarcity of longitudinal data to help characterize CRND, determine its clinical course, and evaluate the benefits of behavioral and/or pharmacologic interventions to treat this adverse condition for patients and survivors. Furthermore, the very limited available data in non-CNS cancer populations other than breast cancer does not allow for comparisons of CRND incidence, prevalence, and severity among different cancer types. The negative impact of cancer and its treatments on brain structures and cognitive functioning are increasingly being recognized, as evidenced by recent inclusion of CRND in the National Comprehensive Cancer Network Survivorship Guidelines (NCCN, 2016). The growing recognition of CRND will facilitate the development and testing of reliable assessment methods and interventions to treat CRND. These clinical research endeavors will also necessitate the systematic application of rigorous methodologic and statistical analytic tools and techniques in CRND research.

Incidence of CRND

As previously mentioned, reported incidence of CRND ranges from 17% to 75% across non-CNS malignancies (e.g., breast, prostate, head and neck, cervical, and colorectal cancers) and treatment modalities (e.g., chemotherapy, radiotherapy, immunotherapy, hormonal therapy, or combined treatments) (Tannock et al., 2004; Wefel et al., 2004; Correa and Ahles, 2008). This reported range of CRND occurrence is based on multiple single-site studies in cancer patients and survivors. The scope of CRND in the USA is not yet known. However, researchers have approximated the magnitude of CRND in an epidemiologic study using data from a US, stratified, multistage probability sample of civilian noninstitutionalized individuals who completed the 2001–2006 National Health and Nutrition Examination Survey, and reported a much greater likelihood of self-reported memory problems among individuals with a history of cancer than those without any history of cancer (Jean-Pierre et al., 2012). Specifically, their analysis revealed that, adjusting for common predictors of memory impairments such as older age, sex, race/ethnicity, educational achievement, poverty level, and poor general health, individuals with a history of cancer or cancer treatment were 40% more likely to report experiencing memory problems that interfere with their activities of daily functioning (Jean-Pierre et al., 2012).

Prevalence of CRND

Previous studies have reported that CRND is detectable in cancer survivors more than a decade following completion of cancer treatment (Koppelmans et al., 2012). Recent advances in cancer therapeutics, early screening, and timely initiation and completion of cancer treatment have contributed to a substantial increase in the number of cancer survivors in the USA. As of 2014, nearly 14.5 million people were estimated to be living with a history of cancer and cancer treatment in the USA (DeSantis et al., 2014). The number of cancer survivors in the USA is expected to increase from 14.5 million to almost 19 million by the year 2024 (DeSantis et al., 2014). Based on this projection, about 14.2 million (up to 75% of the projected 19 million) cancer patients and survivors could be expected to develop CRND symptoms at some point during their cancer treatment by the year 2024 (American Cancer Society, 2014; DeSantis et al., 2014). This equates to roughly 4% of the projected total US population. Cancer survivors are living longer and are experiencing greater exposure to adverse conditions such as CRND that negatively impact their psychosocial functioning and overall quality of life.

Mortality of CRND

Even though CRND has been reported in cancer survivors 20 years post completion of cancer treatment, studies that track data on CRND-related mortality are nonexistent. Of particular interest in this area is the degree to which a history of cancer and treatment may increase risk for cognitive disorders of aging (Ahles

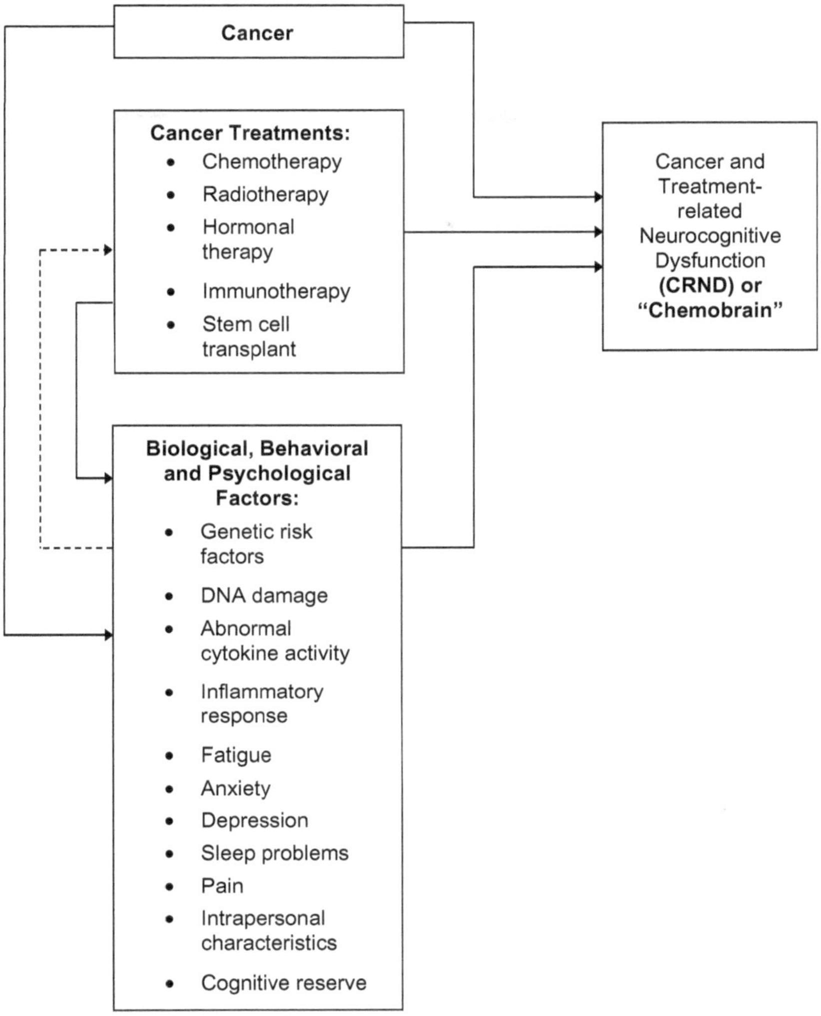

Fig. 17.1. Hypothesized convergent and indirect causal structure of cancer and treatment-related neurocognitive dysfunction.

et al., 2012; Mandelblatt et al., 2013). Strategies for reliable assessment and monitoring of CRND are usually limited in their comprehensiveness and accuracy. Therefore, a definitive diagnosis of CRND may not generally be determined. Although not yet realized, studies correlating symptomatology with postmortem examination of brain changes in cancer patients and survivors with or without CRND may provide important insights on this debilitating adverse condition in the future.

PATHOPHYSIOLOGY OF CRND

Even though the underlying biologic and neuropsychologic mechanisms of CRND are not yet clearly described, its etiology is generally attributed to numerous factors that can independently or synergistically impact fundamental aspects of mental processes and cognitive outcomes, such as performance in different types of attention and memory tasks (Fig. 17.1) (Ahles and Saykin, 2007). Some of the suggested causes of CRND include the tumor malignancy itself (symptoms related to the cancer disease process itself might be conceptualized as part of a paraneoplastic syndrome), brain metastases, genetic predisposition to cognitive impairments, failure of the integrity of the blood–brain barrier, direct or indirect treatment-related brain toxicities, deoxyribonucleic acid (DNA) damage, neuronal impairment, microvasculature obstruction, infarction of brain tissue, and changes in hormonal levels such as abnormal cytokine activities (e.g., increases in proinflammatory cytokines) (Rolig and McKinnon, 2000; Trask et al., 2000; Reichenberg et al., 2001; Kelley et al., 2003; Verstappen et al., 2003; Blasiak et al., 2004; Scheibel et al., 2004; Dietrich et al., 2006). Other possible causes of CRND include cancer- and treatment-related anemia, stress, poor nutrition (e.g., low levels of iron, folic acid, or other vitamins), psychologic distress (e.g., anxiety and depression), sleep dysfunction, and cancer/treatment-related fatigue (Meyers, 1997, 2000; Bender et al., 2001, 2007; Ahles et al., 2002; O'Shaughnessy, 2002, 2003; Weiss, 2003; Tannock et al., 2004; Meyers et al., 2005; Staat and Segatore, 2005).

The conditions normally observed in cancer patients and survivors with CRND include mild to moderate difficulties concentrating, paying attention, remembering information, and carrying out planned sets of cognitive-behavioral activities to accomplished desired goals. These cognitive difficulties have been both self-reported and determined by neuropsychologic testing (despite occurrence of inconsistencies of outcomes) and neuroimaging assessment. Even though knowledge of the deleterious effects of CRND on brain structures and cognitive functioning is well evidenced, the underlying neurobiologic processes or mechanisms influencing the development and progression of neurocognitive dysfunction in cancer patients and survivors are not yet systematically characterized.

Additionally, paraneoplastic syndromes resulting from cancer-fighting agents of the immune system attacking parts of the CNS may be implicated in CRND among cancer patients and survivors with non-CNS malignancies (e.g., breast, lungs, ovarian and blood cancers). When they occur, paraneoplastic syndromes can manifest as problems in various domains of cognition, including thinking and remembering, as well as complications in endocrine, neuromuscular, cardiovascular, hematologic, and other symptoms. Signs and symptoms of neoplastic syndromes include difficulties ranging from motor function (e.g., problems with balance, coordination, and walking, as well as fine motor skills), muscular issues (e.g., loss of muscle tone), and cognitive processes (e.g., sensory, speech, sleep, and memory problems). Nevertheless, studies that examined the effects of paraneoplastic syndromes on cognition in cancer patients and survivors are still lacking. Future studies on the characterization of CRND should consider the potential contributions of paraneoplastic syndromes. These studies could focus on and test for certain factors (e.g., inflammation, and immune system problems), which are already proposed as plausible causes of cognitive impairments in the context of cancer and its treatments. Specifically, inflammation of the limbic system (e.g., limbic encephalitis) is known to affect emotion, behavior, and certain types of memory. Additionally, investigation of overall CNS inflammation such as encephalomyelitis and peripheral neuropathy could hold promises to understanding CRND. Furthermore, future studies could test for certain antibodies and white blood cells such as T cells that may be implicated in the immune system attacking both normal cells and cancer cells, thereby leading to paraneoplastic syndromes and CRND.

GENETIC EPIDEMIOLOGY OF CRND

Genetic epidemiology studies may provide valuable data to enhance our understanding of CRND and facilitate the development of more cogent description of this adverse condition across cancer populations (Yokoyama et al., 2014; Davies et al., 2015; Hibar et al., 2015). Such studies could propose and test genomic hypotheses to aid in accurate prediction and identification of cancer patients at risk of developing cognitive impairments. Previous studies have described possible roles of genetic factors in the development of CRND. For example, Ahles and colleagues (2003) reported that long-term cancer survivors treated with chemotherapy with at least one apolipoprotein ε4 (APOE ε4) allele scored statistically lower on various tests of cognitive functioning compared to cancer survivors without the ε4 allele. Koleck and colleagues (2014) also examined the relationship between APOE ε4 and cognitive functioning in postmenopausal women with early-stage breast cancer at three time points prior to and during cancer treatment in a sample of 37 women receiving chemotherapy and anastrozole, 41 women receiving anastrozole alone, and 50 healthy control women. Koleck et al. (2014) also reported a statistically significant effect of APOE ε4 on cognitive domains of attention, verbal and visual learning and memory, and executive function measured by scores on a comprehensive neuropsychologic test battery. Another study on the role of genetic factors in CRND examined the association of cognitive functioning with catechol-*O*-methyltransferase (COMT). Small et al. (2011) studied breast cancer patients treated with and without chemotherapy and healthy control women, and compared individuals who had at least one COMT Val allele to those homozygous for the Met allele. These authors found that Val carriers had poorer performance on tests of attention, verbal fluency, and motor speed as compared to those without this allele. In addition, breast cancer patients treated with chemotherapy who were Val carriers showed poorer performance on measures of attention than healthy individuals who were Val carriers.

Genomewide association studies (GWAS) have identified candidate genes related to different aspects of cognitive functioning in healthy individuals (Barral et al., 2014; Debette et al., 2015; Ibrahim-Verbaas et al., 2016) but these approaches have not yet been applied to CRND. Systematic studies on the genetic etiology of CRND – ranging from studies that examine gene variant(s) and gene expression to studies on gene product(s) and altered physiology in relation to cognitive dysfunction in cancer patients and survivors – are urgently needed. These types of genetic studies can be applied to CRND to facilitate more precise characterization, prediction of risks, and development and testing of reliable interventions to treat neurocognitive impairments for cancer patients and survivors.

Genetic epidemiology studies of CRND can apply different designs, including family aggregation, pattern of inheritance, linkage and association studies, to help determine gene–environment relationships, and evaluate their roles on the development and characteristics of this adverse condition. Specifically, family aggregation studies could help establish clustering of CRND or risk factors for this debilitating condition in families of specific cancer patients and survivors. This is an important approach as family aggregation for CRND could be due to genetic and/or exposures to similar environmental factors, which may help explain some of the variance in cognitive impairments among individuals with non-CNS malignancies throughout the continuum of their cancer diagnosis, treatment, and survivorship experience.

Some of the key factors that could be examined in familial aggregation studies include: (1) the methods of determining the presence and severity of CRND (e.g., patient self-reported subjective experience of CRND or verified evidence of neurocognitive impairments through neuropsychologic tests and brain imaging); (2) age of onset of CRND (within the context of cancer and cancer treatment); (3) degree of relatedness of affected relatives; and (4) number of relatives affected with CRND. This knowledge will facilitate the development of an operational definition of a positive family history for CRND, and will also provide data to help calculate the relative risk ratio (λ) for CRND (where λ = prevalence of CRND in relatives of affected persons/prevalence of CRND in the general population). Information about heritability and concordance of CRND is not known in the field. Familial aggregation studies comparing twin (monozygous and dizygous) pairs and siblings to measure the degree of genetic contribution will be an innovative and useful approach to characterize CRND. If a genetic influence is identified for CRND, then segregation, linkage, and association studies can be conducted to determine patterns of inheritance and the relationships among alleles in term of their expression over one another (e.g., dominant vs. recessive), on which part of which chromosome a possible gene related to CRND is located, and which allele of which gene is associated with CRND.

Knowledge about the genetic etiology of CRND is still lacking. A genetic epidemiology framework to identify candidate genes for this condition and genomewide association research on CRND could prove beneficial to treating this problem for cancer patients and survivors. Such an approach to CRND would help answer important questions about family aggregation, patterns of inheritance, and information about CRND-related genes and specific chromosomes that carry these genes, as well as enhancing oncology researchers' knowledge of specific gene variants and gene products as potential direct or indirect cause(s) of CRND in cancer patients and survivors.

FACTORS INFLUENCING THE CHARACTERIZATION OF CRND

Reliable assessment of incidence and severity of CRND remains a clinical research and practice challenge for several reasons. First, there are no "gold-standard" measures that were systematically developed and psychometrically validated specifically for cancer populations. The lack of "gold-standard" measures of cancer-specific neurocognitive dysfunction is probably due, at least partly, to a lack of consensus development studies about most appropriate and reliable methods for assessing the presence and severity of CRND (Jean-Pierre, 2010). Consequently, oncology researchers generally use readily available objective and validated neuropsychologic tests as well as patients' and survivors' self-reported subjective complaints of cognitive difficulties to establish the presence of CRND (Kayl et al., 2008). Thus, the presence and severity of CRND are normally based on comparative analyses of neurocognitive performance of cancer patients and survivors to that of healthy reference control groups (i.e., individuals without a history of cancer or cancer treatment), using scores on conveniently obtainable neuropsychologic measures. This *modus operandi* persists in behavioral and clinical oncology research, despite our growing understanding and recognition that the ecologic validity of established neuropsychologic measures may still need to be determined for cancer populations, where ability to detect more subtle cognitive changes may be needed (Jean-Pierre, 2010).

Second, normal cerebral operations involve complex interactions among neural substrates of cognition and mental processing systems that allow a person to discriminate between relevant and irrelevant information, selectively react to stimuli based on *a priori* knowledge in the context of internal drives and intentions, maintain and manipulate mental information, and monitor and control behavioral outcome responses. These neurocognitive and mental processes also involve intricate and essential aspects of different types of attention and memory that are difficult to evaluate and describe even in healthy people (i.e., individuals with a noncompromised CNS). Ascertaining specific alterations in these neurocognitive and mental processes is expected to be more challenging in cancer patients and survivors with compromised neurocognitive systems.

Systematic description of the relationships among complicated brain architecture, functional neuroanatomy, and mental processes and behavioral outcomes (e.g., different types of attention and memory procedures, cognitive information processing speed, and executive functioning) has been a topic of increasing interest in both basic and clinical research. Despite growing interest in understanding and explicating brain–behavior relationships, reliable models that can identify and cogently describe the underlying neurobiologic mechanisms influencing cognitive functioning in cancer populations are still lacking. Understanding the complexities of brain–behavior relationships is particularly challenging in situations where endogenous (e.g., brain dysfunction due to non-CNS malignancies) or exogenous (e.g., chemotherapy for non-CNS malignancies that cross the blood–brain barrier, radiotherapy-related systemic inflammation, and other effects of cancer and its various types of treatment) stress is involved. These challenges highlight the need for an integrative assessment and management approach to facilitate the development of more methodical descriptions of interactive cognitive systems and complex brain–behavior interrelationships in the context of non-CNS malignant tumors, their treatments, and the resulting neurotoxicities (Fig. 17.2).

CLINICAL TRIALS TO TREAT CRND

Even though the precise etiology of CRND is not yet clearly understood, cancer control researchers and clinical oncology professionals have recognized the growing need to develop and test effective interventions to help alleviate the burden and suffering of this adverse condition for patients and survivors. The attribution of the cause of CRND is generally multifactorial, and includes divergent biologic, genetic, psychologic, behavioral, and cancer and treatment-related components, as illustrated in Figure 17.1. Development and testing of interventions to treat CRND have been delayed by difficulties characterizing and reliably assessing its presence and severity. Consequently, relatively few studies on the treatment of CRND have been conducted, though this literature is now growing.

Theories regarding potential causes (e.g., biologic, psychosocial, biobehavioral, changes in hormonal activities, sleep impairments, and fatigue) of CRND have influenced research questions and examination of interventions to mitigate this adverse condition for cancer patients and survivors. For instance, Gagnon and colleagues (2005) examined the benefits of methylphenidate hydrochloride to cognitive functioning in patients with advanced-stage cancers and hypoactive delirium in inpatient and outpatient palliative care settings, and reported improvement in overall cognitive functioning and psychomotor activities following methylphenidate treatment. Other researchers have reported inconsistent outcomes of the effects of epoetin alfa on cognitive functioning in cancer patients undergoing chemotherapy (O'Shaughnessy, 2002; Chang et al., 2004; Iconomou et al., 2008). Jean-Pierre and colleagues (2009) examined the neuroprotective effect of paroxetine hydrochloride among 781 cancer patients receiving chemotherapy and reported that, controlling for depression, cancer patients receiving paroxetine scored better on cognitive functioning than those on placebo. Researchers have also reported statistically significant effects of modafinil on attention and psychomotor functioning in patients with advanced cancers (Lundorff et al., 2009).

Other researchers have examined the benefits of behavioral and psychologic interventions to treat CRND. Ferguson et al. (2007, 2012) have examined the utility of a Memory and Attention Adaptation Training (MAAT) intervention in chemotherapy-treated breast cancer patients. Outcomes of wait-list and randomized controlled trials revealed improvement in self-reported cognitive functioning and quality of life, as well as performance on standardized neuropsychologic testing. Improvements in patients receiving MAAT were apparent at posttreatment, and persisted at follow-up assessment 6 months posttreatment. Other researchers have reported plausible benefits of group cognitive training on cognitive functioning in cancer patients (Von Ah et al., 2012; Cherrier et al., 2013; Von Ah, 2015).

Limited understanding of the etiology and precise underlying mechanisms of CRND has resulted in a lack of guidelines for the assessment and management of this problem for cancer patients and survivors. Increases in survival due to improved screening and advancements in cancer therapeutics highlight the urgent need to further understand and develop strategies to reliably prevent, assess, treat, and continuously monitor the negative effects of cancer and its treatments on the brain and its function for patients and survivors.

The literature on interventions to treat CRND is still in its infancy. Researchers are implementing pilot evaluations of different biologic, neuropsychologic, or combined modality interventions to assess and treat CRND, as evidenced on registry and results databases of publicly and privately supported clinical studies of human subjects (e.g., www.clinicaltrials.gov). While projects in the pipeline (e.g., studies registered on www.clinicaltrials.gov, and personal communications with various cancer control research labs) seem encouraging, the reliability, efficacy, and effectiveness of these interventions to address CRND across cancer populations are still uncertain. Phase III clinical trials to determine efficacy, effectiveness, and safety of behavioral/psychologic or drug interventions to affect CRND may still be implausible given the lack of understanding of the precise mechanistic pathways through which cancer and its treatments impact different brain

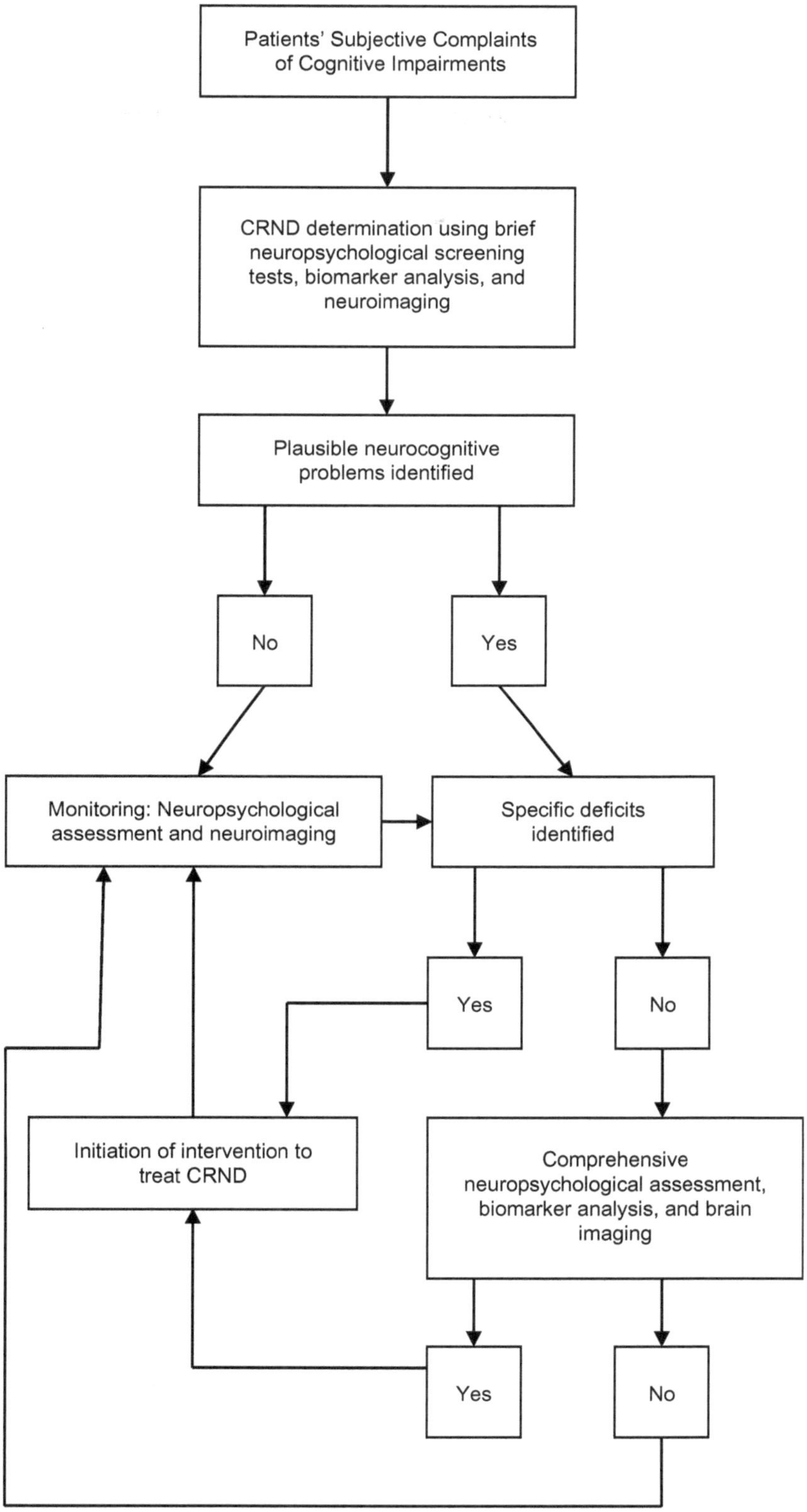

Fig. 17.2. Flow diagram for the assessment and management of cancer and treatment-related neurocognitive dysfunction (CRND).

structures, unconscious mental processes, and behavioral outcomes, as evidenced in different aspects of attention, memory, information-processing speed, and executive functioning.

IMPLICATIONS OF CRND FOR SURVIVORSHIP AND QUALITY OF LIFE

Neurocognitive impairments (e.g., difficulties in different aspects of attention, memory, and other aspects of mental processes and behavioral outcomes) can be serious adverse effects of cancer and its treatments. These deleterious disease- and treatment-related side-effects can be enduring and can impact cancer patients and survivors on multiple dimensions of psychologic, behavioral, sociorelational, cognitive performance, and quality-of-life performance. Although not yet systematically studied, CRND might also influence treatment outcomes in some cases, particularly if problems in attention, memory, and other aspects of cognitive functioning interfere with patients' or survivors' abilities to follow through with a recommended cancer treatment regimen or planned follow-up activities.

Strategies to improve neurocognitive functions can help alleviate cancer burden, improve psychosocial and cognitive functioning, and enhance overall quality of life for individuals with a history of cancer and cancer treatments. Control of CRND can adopt a stepwise or multiphase approach ranging on a continuum from prevention, through timely initiation of treatment, and then follow-up monitoring. At the prevention phase, we can focus our efforts on identifying and managing conditions that can influence brain function and cognitive processes, including psychologic distress and behavioral problems (e.g., depression, anxiety, adjustment issues, pain, fatigue, and sleep problems), maintenance of good nutrition (e.g., a healthy and balanced diet that includes consumption of fruits, vegetables, and recommended daily supplements of fiber, carbohydrate, minerals, iron, folic acid, vitamins, fats, and protein), and engagement in mental activities (e.g., reading, journaling, note taking, using a daily calendar, engaging in interpersonal interactions and communication with others, and utilization of structured cognitive training or rehabilitation programs) that promote and enhance cognitive reserve. The treatment phase will involve the use of targeted behavioral (psychotherapy), pharmacotherapy (e.g., psychostimulant and antidepressant medications), nutritional evaluation and management, and physical therapy and improvement of overall physical strength and energy levels. Monitoring at the survivorship phase will include continuing evaluation and application of the activities described in the previous phases to promote healthy and functioning cognitive performance.

CONCLUSION

Issues related to cancer survivorship concerns are a frequent and rapidly growing public health concern in the USA and worldwide. Advancements in cancer prevention and control, biomedical research, and cancer drug development have contributed to improved cancer therapies, as evidenced by increases in cancer survivorship. However, increased survivorship is accompanied by corresponding rises in cancer and cancer treatment-related side-effects that can affect activities of daily functioning for cancer patients and survivors. Previous studies on cancer and treatment adverse effects focused primarily on other issues, including pain, nausea, and emesis, fatigue, psychologic distress, and sleep dysfunction. A growing literature, however, has focused on CRND. It is important to understand and characterize CRND in order to facilitate the development of effective strategies for assessing and treating this detrimental condition for patients and survivors.

Difficulties determining the precise etiology of CRND and describing the underlying mechanistic pathways through which cancer and its treatments can lead to cognitive impairments have delayed progress in the development and testing of reliable interventions to treat this adverse condition. A multistep approach to the assessment and management of CRND seems promising. Future studies need to focus on integrating data from patients, objective neuropsychologic measures, imaging, and biologic measures of variables known to be associated with cognitive deficits. Oncology professionals need to systematically assess and encourage patients and survivors to discuss cognitive problems, and be prepared to treat CRND.

ACKNOWLEDGMENTS

This work was supported in part by grants from the National Institutes of Health, National Center for Advancing Translational Sciences, Clinical and Translational Sciences Award Grant Number KL2 TR000163, the Walther Cancer Foundation, the Eunice Kennedy Shriver National Institute of Child Health and Human Development of the National Institutes of Health (U54 HD062484), and the Indiana University Melvin and Bren Simon Cancer Center American Cancer Society Institutional grant.

References

Ahles TA, Saykin A (2001). Cognitive effects of standard-dose chemotherapy in patients with cancer. Cancer Invest 19: 812–820.

Ahles TA, Saykin AJ (2007). Candidate mechanisms for chemotherapy-induced cognitive changes. Nat Rev Cancer 7: 192–201.

Ahles TA, Saykin AJ, Furstenberg CT et al. (2002). Neuropsychologic impact of standard-dose systemic chemotherapy in long-term survivors of breast cancer and lymphoma. J Clin Oncol 20: 485–493.

Ahles TA, Saykin AJ, Noll WW et al. (2003). The relationship of APOE genotype to neuropsychological performance in long-term cancer survivors treated with standard dose chemotherapy. Psychooncology 12: 612–619.

Ahles TA, Root JC, Ryan EL (2012). Cancer- and cancer treatment-associated cognitive change: an update on the state of the science. J Clin Oncol 30: 3675–3686.

American Cancer Society (2014). Cancer Facts and Figures 2014, American Cancer Society, Atlanta, GA.

American Cancer Society (2015). Global Cancer Facts and Figures, 3rd edn. American Cancer Society, Atlanta, GA.

Barral S, Cosentino S, Christensen K et al. (2014). Common genetic variants on 6q24 associated with exceptional episodic memory performance in the elderly. JAMA Neurol 71: 1514–1519.

Behin A, Delattre J-Y (2004). Complications of radiation therapy on the brain and spinal cord. Semin Neurol 24: 405–417.

Bender CM, Paraska KK, Sereika SM et al. (2001). Cognitive function and reproductive hormones in adjuvant therapy for breast cancer: a critical review. J Pain Symptom Manage 21: 407–424.

Bender CM, Sereika SM, Brufsky AM et al. (2007). Memory impairments with adjuvant anastrozole versus tamoxifen in women with early-stage breast cancer. Menopause 14: 995–998.

Blasiak J, Arabski M, Krupa R et al. (2004). Basal, oxidative and alkylative DNA damage, DNA repair efficacy and mutagen sensitivity in breast cancer. Mutat Res 554: 139–148.

Boykoff N, Moieni M, Subramanian SK (2009). Confronting chemobrain: an in-depth look at survivors' reports of impact on work, social networks, and health care response. J Cancer Surviv 3: 223–232.

Brezden CB, Phillips KA, Abdolell M et al. (2000). Cognitive function in breast cancer patients receiving adjuvant chemotherapy. J Clin Oncol 18: 2695–2701.

Chang J, Couture FA, Young SD et al. (2004). Weekly administration of epoetin alfa improves cognition and quality of life in patients with breast cancer receiving chemotherapy. Support Canc Ther 2: 52–58.

Cherrier MM, Anderson K, David D et al. (2013). A randomized trial of cognitive rehabilitation in cancer survivors. Life Sci 93: 617–622.

Cleeland CS, Zhao F, Chang VT et al. (2013). The symptom burden of cancer: evidence for a core set of cancer-related and treatment-related symptoms from the Eastern Cooperative Oncology Group Symptom Outcomes and Practice Patterns study. Cancer 119: 4333–4340.

Correa DD, Ahles TA (2008). Neurocognitive changes in cancer survivors. Cancer J 14: 396–400.

Davies G, Armstrong N, Bis JC et al. (2015). Genetic contributions to variation in general cognitive function: a meta-analysis of genome-wide association studies in the CHARGE consortium (N = 53949). Mol Psychiatry 20: 183–192.

Debette S, Ibrahim Verbaas CA, Bressler J et al. (2015). Genome-wide studies of verbal declarative memory in nondemented older people: the Cohorts for Heart and Aging Research in Genomic Epidemiology consortium. Biol Psychiatry 77: 749–763.

DeSantis CE, Lin CC, Mariotto AB et al. (2014). Cancer treatment and survivorship statistics, 2014. CA Cancer J Clin 64: 252–271.

Dietrich J, Han R, Yang Y et al. (2006). CNS progenitor cells and oligodendrocytes are targets of chemotherapeutic agents in vitro and in vivo. J Biol 5: 22.

Dietrich J, Monje M, Wefel J et al. (2008). Clinical patterns and biological correlates of cognitive dysfunction associated with cancer therapy. Oncologist 13: 1285–1295.

Faller H, Brahler E, Harter M et al. (2015). Performance status and depressive symptoms as predictors of quality of life in cancer patients. A structural equation modeling analysis. Psychooncology 24: 1456–1462.

Falleti MG, Sanfilippo A, Maruff P et al. (2005). The nature and severity of cognitive impairment associated with adjuvant chemotherapy in women with breast cancer: a meta-analysis of the current literature. Brain Cognit 59: 60–70.

Ferguson RJ, Ahles TA, Saykin AJ et al. (2007). Cognitive-behavioral management of chemotherapy-related cognitive change. Psychooncology 16: 772–777.

Ferguson RJ, McDonald BC, Rocque MA et al. (2012). Development of CBT for chemotherapy-related cognitive change: results of a waitlist control trial. Psychooncology 21: 176–186.

Ferlay J, Soerjomataram I, Dikshit R et al. (2015). Cancer incidence and mortality worldwide: sources, methods and major patterns in GLOBOCAN 2012. Int J Cancer 136: E359–E386.

Gagnon B, Low G, Schreier G (2005). Methylphenidate hydrochloride improves cognitive function in patients with advanced cancer and hypoactive delirium: a prospective clinical study. J Psychiatry Neurosci 30: 100–107.

Ganz PA (2012). Doctor, will the treatment you are recommending cause chemobrain? J Clin Oncol 30: 229–231.

Gottschalk LA, Kunkel R, Wohl TH et al. (1969). Total and half body irradiation. Effect on cognitive and emotional processes. Arch Gen Psychiatry 21: 574–580.

Hibar DP, Stein JL, Renteria ME et al. (2015). Common genetic variants influence human subcortical brain structures. Nature 520: 224–229.

Hurria A, Somlo G, Ahles T (2007). Renaming "chemobrain". Cancer Invest 25: 373–377.

Ibrahim-Verbaas CA, Bressler J, Debette S et al. (2016). GWAS for executive function and processing speed suggests involvement of the CADM2 gene. Mol Psychiatry 21: 189–197.

Iconomou G, Koutras A, Karaivazoglou K et al. (2008). Effect of epoetin alpha therapy on cognitive function in anaemic patients with solid tumours undergoing chemotherapy. Eur J Canc Care (Engl) 17: 535–541.

Jean-Pierre P (2010). Management of cancer-related cognitive dysfunction-conceptualization challenges and implications for clinical research and practice. US Oncol 6: 9–12.

Jean-Pierre P, Mohile S, Morrow G et al. (2009). Neuroprotective effect of SSRI among 781 cancer patients receiving chemotherapy: a URCC CCOP study. J Clin Oncol 27 (suppl). abstr 9512.

Jean-Pierre P, Winters PC, Ahles TA et al. (2012). Prevalence of self-reported memory problems in adult cancer survivors: a national cross-sectional study. J Oncol Pract/Am Soc Clin Oncol 8: 30–34.

Kayl AE, Collins R, Wefel JS (2008). Neuropsychological assessment of adults with cancer. In: CA Meyers, JR Perry (Eds.), Cognition and Cancer, Cambridge University Press, Cambridge, MA.

Kelley KW, Bluthe R-M, Dantzer R et al. (2003). Cytokine-induced sickness behavior. Brain Behav Immun 17 (Suppl 1): S112–S118.

Kenzik K, Pisu M, Johns SA et al. (2015). Unresolved pain interference among colorectal cancer survivors: implications for patient care and outcomes. Pain Med 16: 1410–1425.

Koleck TA, Bender CM, Sereika SM et al. (2014). Apolipoprotein E genotype and cognitive function in postmenopausal women with early-stage breast cancer. Oncol Nurs Forum 41: E313–E325.

Koppelmans V, Breteler MMB, Boogerd W et al. (2012). Neuropsychological performance in survivors of breast cancer more than 20 years after adjuvant chemotherapy. J Clin Oncol 30: 1080–1086.

Kurita GP, Sjogren P (2015). Pain management in cancer survivorship. Acta Oncol 54: 629–634.

Luckett T, Goldstein D, Butow PN et al. (2011). Psychological morbidity and quality of life of ethnic minority patients with cancer: a systematic review and meta-analysis. Lancet Oncol 12: 1240–1248.

Lundorff LE, Jonsson BH, Sjogren P (2009). Modafinil for attentional and psychomotor dysfunction in advanced cancer: a double-blind, randomised, cross-over trial. Palliat Med 23: 731–738.

Mackenzie CR (2015). Breast cancer survivors' experiences of partner support and physical activity participation. Psychooncology 24: 1197–1203.

Mandelblatt JS, Hurria A, McDonald BC et al. (2013). Cognitive effects of cancer and its treatments at the intersection of aging: what do we know; what do we need to know? Semin Oncol 40: 709–725.

Meyers CA (1997). Issues of quality of life in neuro-oncology. In: CJ Vecht (Ed.), Handbook of Clinical Neurology 23 Neuro-Oncology, Part 1. Brain Tumors: Principles of Biology, Diagnosis and Therapy, Elsevier Science, Amsterdam.

Meyers CA (2000). Neurocognitive dysfunction in cancer patients. Oncology (Williston Park) 14: 75–85.

Meyers CA, Albitar M, Estey E (2005). Cognitive impairment, fatigue, and cytokine levels in patients with acute myelogenous leukemia or myelodysplastic syndrome. Cancer 104: 788–793.

Myers JS (2012). Chemotherapy-related cognitive impairment: the breast cancer experience. Oncol Nurs Forum 39: E31–E40.

National Cancer Institute (2016). SEER Stat Fact Sheets: Cancer of Any Site. [Online]. Bethesda, MD, http://seer.cancer.gov/statfacts/html/all.html [Accessed 04/28/2016].

NCCN (2016). Clinical Practice Guidelines in Oncology (NCCN Guidelines) Survivorship, Version 1. National Comprehensive Cancer Network, Inc. 2016.

O'Shaughnessy JA (2002). Effects of epoetin alfa on cognitive function, mood, asthenia, and quality of life in women with breast cancer undergoing adjuvant chemotherapy. Clin Breast Cancer 3 (Suppl 3): S116–S120.

O'Shaughnessy JA (2003). Chemotherapy-induced cognitive dysfunction: a clearer picture. Clin Breast Cancer 4 (Suppl 2): S89–S94.

Paraska K, Bender CM (2003). Cognitive dysfunction following adjuvant chemotherapy for breast cancer: two case studies. Oncol Nurs Forum 30: 473–478.

Reichenberg A, Yirmiya R, Schuld A et al. (2001). Cytokine-associated emotional and cognitive disturbances in humans. Arch Gen Psychiatry 58: 445–452.

Reitan RM (1958). Validity of the Trail Making Test as an indicator of organic brain damage. Percept Mot Skills 8: 271–276.

Rolig RL, McKinnon PJ (2000). Linking DNA damage and neurodegeneration. Trends Neurosci 23: 417–424.

Saita E, Acquati C, Kayser K (2015). Coping with early stage breast cancer: examining the influence of personality traits and interpersonal closeness. Front Psychol 6: 88.

Sales PMG, Carvalho AF, McIntyre RS et al. (2014). Psychosocial predictors of health outcomes in colorectal cancer: a comprehensive review. Cancer Treat Rev 40: 800–809.

Saykin AJ, Ahles TA, McDonald BC (2003). Mechanisms of chemotherapy-induced cognitive disorders: neuropsychological, pathophysiological, and neuroimaging perspectives. Semin Clin Neuropsychiatry 8: 201–216.

Schagen SB, van Dam FS, Muller MJ et al. (1999). Cognitive deficits after postoperative adjuvant chemotherapy for breast carcinoma. Cancer 85: 640–650.

Scheibel RS, Valentine AD, O'Brien S et al. (2004). Cognitive dysfunction and depression during treatment with interferon-alpha and chemotherapy. J Neuropsychiatry Clin Neurosci 16: 185–191.

Schiff D, Lee EQ, Nayak L et al. (2015). Medical management of brain tumors and the sequelae of treatment. Neuro Oncol 17: 488–504.

Small BJ, Rawson KS, Walsh E et al. (2011). Catechol-*O*-methyltransferase genotype modulates cancer treatment-related cognitive deficits in breast cancer survivors. Cancer 117: 1369–1376.

Staat K, Segatore M (2005). The phenomenon of chemo brain. Clin J Oncol Nurs 9: 713–721.

Stewart BW, Wild CP (Eds.), (2014). World Cancer Report 2014, International Agency for Research on Cancer, World Health Organization, Lyon, France.

Tannock IF, Ahles TA, Ganz PA et al. (2004). Cognitive impairment associated with chemotherapy for cancer: report of a workshop. J Clin Oncol 22: 2233–2239.

Tchen N, Juffs HG, Downie FP et al. (2003). Cognitive function, fatigue, and menopausalmptoms in women receiving adjuvant chemotherapy for breast cancer. J Clin Oncol 21: 4175–4183.

Trask PC, Esper P, Riba M et al. (2000). Psychiatric side effects of interferon therapy: prevalence, proposed mechanisms, and future directions. J Clin Oncol 18: 2316–2326.

van Dam FS, Schagen SB, Muller MJ et al. (1998). Impairment of cognitive function in women receiving adjuvant treatment for high-risk breast cancer: high-dose versus standard-dose chemotherapy. J Natl Cancer Inst 90: 210–218.

Vardy J, Tannock I (2007). Cognitive function after chemotherapy in adults with solid tumours. Crit Rev Oncol Hematol 63: 183–202.

Verstappen CCP, Heimans JJ, Hoekman K et al. (2003). Neurotoxic complications of chemotherapy in patients with cancer: clinical signs and optimal management. Drugs 63: 1549–1563.

Von Ah D (2015). Cognitive changes associated with cancer and cancer treatment: state of the science. Clin J Oncol Nurs 19: 47–56.

Von Ah D, Carpenter JS, Saykin A et al. (2012). Advanced cognitive training for breast cancer survivors: a randomized controlled trial. Breast Cancer Res Treat 135: 799–809.

Wefel JS, Lenzi R, Theriault R et al. (2004). 'Chemobrain' in breast carcinoma? A prologue. Cancer 101: 466–475.

Weiss MJ (2003). New insights into erythropoietin and epoetin alfa: mechanisms of action, target tissues, and clinical applications. Oncologist 8 (Suppl 3): 18–29.

Wieneke MH, Dienst ER (1995). Neuropsychological assessment of cognitive functioning following chemotherapy for breast cancer. Psychooncology 4: 61–66.

Yokoyama JS, Evans DS, Coppola G et al. (2014). Genetic modifiers of cognitive maintenance among older adults. Hum Brain Mapp 35: 4556–4565.

Handbook of Clinical Neurology, Vol. 138 (3rd series)
Neuroepidemiology
C. Rosano, M.A. Ikram, and M. Ganguli, Editors
http://dx.doi.org/10.1016/B978-0-12-802973-2.00018-5

Chapter 18

Sickle cell disease

J. STROUSE*
Division of Hematology, Department of Medicine and Division of Pediatric Hematology/Oncology, Department of Pediatrics, Duke University, Durham, NC, USA

Abstract

Sickle cell disease (SCD) is an inherited hemoglobinopathy caused by a mutation in the sixth amino acid of the β-globin gene (*HBB*). It is the most common serious genetic diseases in childhood, affecting approximately 1 in 2500 births and 100 000 individuals in the USA, in addition to 300 000 new cases globally each year. Central nervous system injury is the most debilitating frequent complication of SCD and includes stroke, silent cerebral infarct (SCI), and cognitive impairment. Among children with sickle cell anemia (HbSS), 11% had a stroke by age 18 years before the implementation of transcranial Doppler screening. SCI is identified in 27% of children with HbSS by their 5th birthday. Children who develop SCI have greater cognitive impairment compared with either children with HbSS without SCI or siblings without SCD. A recent study of adults demonstrated significant cognitive dysfunction, even in participants with apparently mild SCD.

OVERVIEW

Stroke, silent cerebral infarct (SCI), and cognitive impairment are frequent complications in children and adults with sickle cell disease (SCD) (Fig. 18.1). SCD is an inherited hemoglobinopathy caused by a mutation in the sixth amino acid of the β-globin gene (*HBB*). Both the homozygous mutation, sickle cell anemia (HbSS) and compound heterozygous mutations with hemoglobin C (HbSC), β-thalassemia, or rarer mutations in *HBB* cause a hemolytic anemia and vaso-occlusion (Rees et al., 2010). The vaso-occlusion results from increased blood viscosity from decreased deformability of red blood cells, elevated levels of cell adhesion proteins on red and white blood cells and endothelial cells, and activation of platelets and the coagulation system. Decreased bioavailability of nitric oxide secondary to hemolysis and the release of free hemoglobin and arginase from red blood cells also affects blood flow and contributes to the activation of platelets and coagulation (Vercellotti and Belcher, 2014).

SCD affects approximately 100 000 people in the USA, with about 2000 infants diagnosed each year by state newborn screening programs (Brousseau et al., 2010). Most children receive comprehensive care from pediatric hematologists, while care for adults is often fragmented, with few comprehensive sickle cell programs for adults. Worldwide, over 300 000 children are born each year with HbSS, with the vast majority in Africa (237 000), India, and the Middle East (49 000) (Piel et al., 2013). These patients have limited access to diagnostic testing or care for SCD and mortality is high during childhood (Makani et al., 2011). Most studies of stroke and other neurologic complications only include patients from North America and Europe, and there are few studies that focus on India, the Middle East, and Sub-Saharan Africa.

STROKE

SCD confers a greatly increased risk of stroke, defined as a sudden focal neurologic deficit, of any duration (Easton

*Correspondence to: John Strouse, MD, PhD, Division of Hematology, Department of Medicine and Division of Pediatric Hematology/Oncology, Department of Pediatrics, Duke University, DUMC 3939, Durham NC 27710, USA. Tel: +1-919-684-0628, Fax: +1-919-6816-74, E-mail: john.strouse@duke.edu

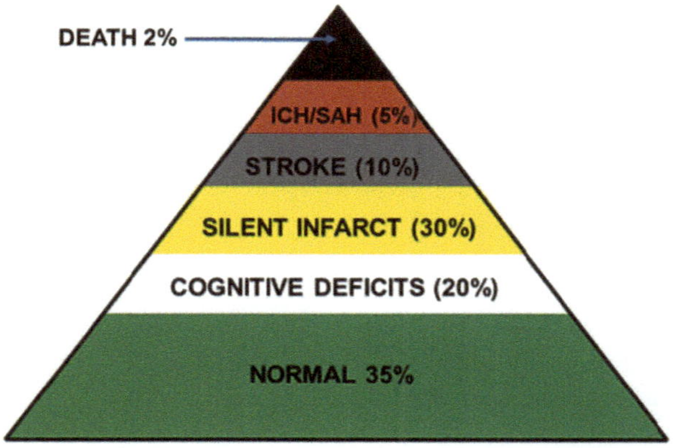

Fig. 18.1. Central nervous complications of sickle cell disease. ICH, intracerebral hemorrhage; SAH, subarachnoid hemorrhage.

et al., 2009), due to focal brain infarction or hemorrhage, in both children and adults. Stroke includes both arterial ischemic stroke with or without hemorrhagic conversion and primary hemorrhagic stroke. Traumatic intracranial hemorrhages are excluded. Common symptoms and signs included hemiplegia, aphasia, sensory deficits, visual field defects, dizziness, and headache. Seizures, meningismus, and impaired consciousness are also seen, but less frequently.

Epidemiology

This association between stroke and SCD was first described by Sydentricker et al. (1923) in a 3-year-old child with left hemiparesis and presumably HbSS and has been demonstrated in both children and adults. The incidence of first stroke in children with HbSS is over 200 times that of the general population, with an incidence of 1020 per 100 000 person-years in children between 2 and 5 years of age, 790 per 100 000 person-years in children 6–9 years of age, and 410 per 100 000 person-years in children 10–19 years of age. Rates of stroke are much lower for children with the other common genotypes of SCD, with 100 per 100 000 person-years for children with HbSC and no strokes reported in a large cohort study that include several hundred children with sickle β-thalassemia (Ohene-Frempong et al., 1998). The rate of stroke is also greatly increased in adults with SCD. Several cohort studies estimated the incidence of first stroke as 500–1280 per 100 000 person-years in adults with HbSS and 360–1160 for all adults with SCD compared to 12 per 100 000 person-years in African Americans less than 35 years old and 202 in those 35–54 years old (Powars et al., 1978, 2005; Ohene-Frempong et al., 1998; Kissela et al., 2004).

An analysis of administrative data from California identified the greatest absolute number of strokes and the highest incidence rates of ischemic stroke in adults 35–64 years old (740/100 000 person-years) and >65 years old (3500/100 000 person-years) (Strouse et al., 2009). These estimates are much higher than the incidence of ischemic stroke in the general African American population (270/100 000 person years for those 35–64 years old and 1500/100 000 person-years for those 65–74 (Kissela et al., 2004). Powars et al. identified an increased risk of stroke (all types) in a prospective cohort of 1056 children and adults with HbSS, in those with chronic lung disease (odds ratio (OR) 3.2), avascular necrosis (OR 7.4), retinopathy (OR 2.5), and renal failure (OR 7.3), and a decreased risk with acute chest syndrome (OR 0.5) (Powars et al., 2005). The results from these studies need to be compared with caution as cross-sectional brain imaging was not widely available during the earliest cohort studies and because of the limitations of analyses using administrative data. These include misclassification of diagnosis and the estimation of the number of adults with SCD based on birth incidence by race and survival to calculate rates. For these reasons, contemporary prospective cohort studies are needed to define the epidemiology of stroke in adults with SCD. These studies will benefit from the inclusion of modern neuroimaging and will likely identify differences in incidence rates and risk factors resulting from the routine implementation of treatment for the primary prevention of ischemic stroke in children with HbSS and greatly increased survival to adulthood (Enninful-Eghan et al., 2010).

Data on the epidemiology of stroke in less developed countries is limited. In Cameroon a small cross-sectional study reported a stroke prevalence of 13% (3/24) of adults with SCD (Njamnshi et al., 2006) and in Nigeria a larger cross-sectional study without neuroimaging identified focal weakness in 1.7% of adults compared to 6.2% of adolescents. A prospective study that included a comprehensive evaluation by a pediatric neurologist of 214 Nigerian children with SCD identified a prevalence of stroke of 9.1% in those with HbSS and 3.7% in children with HbSC (Lagunju and Brown, 2012). This is in marked contrast to the low prevalence (0.7% of children and 1.8% of adults) identified in a survey-based study of stroke prevalence in 5,721 patients with SCD (proportion with HbSS not reported) registered in 14 sickle-cell clinics at tertiary health institutions in Nigeria (Jude et al., 2014). The low prevalence in some studies likely reflects incomplete ascertainment or decreased survival of children and adolescents with stroke in Nigeria (Kehinde et al., 2008). A large prospective cohort of 310 children with HbSS followed from birth in Jamaica identified stroke in 17 children with a cumulative stroke incidence of 7.8% at 14 years of age, but data for adults have not been reported (Balkaran et al., 1992).

ARTERIAL ISCHEMIC STROKE

An arterial ischemic stroke is a sudden, focal neurologic deficit secondary to arterial infarction of central nervous system tissue on brain imaging or autopsy with or without hemorrhagic conversion (Adams et al., 2007). It excludes venous infarction secondary to cerebral sinus thrombosis, but does include cerebral fat embolization syndrome, a complication seen more frequently in people with SCD. Common symptoms and signs include sudden onset of hemiplegia or hemiparesis, aphasia or dysarthria, sensory deficits, visual field defects, decreased level of consciousness, and seizures. Headaches, emesis, and hypertension, and bradycardia are less common (Strouse et al., 2006b).

Epidemiology

Arterial ischemic stroke is a common complication of SCD, with a bimodal distribution of highest incidence in young children (2–5 years of age) and older adults with SCD. Most first arterial ischemic strokes occur in children with HbSS between the ages of 2 and 5 (incidence 700 per 100 000 person-years) and 6 and 9 years (510 per 100 000 person-years) (Ohene-Frempong et al., 1998). In adults with HbSS, the risk of first ischemic stroke is relatively low in young adults 20–29 years old (40 per 100 000 person-years, and increased moderately in middle age (370 per 100 000 person-years in those 30–39 years old and 240 per 100 000 person-years in those 40–49 years old). Adults greater than 50 years old have a higher rate (620 per 100 000 person-years) (Strouse et al., 2009).

Numerous clinical and genetic risk factors for ischemic stroke in SCD have been identified, mostly from pediatric studies that are summarized in Tables 18.1–18.3. More recent studies have focused on abnormally elevated cerebral blood flow velocity and cerebral vasculopathy by magnetic resonance angiography as endpoints instead of ischemic stroke, as the rate of stroke has fallen with the widespread implementation of transcranial Doppler (TCD) ultrasound to screen for increased risk of stroke in children with HbSS.

Table 18.1

Clinical and biologic risk factors for ischemic stroke in sickle cell disease

Risk factor	Odds ratio (95% CI)	*p*-value	Participants	Comments
Homocysteine (> median) (Houston et al., 1997)	3.5 (1.1–12)	0.03	16 with stroke 83 without stroke	50% adults, corrected for age, stroke type not specified
Silent cerebral infarct (Miller et al., 2001)	14	0.006	248 children	Infant cohort CSSCD
Nocturnal SaO_2 (for every 1% increase) (Kirkham et al., 2001)	Hazard ratio 0.82 (0.7–0.9)	0.003	19 with CNS events 76 without CNS events	7 strokes, 8 TIAs, 4 seizures
Elevated MCA/dICA CBFV	Risk ratio	0.16	315 children	Medical College of Georgia cohort
170–199 cm/s	3.5 (0.7–17)	<0.0001		
≥200 cm/s (Adams et al., 1997)	17 (6.9–40)			
Isolated elevated ACA CBFV (>170 cm/s) (Kwiatkowski et al., 2006)	10.5	<0.001	54 strokes in 1975 children	Analysis of screening TCDs from STOP
Aplastic crisis (Wierenga et al., 2001)	58		5 strokes in 346 aplastic crises	Children with HbSS in Kingston, Jamaica Sickle Clinic
Prior TIA (Ohene-Frempong et al., 1998)	56 (12–285)	<0.001	2436 children and adults with HbSS	Did not report children and adults separately
Steady-state Hb (per g/dL)	1.9 (1.3–2.6)	<0.001		
Acute chest syndrome (ACS) within 2 weeks	7 (1.8–27)	0.001		
	2.4 (1.3–4.5)	0.005		
ACS rate (event/year)	1.3 (1.03–1.7)	0.033		
SBP (10-mm increase)				
Hypertension (Strouse et al., 2009)	4.1 (2.9–5.7)	<0.0001	255 acute strokes among 69 586 discharges with diagnosis of sickle cell disease	
Diabetes mellitus	2.2 (1.2–3.9)	<0.05		
Hyperlipidemia	6.9 (2.9–14)	<0.0001		
Renal disease	4.2 (2.4–6.8)	<0.0001		
Atrial fibrillation	4.9 (2.2–9.5)	<0.0005		

95% CI, 95% confidence interval; CSSCD, Cooperative Study of Sickle Cell Disease; $Sa\text{O}_2$, oxygen saturation; CNS, central nervous system; TIA, transient ischemic attack; MCA, middle cerebral artery; dICA, distal internal carotid artery; CBFV, cerebral blood flow velocity; ACA, anterior cerebral artery; TCD, transcranial Doppler ultrasound; STOP, Stroke Prevention Study; HbSS, sickle cell anemia; Hb, hemoglobin; ACS, SBP, systolic blood pressure.

Table 18.2

Risk factors for hemorrhagic stroke in sickle cell disease

Risk factor	Odds ratio (95% CI)	*p*-value	Participants	Comments
Steady-state Hb (for every 1 g/dL decrease) (Ohene-Frempong et al., 1998)	1.6 (1.1–2.4)	0.013	2436 children and adults with HbSS	Did not report children and adults separately
Steady-state leukocyte count for every 5000/μL increase)	1.9 (1.7–2.2)	0.026		
Hypertension (Strouse et al., 2006b)	NC (1.7–NC)	<0.05	15 children with hemorrhagic and 29 with ischemic stroke	Case-control study
Events in the last 14 days				
Transfusion of RBCs	35 (4.9–289)	<0.001		
Corticosteroids	20 (2.9–217)	<0.005		
NSAIDs	4.4 (0.9–21)	<0.05		
Transfusion in last 14 days (Strouse et al., 2008)	15 (1.5–708)	<0.01	20 adults with hemorrhagic, 29 with ischemic stroke	Case-control study
Hypertension (Strouse et al., 2009)	7.7 (4.7–13)	<0.0001	255 acute strokes among 69 586 discharges with sickle cell disease	Did not report children and adults separately
Renal disease	7.2 (3.4–14)	<0.0001		
Coagulopathy	9.1 (2.8–23)	<0.0005		
Atrial fibrillation	4.3 (0.9–13)	<0.05		

95% CI, 95% confidence interval; Hb, hemoglobin; HbSS, sickle cell anemia; RBCs, red blood cells; NSAIDs, nonsteroidal anti-inflammatory drugs; NC, not calculated.

Table 18.3

Genetic risk factors for stroke in children with sickle cell disease

Gene	Odds ratio (95% CI)	*p*-value	Participants	Comments
ANXA2 Annexin a2 (Flanagan et al., 2011)	2.7 (1.3–5.8)	0.007		All ischemic
TGFBR3 Transforming growth factor-β receptor III	2.5 (1.3–5.0)	0.005		
	2.2 (1.1–4.2)	0.016	130 with stroke 103 without stroke	
TEK Tek tyrosine kinase	0.47 (0.3–0.8)	0.003		
ADCY9 Adenylate cyclase 9				
HbA2 α-thalassemia 3.7 kb deletion (Flanagan et al., 2011)	0.45 (0.2–0.8)	0.009		
HbA2 α-thalassemia deletion (Adams et al., 1994)	0.42 (0.2–0.9)	0.02	44 with stroke 256 without stroke	Unspecified stroke type
HbA2 α-thalassemia deletion (Sebastiani et al., 2005)	Bayesian network	NA	92 stroke 1306 without stroke	Hemorrhagic and ischemic
BMP6 Bone morphogenetic protein				
SELP Selectin P				
MET proto-oncogene				
ADRB2 Angiotensin receptor 27E (Hoppe et al., 2004)	0.53	0.033		
HLA-A Human leukocyte antigen A	7.7	0.013	36 ischemic stroke 159 with normal MRI	CSSCD Infant cohort
IL4R Interleukin-4 receptor (Hoppe et al., 2004)	2.5	0.006		
VCAM1 vascular cell adhesion molecule 1 VCAM1(-1594) C (Hoppe et al., 2004)	1.1	0.72		
VCAM1 1238G>C (Taylor et al., 2002)	0.4 (0.2–0.8)	0.02	51 with and without stroke	Unspecified stroke type
VCAM1 1238G>C (Belisario et al., 2015a)	1.7 (0.6, 5.0)	0.35	21 with stroke 365 without stroke	Ischemic stroke
TNF-α -308G>A	2.5 (1.1, 5.9)	0.03		
HbA2 $\alpha^{3.7}$-thalassemia deletion	7.2 (1,54)	0.5 NA		

Table 18.3

Continued

Gene	Odds ratio (95% CI)	*p*-value	Participants	Comments
HbA2 α-thalassemia deletion	Bayesian network		92 stroke 1306 without stroke	Hemorrhagic and ischemic
BMP6 Bone morphogenetic protein				
SELP Selectin P				
MET proto-oncogene (Sebastiani et al., 2005)				
AGT Angiotensinogen GT repeat Allele 3 or 4 (Tang et al., 2001)	4 (1.3–13)	0.02	21 with stroke 42 without stroke	Unspecified stroke type
Allele 3 or 4 (Romana et al., 2004)	0.2 (0.03–1)	0.06	8 with stroke	Unspecified stroke type
Allele 1	3.9 (1.01–15)	0.047	148 without stroke	
GDTT1 null	0.6 (0.1, 2.6)	0.45	31 with stroke	
GSTM1 null (de Oliveira Filho et al., 2013)	3.9 (1.7, 8.6)	0.0005	175 without stroke	
HIMP-2 and BCLL11A (Leonardo et al., 2016)	0.27 (0.13, 0.57)	0.0006	25 with stroke* 180 without stroke	Unspecified stroke type
B-globin Central African haplotype	2.9 (1.6, 4.3)	0.003	67 with stroke	62 ischemic
HbA2 $\alpha^{3.7}$-thalassemia deletion	0.4 (0.1, 1.04)	0.06	194 without stroke	3 SCI
G6PD A-variant (Domingos et al., 2014)	0.5 (0.2, 1.9)	0.35		2 hemorrhagic
eNOS4a Endothelial nitric oxide synthase (Tantawy et al., 2015)	Not provided	0.029	13 with stroke 38 without stroke	Unspecified stroke type
GOLGB1 Giantin Y1212C	0.2 (0.1, 0.4)	<.0001	120 stroke	Overt stroke
ENPP1 K173Q	0.4 (0.3, 0.7)	<.0011	104 without stroke	Discovery
GOLGB1 Y1212C	0.4 (0.1, 0.99)	0.035	57 stroke	Overt stroke
ENPP1 K173Q (Flanagan et al., 2013)	0.6 (0.3,0.99)	0.031	231 without stroke	Validation
ENPP1 K173 variant (Belisario et al., 2015b)	Protective	0.037	23 with stroke 372 without stroke	Ischemic stroke

*Mostly adults (age range 12–68 years).
95% CI, 95% confidence interval; NA, not applicable; MRI, magnetic resonance imaging; CSSCD, Cooperative Study of Sickle Cell Disease; SCI, silent cerebral infarct.

The most consistently identified clinical risk factors for first arterial ischemic stroke in children are genotype (highest for HbSS), age, recent acute chest syndrome, elevated cerebral blood flow velocity in the distal internal carotid and middle cerebral arteries, history of SCI, low hemoglobin, acute anemic events, including aplastic crisis secondary to parvovirus, and cerebral vasculopathy. Clinical risk factors in adults include genotype (risk is greatest for HbSS), increasing age, increased systolic blood pressure or hypertension, and lower baseline hemoglobin (Powars et al., 1978; Pegelow et al., 1997; Ohene-Frempong et al., 1998; Strouse et al., 2009). Other risk factors for ischemic stroke in the general population (diabetes mellitus, atrial fibrillation, hyperlipidemia, and renal disease) may also contribute to increased risk in adults with SCD, but have only been identified in a single study (Strouse et al., 2009).

Ischemic stroke recurs frequently in untreated children with SCD (67%) and most recurrent strokes occur within 24 months (Powars et al., 1978). The risk of recurrent stroke is likely lower in those with hypertension within 24 hours, acute chest syndrome, fever, exchange transfusion, or acute anemic event requiring transfusion within 2 weeks of the initial stroke, based on a large retrospective study of children and young adults on chronic transfusions for secondary stroke prevention (Scothorn et al., 2002). Moyamoya is associated with a greatly increased risk of recurrent transient ischemic attack and stroke (58%) compared to children without moyamoya (28%) despite treatment with chronic transfusion (Dobson et al., 2002).

Genetic modifiers of disease severity in SCD including stroke has been an area of intense interest for over 15 years. Candidate gene approaches have been generally used to evaluate the association between arterial ischemic stroke and/or cerebral vasculopathy of the carotid or proximal middle and anterior cerebral arteries (Table 18.4). Multiple single-nucleotide polymorphisms in candidate genes have been associated with stroke risk and some of these were incorporated into a Bayesian network model to predict stroke. This model was developed and then validated in a smaller second cohort with excellent predictive power; however, it has not been replicated or implemented for clinical use (Sebastiani et al., 2005).

Table 18.4

Risk factors for silent cerebral infarct in children with sickle cell disease

Risk factor	Odds ratio (95% CI)	p-value	Participants	Comments
Painful event rate	0.5 (0.3, 0.95)	0.034		
Seizure history	14 (1.5, 141)	0.023	43 silent infarcts	Some MRIs obtained on
Leukocyte count <11 800/μL	3.2 (1.2, 14)	0.016	188 normal MRI	0.6 and 1 T scanners
SEN β^S-globin haplotype (Kinney et al., 1999)	2.5 (1.03, 6.2)	0.044		
Baseline hemoglobin				
<7.6 g/dL	2.1 (1.5, 3.1)	<.001	251 silent infarcts	Silent Infarct Transfusion
7.6–8.5 g/dL	1.2 (0.8, 1.8)	0.37	563 normal MRI	(SIT) trial
Systolic blood pressure				Limited to children age
104–112 mmHg				5–15 years
>112 mmHg	1.6 (1.1, 2.3)	0.025		
Male sex (DeBaun et al., 2012b)	1.7 (1.2, 2.5)	0.005		
	1.4 (1.04, 1.9)	0.026		
Cerebral artery stenoses (Arkuszewski et al., 2014)	1.7 (1.2, 2,4)	0.0006	26 silent infarcts 41 normal MRI	
MRA-defined vasculopathy (Thangarajh et al., 2012)	2.8 (1.6, 5.2)	0.0007	214 silent infarcts 302 normal MRI	Subset of SIT trial
Baseline Hb <7 g/dL by age 3 years	2.9 (1.1, 7.9)	0.039		
Acute anemic events	3.4 (1.01, 11)	0.048	34 silent infarcts	
Isolated extracranial internal carotid artery stenosis (Bernaudin et al., 2015)	3.1 (1.1, 8.8)	0.033	155 normal MRI	

95% CI, 95% confidence interval; MRI, magnetic resonance imaging; MRA, magnetic resonance angiography; Hb, hemoglobin.

Pathophysiology

There are multiple factors that contribute to the development of arterial ischemic stroke in children and adults with SCD. Children with HbSS have moderate to severe anemia, resulting in increased cerebral blood flow velocities, increased cerebral blood flow, and, in a subset of children, cerebral vasculopathy with intimal proliferation and narrowing or occlusion of distal internal carotid and anterior and middle cerebral arteries (Stockman et al., 1972; Brass et al., 1991; Adams et al., 1992; Strouse et al., 2006a). This may occur secondary to turbulent flow and intimal injury by dense sickled cells or from decreased bioavailability of nitric oxide, as nitric oxide is reduced by the intravascular hemolysis from SCD (Schechter and Gladwin, 2003). Activation of the coagulation system with elevated levels of D-dimer and chronic inflammation contribute to an increased risk of venous thromboembolism and likely arterial ischemic stroke in SCD (Ataga et al., 2012; Naik et al., 2014).

Primary stroke prevention

The majority of arterial ischemic strokes can be prevented in children with HbSS. Screening with TCD ultrasound identifies children with HbSS at greatly increased risk of stroke (those with time-averaged mean maximal blood flow velocities in the distal internal carotid and middle cerebral arteries of >200 cm/s) and the Stroke Prevention Trial in Sickle Cell Anemia showed that transfusions of sickle-negative blood, typically every 4 weeks, to achieve a trough hemoglobin S<30%, decreased the absolute risk of stroke from 30% over 30 months to 3% (Adams et al., 1992, 1998).

A randomized trial comparing continued transfusions vs. stopping transfusions in patients who had normalization of their cerebral blood flow velocities on transfusions was stopped early because of an increased rate of stroke and increased velocities in the group that stopped. In 2016, results from a similar trial that randomized patients with abnormal TCDs after at least a year of transfusions to continued transfusions vs. hydroxyurea in patients with abnormal TCDs, demonstrated noninferiority of hydroxyurea compared to transfusion (Ware et al., 2016).

Several studies have evaluated TCD in adults with HbSS, but have not identified the greatly elevated velocities associated with increased risk of stroke in children with HbSS (Silva et al., 2006; Valadi et al., 2006). One prospective study evaluated the relationship between TCD and intracranial stenosis in adults. The authors reported that a velocity of 123.5 cm/s or higher in the internal carotid or middle cerebral artery had excellent sensitivity (100%) and moderate specificity (73%) to detect intracranial stenosis of the internal carotid or middle cerebral artery, but this small study has not been validated or demonstrated to predict stroke risk (Silva et al., 2009). No potential interventions to reduce stroke risk in adults have been rigorously evaluated, although, based on the general stroke literature, it is reasonable to aggressively treat known risk factors for stroke, including hypertension, atrial fibrillation, tobacco use, diabetes mellitus, and hypercholesterolemia, We also often consider interventions to increase a low baseline hemoglobin (by hydroxyurea, transfusion, or hematopoietic stem cell transplantation) (Strouse et al., 2011). The increased use of hydroxyurea likely is reducing the number of children with SCD that develop abnormal TCDs and need to initiate transfusion. The rate of stroke has also clearly decreased with the effective implementation of TCD screening for increased stroke risk and treatment with transfusion and hydroxyurea, as described above (Armstrong-Wells et al., 2009).

Secondary stroke prevention

Regular transfusion of sickle-negative blood to maintain a trough hemoglobin S less than 30% of total hemoglobin is the most commonly used therapy for the secondary prevention of stroke. Recurrence rates after first ischemic stroke are as high as 67%, but decrease to 2–4 per 100 person-years with regular transfusions of sickle-negative blood (Wilimas et al., 1980; Wang et al., 1991; Balkaran et al., 1992). Transfusion is the most frequently used therapy in children, but there is some evidence that hydroxyurea also reduces stroke recurrence. However, it is less efficacious than regular transfusions and is typically used as an alternative in patients who are difficult to transfuse or do not have access to safe transfusion (Ware et al., 1999, 2004, 2014). Hydroxyurea is used more frequently for secondary stroke prevention in adults, as there are very limited data to guide secondary stroke prevention in adults with SCD and some adults have a strong preference to stop transfusions (Strouse et al., 2011; Kassim et al., 2015).

PRIMARY HEMORRHAGIC STROKE

This is the global term for acute onset of bleeding into the brain that can be intraparenchymal, intraventricular, or subarachnoid blood. This is not due to trauma or hemorrhagic conversion of an ischemic stroke. Clinical presentation is nontraumatic abrupt onset of severe headache, altered level of consciousness, and/or focal neurologic deficit that is associated with a focal collection of blood within the brain parenchyma, the ventricles, or the subarachnoid space. Blood will be seen on imaging or at autopsy and is not due to trauma or hemorrhagic conversion of an ischemic stroke.

Epidemiology

The only large prospective study of hemorrhagic stroke in SCD, the Cooperative Study of Sickle Cell Disease, enrolled over 4000 children and adults. There were 11 first primary hemorrhagic strokes in children and 14 in adults. Hemorrhagic stroke occurred frequently in children 2–5 years of age (150 per 100 000 person-years), 6–9 years (250 per 100 000 person-years), 10–19 years (140 per 100 000 person-years) and most frequently in young adults, 20–29 years (440 per 100 000 person-years). This is over 200 times the rate in children overall and over 30 times the rate (14 per 100 000 person-years) seen in African Americans 20–44 years old in the Manhattan Stroke Study (Earley et al., 1998; Ohene-Frempong et al., 1998; Jacobs et al., 2002). A similar rate of hemorrhagic stroke (330 per 100 000 person-years) in adults with SCD 19–34 years old was also identified in a study using California discharge data (Strouse et al., 2009). Mortality within the first 15 days was 24% of children and adults with hemorrhagic stroke in the Cooperative Study of Sickle Cell Disease (Ohene-Frempong et al., 1998). Both children and adults were more likely to die with intracerebral hemorrhage (50–80%) than subarachnoid hemorrhage (0–27%) (Powars et al., 1978; Strouse et al., 2006b).

Risk factors

The Cooperative Study of Sickle Cell Disease identified low steady-state hemoglobin and high steady-state leukocyte count (Ohene-Frempong et al., 1998) as clinical risk factors for hemorrhagic stroke in children and adults with SCD. Several risk factors for hemorrhagic stroke in the general population (renal disease, hypertension, coagulopathy) were identified in a discharge database of patients with SCD from California (Strouse et al., 2009). A pair of case-control studies of 15 children and 20 adults with hemorrhagic stroke and 29 children and 18 adults with ischemic stroke identified strong associations between hemorrhagic stroke and a history of hypertension, coagulopathy, and recent (in the last 14 days) transfusion, treatment with corticosteroids, or acute chest syndrome in children (Strouse et al., 2006b). In adults, only transfusion in the last 14 days was significantly associated with hemorrhagic stroke (Strouse et al., 2008).

Pathophysiology

The pathophysiology of primary hemorrhagic stroke in SCD includes several likely mechanisms. Most common, especially in adults, is bleeding from cerebral aneurysms. Cerebral aneurysms are much more common in people with SCD and multiple aneurysms are more frequent. A recent study identified saccular aneurysms in 5 of 55 (9%) adults with HbSS and no prior stroke (Kassim et al., 2016). The increased prevalence of aneurysms in people with SCD may be secondary to cerebral vasculopathy, with injury to the intima and media leading to the formation of aneurysms (Oyesiku et al., 1991). Moyamoya disease can lead to bleeding from collateral cerebral blood vessels. These fragile blood vessels proliferate in response to obstructive cerebral vasculopathy and are associated with bleeding in other diseases, more commonly in adults (Kainth et al., 2013). Underlying coagulopathy from liver disease or anticoagulation for venous thromboembolism (a frequent occurrence in people with SCD) also likely contributes to the increased risk of hemorrhage (Strouse et al., 2007; Naik et al., 2014). Finally, people with SCD are more likely to develop posterior reversible encephalopathy syndrome and may develop primary hemorrhagic stroke as a complication of the hypertension and inadequate cerebral autoregulation (Solh et al., 2016).

Primary and secondary prevention

There are no validated methods for the primary prevention of hemorrhagic stroke in children or adults with SCD and these studies would be challenging to undertake, given the lower incidence of hemorrhagic stroke and the rarity of SCD outside of countries with limited healthcare resources. There are only a few small studies that address recurrent stroke after a first hemorrhagic stroke in SCD patients. These include two small retrospective cohort studies that included 15 children and 19 adults with both intracranial and subarachnoid hemorrhage. They did not identify recurrent strokes or transient ischemic events, including several patients that did not receive chronic transfusions or hydroxyurea, but had limited follow-up time (29 and 25 person-years) (Strouse et al., 2006b, 2008). A case series evaluated treatment of ruptured aneurysm by craniotomy and clip ligation in 15 patients 18–43 years old after subarachnoid hemorrhage. Eleven patients survived, 8 with good recovery and 3 with moderate disability. Two died before surgery (Oyesiku et al., 1991). More recently, coil embolization has been successfully used as a treatment of ruptured and unruptured aneurysms in children and adults with SCD (McConachie et al., 1998; Lo Presti et al., 2015). It is difficult to provide guidance based on this weak evidence, but some experts recommend hydroxyurea or observation alone after primary hemorrhagic stroke in patients with SCD without another indication for transfusion or hematopoietic stem cell transplantation. It is reasonable to implement treatments shown to reduce the risk of recurrent stroke in the general population with primary hemorrhagic stroke. These depend on the identified cause and may include meticulous control of blood pressure if the cause is hypertension, or the discontinuation or modification of anticoagulation. For aneurysmal subarachnoid hemorrhage, treatment of the aneurysm by coiling or surgical clipping is effective (Connolly et al., 2012).

SILENT CEREBRAL INFARCT

Silent cerebral infarct is defined as cerebral infarction on imaging or autopsy in the absence of transient ischemic attack or stroke. There are two competing definitions based on the imaging findings that have been used in multicenter studies of SCD: (1) a magnetic resonance imagin (MRI) lesion measuring at least 3 mm in greatest linear dimension, visible in at least two planes of T2-weighted images (Casella et al., 2010); (2) an MRI lesion measuring at least 5 mm in greatest linear dimension in the T2-weighted image, with corresponding hypointensity on the T1-weighted image (Vichinsky et al., 2010). There should not be a history, physical findings of a corresponding focal neurologic deficit, a recognized stroke syndrome, or transient ischemic attack (Vermeer et al., 2007). However, children with SCI have poorer performance on cognitive testing, including full-scale IQ and tests of attention and executive function compared with children with SCD and normal MRI of the brain (Wang et al., 2001).

Epidemiology

The highest incidence of SCI is in the first 4 years of life, as the cumulative prevalence increases (Fig. 18.2) from 12% at 15 months (Wang et al., 1998, 2008) to 28% at 3.7 years (Kwiatkowski et al., 2009), 31% at 9.1 years, and 37.4% at 14 years (Bernaudin et al., 2011). In adults, two small studies have identified point prevalence of 20–53% in adults with HbSS and HbSβnull thalassemia and a larger study of 104 adults identified a prevalence of 53% in those with HbSβnull thalassemia and 31% in those with HbSB+ thalassemia (Marouf et al., 2003; Rigano et al., 2013; Kassim et al., 2016). Specific neurologic morbidities associated with SCI include decrements in cognition, poor academic attainment, progression to overt stroke, and new or progressive SCI on MRI (Miller et al., 2001; Wang et al., 2001; Pegelow et al., 2002; Schatz, 2004; Debaun et al., 2012a; Cancio et al., 2015).

Multiple risk factors have been identified for SCI in children with SCD, including genotype (much higher prevalence in children with HbSS than HbSC (5.8% in the Cooperative Study of Sickle Cell Disease vs. 13.5% in a recent single institution study), history of seizure, increased leukocyte count, early splenic dysfunction, Senegal haplotype, low hemoglobin concentration, acute

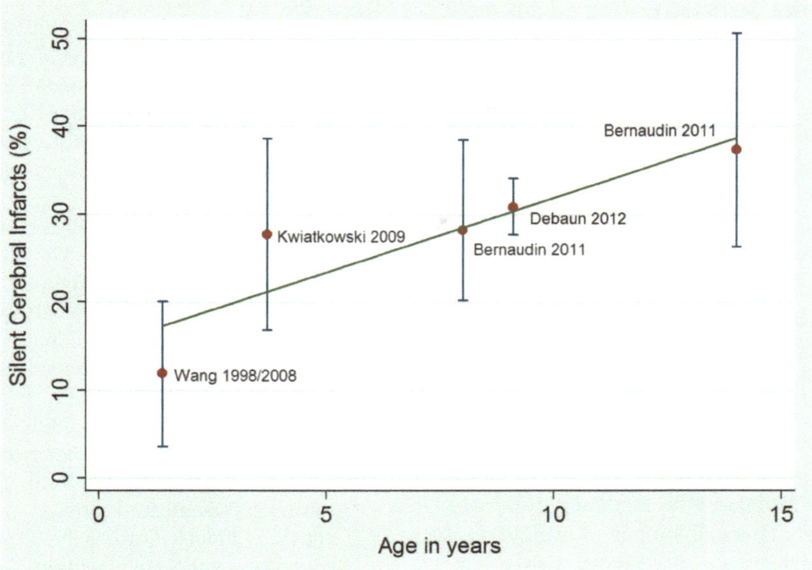

Fig. 18.2. Prevalence of silent cerebral infarcts by age.

anemic events, vasculopathy, glucose-6-phosphate dehydrogenase deficiency, hypertension, and male sex (Kinney et al., 1999; DeBaun et al., 2012b; Guilliams et al., 2015).

Recurrence was frequent in the Cooperative Study of Sickle Cell Disease, the first large study of SCI in children with SCD, with a rate of 8 per 100 person-years. However, rate of recurrence (2.4 per 100 person-years) was much lower in the Silent Infarct Transfusion Trial in the usual treatment arm (Pegelow et al., 2002; DeBaun et al., 2014).

Pathophysiology

The exact pathophysiology of SCI is unknown, but it likely occurs in periods of increased cerebral oxygen demand or decreased delivery based on studies of risk factors and cerebral blood flow in SCD (Debaun et al., 2012a). A single autopsy study supports an important role for increased adhesion of sickled red blood cells in the postcapillary venules (Rothman et al., 1986).

Primary prevention

Currently there are no validated approaches for the primary prevention of SCI. Both hydroxyurea and hematopoietic stem cell transplantation have been proposed and transfusion would likely be an effective approach (DeBaun and Casella, 2014; DeBaun and Kirkham, 2016).

Secondary prevention

The Silent Cerebral Infract Trial randomized 196 children with HbSS or sickle β-null thalassemia and SCI to usual care or regular transfusions of sickle-negative blood to maintain a trough hemoglobin S less than 30% (same treatment as used for secondary stroke prevention in children with HbSS) for 3 years. The group randomized to transfusion had a reduction in the combined endpoint of new or progressive SCI or stroke to 2 per 100 person-years, compared to 4.8 per 100 person-years in the group assigned to routine care (DeBaun et al., 2014). The efficacy of hydroxyurea to prevent recurrent SCI is unknown.

COGNITIVE IMPAIRMENT

Children with SCD often present with difficulty in school and school failure as the organization and academic demands increase. In a large study of school-aged children with HbSS, 17% had been retained at least one grade and 18% received special education services (King et al., 2014b). Adults may have difficulty with employment and organization of their medical care (Sanger et al., 2016).

Epidemiology

Cognitive impairment is highly prevalent in people with SCD, with mean intelligent quotient (IQ) in those without stroke or SCI substantially lower (7 points) than healthy controls (Kawadler et al., 2016). In the infant cohort of the Cooperative Study of Sickle Cell Disease, children with overt stroke or SCI had full-scale IQ 7 points lower than those with normal MRI of the brain (Wang et al., 2001). Several studies have associated increased cerebral blood flow velocity or cerebral blood flow and lower hemoglobin concentration with cognitive impairment in children with SCD (Bernaudin et al., 2000; Kral et al., 2003; Steen et al., 2003; Strouse et al., 2006a). Other factors associated with lower IQ include lower hemoglobin oxygen saturation, increased age, male sex, growth delay, and measures of socioeconomic status (Knight et al., 1995; Wang et al., 2001; Puffer et al., 2010; King et al., 2014a). The single rigorous study of cognition in adults, a cross-sectional study of 149 neurologically asymptomatic people with HbSS, identified significantly lower performance IQ in older patients (greater than 40 years) with severe anemia (Hb < 7.6 g/dL) (Vichinsky et al., 2010).

Children and adults with SCD also have lower average performance for specific domains of cognitive function. In particular, performance on measures of executive function and attention, including working memory, is impaired, especially in those with evidence of silent or overt cerebral infarcts (Berkelhammer et al., 2007). In adults without infarcts, processing speed was notably slower compared to matched controls without SCD from the same community. The same study also identified poorer performance on tests of executive function (Vichinsky et al., 2010).

Pathophysiology

Potential mechanisms of cognitive impairment include decreased cerebral oxygen delivery secondary to severe anemia and hypoxemia, SCI, and ischemic and hemorrhagic strokes (Wang et al., 2001; King et al., 2014a). In addition, chronic inflammation may contribute to cognitive impairment (Andreotti et al., 2015). Multiple studies in children and adults support the association of severe anemia to cognitive impairment in the absence of cerebral ischemia, and several studies have demonstrated low cerebral tissue hemoglobin saturation in children and adults with SCD (Bernaudin et al., 2000; Nahavandi et al., 2004; Raj et al., 2004; Tavakkoli et al., 2005; Vichinsky et al., 2010; Quinn and Dowling, 2012).

Primary and secondary prevention

There have been no studies that have identified sustained effective strategies to prevent cognitive impairment in SCD. Cognitive rehabilitation has been evaluated in pilot studies in children and may improve memory (Yerys et al., 2003; King et al., 2007). In the BABYHUG Trial that compared hydroxyurea vs. placebo for 2 years in young children with HbSS, the mental development index from the Bayley Infant Scales of Development was similar between the two arms. However, all 5 children with mental development index score <70 (2 standard deviations below the mean) were in the placebo arm of the trial (Wang et al., 2011).

References

Adams R, McKie V, Nichols F et al. (1992). The use of transcranial ultrasonography to predict stroke in sickle cell disease. N Engl J Med 326: 605–610.

Adams RJ, Kutlar A, McKie V et al. (1994). Alpha thalassemia and stroke risk in sickle cell anemia. Am J Hematol 45: 279–282.

Adams RJ, McKie VC, Carl EM et al. (1997). Long-term stroke risk in children with sickle cell disease screened with transcranial Doppler. Ann Neurol 42: 699–704.

Adams RJ, McKie VC, Hsu L et al. (1998). Prevention of a first stroke by transfusions in children with sickle cell anemia and abnormal results on transcranial Doppler ultrasonography. N Engl J Med 339: 5–11.

Adams Jr HP, del Zoppo G, Alberts MJ et al. (2007). Guidelines for the early management of adults with ischemic stroke: a guideline from the American Heart Association/American Stroke Association Stroke Council, Clinical Cardiology Council, Cardiovascular Radiology and Intervention Council, and the Atherosclerotic Peripheral Vascular Disease and Quality of Care Outcomes in Research Interdisciplinary Working Groups: The American Academy of Neurology affirms the value of this guideline as an educational tool for neurologists. Circulation 115: e478–e534.

Andreotti C, King AA, Macy E et al. (2015). The association of cytokine levels with cognitive function in children with sickle cell disease and normal MRI studies of the brain. J Child Neurol 30: 1349–1353.

Arkuszewski M, Krejza J, Chen R et al. (2014). Sickle cell anemia: intracranial stenosis and silent cerebral infarcts in children with low risk of stroke. Adv Med Sci 59: 108–113.

Armstrong-Wells J, Grimes B, Sidney S et al. (2009). Utilization of TCD screening for primary stroke prevention in children with sickle cell disease. Neurology 72: 1316–1321.

Ataga KI, Brittain JE, Desai P et al. (2012). Association of coagulation activation with clinical complications in sickle cell disease. PLoS One 7: e29786.

Balkaran B, Char G, Morris JS et al. (1992). Stroke in a cohort of patients with homozygous sickle cell disease. J Pediatr 120: 360–366.

Belisario AR, Nogueira FL, Rodrigues RS et al. (2015a). Association of alpha-thalassemia, TNF-alpha (308G > A) and VCAM-1 (c.1238G > C) gene polymorphisms with cerebrovascular disease in a newborn cohort of 411 children with sickle cell anemia. Blood Cells Mol Dis 54: 44–50.

Belisario AR, Sales RR, Toledo NE et al. (2015b). Association between ENPP1 K173Q and stroke in a newborn cohort of 395 Brazilian children with sickle cell anemia. Blood 126: 1259–1260.

Berkelhammer LD, Williamson AL, Sanford SD et al. (2007). Neurocognitive sequelae of pediatric sickle cell disease: a review of the literature. Child Neuropsychol 13: 120–131.

Bernaudin F, Verlhac S, Freard F et al. (2000). Multicenter prospective study of children with sickle cell disease:

radiographic and psychometric correlation. J Child Neurol 15: 333–343.
Bernaudin F, Verlhac S, Arnaud C et al. (2011). Impact of early transcranial Doppler screening and intensive therapy on cerebral vasculopathy outcome in a newborn sickle cell anemia cohort. Blood 117: 1130–1140.
Bernaudin F, Verlhac S, Arnaud C et al. (2015). Chronic acute anemia and extracranial internal carotid stenosis are risk factors for silent cerebral infarcts in sickle cell anemia. Blood 125: 1653–1661.
Brass LM, Prohovnik I, Pavlakis SG et al. (1991). Middle cerebral artery blood velocity and cerebral blood flow in sickle cell disease. Stroke 22: 27–30.
Brousseau DC, Panepinto JA, Nimmer M et al. (2010). The number of people with sickle-cell disease in the United States: national and state estimates. Am J Hematol 85: 77–78.
Cancio MI, Helton KJ, Schreiber JE et al. (2015). Silent cerebral infarcts in very young children with sickle cell anaemia are associated with a higher risk of stroke. Br J Haematol 171: 120–129.
Casella JF, King AA, Barton B et al. (2010). Design of the Silent Infarct Transfusion (SIT) trial. Pediatr Hematol Oncol 27: 69–89.
Connolly Jr ES, Rabinstein AA, Carhuapoma JR et al. (2012). Guidelines for the management of aneurysmal subarachnoid hemorrhage: a guideline for healthcare professionals from the American Heart Association/American Stroke Association. Stroke 43: 1711–1737.
de Oliveira Filho RA, Silva GJ, de Farias Domingos I et al. (2013). Association between the genetic polymorphisms of glutathione S-transferase (GSTM1 and GSTT1) and the clinical manifestations in sickle cell anemia. Blood Cells Mol Dis 51: 76–79.
DeBaun MR, Casella JF (2014). Transfusions for silent cerebral infarcts in sickle cell anemia. N Engl J Med 371: 1841–1842.
DeBaun MR, Kirkham FJ (2016). Central nervous system complications and management in sickle cell disease: a review. Blood 127: 828–838.
DeBaun MR, Armstrong FD, McKinstry RC et al. (2012a). Silent cerebral infarcts: a review on a prevalent and progressive cause of neurological injury in sickle cell anemia. Blood 119: 4587–4596.
DeBaun MR, Sarnaik SA, Rodeghier MJ et al. (2012b). Associated risk factors for silent cerebral infarcts in sickle cell anemia: low baseline hemoglobin, gender and relative high systolic blood pressure. Blood 119: 3684–3690.
DeBaun MR, Gordon M, McKinstry RC et al. (2014). Controlled trial of transfusions for silent cerebral infarcts in sickle cell anemia. N Engl J Med 371: 699–710.
Dobson SR, Holden KR, Nietert PJ et al. (2002). Moyamoya syndrome in childhood sickle cell disease: a predictive factor for recurrent cerebrovascular events. Blood 99: 3144–3150.
Domingos IF, Falcao DA, Hatzlhofer BL et al. (2014). Influence of the beta haplotype and alpha-thalassemia on stroke development in a Brazilian population with sickle cell anaemia. Ann Hematol 93: 1123–1129.
Earley CJ, Kittner SJ, Feeser BR et al. (1998). Stroke in children and sickle-cell disease: Baltimore-Washington Cooperative Young Stroke Study. Neurology 51: 169–176.
Easton JD, Saver JL, Albers GW et al. (2009). Definition and evaluation of transient ischemic attack: a scientific statement for healthcare professionals from the American Heart Association/American Stroke Association Stroke Council; Council on Cardiovascular Surgery and Anesthesia; Council on Cardiovascular Radiology and Intervention; Council on Cardiovascular Nursing; and the Interdisciplinary Council on Peripheral Vascular Disease. The American Academy of Neurology affirms the value of this statement as an educational tool for neurologists. Stroke 40: 2276–2293.
Enninful-Eghan H, Moore RH, Ichord R et al. (2010). Transcranial Doppler ultrasonography and prophylactic transfusion program is effective in preventing overt stroke in children with sickle cell disease. J Pediatr 2010: 29.
Flanagan JM, Frohlich DM, Howard TA et al. (2011). Genetic predictors for stroke in children with sickle cell anemia. Blood 117: 6681–6684.
Flanagan JM, Sheehan V, Linder H et al. (2013). Genetic mapping and exome sequencing identify 2 mutations associated with stroke protection in pediatric patients with sickle cell anemia. Blood 121: 3237–3245.
Guilliams KP, Fields ME, Hulbert ML (2015). Higher-than-expected prevalence of silent cerebral infarcts in children with hemoglobin SC disease. Blood 125: 416–417.
Hoppe C, Klitz W, Cheng S et al. (2004). Gene interactions and stroke risk in children with sickle cell anemia. Blood 103: 2391–2396.
Houston PE, Rana S, Sekhsaria S et al. (1997). Homocysteine in sickle cell disease: relationship to stroke. Am J Med 103: 192–196.
Jacobs BS, Boden-Albala B, Lin I-F et al. (2002). Stroke in the young in the Northern Manhattan Stroke Study. Stroke 33: 2789–2793.
Jude MA, Aliyu GN, Nalado AM et al. (2014). Stroke prevalence amongst sickle cell disease patients in Nigeria: a multi-centre study. Afr Health Sci 14: 446–452.
Kainth D, Chaudhry SA, Kainth H et al. (2013). Epidemiological and clinical features of moyamoya disease in the USA. Neuroepidemiology 40: 282–287.
Kassim AA, Galadanci NA, Pruthi S et al. (2015). How I treat and manage strokes in sickle cell disease. Blood 125: 3401–3410.
Kassim AA, Pruthi S, Day M et al. (2016). Silent cerebral infarcts and cerebral aneurysms are prevalent in adults with sickle cell anemia. Blood 127: 2038–2040.
Kawadler JM, Clayden JD, Clark CA et al. (2016). Intelligence quotient in paediatric sickle cell disease: a systematic review and meta-analysis. Dev Med Child Neurol. http://dx.doi.org/10.1111/dmcn.13113 (epub ahead of print).
Kehinde MO, Temiye EO, Danesi MA (2008). Neurological complications of sickle cell anemia in Nigerian Africans – a case-control study. J Natl Med Assoc 100: 394–399.
King AA, White DA, McKinstry RC et al. (2007). A pilot randomized education rehabilitation trial is feasible in sickle cell and strokes. Neurology 68: 2008–2011.

King AA, Strouse JJ, Rodeghier MJ et al. (2014a). Parent education and biologic factors influence on cognition in sickle cell anemia. Am J Hematol 89: 162–167.

King AA, Rodeghier MJ, Panepinto JA et al. (2014b). Silent cerebral infarction, income and grade retention among students with sickle cell. Am J Hematol 89: E188–E192.

Kinney TR, Sleeper LA, Wang WC et al. (1999). Silent cerebral infarcts in sickle cell anemia: a risk factor analysis. The Cooperative Study of Sickle Cell Disease. Pediatrics 103: 640–645.

Kirkham FJ, Hewes DK, Prengler M et al. (2001). Nocturnal hypoxaemia and central-nervous-system events in sickle-cell disease. Lancet 357: 1656–1659.

Kissela B, Schneider A, Kleindorfer D et al. (2004). Stroke in a biracial population: the excess burden of stroke among blacks. Stroke 35: 426–431.

Knight S, Singhal A, Thomas P et al. (1995). Factors associated with lowered intelligence in homozygous sickle cell disease. Arch Dis Child 73: 316–320.

Kral MC, Brown RT, Nietert PJ et al. (2003). Transcranial Doppler ultrasonography and neurocognitive functioning in children with sickle cell disease. Pediatrics 112: 324–331.

Kwiatkowski JL, Granger S, Brambilla DJ et al. (2006). Elevated blood flow velocity in the anterior cerebral artery and stroke risk in sickle cell disease: extended analysis from the STOP trial. Br J Haematol 134: 333–339.

Kwiatkowski JL, Zimmerman RA, Pollock AN et al. (2009). Silent infarcts in young children with sickle cell disease. Br J Haematol 146: 300–305.

Lagunju IA, Brown BJ (2012). Adverse neurological outcomes in Nigerian children with sickle cell disease. Int J Hematol 96: 710–718.

Leonardo FC, Brugnerotto AF, Domingos IF et al. (2016). Reduced rate of sickle-related complications in Brazilian patients carrying HbF-promoting alleles at the BCL11A and HMIP-2 loci. Br J Haematol 173: 456–460.

Lo Presti A, Weil AG, Fallah A et al. (2015). Treatment of a cerebral pial arteriovenous fistula in a patient with sickle cell disease-related moyamoya syndrome: case report. J Neurosurg Pediatr 16: 207–211.

Makani J, Cox SE, Soka D et al. (2011). Mortality in sickle cell anemia in Africa: a prospective cohort study in Tanzania. PLoS One 6: e14699.

Marouf R, Gupta R, Haider MZ et al. (2003). Silent brain infarcts in adult Kuwaiti sickle cell disease patients. Am J Hematol 73: 240–243.

McConachie NS, Jaspan T, Hatrick AG et al. (1998). Complications of sickle cell disease: intracranial aneurysms and their treatment. Clin Radiol 53: 388–389.

Miller ST, Macklin EA, Pegelow CH et al. (2001). Silent infarction as a risk factor for overt stroke in children with sickle cell anemia: a report from the Cooperative Study of Sickle Cell Disease. J Pediatr 139: 385–390.

Nahavandi M, Tavakkoli F, Hasan SP et al. (2004). Cerebral oximetry in patients with sickle cell disease. Eur J Clin Invest 34: 143–148.

Naik RP, Streiff MB, Haywood Jr C et al. (2014). Venous thromboembolism incidence in the cooperative study of sickle cell disease. J Thromb Haemost 12: 2010–2016.

Njamnshi AK, Mbong EN, Wonkam A et al. (2006). The epidemiology of stroke in sickle cell patients in Yaounde, Cameroon. J Neurol Sci 250: 79–84.

Ohene-Frempong K, Weiner SJ, Sleeper LA et al. (1998). Cerebrovascular accidents in sickle cell disease: rates and risk factors. Blood 91: 288–294.

Oyesiku NM, Barrow DL, Eckman JR et al. (1991). Intracranial aneurysms in sickle-cell anemia: clinical features and pathogenesis. J Neurosurg 75: 356–363.

Pegelow CH, Colangelo L, Steinberg M et al. (1997). Natural history of blood pressure in sickle cell disease: risks for stroke and death associated with relative hypertension in sickle cell anemia. Am J Med 102: 171–177.

Pegelow CH, Macklin EA, Moser FG et al. (2002). Longitudinal changes in brain magnetic resonance imaging findings in children with sickle cell disease. Blood 99: 3014–3018.

Piel FB, Patil AP, Howes RE et al. (2013). Global epidemiology of sickle haemoglobin in neonates: a contemporary geostatistical model-based map and population estimates. Lancet 381: 142–151.

Powars D, Wilson B, Imbus C et al. (1978). The natural history of stroke in sickle cell disease. Am J Med 65: 461–471.

Powars DR, Chan LS, Hiti A et al. (2005). Outcome of sickle cell anemia: a 4-decade observational study of 1056 patients. Medicine (Baltimore) 84: 363–376.

Puffer ES, Schatz JC, Roberts CW (2010). Relationships between somatic growth and cognitive functioning in young children with sickle cell disease. J Pediatr Psychol 35: 892–904.

Quinn CT, Dowling MM (2012). Cerebral tissue hemoglobin saturation in children with sickle cell disease. Pediatr Blood Cancer 59: 881–887.

Raj A, Bertolone SJ, Mangold S et al. (2004). Assessment of cerebral tissue oxygenation in patients with sickle cell disease: effect of transfusion therapy. J Pediatr Hematol Oncol 26: 279–283.

Rees DC, Williams TN, Gladwin MT (2010). Sickle-cell disease. Lancet 376: 2018–2031.

Rigano P, Pecoraro A, Calvaruso G et al. (2013). Cerebrovascular events in sickle cell-beta thalassemia treated with hydroxyurea: a single center prospective survey in adult Italians. Am J Hematol 88: E261–E264.

Romana M, Diara JP, Doumbo L et al. (2004). Angiotensinogen gene associated polymorphisms and risk of stroke in sickle cell anemia: additional data supporting an association. Am J Hematol 76: 310–311.

Rothman SM, Fulling KH, Nelson JS (1986). Sickle cell anemia and central nervous system infarction: a neuropathological study. Ann Neurol 20: 684–690.

Sanger M, Jordan L, Pruthi S et al. (2016). Cognitive deficits are associated with unemployment in adults with sickle cell anemia. J Clin Exp Neuropsychol 38: 661–671.

Schatz J (2004). Brief report: Academic attainment in children with sickle cell disease. J Pediatr Psychol 29: 627–633.

Schechter AN, Gladwin MT (2003). Hemoglobin and the paracrine and endocrine functions of nitric oxide. N Engl J Med 348: 1483–1485.

Scothorn DJ, Price C, Schwartz D et al. (2002). Risk of recurrent stroke in children with sickle cell disease receiving blood transfusion therapy for at least five years after initial stroke. J Pediatr 140: 348–354.

Sebastiani P, Ramoni MF, Nolan V et al. (2005). Genetic dissection and prognostic modeling of overt stroke in sickle cell anemia. Nat Genet 37: 435–440.

Silva G, Vicari P, Figueiredo M et al. (2006). Migraine-mimicking headache and sickle cell disease: a transcranial Doppler study. Cephalalgia 26: 678–683.

Silva GS, Vicari P, Figueiredo MS et al. (2009). Brain magnetic resonance imaging abnormalities in adult patients with sickle cell disease correlation with transcranial Doppler findings. Stroke 40: 2408–2412.

Solh Z, Taccone MS, Marin S et al. (2016). Neurological PRESentations in sickle cell patients are not always stroke: a review of posterior reversible encephalopathy syndrome in sickle cell disease. Pediatr Blood Cancer 63: 983–989.

Steen RG, Miles MA, Helton KJ et al. (2003). Cognitive impairment in children with hemoglobin SS sickle cell disease: relationship to MR imaging findings and hematocrit. AJNR Am J Neuroradiol 24: 382–389.

Stockman JA, Nigro MA, Mishkin MM et al. (1972). Occlusion of large cerebral vessels in sickle-cell anemia. N Engl J Med 287: 846–849.

Strouse JJ, Cox CS, Melhem ER et al. (2006a). Inverse correlation between cerebral blood flow measured by continuous arterial spin-labeling (CASL) MRI and neurocognitive function in children with sickle cell anemia (SCA). Blood 108: 379–381.

Strouse JJ, Hulbert ML, DeBaun MR et al. (2006b). Primary hemorrhagic stroke in children with sickle cell disease is associated with recent transfusion and use of corticosteroids. Pediatrics 118: 1916–1924.

Strouse JJ, Jordan L, Casella JF (2007). The Excess Burden of Stroke Among Adults with Homozygous Sickle Cell Disease. International Stroke Meeting, San Francisco, CA, February 2007. Stroke 38: 460–461.

Strouse JJ, Field J, Crawford RD et al. (2008). Antecedent transfusion and primary hemorrhagic stroke in adults with sickle cell disease. ASH Annual Meeting Abstracts 112: 1437.

Strouse JJ, Jordan LC, Lanzkron S et al. (2009). The excess burden of stroke in hospitalized adults with sickle cell disease. Am J Hematol 84: 548–552.

Strouse JJ, Lanzkron S, Urrutia V (2011). The epidemiology, evaluation and treatment of stroke in adults with sickle cell disease. Expert Rev Hematol 4: 597–606.

Sydentricker V, Mulherin W, Houseal R (1923). Sickle cell anemia: report of two cases in children with necropsy in one case. Am J Dis Child 26: 132–154.

Tang DC, Prauner R, Liu W et al. (2001). Polymorphisms within the angiotensinogen gene (GT-repeat) and the risk of stroke in pediatric patients with sickle cell disease: a case-control study. Am J Hematol 68: 164–169.

Tantawy AA, Adly AA, Ismail EA et al. (2015). Endothelial nitric oxide synthase gene intron 4 VNTR polymorphism in sickle cell disease: relation to vasculopathy and disease severity. Pediatr Blood Cancer 62: 389–394.

Tavakkoli F, Nahavandi M, Wyche MQ et al. (2005). Effects of hydroxyurea treatment on cerebral oxygenation in adult patients with sickle cell disease: an open-label pilot study. Clin Ther 27: 1083–1088.

Taylor JG, Tang DC, Savage SA et al. (2002). Variants in the VCAM1 gene and risk for symptomatic stroke in sickle cell disease. Blood 100: 4303–4309.

Thangarajh M, Yang G, Fuchs D et al. (2012). Magnetic resonance angiography-defined intracranial vasculopathy is associated with silent cerebral infarcts and glucose-6-phosphate dehydrogenase mutation in children with sickle cell anaemia. Br J Haematol 159: 352–359.

Valadi N, Silva GS, Bowman LS et al. (2006). Transcranial Doppler ultrasonography in adults with sickle cell disease. Neurology 67: 572–574.

Vercellotti GM, Belcher JD (2014). Not simply misshapen red cells: multimolecular and cellular events in sickle vaso-occlusion. Clinical Investigation 124: 1462–1465.

Vermeer SE, Longstreth Jr WT, Koudstaal PJ (2007). Silent brain infarcts: a systematic review. Lancet Neurol 6: 611–619.

Vichinsky EP, Neumayr LD, Gold JI et al. (2010). Neuropsychological dysfunction and neuroimaging abnormalities in neurologically intact adults with sickle cell anemia. Neurocognitive complications of sickle cell anemia in adults. JAMA 303: 1823–1831.

Wang WC, Kovnar EH, Tonkin IL et al. (1991). High risk of recurrent stroke after discontinuance of five to twelve years of transfusion therapy in patients with sickle cell disease. J Pediatr 118: 377–382.

Wang WC, Langston JW, Steen RG et al. (1998). Abnormalities of the central nervous system in very young children with sickle cell anemia. J Pediatr 132: 994–998.

Wang W, Enos L, Gallagher D et al. (2001). Neuropsychologic performance in school-aged children with sickle cell disease: a report from the Cooperative Study of Sickle Cell Disease. J Pediatr 139: 391–397.

Wang WC, Pavlakis SG, Helton KJ et al. (2008). MRI abnormalities of the brain in one-year-old children with sickle cell anemia. Pediatr Blood Cancer 51: 643–646.

Wang WC, Ware RE, Miller ST et al. (2011). Hydroxycarbamide in very young children with sickle-cell anaemia: a multicentre, randomised, controlled trial (BABY HUG). Lancet 377: 1663–1672.

Ware RE, Zimmerman SA, Schultz WH (1999). Hydroxyurea as an alternative to blood transfusions for the prevention of recurrent stroke in children with sickle cell disease. Blood 94: 3022–3026.

Ware RE, Zimmerman SA, Sylvestre PB et al. (2004). Prevention of secondary stroke and resolution of transfusional iron overload in children with sickle cell anemia using hydroxyurea and phlebotomy. J Pediatr 145: 346–352.

Ware RE, Schultz WH, Yovetich N et al. (2014). Stroke with transfusions changing to hydroxyurea (SWiTCH): a phase III randomized clinical trial for treatment of children with sickle cell anemia, stroke, and iron overload. Pediatr Blood Cancer 124 (6): 891–898.

Ware RE, Davis BR, Schultz WH et al. (2016). Hydroxycarbamide versus chronic transfusion for maintenance of transcranial Doppler flow velocities in children with sickle cell anaemia—TCD With Transfusions Changing to Hydroxyurea (TWiTCH): a multicentre, open-label, phase 3, non-inferiority trial. Lancet 387: 661–670.

Wierenga KJ, Serjeant BE, Serjeant GR (2001). Cerebrovascular complications and parvovirus infection in homozygous sickle cell disease. J Pediatr 139: 438–442.

Wilimas J, Goff JR, Anderson Jr HR et al. (1980). Efficacy of transfusion therapy for one to two years in patients with sickle cell disease and cerebrovascular accidents. J Pediatr 96: 205–208.

Yerys BE, White DA, Salorio CF et al. (2003). Memory strategy training in children with cerebral infarcts related to sickle cell disease. J Pediatr Hematol Oncol 25: 495–498.

Index

NB: Page numbers in *italics* refer to figures and tables.

E

P

Q

R

FSC
www.fsc.org
MIX
Papier | Fördert
gute Waldnutzung
FSC® C083411

Kolibri 360 GmbH
Ferdinand-Jühlke-Straße 7
99095 Erfurt, Deutschland
produktsicherheit@kolibri360.de